P9-AGV-110

ORDINARY DIFFERENTIAL EQUATIONS

PURE AND APPLIED MATHEMATICS

A Series of Texts and Monographs

Edited by: R. COURANT · L. BERS · J. J. STOKER

VOLUME XXI

To my wife

ORDINARY DIFFERENTIAL EQUATIONS

JACK K. HALE

ROBERT E. KRIEGER PUBLISHING COMPANY
MALABAR, FLORIDA

NY, INC.

Copyright © 1969 (Original Material) by
JOHN WILEY & SONS, INC.
Copyright © 1980 (New Material) by
ROBERT E. KRIEGER PUBLISHING COMPANY, INC.

Printed in the United States of America

Library of Congress Cataloging in Publication Data

Hale, Jack K.
    Ordinary differential equations.

    Second edition of original published by Wiley-Inter-
science, New York, which was issued as v. 21 of Pure and
applied mathematics.
    Bibliography: p.
    Includes index.
    1. Differential equations.   I. Title.
[QA372.H184 1980]   515'.352   79-17238
ISBN 0-89874-011-8

Preface

This book is the outgrowth of a course given for a number of years in the Division of Applied Mathematics at Brown University. Most of the students were in their first and second years of graduate study in applied mathematics, although some were in engineering and pure mathematics. The purpose of the book is threefold. First, it is intended to familiarize the reader with some of the problems and techniques in ordinary differential equations, with the emphasis on nonlinear problems. Second, it is hoped that the material is presented in a way that will prepare the reader for intelligent study of the current literature and for research in differential equations. Third, in order not to lose sight of the applied side of the subject, considerable space has been devoted to specific analytical methods which are presently widely used in the applications.

Since the emphasis throughout is on nonlinear phenomena, the global theory of two-dimensional systems has been presented immediately after the fundamental theory of existence, uniqueness, and continuous dependence. This also has the advantage of giving the student specific examples and concepts which serve to motivate study of later chapters. Since a satisfactory global theory for general n-dimensional systems is not available, we naturally turn to local problems and, in particular, to the behavior of solutions of differential equations near invariant sets. In the applications it is necessary not only to study the effect of variations of the initial data but also in the vector field. These are discussed in detail in Chapters III and IV in which the invariant set is an equilibrium point. In this way many of the basic and powerful methods in differential equations can be examined at an elementary level. The analytical methods developed in these chapters are immediately applicable to the most widely used technique in the practical theory of nonlinear oscillations, the method of averaging, which is treated in Chapter V. When the invariant set corresponds to a periodic orbit and only autonomous perturbations in the vector field are permitted, the discussion is similar to that for an equilibrium point and is given in Chapter VI. On the other hand, when the perturbations in the vector field are nonautonomous or the invariant set is a closed curve with equilibrium points, life is not so simple. In Chapter VII an attempt has been made to present this more complicated

and important subject in such a way that the theory is a natural generalization of the theory in Chapter IV. Chapter VIII is devoted to a general method
for determining when a periodic differential equation containing a small
parameter has a periodic solution. The reason for devoting a chapter to this
subject is that important conclusions are easily obtained for Hamiltonian
systems in this framework and the method can be generalized to apply to
problems in other fields such as partial differential, integral, and functional
differential equations. The abstract generalization is made in Chapter IX
with an application to analytic solutions of linear systems with a singularity,
but space did not permit applications to other fields. The last chapter is
devoted to elementary results and applications of the direct method of
Lyapunov to stability theory. Except for Chapter I this topic is independent
of the remainder of the book and was placed at the end to preserve continuity
of ideas.

For the sake of efficiency and to acquaint the student with concrete
applications of elementary concepts from functional analysis, I have presented the material with an element of abstraction. Relevant background
material appears in Chapter 0 and in the appendix on almost periodic
functions, although I assume that the reader has had a course in advanced
calculus. A one-semester course at Brown University usually covers the
saddlepoint property in Chapter III; the second semester is devoted to
selections from the remaining chapters. Throughout the book I have made
suggestions for further study and have provided exercises, some of which are
difficult. The difficulty usually arises because the exercises are introduced
when very little technique has been developed. This procedure was followed
to permit the student to develop his own ideas and intuition. Plenty of time
should be allowed for the exercises and appropriate hints should be given
when the student is prepared to receive them.

No attempt has been made to cover all aspects of differential equations.
Lack of space, however, forced the omission of certain topics that contribute
to the overall objective outlined above; for example, the general subject of
boundary value problems and Green's functions belong in the vocabulary of
every serious student of differential equations. This omission is partly justi-
fied by the fact that this topic is usually treated in other courses in applied
mathematics and, in addition, excellent presentations are available in the
literature. Also, specific applications had to be suppressed, but individuals
with special interest can easily make the correlation with the theoretical
results herein.

I have received invaluable assistance in many conversations with my
colleagues and students at Brown University. Special thanks are due to
C. Olech for his direct contribution to the presentation of two-dimensional
systems, to M. Jacobs for his thought-provoking criticisms of many parts of

the original manuscript, and to W. S. Hall and D. Sweet for their comments. I am indebted to K. Nolan for her endurance in the excellent preparation of the manuscript. I also wish to thank the staff of Interscience for being so efficient and cooperative during the production process.

Jack K. Hale

Providence, Rhode Island
September, 1969

Preface to Revised Edition

For this revised edition, I am indebted to several colleagues for their assistance in the elimination of misprints and the clarification of the presentation. The section on integral manifolds has been enlarged to include a more detailed discussion of stability. In Chapter VIII, new material is included on Hopf bifurcation, bifurcation with several independent parameters and subharmonic solutions. A new section in Chapter X deals with Ważewski's principle. The Appendix on almost periodic functions has been completely rewirtten using the modern definition of Bochner.

Jack K. Hale

April 1980

Contents

ORDINARY DIFFERENTIAL EQUATIONS

CHAPTER 0

Mathematical Preliminaries

In this chapter we collect a number of basic facts from analysis which play an important role in the theory of differential equations.

0.1. Banach Spaces and Examples

Set intersection is denoted by $\cap$, set union by $\cup$, set inclusion by $\subset$ and $x \in S$ denotes x is a member of the set S. R (or C) will denote the real (or complex) field. An *abstract linear vector space* (or *linear space*) $\mathscr{X}$ over R (or C) is a collection of elements $\{x, y, \ldots\}$ such that for each x, y in $\mathscr{X}$, the sum $x + y$ is defined, $x + y \in \mathscr{X}$, $x + y = y + x$ and there is an element 0 in $\mathscr{X}$ such that $x + 0 = x$ for all $x \in \mathscr{X}$. Also for any number $a, b \in R$ (or C), scalar multiplication ax is defined, $ax \in \mathscr{X}$ and $1 \cdot x = x$, $(ab)x = a(bx) = b(ax)$, $(a + b)x = ax + by$ for all $x, y \in \mathscr{X}$. A linear space $\mathscr{X}$ is a *normed linear space* if to each x in $\mathscr{X}$, there corresponds a real number $|x|$ called the *norm* of x which satisfies

(i) $|x| > 0$ for $x \neq 0$, $|0| = 0$;
(ii) $|x + y| \leq |x| + |y|$ (triangle inequality);
(iii) $|ax| = |a| \cdot |x|$ for all a in R (or C) and x in X.

When confusion may arise, we will write $|\cdot|_{\mathscr{X}}$ for the norm function on $\mathscr{X}$. A sequence $\{x_n\}$ in a normed linear space $\mathscr{X}$ converges to x in $\mathscr{X}$ if $\lim_{n \to \infty} |x_n - x| = 0$. We shall write this as $\lim x_n = x$. A sequence $\{x_n\}$ in $\mathscr{X}$ is a *Cauchy sequence* if for every $\varepsilon > 0$, there is an $N(\varepsilon) > 0$ such that $|x_n - x_m| < \varepsilon$ if $n, m \geq N(\varepsilon)$. The space $\mathscr{X}$ is *complete* if every Cauchy sequence in $\mathscr{X}$ converges to an element of $\mathscr{X}$. A complete normed linear space is a *Banach space*. The ε-*neighborhood* of an element x of a normed linear space $\mathscr{X}$ is $\{y$ in $\mathscr{X} : |y - x| < \varepsilon\}$. A set S in $\mathscr{X}$ is open if for every $x \in S$, an ε-neighborhood of x is also contained in $\mathscr{X}$. An element x is a *limit point* of a set S if each ε-neighborhood of x contains points of S. A set S is *closed* if it contains its limit points. The *closure* of a set S is the union of S and its limit points. A set S is *dense* in $\mathscr{X}$ if the closure of S is $\mathscr{X}$. If S is a subset of $\mathscr{X}$,

1

A is a subset of R and V_a, $a \in A$ is a collection of open sets of $\mathscr{X}$ such that $\bigcup_{a \in A} V_a \supset S$, then the collection V_a is called an *open covering of S*. A set S in $\mathscr{X}$ is *compact* if every open covering of S contains a finite number of open sets which also cover S. For Banach spaces, this is equivalent to the following: a set S in a Banach space is compact if every sequence $\{x_n\}$, $x_n \in S$, contains a subsequence which converges to an element of S. A set S in $\mathscr{X}$ is *bounded* if there exists an $r > 0$ such that $S \subset \{x \in \mathscr{X}: |x| < r\}$.

Example 1.1. Let $R^n(C^n)$ be the space of real (complex) n-dimensional column vectors. For a particular coordinate system, elements x in $R^n(C^n)$ will be written as $x = (x_1, \ldots, x_n)$ where each x_j is in $R(C)$. If $x = (x_1, \ldots, x_n)$, $y = (y_1, \ldots, y_n)$ are in $R^n(C^n)$, then $ax + by$ for a, b in $R(C)$ is defined to be $(ax_1 + by_1, \ldots, ax_n + by_n)$. The space $R^n(C^n)$ is clearly a linear space. It is a Banach space if we choose $|x|$, $x = \mathrm{col}(x_1, \ldots, x_n)$, to be either $\sup_i |x_i|$, $\sum_i |x_i|$ or $[\sum_i |x_i|^2]^{1/2}$. Each of these norms is equivalent in the sense that a sequence converging in one norm converges in any of the other norms. $R^n(C^n)$ is complete because convergence implies coordinate wise convergence and $R(C)$ is complete.

A set S in $R^n(C^n)$ is compact if and only if it is closed and bounded.

EXERCISE 1.1. If E is a finite dimensional linear vector space and $|\cdot|$, $\|\cdot\|$ are two norms on E, prove there are positive constants m, M such that $m|x| \leq \|x\| \leq M|x|$ for all x in E.

Example 1.2. Let D be a compact subset of R^m [or C^m] and $\mathscr{C}(D, R^n)$ [or $\mathscr{C}(D, C^n)$] be the linear space of continuous functions which take D into R^n [or C^n]. A sequence of functions $\{\phi_n, n = 1, 2, \ldots\}$ in $\mathscr{C}(D, R^n)$ is said to *converge uniformly* on D if there exists a function ϕ taking D into R^n such that for every $\varepsilon > 0$ there is an $N(\varepsilon)$ (independent of n) such that $|\phi_n(x) - \phi(x)| < \varepsilon$ for all $n \geq N(\varepsilon)$ and x in D. A sequence $\{\phi_n\}$ is said to be uniformly bounded if there exists an $M > 0$ such that $|\phi_n(x)| < M$ for all x in D and all $n = 1, 2, \ldots$. A sequence $\{\phi_n\}$ is said to be *equicontinuous* if, for every $\varepsilon > 0$, there is a $\delta > 0$ such that

$$|\phi_n(x) - \phi_n(y)| < \varepsilon, \qquad n = 1, 2, \ldots,$$

if $|x - y| < \delta$, x, y in D. A function f in $\mathscr{C}(D, R^n)$ is said to be *Lipschitzian* in D if there is a constant K such that $|f(x) - f(y)| \leq K|x - y|$ for all $x, y,$ in D. The most frequently encountered equicontinuous sequences in $\mathscr{C}(D, R^n)$ are sequences $\{\phi_n\}$ which are Lipschitzian with a Lipschitz constant independent of n.

LEMMA 1.1. (Ascoli-Arzela). Any uniformly bounded equicontinuous sequence of functions in $\mathscr{C}(D, R^n)$ has a subsequence which converges uniformly on D.

LEMMA 1.2. If a sequence in $\mathscr{C}(D, R^n)$ converges uniformly on D, then the limit function is in $\mathscr{C}(D, R^n)$.

If we define

$$|\phi| = \max_{x \in D}|\phi(x)|,$$

then one easily shows this is a norm on $\mathscr{C}(D, R^n)$ and the above lemmas show that $\mathscr{C}(D, R^n)$ is a Banach space with this norm. The same remarks apply to $\mathscr{C}(D, C^n)$.

EXERCISE 1.2. Suppose $m = n = 1$. Show that $\mathscr{C}(D, R)$ is a normed linear space with the norm defined by

$$\|\phi\| = \int_D |\phi(x)| \, dx.$$

Give an example to show why this space is not complete. What is the completion of this space?

0.2. Linear Transformations

A function taking a set A of some space into a set B of some space will be referred to as a *transformation* or *mapping* of A into B. A will be called the *domain* of the mapping and the set of values of the mapping will be called the *range* of the mapping. If f is a mapping of A into B, we simply write $f: A \to B$ and denote the range of f by $f(A)$. If $f: A \to B$ is one to one and continuous together with its inverse, then we say f is a *homeomorphism* of A onto B. If $\mathscr{X}$, $\mathscr{Y}$ are real (or complex) Banach spaces and $f: \mathscr{X} \to \mathscr{Y}$ is such that $f(a_1 x_1 + a_2 x_2) = a_1 f(x_1) + a_2 f(x_2)$ for all x_1, x_2 in $\mathscr{X}$ and all real (or complex) numbers a_1, a_2, then f is called a *linear mapping*. A linear mapping f of $\mathscr{X}$ into $\mathscr{Y}$ is said to be *bounded* if there is a constant K such that $|f(x)|_{\mathscr{Y}} \leq K|x|_{\mathscr{X}}$ for all x in $\mathscr{X}$.

LEMMA 2.1. Suppose $\mathscr{X}$, $\mathscr{Y}$ are Banach spaces. A linear mapping $f: \mathscr{X} \to \mathscr{Y}$ is bounded if and only if it is continuous.

EXERCISE 2.1. Prove this lemma.

EXERCISE 2.2. Show that each linear mapping of R^n (or C^n) into R^m (or C^m) can be represented by an $m \times n$ real (or complex) matrix and is therefore necessarily continuous.

The *norm* $|f|$ of a continuous linear mapping $f: \mathscr{X} \to \mathscr{Y}$ is defined as

$$|f| = \sup\{|fx|_{\mathscr{Y}}: |x|_{\mathscr{X}} = 1\}.$$

It is easy to show that $|f|$ defined in this way satisfies the properties (i)–(iii) in the definition of a norm and also that

$$|fx|_{\mathscr{Y}} \leq |f| \cdot |x|_{\mathscr{X}} \qquad \text{for all } x \text{ in } \mathscr{X}.$$

If a linear map taking an n-dimensional linear space into an m-dimensional linear space is defined by an $m \times n$ matrix A, we write its norm as $|A|$.

EXERCISE 2.3.   If $\mathscr{X}$, $\mathscr{Y}$ are Banach spaces, let $L(\mathscr{X}, \mathscr{Y})$ be the set of bounded linear operators taking $\mathscr{X}$ into $\mathscr{Y}$. Prove the $L(\mathscr{X}, \mathscr{Y})$ is a Banach space with the norm defined above.

Example 2.1.   Define $f: \mathscr{C}([0, 1], R) \to R$ by $f(\phi) = \int_0^1 \phi(s)\, ds$. The map f is linear and continuous with $|f| = 1$.

Example 2.2.   Define $S = \{\phi$ in $\mathscr{C}([0, 1], R)$ which have a continuous first derivative$\}$. S is dense in $\mathscr{C}([0, 1], R^n)$. For any ϕ in S, define $f\phi(t) = d\phi(t)/dt$, $0 \leq t \leq 1$. The function f is linear but not bounded. In fact, the sequence of functions $\phi_n(t) = t^n$, $0 \leq t \leq 1$, satisfies $\|\phi_n\| = 1$, but $\|f\phi_n\| = n$. Another way to show unboundedness is to prove f is not continuous. Consider the functions $\phi_n(t) = t^n - t^{n+1}$, $0 \leq t \leq 1$, $n \geq 1$. In $\mathscr{C}([0, 1], R)$, $\phi_n \to 0$ as $n \to \infty$, but $f\phi_n(t) = t^{n-1}[n - (n+1)t]$ and $f\phi_n(1) = -1$, which does not approach zero as $n \to \infty$.

Another very important tool from functional analysis which can be used frequently in differential equations is the principle of uniform boundedness. In this book, we have chosen to circumvent this principle by using more elementary proofs except in one instance in Chapter IV.

Principle of uniform boundedness

Suppose $\mathscr{A}$ is an index set and T_α, α in $\mathscr{A}$, are bounded linear maps from a Banach space $\mathscr{X}$ to a Banach space $\mathscr{Y}$ such that for each x in $\mathscr{X}$, $\sup_{\alpha \text{ in } \mathscr{A}}|T_\alpha x| < \infty$. Then $\sup_{\alpha \text{ in } \mathscr{A}}|T_\alpha| < \infty$.

0.3 Fixed Point Theorems

A fixed point of a transformation $T: \mathscr{X} \to \mathscr{X}$ is a point x in $\mathscr{X}$ such that $Tx = x$. Theorems concerning the existence of a fixed point of a transformation are very convenient in differential equations even though not absolutely necessary. Such theorems should be considered as a tool which avoids the repetition of standard arguments and permits one to concentrate on the essential elements of the problem.

One standard tool in analysis is successive approximations. The basic elements of the method of successive approximations have been abstracted into the so called contraction mapping principle by Banach and Cacciopoli. If $\mathscr{F}$ is a subset of a Banach space $\mathscr{X}$ and T is a transformation taking $\mathscr{F}$ into a Banach space $\mathscr{B}$ (written as $T\colon \mathscr{F} \to \mathscr{B}$), then T is a *contraction* on $\mathscr{F}$ if there is a λ, $0 \leq \lambda < 1$, such that

$$|Tx - Ty| \leq \lambda |x - y|, \qquad x,\, y \in \mathscr{F}.$$

The constant λ is called the *contraction constant* for T on $\mathscr{F}$.

THEOREM 3.1. (*Contraction mapping principle of Banach-Cacciopoli*). If $\mathscr{F}$ is a closed subset of a Banach space $\mathscr{X}$ and $T\colon \mathscr{F} \to \mathscr{F}$ is a contraction on $\mathscr{F}$, then T has a unique fixed point $\bar{x}$ in $\mathscr{F}$. Also, if x_0 in $\mathscr{F}$ is arbitrary, then the sequence $\{x_{n+1} = Tx_n,\ n = 0,\ 1,\ 2,\ \ldots\}$ converges to $\bar{x}$ as $n \to \infty$ and $|\bar{x} - x_n| \leq \lambda^n |x_1 - x_0|/(1 - \lambda)$, where $\lambda < 1$ is the contraction constant for T on $\mathscr{F}$.

PROOF. *Uniqueness.* If $0 \leq \lambda < 1$ is the contraction constant for T on $\mathscr{F}$, $x = Tx$, $y = Ty$, $x,\, y \in \mathscr{F}$, then $|x - y| = |Tx - Ty| \leq \lambda |x - y|$. This implies $|x - y| \leq 0$ and thus $|x - y| = 0$ and $x = y$.

Existence. Let x_0 be arbitrary, and $x_{n+1} = Tx_n$, $n = 0,\ 1,\ 2,\ \ldots$. By hypotheses, each x_n, $n = 0,\ 1,\ \ldots$, is in $\mathscr{F}$. Also, $|x_{n+1} - x_n| \leq \lambda |x_n - x_{n-1}|$ $\leq \cdots \leq \lambda^n |x_1 - x_0|$, $n = 0,\ 1,\ \ldots$. Thus, for $m > n$,

$$|x_m - x_n| \leq |x_m - x_{m-1}| + |x_{m-1} - x_{m-2}| + \cdots + |x_{n+1} - x_n|$$

$$\leq [\lambda^{m-1} + \lambda^{m-2} + \cdots + \lambda^n] |x_1 - x_0|$$

$$= \lambda^n [1 + \lambda + \cdots + \lambda^{m-n-1}] |x_1 - x_0|$$

$$= \frac{\lambda^n [1 - \lambda^{m-n}]}{1 - \lambda} |x_1 - x_0| \leq \frac{\lambda^n}{1 - \lambda} |x_1 - x_0|.$$

Thus the sequence $\{x_n\}$ forms a Cauchy sequence and there is an $\bar{x}$ in $\mathscr{X}$ such that $\lim_{n\to\infty} x_n = \bar{x}$. Since $\mathscr{F}$ is closed, $\bar{x}$ is in $\mathscr{F}$. Since T is continuous and $|\cdot|$ is continuous (the latter because $|x| - |x_n - x| \leq |x_n| \leq |x_n - x| + |x|$), it follows that

$$0 = \lim_{m \to \infty} |x_{m+1} - Tx_m| = |\lim_{m \to \infty} [x_{m+1} - Tx_m]| = |\bar{x} - T\bar{x}|,$$

which implies $T\bar{x} = \bar{x}$. This gives the existence of a fixed point.

To prove the last estimate, take the limit as $m \to \infty$ in the previous estimate of $|x_m - x_n|$. This completes the proof of the theorem.

EXERCISE 3.1. Suppose $\mathscr{X}$ is a Banach space, $T\colon \mathscr{X} \to \mathscr{X}$ is a continuous linear operator with $|T| < 1$. Show that $I - T$, I the identity operator, has a bounded inverse; that is, prove that the equation $(I - T)x = y$ has a unique solution x in $\mathscr{X}$ which depends continuously upon y in $\mathscr{X}$.

Let $\mathscr{X}$, $\mathscr{Y}$ be Banach spaces and D be an open set in $\mathscr{X}$. A function $f\colon D \to \mathscr{Y}$ is said to be (*Frechet*) *differentiable* at a point x in D if there is a bounded linear operator $A(x)$ taking $\mathscr{X} \to \mathscr{Y}$ such that for every $h \in \mathscr{X}$ with $x + h \in D$,

$$|f(x+h) - f(x) - A(x)h| \leq \rho(|h|, x),$$

where $\rho(|h|, x)$ satisfies $\rho(|h|, x)/|h| \to 0$ as $|h| \to 0$. The linear operator $A(x)$ is called the *derivative of f at x* and $A(x)h$ the *differential of f at x*.

EXERCISE 3.2. Suppose $\mathscr{X}$, $\mathscr{Y}$ are Banach spaces, $\mathscr{A}$ is an open subset of $\mathscr{X}$, $f\colon \mathscr{A} \to \mathscr{Y}$, and

$$\lim_{t \to 0} \frac{f(x_0 + th) - f(x_0)}{t} = \omega(x_0)h$$

exists for every $x_0 \in \mathscr{A}$, $h \in \mathscr{X}$, where the limit is taken for t real. Suppose $\omega(x_0)$ is a continuous linear mapping for all $x_0 \in \mathscr{A}$ and suppose ω considered as a mapping from $\mathscr{A}$ into $L(\mathscr{X}, \mathscr{Y})$ is continuous, where $L(\mathscr{X}, \mathscr{Y})$ is defined in Exercise 2.3. Prove that f is (Frechet) differentiable with derivative $\omega(x_0)$ at $x_0 \in \mathscr{A}$.

EXERCISE 3.3. Prove that $\omega(x_0)$ of the previous exercise can exist for every h, be a continuous linear mapping and not be the Frechet derivative of f at $x_0 \in \mathscr{A}$.

EXERCISE 3.4. Let $\mathscr{C}^1([0, 1], R^n)$ denote the space of continuously differentiable functions $x\colon [0, 1] \to R^n$, with addition and scalar multiplication defined in the usual way and let $|x| = \sup_{0 \leq t \leq 1} |x(t)| + \sup_{0 \leq t \leq 1} |\dot{x}(t)|$ where $\dot{x}(t) = dx/dt$. Prove $\mathscr{C}^1([0, 1], R^n)$ is a Banach space.

EXERCISE 3.5. Let $w\colon \mathscr{C}^1([0, 1], R^n) \times [0, 1] \to R^n$ be the evaluation mapping, $w(x, t) = x(t)$. Prove that w is continuously differentiable and compute its derivative. Do this exercise two ways, using the definition of the derivative and also Exercise 3.2.

If $f\colon R^m \to R^n$ is differentiable at a point x, then $A(x) = \partial f(x)/\partial x = (\partial f_i(x)/\partial x_j, i = 1, 2, \ldots, n, j = 1, 2, \ldots, m)$, is the Jacobian matrix of f with respect to x.

For later reference, it is convenient to have notation for order relations. We shall say a function $f(x) = O(|x|)$ as $|x| \to 0$ if $|f(x)|/|x|$ is bounded for x in a neighborhood of zero and $f(x) = o(|x|)$ as $|x| \to 0$ if $|f(x)|/|x| \to 0$ as $|x| \to 0$.

Suppose $\mathscr{F}$ is a subset of a Banach space $\mathscr{X}$, $\mathscr{G}$ is a subset of a Banach space $\mathscr{Y}$ and $\{T_y, y \in \mathscr{G}\}$ is a family of operators taking $\mathscr{F} \to \mathscr{X}$. The operator T_y is said to be a *uniform contraction* on $\mathscr{F}$ if $T_y\colon \mathscr{F} \to \mathscr{F}$ and there is a λ, $0 \leq \lambda < 1$ such that

$$|T_y x - T_y \bar{x}| \leq \lambda |x - \bar{x}| \quad \text{for all } y \text{ in } \mathscr{G}, x, \bar{x} \text{ in } \mathscr{F}.$$

In other words, T_y is a contraction for each y in $\mathscr{G}$ and the contraction constant can be chosen independent of y in $\mathscr{G}$.

THEOREM 3.2. If $\mathscr{F}$ is a closed subset of a Banach space $\mathscr{X}$, $\mathscr{G}$ is a subset of a Banach space $\mathscr{Y}$, $T_y\colon \mathscr{F} \to \mathscr{F}$, y in $\mathscr{G}$ is a uniform contraction on $\mathscr{F}$ and $T_y x$ is continuous in y for each fixed x in $\mathscr{F}$, then the unique fixed point $g(y)$ of T_y, y in $\mathscr{G}$, is continuous in y. Furthermore, if $\mathscr{F}$, $\mathscr{G}$ are the closures of open sets $\mathscr{F}^\circ$, $\mathscr{G}^\circ$ and $T_y x$ has continuous first derivatives $A(x, y)$, $B(x, y)$ in y, x, respectively, then $g(y)$ has a continuous first derivative with respect to y in $\mathscr{G}^\circ$.

COROLLARY 3.1. Suppose $\mathscr{X}$ is a Banach space, $\mathscr{G}$ is a subset of a Banach space $\mathscr{Y}$, $A_y\colon \mathscr{X} \to \mathscr{X}$ are continuous linear operators for each y in $\mathscr{G}$, $|A_y| \leq \delta < 1$ for all y in G and $A_y x$ is continuous in y for each x in X. Then the operator $I - A_y$ has a bounded inverse which depends continuously upon y, $|(I - A_y)^{-1}| \leq (1 - \delta)^{-1}$.

PROOFS. Since $T_y\colon \mathscr{F} \to \mathscr{F}$ is a uniform contraction, there is a λ, $0 \leq \lambda < 1$ such that $|T_y x - T_y \bar{x}| \leq \lambda |x - \bar{x}|$ for all y in $\mathscr{G}$, x, $\bar{x}$ in $\mathscr{F}$. Let $g(y)$ be the unique fixed point of T_y in $\mathscr{F}$ which exists from Theorem 3.1. Then

$$g(y + h) - g(y) = T_{y+h} g(y + h) - T_y g(y)$$
$$= T_{y+h} g(y + h) - T_{y+h} g(y) + T_{y+h} g(y) - T_y g(y),$$

and

$$|g(y + h) - g(y)| \leq \lambda |g(y + h) - g(y)| + |T_{y+h} g(y) - T_y g(y)|.$$

This implies

$$|g(y + h) - g(y)| \leq (1 - \lambda)^{-1} |T_{y+h} g(y) - T_y g(y)|.$$

Since $T_y x$ is continuous in y for each fixed x in $\mathscr{F}$, we see that $g(y)$ is continuous. This proves the first part of the theorem.

The proof of Corollary 3.1 is now almost immediate. In fact, we need to show that the equation $x - A_y x = z$ has a unique solution for each z in $\mathscr{X}$ and this solution depends continuously upon y, z. This is equivalent to finding the fixed points of the operator $T_{y, z}$ defined by $T_{y, z} x = A_y x + z$, x in $\mathscr{X}$. Since A_y is a uniform contraction, $(I - A_y)^{-1}$ exists, is bounded and continuous in y. Also, if $(I - A_y)x = y$, then $|y| \geq (1 - \delta)|x|$ and the proof of Corollary 3.1 is complete.

To prove the last part of the theorem, suppose $T_y x$ has continuous first derivatives $A(x, y)$, $B(x, y)$ with respect to y, x, respectively, for $y \in \mathscr{G}^\circ$, $x \in \mathscr{F}^\circ$. Let us use the fact that $g(y) = T_y g(y)$ and try to find the equation that the differential $z = C(y)h$, h in $\mathscr{Y}$ of g will have to satisfy if g has a derivative $C(y)$. If the chain rule of differentiation is valid then

(3.1) $$z = B(g(y), y)z + A(g(y), y)h$$

where h is an arbitrary element of $\mathcal{Y}$. It is easy to show that T_y being a uniform contraction implies $|B(x,y)| < \delta < 1$ for x in $\mathcal{F}^\circ$, y in $\mathcal{G}^\circ$.

Since $|B(x, y)| < \delta < 1$ for x in $\mathcal{F}^\circ$, y in $\mathcal{G}^\circ$, an application of Corollary 3.1 implies, for each y in $\mathcal{G}^\circ$, h in $\mathcal{Y}$, the existence of a unique solution $z(y, h)$ of (3.1) which is continuous in y, h. From uniqueness, one observes that $z(y, \alpha h + \beta u) = \alpha z(y, h) + \beta z(y, u)$ for all scalars α, β and h, u in $\mathcal{Y}$; that is, $z(y, h)$ is linear in h and may be written as $C(y)h$, where $C(y)\colon \mathcal{Y} \to \mathcal{X}$ is a continuous linear operator for each y is also continuous in y. To show that $C(y)$ is the derivative of $g(y)$, let $w = g(y + h) - g(y)$, $B(g(y),y) = B(y)$, $A(g(y),y) = A(y)$ and observe that w satisfies the equation

$$w - B(y)w - A(y)h + f(w,h,y) = 0$$

where, for any $\epsilon > 0$, there is a $\nu > 0$ such that $|f(w,h,y)| < \epsilon(|w| + |h|)$ for $|h| < \nu$, y in $\mathcal{G}^0$. From Corollary 3.1,

$$w - [I - B(y)]^{-1}A(y)w + F(w,h,y) = 0$$

where $|F(w,h,y)| < \epsilon(1 - \delta)^{-1}(|w| + |h|)$ for $|h| < \nu, y$ in $\mathcal{G}^0$. But, this implies

$$|w| \le \mu|h|, \qquad \mu = 2|[I - B(y)]^{-1}A(y)| + 1$$

$$|y - [I - B(y)]^{-1}A(y)h| \le \frac{\epsilon(1 + \mu)}{1 - \delta}|h|.$$

for $|h| < \nu$. This shows that $g(y)$ is continuously differentiable in y and the derivative is given by $C(y)$ satisfying Equation (3.1). This completes the proof of the theorem.

To illustrate the contraction principle, we prove the following important theorem of implicit functions. In the statement of this result det A for an $m \times m$ matrix A denotes the determinant of A.

THEOREM 3.3. (of *Implicit Functions*). Suppose $F\colon R^m \times R^n \to R^m$ has continuous first partial derivatives and $F(0, 0) = 0$. If the Jacobian matrix $\partial F(x, y)/\partial x$ of F with respect to x satisfies det $\partial F(0, 0)/\partial x \neq 0$, then there exist neighborhoods U, V of 0 in R^m, R^n, respectively, such that for each fixed y in V the equation $F(x, y) = 0$ has a unique solution x in U. Furthermore, this solution can be given as $x = g(y)$, where g has continuous first derivatives and $g(0) = 0$.

PROOF: Let us write

$$F(x, y) = Ax - N(x, y),$$

$$A = \frac{\partial F(0, 0)}{\partial x},$$

$$N(x, y) = \frac{\partial F(0, 0)}{\partial x} x - F(x, y), \ N(0, 0) = 0.$$

From the expression for N, we have

$$\frac{\partial N(x, y)}{\partial x} = \frac{\partial F(0, 0)}{\partial x} - \frac{\partial F(x, y)}{\partial x}.$$

The hypothesis of continuity of $\partial F(x, y)/\partial x$ implies $\partial N(x, y)/\partial x \to 0$ as $x \to 0$, $y \to 0$. We therefore have the existence of a function $k(y, \rho)$ which is continuous in $y \in R^n$ and $\rho \geqq 0$ such that $k(0, 0) = 0$ and

$$|N(x, y) - N(\bar{x}, y)| \leqq k(y, \rho)|x - \bar{x}|$$

for all y in R^n and $x, \bar{x}$ with $|x|, |\bar{x}| \leqq \rho$. Since the matrix A is assumed to be nonsingular, finding a solution to $F(x, y) = 0$ is equivalent to finding a solution of the equation $x = A^{-1}N(x, y)$, where A^{-1} is the inverse of A. This is equivalent to finding a fixed point of the operator $T_y: R^m \to R^m$ defined by $T_y x = A^{-1}N(x, y)$. We now show that T_y is a contraction on an appropriate set. There is a constant K (see Exercise 2.2) such that $|A^{-1}x| \leqq K|x|$ for all x in R^m and therefore

$$|T_y x| = |A^{-1}N(x, y)| \leqq K |N(x, y)|$$
$$= K |N(x, y) - N(0, y) + (0, y)|$$
$$\leqq Kk(y, \rho) |x| + K |N(0, y)|$$
$$|T_y x - T_y \bar{x}| \leqq Kk(y, \rho) |x - \bar{x}|$$

for $|x|, |\bar{x}| \leqq \rho$ and all y. Choose ε, δ positive and so small that

$$Kk(y, \rho)\rho + K |N(0, y)| < \varepsilon \quad \text{for} \quad |y| \leqq \delta, \rho \leqq \varepsilon,$$
$$\sup\{Kk(y, \rho), |y| \leqq \delta, \rho \leqq \varepsilon\} < 1,$$

and let $U = \{x \text{ in } R^m: |x| < \varepsilon\}$, $V = \{y \text{ in } R^n: |y| < \delta\}$. It follows that T_y is a uniform contraction of U into U for y in V_0. Therefore T_y has a unique fixed point $g(y)$ in U from Theorem 3.1. It is clear that $g(0) = 0$. Since $T_y x$ is continuous in y for each x it follows from Theorem 3.2 that $g(y)$ is continuous in y. Also $T_y x$ has continuous first derivatives with respect to x and y with the derivative with respect to x being given by $A^{-1} \partial N(x, y)/\partial x$. Theorem 3.2 implies therefore the continuous differentiability of $g(y)$ in y and completes the proof of the implicit function theorem.

EXERCISE 3.6. State and prove a generalization of Theorem 3.3 for Banach spaces. Hint: Redo the steps in the proof of Theorem 3.3 for Banach spaces making appropriate changes and hypotheses where necessary.

The contraction mapping principle can be regarded as a fixed point theorem. Other fixed point theorems of a more sophisticated type are very useful in differential equations. We formulate two more which are used in this book.

An obvious fixed point theorem in one dimension is the following: any continuous mapping of the closed interval [0, 1] into itself must have a fixed point. The proof is obvious if one simply observes that the existence of a fixed point is equivalent to saying that the graph of the function in 2-space must cross the diagonal of the unit square with vertices at (0, 0), (1, 0), (0, 1), (1, 1). After some thought it seems plausible that a similar result should hold in higher dimensions but the proof is difficult. This is the celebrated

BROUWER FIXED POINT THEOREM. Any continuous mapping of the closed unit ball in R^n into itself must have a fixed point.

If a subset A of R^n is homeomorphic to the closed unit ball in R^n and f is a continuous mapping of A into A, then the Brouwer Fixed Point Theorem implies f has a fixed point in A.

Suppose $f: R^n \to R^n$ is a continuous mapping. The zeros of the function f coincide with the fixed points of the mapping g defined by $g(x) = x + f(x)$. If we can show that there is a set D in R^n which is homeomorphic to the closed unit ball in R^n such that g takes D into D, then the Brouwer fixed point theorem implies that g has a fixed point in D and f has a zero in D. This is a very important application of the Brouwer fixed point theorem.

The Brouwer fixed point theorem has been generalized to Banach spaces by Schauder and even more general spaces by Tychonov. We formulate this result for Banach spaces. Recall that a subset $\mathscr{A}$ of a Banach space is compact if any sequence $\{\phi_n\}$, $n = 1, 2, \ldots$ in $\mathscr{A}$ has a subsequence which converges to an element of $\mathscr{A}$. A subset $\mathscr{A}$ is *convex* if for x, y in $\mathscr{A}$ it follows that $tx + (1-t)y$ is in $\mathscr{A}$ for $0 \leq t \leq 1$; that is, $\mathscr{A}$ contains the "line segment" joining x and y. A mapping f of a Banach space $\mathscr{X}$ into a Banach space $\mathscr{Y}$ is said to be *compact* if for every bounded set $\mathscr{A}$ in $\mathscr{X}$ the closure of the set $\{f(x), x \text{ in } \mathscr{A}\}$ is compact. If, in addition, f is continuous, it is called *completely continuous*.

SCHAUDER FIXED POINT THEOREM. If $\mathscr{A}$ is a convex, compact subset of a Banach space $\mathscr{X}$ and $f: \mathscr{A} \to \mathscr{A}$ is continuous, then f has a fixed point in $\mathscr{A}$.

COROLLARY. If $\mathscr{A}$ is a closed, bounded, convex subset of a Banach space $\mathscr{X}$ and $f: \mathscr{A} \to \mathscr{A}$ is completely continuous, then f has a fixed point in $\mathscr{A}$.

The proof of the corollary proceeds as follows: Since $f(\mathscr{A}) \subset \mathscr{A}$ and $\mathscr{A}$ is closed, the closure of $f(\mathscr{A})$ belongs to $\mathscr{A}$ and is compact by hypothesis. Furthermore, the convex closure $\mathscr{B}$ of $f(\mathscr{A})$ [the smallest closed convex set containing $f(\mathscr{A})$] belongs to $\mathscr{A}$ since $\mathscr{A}$ is convex. A theorem of Mazur states that $\mathscr{B}$ is compact. Since $\mathscr{B} \subset \mathscr{A}$, $f(\mathscr{B}) \subset f(\mathscr{A}) \subset \mathscr{B}$ and the previous result implies the existence of a fixed point in $\mathscr{B} \subset \mathscr{A}$.

As remarked earlier, the Schauder theorem was extended to more general spaces (locally convex linear topological spaces) by Tychonov. We do not wish to introduce all of the terminology of locally convex linear topological spaces. In fact, the only such space for which we need the extended form of this theorem is for the space of continuous functions $f: R \to C^n$ for which convergence in the space is equivalent to uniform convergence on compact subsets. A set is bounded in this space if all elements of the set are uniformly bounded continuous functions. The statement of the extended form of this theorem is now exactly the same as before. We will refer to the fixed point theorem in this situation as the *Schauder-Tychonov theorem*.

EXERCISE 3.7.   Show the Schauder theorem is false if either the compactness or the convexity of A is eliminated.

A very useful compact, convex subset of $\mathscr{C}(I, R^n)$, I a closed bounded interval of R^1, is obtained in the following manner. Suppose M, β are positive constants and $\mathscr{A}$ is the subset of $\mathscr{C}(I, R^n)$ such that ϕ in $\mathscr{A}$ implies $|\phi| \leq \beta$, $|\phi(t) - \phi(\bar{t})| \leq M |t - \bar{t}|$, for $t, \bar{t}$ in I. The set $\mathscr{A}$ is obviously convex and closed. Furthermore, any sequence $\{\phi_n\}$ in $\mathscr{A}$ is uniformly bounded and equicontinuous. Lemmas 1.1 and 1.2 imply the existence of a ϕ in $\mathscr{C}(I, R^n)$ such that $\lim_{n \to \infty} \phi_n = \phi$. But $\mathscr{A}$ is closed so that ϕ belongs to $\mathscr{A}$. This proves compactness of $\mathscr{A}$.

The following books are standard references on analysis and functional analysis.

Dunford, N., and J. T. Schwartz, *Linear Operators, Part I: General Theory*, Interscience, New York, 1964.

Graves, L. M., *The Theory of Functions of Real Variables*, 2nd Edition, McGraw-Hill, New York, 1956.

Hurewicz, W., and H. Walman, *Dimension Theory*, Princeton University Press, Princeton, N.J., 1941.

Liusternik, L. A., and V. J. Sobolev, *Elements of Functional Analysis*, Ungar, New York, 1965.

Rudin, W., *Real and Complex Analysis*, McGraw-Hill, New York, 1966.

Yoshida, K., *Functional Analysis*, Springer-Verlag, Berlin, 1965.

CHAPTER I

General Properties of Differential Equations

The purpose of this chapter is to discuss those properties of differential equations which are not dependent upon the specific form of the vector field. The basic existence theorem of Section 1 shows that a differential equation does define a family of functions and Sections 2 and 3 discuss the dependence of this family upon the initial values and parameters. Section 4 contrasts the concept of stability with the concept of continuous dependence upon initial values. Section 5 is concerned with differential equations with vector fields that are only Lebesgue integrable in t. Section 6 is devoted to differential inequalities and their application to the problem of obtaining upper and lower bounds for solutions of differential equations. Sections 7 and 8 deal with some properties of the solution of differential equations which are characteristic of the fact that the vector field is independent of time; namely, the existence of cylinders of orbits near regular points and the concepts of invariant and minimal sets.

I.1. Existence

Let t be a real scalar; let D be an open set in R^{n+1} with an element of D written as (t, x); let $f \colon D \to R^n$, be continuous and let $\dot{x} = dx/dt$. A differential equation is a relation of the form

(1.1)
$$\dot{x}(t) = f(t, x(t)) \quad \text{or, briefly} \quad \dot{x} = f(t, x).$$

We say x is a *solution* of (1.1) on an interval $I \subset R$ if x is a continuously differentiable function defined on I, $(t, x(t)) \in D$, $t \in I$ and x satisfies (1.1) on I. We refer to f as a *vector field* on D.

Example 1.1. Let $D = R^2$, $f(t, x) = x^2$. The function $\phi(t) = -\dfrac{1}{t + c}$, c an arbitrary real number, $c \neq 0$, is a solution of $\dot{x} = x^2$ for $t \in (-c, \infty)$ if $c > 0$; $t \in (-\infty, -c)$ if $c < 0$. (See Fig. 1.1).

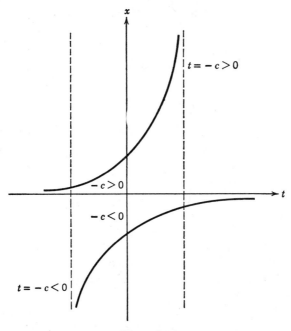

Figure I.1.1

Example 1.2. Let $D = R^2$, $f(t, x) = \sqrt{x}$ for $x \geqq 0$, $= 0$ for $x < 0$. The function $\phi(t) = (t - c)^2/4$, $t \geqq c$, is a solution of $\dot{x} = \sqrt{x}$, $x \geqq 0$. Notice $x = 0$ is also a solution (see Fig. 1.2).

Suppose $(t_0, x_0) \in D$ is given. An *initial value problem for equation (1.1)* consists of finding an interval I containing t_0 and a solution x of (1.1)

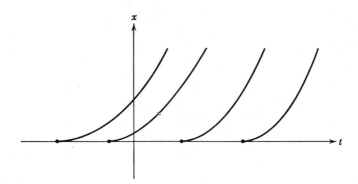

Figure I.1.2

satisfying $x(t_0) = x_0$. We write this problem symbolically as

(1.2) $\dot{x} = f(t, x), x(t_0) = x_0, t \in I.$

If there exists an interval I containing t_0 and an x satisfying (1.2), we refer to this as a solution of (1.1) passing through (t_0, x_0).

For the initial value problem, $\dot{x} = x^2$, $x(0) = -c$, c real, Example 1.1 shows the interval I may depend upon c and may not be the whole real line.

The initial value problem $\dot{x} = \sqrt{x}$, $x \geq 0$, $x(0) = 0$, has the solution $x = 0$ on $(-\infty, \infty)$. The function

$$x(t) = \begin{cases} \dfrac{(t-c)^2}{4}, & t \geq c \geq 0, \\[2mm] 0, & t \leq c. \end{cases}$$

is also a solution. Therefore there need not be a unique solution of (1.2) for every continuous function f.

Our first objective in this chapter is to discuss existence, uniqueness, continuation of solutions, and continuous dependence of solutions on initial data and parameters.

First, notice that consideration of vector equations makes it unnecessary to consider n^{th} order equations. In fact, if y is a scalar, $y^{(j)}$ denotes $d^j y / dt^j$ and

$$y^{(n)} = F(t, y, y', \ldots, y^{(n-1)}),$$
$$y^{(j)}(t_0) = x_{j+1, 0}, \quad j = 0, 1, \ldots, n-1,$$

then, letting $x = (y, y^{(1)}, \ldots, y^{(n-1)})$, $f = (x_2, \ldots, x_n, F)$, we obtain the equivalent problem

$$\dot{x} = f(t, x), \qquad x(t_0) = x_0 = (x_{10}, \ldots, x_{n0}).$$

Also, complex valued differential equations of the real variable t are included in the discussion of (1.1) since one can obtain a real system by taking the real and imaginary parts.

LEMMA 1.1. Problem (1.2) is equivalent to

(1.3) $x(t) = x_0 + \displaystyle\int_{t_0}^{t} f(\tau, x(\tau)) \, d\tau$

provided $f(t, x)$ is continuous.

The proof of this lemma is obvious.

THEOREM 1.1. (Peano) (Existence). If f is continuous in D, then for any $(t_0, x_0) \in D$, there is at least one solution of (1.1) passing through (t_0, x_0).

PROOF. Suppose α, β are positive numbers chosen so that the closed rectangle $B(\alpha, \beta, t_0, x_0) = B(\alpha, \beta) = \{(t, x) : t \in I_\alpha, |x - x_0| \leq \beta\}$, $I_\alpha = I_\alpha(t_0) = $

$\{t: |t - t_0| \leqq \alpha\}$, belongs to D. Let $M = \sup\{|f(t, x)|, (t, x) \in B(\alpha, \beta)\}$. Choose $\bar{\alpha}, \bar{\beta}$ so that $0 < \bar{\alpha} \leqq \alpha$, $0 < \bar{\beta} \leqq \beta$, $M\bar{\alpha} \leqq \bar{\beta}$ and define the set $\mathscr{A} = \mathscr{A}(\bar{\alpha}, \bar{\beta})$ of functions ϕ in $\mathscr{C}(I_{\bar{\alpha}}, R^n)$ which satisfy $\phi(t_0) = x_0$, $|\phi(t) - x_0| \leqq \bar{\beta}$ for all t in $I_{\bar{\alpha}}$. From Chapter 0 the set $\mathscr{A}$ is convex, closed and bounded.

For any ϕ in $\mathscr{A}$ define the function $T\phi$ by the relation

$$T\phi(t) = x_0 + \int_{t_0}^t f(s, \phi(s))\, ds, \quad t \in I_{\bar{\alpha}},$$

From Lemma 1.1 finding fixed points of T is equivalent to solving the above initial value problem for (1.1). We now apply the Schauder fixed point theorem to assert the existence of a fixed point of T in $\mathscr{A}$.

Obviously $T\phi$, ϕ in $\mathscr{A}$, is in $\mathscr{C}(I_{\bar{\alpha}}, R^n)$ and $T\phi(t_0) = x_0$. Also, for $t \in I_{\bar{\alpha}}$,

$$|T\phi(t) - x_0| \leqq \left| \int_{t_0}^t |f(s, \phi(s))|\, ds \right| \leqq M|t - t_0|$$

$$\leqq M\bar{\alpha} \leqq \bar{\beta}$$

since $B(\bar{\alpha}, \bar{\beta}) \subset B(\alpha, \beta)$. Thus $T: \mathscr{A} \to \mathscr{A}$. Also

$$|T\phi(t) - T\phi(\bar{t})| \leqq \left| \int_{\bar{t}}^t |f(s, \phi(s))|\, ds \right| \leqq M|t - \bar{t}|$$

for all t, $\bar{t}$ in $I_{\bar{\alpha}}$. This implies the set $T(\mathscr{A})$ is an equicontinuous family and therefore the closure of $T(\mathscr{A})$ is compact.

Finally, for any ϕ, $\bar{\phi}$ in $\mathscr{A}$, it follows from the uniform continuity of $f(t, x)$ on $B(\alpha, \beta)$ that for any $\varepsilon > 0$ there is a $\delta > 0$ such that

$$|T\phi(t) - T\bar{\phi}(t)| \leqq \left| \int_{t_0}^t |f(s, \phi(s)) - f(s, \bar{\phi}(s))|\, ds \right| \leqq \varepsilon\bar{\alpha},$$

for all t in $I_{\bar{\alpha}}$ provided that $|\phi(s) - \bar{\phi}(s)| \leqq \delta$ for all s in $I_{\bar{\alpha}}$. But this is precisely the statement that T is a continuous mapping; that is, for any $\varepsilon > 0$, there is a $\delta > 0$ such that $|T\phi - T\bar{\phi}| \leqq \varepsilon\bar{\alpha}$ if $|\phi - \bar{\phi}| \leqq \delta$.

All of the conditions of the Schauder theorem are satisfied and we can assert the existence of a fixed point of $\mathscr{A}$. This completes the proof of the theorem.

COROLLARY 1.1. If U is a compact set of D, $U \subset V$, an open set in D with the closure $\bar{V}$ of V in D, then there is an $\alpha > 0$ such that, for any initial value $(t_0, x_0) \in U$, there is a solution of (1.1) through (t_0, x_0) which exists at least on the interval $t_0 - \alpha \leqq t \leqq t_0 + \alpha$.

PROOF. One merely repeats the proof of Theorem 1.1 with the further restriction on $\bar{\alpha}, \bar{\beta}$ to ensure that $B(\bar{\alpha}, \bar{\beta}, t_0, x_0) \subset V$ for all $(t_0, x_0) \in U$.

Other proofs of the Peano theorem can be given without using the Schauder theorem. We illustrate the idea for one special construction, the Euler method of numerical analysis. This method consists of dividing the interval $I_\alpha = \{t\colon |t - t_0| \leq \alpha\}$ into equal segments of length h and then on each of these small segments approximating the "solution" by a straight line. More specifically, for a given h, define the function ϕ^h on I_α by

$$\phi^h(t) = x_0 + f(t_0, x_0)(t - t_0), \qquad t_0 \leq t \leq t_0 + h,$$
$$\phi^h(t) = \phi^h(t_0 + h) + f(t_0 + h, \phi^h(t_0 + h))(t - t_0 - h), \qquad t_0 + h \leq t \leq t_0 + 2h,$$

and so forth. One may have to choose α small in order for $(t, \phi^h(t))$ to be in D. Pictorially, we will have a polygonal function defined which for h small should approximate a solution of (1.1) if it exists. One now chooses a sequence $\{h_k\}$ such that $h_k \to 0$ as $k \to \infty$ and uses the Ascoli-Arzela theorem to show that a subsequence of the sequence ϕ^{h_k} converges to a solution of (1.1).

EXERCISE 1.1. Supply all details of the proof of the Peano theorem of existence using polygonal line segments.

EXERCISE 1.2. State an implicit function theorem whose validity will be implied by the existence Theorem 1.1.

I.2. Continuation of Solutions

If ϕ is a solution of a differential equation on an interval I, we say $\hat{\phi}$ is a *continuation* of ϕ if $\hat{\phi}$ is defined on an interval $\hat{I}$ which properly contains I, $\hat{\phi}$ coincides with ϕ on I and $\hat{\phi}$ satisfies the differential equation on $\hat{I}$. A solution ϕ is *noncontinuable* if no such continuation exists; that is, the interval I is the *maximal interval of existence* of the solution ϕ.

LEMMA 2.1. If D is an open set in R^{n+1}, $f\colon D \to R^n$ is continuous and bounded on D, then any solution $\phi(t)$ of (1.1) defined on an interval (a, b) is such that $\phi(a + 0)$ and $\phi(b - 0)$ exist. If $f(b, \phi(b - 0))$ is or can be defined so that $f(t, x)$ is continuous at $(b, \phi(b - 0))$, then $\phi(t)$ is a solution of (1.1) on $(a, b]$. The same remark applies to the left endpoint a.

PROOF. We first show that the limits $\phi(a + 0)$, $\phi(b - 0)$ exist. For any t_0 in (a, b),

$$\phi(t) = \phi(t_0) + \int_{t_0}^t f(s, \phi(s))\, ds, \qquad a < t < b,$$

and, for $a < t_1 \leq t_2 < b$,

$$|\phi(t_2) - \phi(t_1)| \leq \int_{t_1}^{t_2} |f(s, \phi(s))|\, ds \leq M(t_2 - t_1),$$

where M is a bound of $f(t, x)$ on D. Therefore $\phi(t_2) - \phi(t_1) \to 0$ as $t_1, t_2 \to a + 0$, which implies $\phi(a + 0)$ exists. A similar argument shows that $\phi(b - 0)$ exists.

The last conclusion of the lemma is obvious from the integral equation for ϕ.

THEOREM 2.1. If D is an open set in R^{n+1}, $f: D \to R^n$ is continuous and $\phi(t)$ is a solution of (1.1) on some interval, then there is a continuation of ϕ to a maximal interval of existence. Furthermore, if (a, b) is a maximal interval of existence of a solution x of (1.1), then $(t, x(t))$ tends to the boundary of D as $t \to a$ and $t \to b$.

PROOF. Suppose $x(t)$ is a solution of (1.1) on an interval I. If I is not a maximal interval of existence, then x can be extended to an interval properly containing I. Therefore, we may assume I is closed on one end, say to the right. We first show that x can be extended to a maximal right interval of existence and, therefore, may assume that $I = [a, b]$ and that x has no extension over $[a, \infty)$. The proof of the extension to the left is very similar.

Suppose U is a compact set of D, $U \subset V$, an open set in D with the closure $\bar{V}$ of V in D. From Corollary 1.1 for any initial value in U there is a solution of (1.1) existing over an interval of length α depending only on U, V and the bound of f on V. Therefore, if $x(t)$, $a \leq t \leq b$, belongs to U, then there is an extension of x to an interval $[a, b + \alpha]$. Since U is compact, one can continue this process a finite number of times to conclude there is an extension of $x(t)$ to an interval $[a, b_U]$ such that $(b_U, x(b_U))$ does not belong to U.

Now choose a sequence V_n of open sets in D such that $\bigcup_{n=1}^{\infty} V_n = D$, $\bar{V}_n$ closed, bounded, $\bar{V}_n \subset V_{n+1}$ for $n = 1, 2, \ldots$. For each V_n, there is a monotone increasing sequence $\{b_n\}$ constructed as above so that the solution $x(t)$ of (1.1) on $[a, b]$ has an extension to the interval $[a, b_n]$ and $(b_n, x(b_n))$ is not in $\bar{V}_n$. Since the b_n are bounded above, let $\omega = \lim_{n \to \infty} b_n$. It is clear that x has been extended to the interval $[a, \omega)$ and cannot be extended any further since the sequence $(b_k, x(b_k))$ is either unbounded or has a limit point on the boundary of the domain of definition of f.

If $\omega = \infty$, the last assertion in the theorem is trivial. Suppose ω is finite, U is a compact set of D and there is a sequence $(t_k, x(t_k)), y$ in $R^n, (\omega, y)$ in U, such that $t_k \to w$, $x(t_k) \to y$ as $k \to \infty$. The fact that f is bounded in a neighborhood of (ω, y) implies x is uniformly continuous on $[a, \omega)$ and $x(t) \to y$ as $t \to \omega^-$. Thus, there is an extension of x to the interval $[a, \omega + \alpha]$. Since $\omega + \alpha > \omega$, this is a contradiction and shows there is a t_U such that $(t, x(t))$ is not in U for $t_U < t < \omega$. Since U is an arbitrary compact set, this proves $(t, x(t))$ tends to the boundary of D. The proof of the theorem is complete.

EXERCISE 2.1. For t, x scalars, give an example of a function $f(t, x)$ which is defined and continuous on an open bounded connected set D and yet not every noncontinuable solution ϕ of (1.1) defined on (a, b) has $\phi(a + 0)$, $\phi(b - 0)$ existing.

The above continuation theorem can be used in specific examples to verify that a solution is defined on a large time interval. For example, if it is desired to show that a solution is defined on an interval $[t_0, \infty)$, it is sufficient to proceed as follows. If the function $f(t, x)$ is continuous for t in (t_1, ∞), $t_1 < t_0$, $|x| < \alpha$, and one can by some means ascertain that a certain solution $x(t)$ must always satisfy $|x(t)| \leq \beta < \alpha$ for all values of $t \geq t_0$ for which $x(t)$ is defined, then necessarily $x(t)$ is defined on $[t_0, \infty)$. In fact, choose any $T \geq t_0$ and γ such that $\beta < \gamma < \alpha$ and define the rectangle D_1 as $D_1 = \{(t, x): t_0 \leq t \leq T, |x| \leq \gamma\}$. Then $f(t, x)$ is bounded on D_1 and the continuation theorem implies that the solution $x(t)$ can be continued to the boundary of D_1. But $\gamma > \beta$ implies that $x(t)$ must reach this boundary by reaching the face of the rectangle defined by $t = T$. Therefore $x(t)$ exists for $t_0 \leq t \leq T$. Since T is arbitrary, this proves the assertion.

I.3. Uniqueness and Continuity Properties

A function $f(t, x)$ defined on a domain D in R^{n+1} is said to be *locally lipschitzian* in x if for any closed bounded set U in D there is a $k = k_U$ such that $|f(t, x) - f(t, y)| \leq k |x - y|$ for (t, x), (t, y) in U. If $f(t, x)$ has continuous first partial derivatives with respect to x in D, then $f(t, x)$ is locally lipschitzian in x.

If $f(t, x)$ is continuous in a domain D, then the fundamental existence theorem implies the existence of at least one solution of (1.1) passing through a given point (t_0, x_0) in D. Suppose, in addition, there is only one such solution $x(t, t_0, x_0)$ through a given (t_0, x_0) in D. For any $(t_0, x_0) \in D$, let $(a(t_0, x_0), b(t_0, x_0))$ be the maximal interval of existence of $x(t, t_0, x_0)$ and let $E \subset R^{n+2}$ be defined by

$$E = \{(t, t_0, x_0): a(t_0, x_0) < t < b(t_0, x_0), (t_0, x_0) \in D\}.$$

The *trajectory through* (t_0, x_0) is the set of points in R^{n+1} given by $(t, x(t, t_0, x_0))$ for t varying over all possible values for which (t, t_0, x_0) belongs to E. The set E is called the domain of definition of $x(t, t_0, x_0)$.

The basic existence and uniqueness theorem under the hypothesis that $f(t, x)$ is locally lipschitzian in x is usually referred to as the *Picard-Lindelöf theorem*. This result as well as additional information is contained in

THEOREM 3.1.   If $f(t, x)$ is continuous in D and locally lipschitzian with respect to x in D, then for any (t_0, x_0) in D, there exists a *unique* solution $x(t, t_0, x_0)$, $x(t_0, t_0, x_0) = x_0$, of (1.1) passing through (t_0, x_0). Furthermore, the domain E in R^{n+2} of definition of the function $x(t, t_0, x_0)$ is open and $x(t, t_0, x_0)$ is continuous in E.

PROOF. Define $I_\alpha = I_\alpha(t_0)$ and $B(\alpha, \beta, t_0, x_0)$ as in the proof of Theorem 1.1. For any given closed bounded subset U of D choose positive α, β so that $B(\alpha, \beta, t_0, x_0)$ belongs to D for each (t_0, x_0) in U and if

$$V = \cup\{B(\alpha, \beta, t_0, x_0); (t_0, x_0) \text{ in } U\},$$

then the closure of V is in D. Let $M = \sup\{|f(t, x)|, (t, x) \text{ in } V\}$ and let k be the lipschitz constant of $f(t, x)$ with respect to x on V. Choose $\bar{\alpha}, \bar{\beta}$ so that $0 < \bar{\alpha} \leq \alpha$, $0 < \bar{\beta} \leq \beta$, $M\bar{\alpha} \leq \bar{\beta}$, $k\bar{\alpha} < 1$ and let $\mathscr{F} = \{\phi \text{ in } \mathscr{C}(I_{\bar{\alpha}}(0), R^n): \phi(0) = 0, |\phi(t)| \leq \bar{\beta} \text{ for } t \text{ in } I_{\bar{\alpha}}(0)\}$. For any ϕ in $\mathscr{F}$, define a continuous function $T\phi$ taking $I_{\bar{\alpha}}(0)$ into R^n by

$$(3.1) \qquad T\phi(t) = \int_{t_0}^{t+t_0} f(s, \phi(s - t_0) + x_0)\, ds, t \text{ in } I_{\bar{\alpha}}(0).$$

By Lemma 1.1, the fixed points of T in $\mathscr{F}$ coincide with the solutions $x(t) = \phi(t - t_0) + x_0$ of (1.1) which pass through (t_0, x_0) and are such that $(t, x(t))$ is in $B(\bar{\alpha}, \bar{\beta}, t_0, x_0)$. Obviously, $T\phi(0) = 0$ and it is easy to see that $|T\phi(t)| \leq M\bar{\alpha} \leq \bar{\beta}$ for all t in $I_{\bar{\alpha}}(0)$. Therefore, $T\mathscr{F} \subset \mathscr{F}$. Also, $|T\phi(t) - T\bar{\phi}(t)| \leq k\bar{\alpha}|\phi - \bar{\phi}|$ for t in $I_{\bar{\alpha}}(0)$ or $|T\phi - T\bar{\phi}| \leq k\bar{\alpha}|\phi - \bar{\phi}|$. Since $k\bar{\alpha} < 1$, T is a contraction mapping of $\mathscr{F}$ into $\mathscr{F}$. Since $\mathscr{F}$ is closed, there is a unique fixed point $\phi(t, t_0, x_0)$ of T in $\mathscr{F}$ and therefore there is a unique solution of (1.1) passing through (t_0, x_0) and such that $(t, x(t))$ belongs to $B(\bar{\alpha}, \bar{\beta}, t_0, x_0)$.

If the mapping T is considered as a function of (t_0, x_0); that is, $T = T_{(t_0, x_0)}$, then the above argument shows that $T_{(t_0, x_0)}$ is a uniform contraction on $\mathscr{F}$ for (t_0, x_0) in U. Therefore, $\phi(\cdot, t_0, x_0)$ is continuous in (t_0, x_0) from Theorem 0.3.2. This means $\phi(t, t_0, x_0)$ is continuous in t_0, x_0 uniformly with respect to t. But $\phi(t, t_0, x_0)$ is obviously continuous in t and therefore $\phi(t, t_0, x_0)$ is jointly continuous in (t, t_0, x_0) for t in $I_{\bar{\alpha}}(0)$, (t_0, x_0) in U. Finally, the function $x(t, t_0, x_0) = \phi(t - t_0, t_0, x_0) + x_0$ is a continuous function of (t, t_0, x_0) for t in $I_{\bar{\alpha}}(t_0)$, (t_0, x_0) in U.

The above proof of local uniqueness implies global uniqueness. In fact, if there exist two solutions $x(t), y(t)$ of (1.1) with $x(t_0) = x_0$, $y(t_0) = y_0$ and a number s such that $x(s) = y(s)$ and, in any neighborhood of s, there exist an s such that $x(s) \neq y(s)$, then we can enclose the trajectories defined by $x(t)$, $y(t)$ for t in a neighborhood of s in a compact set U contained entirely in D. The above proof shows there is a unique solution of (1.1) passing through any point in U. This is a contradiction.

Now suppose (s, t_0, x_0) is any given point in the domain of definition E of $x(s, t_0, x_0)$. For ease in notation suppose $s \geq t_0$. The case $s \leq t_0$ is treated in the same manner. This implies the closed set $U = \{(t, x(t, t_0, x_0), t_0 \leq t \leq s\}$ belongs to E. Therefore, we can apply the previous results to see that $x(t, \xi, \eta)$

is a continuous function of (t, ξ, η) for $|t - \xi| \leqq \bar{\alpha}$, (ξ, η) in U. There exists an integer k such that $t_0 + k\bar{\alpha} > s \geqq t_0 + (k - 1)\bar{\alpha}$. From uniqueness, we have $x(t + t_0 + \bar{\alpha}, t_0, x_0) = x(t + t_0 + \bar{\alpha}, t_0 + \bar{\alpha}, x(t_0 + \bar{\alpha}, t_0, x_0))$ for any t. But the previous remarks imply this function is continuous for $|t| \leqq \bar{\alpha}$. Therefore, $x(\xi, t_0, x_0)$ is continuous for $|\xi - t_0| \leqq 2\bar{\alpha}$, (t_0, x_0) in D as long as (ξ, t_0, x_0) is in E. An obvious induction argument proves the continuity of $x(s, t_0, x_0)$ at s. The previous argument also implies E is open. This proves the theorem.

THEOREM 3.2. If in addition to the hypotheses of Theorem 3.1 the function $f = f(t, x, \lambda)$ depends upon a parameter λ in a closed set G in R^k, is continuous for (t, x) in D, λ in G, and has a local lipschitz constant with respect to x independent of λ in G, then for each (t_0, x_0) in D, λ in G, there is a unique solution $x(t, t_0, x_0, \lambda)$, $x(t_0, t_0, x_0, \lambda) = x_0$, passing through (t_0, x_0) and it is a continuous function of (t, t_0, x_0, λ) in its domain of definition.

PROOF. The proof is essentially the same as the proof of Theorem 3.1 except one chooses $M = \sup\{|f(t, x, \lambda)| : (t, x, \lambda) \text{ in } V \times G_1\}$ and k independent of λ to be a local lipschitz constant with respect to x on $V \times G_1$, where G_1 is an arbitrary closed bounded set in G.

Since the contraction principle was used in Theorems 3.1 and 3.2 to prove the local existence and uniqueness of the solution passing through (t_0, x_0) the solution can be obtained by the method of successive approximations

$$(3.2) \qquad x^{(n+1)} = Tx^{(n)}, \qquad n = 0, 1, 2, \ldots,$$

$$Tx(t) = x_0 + \int_{t_0}^{t} f(s, x(s))\, ds, \qquad |t - t_0| \leqq \bar{\alpha},$$

where $x^{(0)}$ is an arbitrary continuous function taking the interval $|t - t_0| \leqq \bar{\alpha}$ into R^n with $|x^{(0)}(t) - x_0| \leqq \bar{\beta}$ for $|t - t_0| \leqq \bar{\alpha}$. The constants $\bar{\alpha}$, $\bar{\beta}$ were chosen so that $M\bar{\alpha} \leqq \bar{\beta}$ and $k\bar{\beta} < 1$ where M and k were bounds on $f(t, x)$ and its lipschitz constant with respect to x over a certain compact set. An obvious choice for $x^{(0)}(t)$ is the constant function x_0. Now, if one works directly with the successive approximations (3.2), one can prove the iterations converge and the equation (1.1) has a unique solution on the interval $|t - t_0| \leqq \bar{\alpha}$ under *only* the assumption $M\bar{\alpha} \leqq \bar{\beta}$ with the restriction $k\bar{\alpha} < 1$ being unnecessary. This makes it appear that the contraction principle is not as effective as successive approximations whereas we implied in Chapter 0 that this principle was introduced so as to replace successive approximations. The discrepancy is resolved by simply observing there are many equivalent norms on the space of continuous functions satisfying $|x(t) - x_0| \leqq \bar{\beta}$ for t in $[t_0, t_0 + \bar{\alpha}]$. In fact, with the constants as above, the norm $|\cdot|$ defined by

$$|x| = \sup_{t_0 \leqq t \leqq t_0 + \bar{\alpha}} \{e^{-2k(t - t_0)}|x(t)|\},$$

is equivalent to the supremum norm used earlier. In this norm, one shows easily that T is a contraction only under the assumption $M\bar{\alpha} \leq \bar{\beta}$, which is the same result obtained by successive approximations. How do you resolve the above discrepancy on $[t_0 - \bar{\alpha}, t_0]$?

Another remark on successive approximations is the fact that one can obtain convergence of the sequence defined by (3.2) for any initial $x^{(0)}$ satisfying appropriate bounds. The sequence obviously would converge more rapidly by a more clever choice of $x^{(0)}(t)$ depending upon f since choosing $x^{(0)}(t)$ as the solution itself would yield the fixed point in one iteration. Effective use of such a procedure would require tremendous ingenuity.

EXERCISE 3.1. Prove directly that the sequence of successive approximations (3.2) converges for $M\bar{\alpha} \leq \bar{\beta}$ where M is a bound for f on $\{(t, x): |t - t_0| \leq \bar{\alpha}, |x - x_0| \leq \bar{\beta}\}$.

In the following, we need other results on the dependence of solutions on parameters and initial data.

THEOREM 3.3. If $f(t, x, \lambda)$ has continuous first derivatives with respect to x, λ for $(t, x) \in D$, λ in an open set $G \subset R^k$, then the solution $x(t, t_0, x_0, \lambda)$, $x(t_0, t_0, x_0, \lambda) = x_0$, of

$$(3.3) \qquad \dot{x} = f(t, x, \lambda),$$

is continuously differentiable with respect to t, t_0, x_0, λ in its domain of definition. The matrix $\partial x(t, t_0, x_0, \lambda)/\partial\lambda$, $\partial x(t_0, t_0, x_0, \lambda)/\partial\lambda = 0$, satisfies the matrix differential equation

$$(3.4) \qquad \dot{y} = \frac{\partial f(t, x(t, t_0, x_0, \lambda), \lambda)}{\partial x} y + \frac{\partial f(t, x(t, t_0, x_0, \lambda), \lambda)}{\partial \lambda}.$$

The matrix $\partial x(t, t_0, x_0, \lambda)/\partial x_0$, $\partial x(t_0, t_0, x_0, \lambda)/\partial x_0 = I$, the identity, satisfies the *linear variational equation*,

$$(3.5) \qquad \dot{y} = \frac{\partial f(t, x(t, t_0, x_0, \lambda), \lambda)}{\partial x} y.$$

Furthermore,

$$(3.6) \qquad \frac{\partial x(t, t_0, x_0, \lambda)}{\partial t_0} = -\frac{\partial x(t, t_0, x_0, \lambda)}{\partial x_0} f(t_0, x_0, \lambda).$$

PROOF. As in Theorem 3.2, we apply Theorem 3.2, Chapter 0. If $\gamma = (x_0, \lambda)$, and $T = T_\gamma$ is defined by (3.1), then we must compute the derivatives $A(\gamma, \phi)$, $B(\gamma, \phi)$ of $T_\gamma \phi$ with respect to γ, ϕ, respectively. One shows easily that $A(\gamma, \phi)$ can be represented by the $n \times (n + k)$ matrix

$$A(\gamma, \phi)(t) = \left[\int_{t_0}^{t_0+t} \left(\frac{\partial f(s, \phi(s - t_0) + x_0, \lambda)}{\partial x} \right) ds, \right.$$
$$\left. \int_{t_0}^{t_0+t} \left(\frac{\partial f(s, \phi(s - t_0) + x_0), \lambda)}{\partial \lambda} \right) ds \right],$$

and the differential $B(\gamma, \phi)\psi$ is given by

$$[B(\gamma, \phi)\psi](t) = \int_{t_0}^{t_0+t} \frac{\partial f(s, \phi(s-t_0)+x_0, \lambda)}{\partial x} \psi(s)\, ds.$$

As in the proof of Theorem 3.2, the constants M, k are chosen as before relative to the set $V \times G_1$ with G_1 a closed bounded set in G. The proof of the differentiability of $\phi(t, t_0, x_0, \lambda)$ with respect to x_0, λ now proceeds exactly as in the proof of Theorem 3.1. The function $x(t, t_0, x_0, \lambda) = \phi(t-t_0, t_0, x_0, \lambda) + x_0$ is obviously differentiable in (t, x_0, λ).

Knowing the differentiability immediately implies relations (3.4) and (3.5). To obtain relation (3.6) observe that uniqueness of the solution implies $x(t, t_0, x_0, \lambda) = x(t, t_0+h, x(t_0+h, t_0, x_0, \lambda), \lambda)$ for any h sufficiently small since at $t_0 + h$, these two functions satisfy the equation and take on the same values. Therefore,

$$x(t, t_0+h, x_0, \lambda) - x(t, t_0, x_0, \lambda)$$
$$= x(t, t_0+h, x_0, \lambda) - x(t, t_0+h, x(t_0+h, t_0, x_0, \lambda), \lambda)$$

and this implies $x(t, t_0, x_0, \lambda)$ is differentiable in t_0 and relation (3.6). The proof of the theorem is complete.

By repeated application of the above theorem, one can obtain higher order derivatives of $x(t, t_0, x_0, \lambda)$ under appropriate hypotheses of f.

EXERCISE 3.2.   If $f(t, x)$ has continuous partial derivatives with respect to x up through order k, show that the solution $x(t, t_0, x_0)$, $x(t_0, t_0, x_0) = x_0$, of (1.1) has continuous derivatives of order k with respect to x_0. Find the differential equation for the jth derivatives with respect to x_0 and observe that the Taylor series for $x(t, t_0, y)$ in y in a neighborhood of x_0 is obtained by solving only nonhomogeneous linear equations.

EXERCISE 3.3.   If $f(t, x)$ is analytic in x in a neighborhood of x_0, show there is an interval around t_0 such that the function $x(t, t_0, x_0)$ of Exercise 3.2 is analytic in a neighborhood of x_0.

EXERCISE 3.4.   If $f(t, x, \lambda)$ has continuous partial derivatives with respect to x, λ up through order k, show that the solution $x(t, t_0, x_0, \lambda)$, $x(t_0, t_0, x_0, \lambda) = x_0$, of (3.3) has continuous derivatives of order k with respect to x_0, λ. Find the differential equations for the partial derivatives with respect to λ. Discuss the determination of the Taylor series in the neighborhood of some point.

EXERCISE 3.5.   As in Exercise 3.2, discuss the analyticity properties of $x(t, t_0, x_0, \lambda)$ in λ when $f(t, x, \lambda)$ satisfies some analyticity conditions.

EXERCISE 3.6. Show that the solutions of the equation $\dot{x} = (A(t)$ $+ \lambda B(t))x$ where $|B(t)|$, $|A(t)|$ are continuous and bounded are entire functions of λ. Can you generalize your result?

EXERCISE 3.7. Suppose $f(t, x, y)$ is continuous and has continuous first derivatives with respect to x, y for $0 \leq t \leq 1$, $x, y \in (-\infty, \infty)$, and the boundary value problem

$$\ddot{x} = f(t, x, \dot{x}), \qquad x(0) = a, \qquad x(1) = b,$$

has a solution $\phi(t)$, $0 \leq t \leq 1$. If $\partial f(t, x, y)/\partial x > 0$ for $t \in [0, 1]$ and all x, y, prove there is an $\varepsilon > 0$ such that the boundary value problem

$$\ddot{x} = f(t, x, \dot{x}), \qquad x(0) = a, \qquad x(1) = \beta,$$

has a solution for $0 \leq t \leq 1$ and $|\beta - b| \leq \varepsilon$. Hint: Consider the solution $\psi(t, \alpha)$ of the initial value problem

$$\ddot{x} = f(t, x, \dot{x}), \qquad x(0) = a, \qquad \dot{x}(0) = \alpha,$$

and let $\psi(t, \alpha_0) = \phi(t)$. For $\alpha - \alpha_0$ sufficiently small, $\psi(t, \alpha)$ exists for $0 \leq t \leq 1$. If $u(t) = \partial \psi(t, \alpha_0)/\partial \alpha$, then

$$(4) \quad \ddot{u} - \frac{\partial f(t, \phi(t), \dot{\phi}(t))}{\partial y} \dot{u} - \frac{\partial f(t, \phi(t), \dot{\phi}(t))}{\partial x} u = 0,$$

with $u(0) = 0$, $\dot{u}(0) = 1$. Show that u is monotone nondecreasing and, thus, $u(1) > 0$. Use the implicit function theorem to solve $\psi(1, \alpha) - \beta = 0$ for α as a function of β.

Equation (3.5) is called the linear variational equation for the following reason. If $x(t) = z(t) + x(t, t_0, x_0, \lambda)$ is a solution of (3.3), then

$$\dot{z}(t) + \frac{\partial x(t, t_0, x_0, \lambda)}{\partial t} = f(t, z(t) + x(t, t_0, x_0, \lambda), \lambda).$$

This implies

$$\dot{z}(t) = f(t, z(t) + x(t, t_0, x_0, \lambda), \lambda) - f(t, x(t, t_0, x_0, \lambda), \lambda)$$

$$= \frac{\partial f(t, x(t, t_0, x_0, \lambda), \lambda)}{\partial x} z(t) + o(|z(t)|)$$

as $|z(t)|$ approaches zero. Equation (3.5) therefore represents the first approximation to the variation $z(t)$ of the true solution of (3.2) from a given solution $x(t, t_0, x_0, \lambda)$. The linear variational equation is obtained by neglecting the terms of order higher than the first in the equation for z.

The conclusions of Theorems 3.1 and 3.2 concerning the continuity properties of $x(t, t_0, x_0, \lambda)$ are valid under much weaker hypotheses than stated. In fact, if we assume the uniqueness of the solution $x(t, t_0, x_0)$ (which is implied when you have a local lipschitz condition in x) then one can prove directly that it must be continuous in (t, t_0, x_0). We need one such result in the study of differential inequalities which we now formulate.

LEMMA 3.1.   Suppose $\{f_n\}$, $n = 1, 2, \ldots$, is a sequence of functions defined and continuous on an open set D in R^{n+1} with $\lim_{n\to\infty} f_n = f_0$ uniformly on compact subsets of D. Suppose (t_n, x_n) is a sequence of points in D converging to (t_0, x_0) in D as $n \to \infty$ and let $\phi_n(t), n = 0, 1, 2, \ldots$, be a solution of the equation $\dot{x} = f_n(t, x)$ passing through the point (t_n, x_n). If $\phi_0(t)$ is defined on $[a, b]$ and is unique, then there is an integer n_0 such that each $\phi_n(t)$, $n \geq n_0$, can be defined on $[a, b]$ and converges to $\phi_0(t)$ uniformly on $[a, b]$.

PROOF.   Let U be a compact subset of D which contains in its interior the set $\{(t, \phi_0(t), a \leq t \leq b\}$ and suppose $|f_0| < M$ on U. Then necessarily $|f_n| < M$ on U if $n \geq n_0$ where n_0 is sufficiently large. Choosing $\bar{\alpha}, \bar{\beta}$ as in the proof of the basic existence theorem such that $M\bar{\alpha} \leq \bar{\beta}$, one can be assured for n sufficiently large that all the $\phi_n(t)$ are defined on $[t_n, t_n + \bar{\alpha}]$. If $[\delta, \gamma] = \bigcap_n [t_n, t_n + \bar{\alpha}]$, then the fact that $t_n \to t_0$ implies $\delta < \gamma$. Also, all the ϕ_n for n sufficiently large are defined on $[\delta, \gamma]$. The sequence $\{\phi_n\}$ is uniformly bounded and equicontinuous since $|f_n| < M$. Therefore, there exists a subsequence which we label again as ϕ_n which converges uniformly to a function $\bar{\phi}$ on $[\delta, \gamma]$. Using the integral equation, we see that $\bar{\phi}$ is ϕ_0. Since every convergent subsequence of ϕ_n on $[\delta, \gamma]$ must also converge to ϕ_0 (by uniqueness of ϕ_0) it follows that ϕ_n converges to ϕ_0 on $[\delta, \gamma]$. Due to the compactness of the set $\{t, \phi(t)\}$, t in $[a, b]$, one completes the proof by successively stepping intervals of length $\gamma - \delta$ until $[a, b]$ is covered.

THEOREM 3.4.   Suppose $f(t, x, \lambda)$ is a continuous function of (t, x, λ) for (t, x) in an open set D and λ in a neighborhood of λ_0 in R^p. If $x(t, t_0, x_0, \lambda_0)$, $x(t_0, t_0, x_0, \lambda_0) = x_0$, is a solution of (3.3) on $[a, b]$ and is unique, then there is a solution $x(t, s, \eta, \lambda)$, $x(s, s, \eta, \lambda) = \eta$, of (3.3) which is defined on $[a, b]$ for all s, η, λ sufficiently near t_0, x_0, λ_0 and is a continuous function of (t, s, η, λ) at (t, t_0, x_0, λ_0).

PROOF.   Lemma 3.1 implies that $x(t, s, \eta, \lambda)$ is a continuous function of s, η, λ at t_0, x_0, λ_0 uniformly with respect to t in $[a, b]$. Thus, for any $\varepsilon > 0$ there is a $\delta_1 > 0$ such that

$$|x(t, s, \eta, \lambda) - x(t, t_0, x_0, \lambda_0)| < \frac{\varepsilon}{2}$$

if $|(s, \eta, \lambda) - (t_0, x_0, \lambda_0)| < \delta_1$. Since $x(t, t_0, x_0, \lambda_0)$ is a continuous function of t for t in $[a, b]$, there is a δ_2 such that

$$\left| x(t, t_0, x_0, \lambda_0) - x(\tau, t_0, x_0, \lambda_0) \right| < \frac{\varepsilon}{2}$$

if $|t - \tau| < \delta_2$. Let $\delta = \min(\delta_1, \delta_2)$. Then

$$\left| x(t, s, \eta, \lambda) - x(\tau, t_0, x_0, \lambda_0) \right|$$
$$\leq \left| x(t, s, \eta, \lambda) - x(t, t_0, x_0, \lambda_0) \right| + \left| x(t, t_0, x_0, \lambda_0) - x(\tau, t_0, x_0, \lambda_0) \right|$$
$$< \varepsilon,$$

provided that $|(s, \eta, \lambda) - (t_0, x_0, \lambda_0)| + |t - \tau| < \delta$. This proves Theorem 3.4.

Under appropriate hypotheses, *the solutions of differential equations define homeomorphisms of subsets of R^n into R^n*. In fact, let us assume for any (t_0, x_0) in D, the equation (1.1) has a unique solution $x(t, t_0, x_0)$ which is jointly continuous in (t, t_0, x_0) in its domain of definition. If $x(t, t_0, x_0)$ exists on an interval $[a, b]$, then there is a neighborhood $U(x_0)$ of x_0 for which the solution $x(t, t_0, x_1)$ exists for t in $[a, b]$ and x_1 in $U(x_0)$. For fixed t_0 and each t in $[a, b]$, the function $x(t, t_0, \cdot)$ can be considered as a mapping T_t of $U(x_0)$ into a neighborhood of $x(t, t_0, x_0)$. The mapping T_t is one to one and continuous by hypothesis and $T_t^{-1} x^* = x(t_0, t, x^*)$. Therefore the inverse function is also continuous which implies T_t is a homeomorphism.

We are now in a position to define a general solution of (1.1). Let U be an open connected set in R^n and $\phi: R \times U \to R^n$ be such that: (a) for any c in U, $\phi(t, c)$ is a solution of (1.1); (b) for any t_0 in R, the mapping $\phi(t_0, \cdot)$: $U \to R^n$ is a homeomorphism on its range. Such a function ϕ will be referred to as a (local) *general solution* of (1.1).

I.4. Continuous Dependence and Stability

Consider the equation $\dot{x} = f(t, x)$ with $f(t, 0) = 0$ for all t in $(-\infty, \infty)$ and the function $f(t, x)$ defined for all (t, x) in R^{n+1}. For any (t_0, x_0) in R^{n+1}, we suppose this equation has a unique solution $x(t, t_0, x_0)$, $x(t_0, t_0, x_0) = x_0$, which is jointly continuous in (t, t_0, x_0) in its region of definition. Since $f(t, 0) = 0$ for all t, it follows that $x = 0$ is a solution on $(-\infty, \infty)$. The hypothesis of continuous dependence implies that given any real numbers t_0 in $(-\infty, \infty)$, $T \geq 0$, there exists a $\delta(T) > 0$ such that the solution $x(t, t_0, x_0)$, $|x_0| \leq \delta(T)$ *exists* on $[t_0, t_0 + T]$. Furthermore, $|x(t, t_0, x_0)| \to 0$ uniformly in t on $[t_0, t_0 + T]$ as $|x_0| \to 0$. In other words, given any $T > 0$ and any $\varepsilon > 0$, there is a $\delta = \delta(\varepsilon, T, t_0)$ such that the solution $x(t, t_0, x_0)$ exists on $I = [t_0, t_0 + T]$, $|x(t, t_0, x_0)| < \varepsilon$ on I provided that $|x_0| < \delta(\varepsilon, T, t_0)$ (see the accompanying figure).

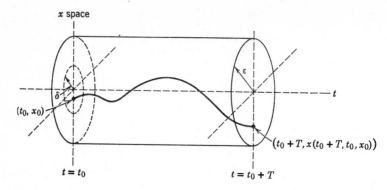

Figure I.4.2

Let us discuss in detail the meaning of continuous dependence on initial data for the equation $\dot{x} = x^2$. Choose $t_0 = 0$ and let $x_0 = c$. A solution $x(t, t_0, x_0) = x(t, 0, c)$ of this equation is $x(t, 0, c) = -c/(ct - 1)$. Uniqueness implies this is the solution with $x(0) = c$. It is clear that $x(t, 0, c)$ is continuous in t, c in the domain of definition of $x(t, 0, c)$. Since $x(t, 0, 0) = 0$, this implies for any $T > 0$ and any $\varepsilon > 0$, there is a $\delta = \delta(\varepsilon, T) > 0$ so that $|x(t, 0, c)| < \varepsilon$ if $|c| < \delta$. The largest value of δ is $\delta/(1 - \delta T) = \varepsilon$. As T increases, δ must decrease and, in fact, δ must approach zero as $T \to \infty$. This implies the continuity of $x(t, 0, c)$ in c is not uniform with respect to t in the infinite interval $[0, \infty)$. For this equation, it is even true that no solution with $x(0) = c > 0$ exists on $[0, \infty)$.

Even though continuous dependence on parameters and initial data is important, it gives information only on finite intervals of time as the above remarks show. An even more important concept is continuous dependence on initial data on infinite intervals of time. This is the concept of stability. In the remainder of this section we assume f smooth enough to ensure existence, uniqueness and continuous dependence on parameters.

Definitions. Suppose $f: [0, \infty) \times R^n \to R^n$. Consider $\dot{x} = f(t, x)$, $f(t, 0) \equiv 0$, $t \in [0, \infty)$. The solution $x = 0$ is called *Liapunov stable*, if for any $\varepsilon > 0$ and any $t_0 \geq 0$, there is a $\delta = \delta(\varepsilon, t_0)$ such that $|x_0| < \delta$ implies $|x(t, t_0, x_0)| < \varepsilon$ for $t \in [t_0, \infty)$. The solution $x = 0$ is *uniformly stable* if it is stable and δ can be chosen independent of $t_0 \geq 0$. The solution $x = 0$ is called *asymptotically stable* if it is stable and there exists a $b = b(t_0)$ such that $|x_0| < b$ implies $|x(t, t_0, x_0)| \to 0$ as $t \to \infty$. The solution $x = 0$ is *uniformly asymptotically stable* if it is uniformly stable, b in the definition of asymptotic stability can be chosen independent of $t_0 \geq 0$, and for every $\eta > 0$ there is a $T(\eta) > 0$ such that $|x_0| < b$ implies $|x(t, t_0, x_0)| < \eta$ if $t \geq t_0 + T(\eta)$. The solution $x = 0$ is *unstable* if it is not stable.

Pictorially, stability is the same as in the above diagram except the solution must remain in the infinite cylinder of radius ε for $t \geq t_0$.

We can discuss the stability and asymptotic stability of any other solution $\bar{x}(t)$ of the equation by replacing x by $\bar{x} + y$ and discussing the zero solution of the equation $\dot{y} = f(t, \bar{x} + y) - f(t, \bar{x})$. The definitions of stability of an arbitrary solution $\bar{x}(t)$ are the same as above except with x replaced by $x - \bar{x}(t)$.

LEMMA 4.1. *If f is either independent of t or periodie in t, then the solution $x = 0$ of (1.1) being stable (asymptotically stable) implies the solution $x = 0$ of (1.1) is uniformly stable (uniformly asymptotically stable).*

EXERCISE 4.1.   Prove Lemma 4.1.

EXERCISE 4.2.   Discuss the stability and asymptotic stability of every solution of the equations $\dot{x} = -x(1 - x)$, $\ddot{x} + x = 0$, and $\ddot{x} + 2^{-1}[x^2 + (x^4 + 4\dot{x}^2)^{1/2}]x = 0$. The latter equation has the family of solutions $x = c \sin(ct + d)$ where c, d are arbitrary constants.

Does stability defined in the above way depend on t_0 in the sense that a solution $x = 0$ may be stable at one value of t_0 and not at another? The answer is no! For $t_1 \leq t_0$, this follows immediately from continuity with respect to initial data. In fact, stability of $x = 0$ implies the existence of a $\delta(t_0, \varepsilon) > 0$ such that $|x_0| \leq \delta(t_0, \varepsilon)$ implies $|x(t, t_0, x_0)| < \varepsilon, t \geq t_0$. Continuity with respect to initial data implies the existence of a $\delta_1 = \delta_1(t_1, \varepsilon, t_0, \delta)$ > 0 so small that $|x_1| < \delta_1(t_1, \varepsilon)$ implies $|x(t, t_1, x_1)| < \delta(t_0, \varepsilon), t_1 \leq t \leq t_0$. Then $|x(t, t_1, x_1)| < \varepsilon$ for $t \geq t_1$, provided that $|x_1| \leq \delta_1(t_1, \varepsilon)$; that is, stability at t_1. For $t_1 \geq t_0$, it is not quite so obvious. Let $V(t_1, \varepsilon) = \{x$ in R^n: $x = x(t_1, t_0, x_0)$ for x_0 in the open ball of radius $\delta(t_0, \varepsilon)$ centered at zero}. Since the mapping $x(t_1, t_0, \cdot)$ is a homeomorphism, there exists a $\delta_1(t_1, \varepsilon)$ such that $\{x: |x| \leq \delta_1(t_1, \varepsilon)\} \subset V(t_1, \varepsilon)$. With this $\delta_1(t_1, \varepsilon)$ we have $|x(t, t_1, x_1)| < \varepsilon$ for $t \geq t_1$ and $|x_1| \leq \delta(t_1, \varepsilon)$; that is, stability at t_1.

EXERCISE 4.3.   In the above definition of asymptotic stability of the solution $x = 0$, we have supposed that $x = 0$ is stable and solutions with initial values in a neighborhood of zero approach zero as $t \to \infty$. Is it possible to have the latter property and also have the solution $x = 0$ unstable? Show this cannot happen if x is a scalar. Give an example in two dimensions where all solutions approach zero as $t \to \infty$ and yet the solution $x = 0$ is unstable. Is it possible to give such an example in two dimensions for an equation whose right hand sides are independent of t?

It is not appropriate at this time to have a detailed discussion of stability, but we will continually bring out more of the properties of this concept.

I.5. Extension of the Concept of a Differential Equation

In Section 1.1, a differential equation was defined for continuous vector fields f. As an immediate consequence, the initial value problem for (1.1) is equivalent to the integral equation

$$(5.1) \qquad x(t) = x_0 + \int_{t_0}^{t} f(s, x(s)) \, ds.$$

For f continuous, any solution of this equation automatically possesses a continuous first derivative. On the other hand, it is clear that (5.1) will be meaningful for a more general class of functions f if it is not required that x have a continuous first derivative. The purpose of this section is to make these notions precise for a class of functions f.

Suppose D is an open set in R^{n+1} and $f : D \to R^n$ is not necessarily continuous. Our problem is to find an absolutely continuous function x defined on a real interval I such that $(t, x(t)) \in D$ for t in I and

$$(5.1) \qquad \dot{x}(t) = f(t, x(t))$$

for all t in I except on a set of Lebesgue measure zero. If such a function x and interval I exist, we say x is a *solution of (5.1)*. A solution of (5.1) through (t_0, x_0) is a solution x of (5.1) with $x(t_0) = x_0$. We will not repeat the phrase "except on a set of Lebesgue measure zero" since it will always be clear that this is understood.

Suppose D is an open set in R^{n+1}. We say that $f : D \to R^n$ satisfies the Carathéodory conditions on D if f is measurable in t for each fixed x, continuous in x for each fixed t and for each compact set U of D, there is an integrable function $m_U(t)$ such that

$$(5.2) \qquad |f(t, x)| \leq m_U(t), \qquad (t, x) \in U.$$

For functions f which satisfy the Carathéodory conditions on a domain D, the conclusions of Sections 1 and 2 carry over without change. If the function $f(t, x)$ is also locally Lipschitzian in x with a measurable Lipschitz function, then the uniqueness property of the solution remains valid. These results are stated below, but only the details of the proof of the existence theorem are given, since the other proofs are essentially the same.

THEOREM 5.1. (Carathéodory). If D is an open set in R^{n+1} and f satisfies the Carathéodory conditions on D, then, for any (t_0, x_0) in D, there is a solution of (5.1) through (t_0, x_0).

PROOF. Suppose α, β are positive numbers chosen so that the rectangle $B(\alpha, \beta) = \{(t, x) : |t - t_0| \leq \alpha, |x - x_0| \leq \beta\}$ is in D. Let $I_\alpha = \{t : |t - t_0| \leq \alpha\}$,

$m = m_{B(\alpha,\,\beta)}$, $M(t) = \int_{t_0}^{t} m(s)\, ds$. Choose $\bar{\alpha}$, $\bar{\beta}$ so that $0 < \bar{\alpha} \leq \alpha$, $0 < \bar{\beta} \leq \beta$, $|M(t)| \leq \bar{\beta}$, $t \in I_{\bar{\alpha}}$. Let $\mathscr{A}$ be the set of functions ϕ in $\mathscr{C}(I_{\bar{\alpha}}, R^n)$ which satisfy $\phi(t_0) = x_0$, $|\phi(t) - x_0| \leq \bar{\beta}$ for all t in $I_{\bar{\alpha}}$. The set $\mathscr{A}$ is a closed, bounded, convex subset of $\mathscr{C}(I_{\bar{\alpha}}, R^n)$.

For any ϕ in $\mathscr{A}$, define the function $T\phi$ by the relation

$$T\phi(t) = x_0 + \int_{t_0}^{t} f(s, \phi(s))\, ds, \qquad t \in I_{\bar{\alpha}}.$$

The fixed points of T in $\mathscr{A}$ coincide with the solutions in $\mathscr{A}$ of (5.1). We now apply the Schauder theorem to prove the existence of a fixed point of T in $\mathscr{A}$.

For any ϕ in $\mathscr{A}$, the operator T is well defined since $f(s, \phi(s))$ is integrable for ϕ in $\mathscr{A}$. Also, $T\phi(t_0) = x_0$, and $T\phi(t)$ is continuous for $t \in I_{\bar{\alpha}}$. Using (5.2), we have

$$|T\phi(t) - x_0| \leq \left| \int_{t_0}^{t} |f(s, \phi(s))|\, ds \right|$$

$$\leq \left| \int_{t_0}^{t} m(s)\, ds \right|$$

$$= |M(t)| \leq \bar{\beta},$$

for all t in $I_{\bar{\alpha}}$. Therefore, $T \colon \mathscr{A} \to \mathscr{A}$.

The operator T is continuous on $\mathscr{A}$. In fact, if $\phi_n \in A$, $\phi_n \to \phi$ in $\mathscr{A}$, then $f(t, x)$ continuous in x for each fixed t implies $f(t, \phi_n(t)) \to f(t, \phi(t))$ as $n \to \infty$ for each t in $I_{\bar{\alpha}}$. Since $|f(t, \phi_n(t))| \leq m(t)$, the Lebesgue dominated convergence theorem implies

$$\int_{t_0}^{t} f(s, \phi_n(s))\, ds \to \int_{t_0}^{t} f(s, \phi(s))\, ds,$$

as $n \to \infty$ for each t in $I_{\bar{\alpha}}$. This proves the assertion.

For any ϕ in $\mathscr{A}$,

$$|T\phi(t) - T\phi(\tau)| \leq |M(t) - M(\tau)|$$

for all t, τ in $I_{\bar{\alpha}}$. Since M is continuous on $I_{\bar{\alpha}}$, it is uniformly continuous and, thus, the set $T\mathscr{A}$ is an equicontinuous set of $\mathscr{C}(I_{\bar{\alpha}}, R^n)$. It is also uniformly bounded. This proves $T\mathscr{A}$ is relatively compact and, thus, T is completely continuous. The Schauder fixed point theorem implies the existence of a fixed point in $\mathscr{A}$ and the theorem is proved.

THEOREM 5.2. *If D is an open set in R^{n+1}, f satisfies the Carathéodory conditions on D and ϕ is a solution of (5.1) on some interval, then there is a continuation of ϕ to a maximal interval of existence. Furthermore, if (a, b) is a maximal interval of existence of (5.1), then $x(t)$ tends to the boundary of D as $t \to a$ and $t \to b$.*

PROOF. The proof is essentially the same as the proof of Theorem 2.1 and is left to the reader.

THEOREM 5.3. Suppose D is an open set in R^{n+1}, f satisfies the Carathéodory conditions on D and for each compact set U in D, there is an integrable function $k_U(t)$ such that

$$(5.3) \qquad |f(t, x) - f(t, y)| \leq k_U(t) |x - y|, \quad (t, x) \in U, \quad (t, y) \in U.$$

Then, for any (t_0, x_0) in U, there exists a *unique* solution $x(t, t_0, x_0)$ of (5.1) passing through (t_0, x_0). The domain E in R^{n+2} of definition of the function $x(t, t_0, x_0)$ is open and $x(t, t_0, x_0)$ is continuous in E.

PROOF. This proof is essentially the same as the proof of Theorem 3.1 and the details are left to the reader. One only needs to choose $M(t)$ as in the proof of Theorem 5.1, let $K(t) = \int_{t_0}^{t} k_{B(\alpha,\beta)}(s)\, ds$ and choose $\bar{\alpha}$, $\bar{\beta}$ so that $0 < \bar{\alpha} \leq \alpha$, $0 < \bar{\beta} \leq \beta$, $|M(t)| \leq \bar{\beta}$, $K|(t)| < 1$ for $t \in I_{\bar{\alpha}}$.

Any linear system

$$(5.4) \qquad \dot{x} = A(t)x + h(t),$$

where $A(t)$ is an $n \times n$ matrix, $h(t)$ is an n-vector whose elements are integrable on every finite interval satisfies the Carathéodory conditions and (5.3). Therefore, the initial value problem has a unique solution.

I.6. Differential Inequalities

Let D_r denote the right hand derivative of a function. If $\omega(t, u)$ is a scalar function of the scalars t, u in some open connected set Ω, we say a function $v(t)$, $a \leq t \leq b$, is a *solution of the differential inequality*

$$(6.1) \qquad D_r v(t) \leq \omega(t, v(t))$$

on $[a, b)$ if $v(t)$ is continuous on $[a, b)$ and has a right hand derivative on $[a, b)$ that satisfies (6.1).

LEMMA 6.1. If $x(t)$ is a continuously differentiable n-vector function on $a \leq t \leq b$, then $D_r|x(t)|$ exists on $a \leq t < b$ and $|D_r(|x(t)|)| \leq |\dot{x}(t)|$, $a \leq t < b$.

PROOF. For any two n-vectors x, u and $0 < \theta \leq 1$, $h > 0$, we have

$$|x + \theta h u| - |\theta x + \theta h u| \leq (1 - \theta)|x|.$$

Therefore,

$$\frac{|x + \theta h u| - |x|}{\theta h} \leq \frac{|x + h u| - |x|}{h};$$

that is, the difference quotient

$$\frac{|x + hu| - |x|}{h}$$

is a nondecreasing function of h. Furthermore, this difference quotient is bounded below by $-|u|$. Consequently,

$$\lim_{h \to 0^+} \frac{|x + hu| - |x|}{h}$$

exists.

If $x(t)$ is continuously differentiable for $a \le t \le b$, then this latter relation implies

$$\lim_{h \to 0^+} \frac{|x(t) + h\dot{x}(t)| - |x(t)|}{h}$$

exists. Since

$$|[|x(t + h)| - |x(t)|] - [|x(t) + h\dot{x}(t)| - |x(t)|]|$$
$$= |[|x(t + h)| - |x(t) + h\dot{x}(t)|]|$$
$$\le |x(t + h) - x(t) - h\dot{x}(t)| = o(h)$$

as $h \to 0^+$, it follows that $D_r(|x(t)|)$ exists and

$$D_r|x(t)| = \lim_{h \to 0^+} \frac{|x(t) + h\dot{x}(t)| - |x(t)|}{h}.$$

It is clear that $|D_r(|x(t)|)| \le |\dot{x}(t)|$ and the lemma is proved.

The same proof also shows that the conclusion of Lemma 6.1 is valid for absolutely continuous functions in the sense that $D_r(|x(t)|)$ exists and satisfies $|D_r(|x(t)|)| \le |\dot{x}(t)|$ almost everywhere for $a \le t \le b$.

THEOREM 6.1. Let $\omega(t, u)$ be continuous on an open connected set $\Omega \subset R^2$ and such that the initial value problem for the scalar equation

(6.2) $\dot{u} = \omega(t, u)$

has a unique solution. If $u(t)$ is a solution of (6.2) on $a \le t \le b$ and $v(t)$ is a solution of (6.1) on $a \le t < b$ with $v(a) \le u(a)$, then $v(t) \le u(t)$ for $a \le t \le b$.

PROOF. Consider the family of equations

(6.3) $\dot{u} = \omega(t, u) + \frac{1}{n}$

for $n = 1, 2, \ldots$. We now apply Lemma 3.1 to (6.3).

If $u_n(t)$ designates the solution of (6.3) with $u_n(a) = u(a)$, then Lemma 3.1 implies there is an n_0 such that $u_n(t)$, $n \geq n_0$, is defined on $[a, b]$ and $u_n(t) \to u(t)$ uniformly on $[a, b]$. We show $v(t) \leq u_n(t)$, for $n \geq n_0$, $a \leq t \leq b$. If this is not so, then there exist $t_2 < t_1$ in (a, b) such that $v(t) > u_n(t)$ on $t_2 < t \leq t_1$, $v(t_2) = u_n(t_2)$. Therefore, $v(t) - v(t_2) > u_n(t) - u_n(t_2)$, $t_2 < t \leq t_1$, which implies

$$D_r v(t_2) \geq \dot{u}_n(t_2) = \omega(t_2, u_n(t_2)) + \frac{1}{n}$$

$$= \omega(t_2, v(t_2)) + \frac{1}{n}$$

$$> \omega(t_2, v(t_2)),$$

which is a contradiction. Consequently, $v(t) \leq u_n(t)$ for t in $[a, b]$, $n \geq n_0$. Since $u_n(t) \to u(t)$ uniformly on $[a, b]$, this proves the theorem.

COROLLARY 6.1. A solution of $D_r v(t) \leq 0$ on $[a, b)$ is nonincreasing on $[a, b)$.

COROLLARY 6.2. Suppose $\omega(t, u)$ and $u(t)$ are as in Theorem 6.1. If $x(t)$ is a continuous n-vector function with a continuous first derivative on $[a, b]$ such that $|x(a)| \leq u(a)$, $(t, |x(t)|) \in \Omega$, $a \leq t \leq b$, and

$$|\dot{x}(t)| \leq \omega(t, |x(t)|), \quad a \leq t \leq b,$$

then $|x(t)| \leq u(t)$ on $a \leq t \leq b$.

PROOF. This is immediate from Lemma 6.1 and Theorem 6.1.

COROLLARY 6.3. Suppose $\omega(t, u)$ satisfies the conditions of Theorem 6.1 for $a \leq t < b$, $u \geq 0$, and let $u(t) \geq 0$ be a solution of (6.2) on $a \leq t < b$. If $f: [a, b) \times R^n \to R^n$ is continuous and

$$|f(t, x)| \leq \omega(t, |x|), \quad a \leq t < b, \quad x \in R^n,$$

then the solutions of

$$\dot{x} = f(t, x), \quad |x(a)| \leq u(a),$$

exist on $[a, b)$ and $|x(t)| \leq u(t)$, t in $[a, b)$.

PROOF. From Corollary 6.2, $|x(t)| \leq u(t)$ as long as $x(t)$ exists. From Theorem 2.1, the solution $x(t)$ can fail to exist on $[a, b)$ only if there is a c, $a < c < b$, such that $x(t)$ is defined on $[a, c)$ and $\overline{\lim} |x(t)| = \infty$ as $t \to c - 0$. On the other hand, this is impossible since $|x(t)| \leq u(t)$, $t \in [a, c)$ and $\lim u(t)$ exists as $t \to c - 0$.

Notice that Corollary 6.3 gives existence of the solution on $[a, b)$, as well as upper bounds on the solutions.

Another simple application of Corollary 6.2 yields

COROLLARY 6.4. Suppose D is an open connected set in R^{n+1}, $f: D \to R^n$ is continuous and $f(t, x)$ is locally lipschitzian in x. If K is any compact set in D and L is the lipschitz constant of f in K, then

$$|x(t, t_0, x_0) - x(t, t_0, x_1)| \leqq e^{L(t-t_0)} |x_0 - x_1|$$

as long as the solutions $x(t, t_0, x_0)$, $x(t, t_0, x_1)$ of (1.1) are such that $(t, x(t, t_0, x_0))$, $(t, x(t, t_0, x_1))$ remain in K.

PROOF. If $z(t) = x(t, t_0, x_0) - x(t, t_0, x_1)$, then $|\dot{z}(t)| \leqq L |z(t)|$ as long as $(t, x(t, t_0, x_0))$, $(t, x(t, t_0, x_1))$ remain in K and Corollary 6.2 gives the result.

As a first illustration of the above results, we prove a result on existence in the large. Suppose $f: (\alpha, \infty) \times R^n \to R^n$ is continuous

$$(6.4) \qquad |f(t, x)| \leqq \phi(t) \psi(|x|),$$

where $\phi(t) \geqq 0$ is continuous for all $t > \alpha$ and $\psi(u)$ is continuous for $u \geqq 0$ and positive for all $u > 0$. Suppose $u(t, t_0, u_0)$, $t_0 > \alpha$, is a solution of $\dot{u} = \phi(t) \psi(u)$, $u(t_0) = u_0 > 0$ and this solution is unique for any $u_0 > 0$. If

$$(6.5) \qquad \int^\infty \frac{du}{\psi(u)} = +\infty,$$

then the solution $u(t, t_0, u_0)$ exists for all $t > \alpha$. In fact, u satisfies the equation

$$\int_{u_0}^u \frac{dv}{\psi(v)} = \int_{t_0}^t \phi(s) \, ds,$$

and if u did not exist for all $t > \alpha$, then the continuation theorem implies there would exist a τ and a sequence $\{t_n\}, t_n \to \tau$ as $n \to \infty$ such that $u(t_n) \to \infty$ as $n \to \infty$. But this is impossible since $\int_{t_0}^\tau \phi < \infty$ and $\int_{u_0}^\infty dv/\psi(v) = +\infty$.

For any given $x_0 \neq 0$ in R^n, choose $u_0 = |x_0|$ and for $x_0 = 0$ choose any $u_0 > 0$. From Corollary 6.3, it follows that any solution

$$x(t, t_0, x_0), \quad x(t_0, t_0, x_0) = x_0, \quad t_0 > \alpha.$$

of $\dot{x} = f(t, x)$ exists and satisfies $|x(t, t_0, x_0)| \leqq u(t, t_0, u_0)$ on $[t_0, \infty)$ provided f satisfies (6.4) and (6.5).

As a special case suppose that $f(t, x) = A(t)x + h(t)$ where $A(t)$, $h(t)$ are continuous for all values of t. Then any solution of the linear equation

$$(6.6) \qquad \dot{x} = A(t)x + h(t)$$

exists on $(-\infty, \infty)$. In fact,

$$|A(t)x + h(t)| \leq |A(t)| \, |x| + |h(t)|$$
$$\leq \max\{|A(t)|, |h(t)|\}(|x| + 1)$$
$$= \phi(t)\psi(|x|),$$
$$\psi(u) = u + 1.$$

Since $\phi(t)$ is continuous and $\int^{\infty} du/\psi(u) = +\infty$, we have the desired result.

Another interesting special case is when $|f(t, x)| \leq K|x|$ for all t in $(-\infty, \infty)$ and x in R^n. For $\phi(t) = 1$ and $\psi(u) = Ku$, the above example shows that all solutions of $\dot{x} = f(t, x)$ exist on $(-\infty, \infty)$.

For later reference, part of these results are summarized in

THEOREM 6.2. If $f: (\alpha, \infty) \times R^n \to R^n$ is continuous, satisfies (6.4) and (6.5) and the solution of $\dot{u} = \phi(t)\psi(u)$, $u(t_0) = u_0 > 0$, $t_0 > \alpha$ exists and is unique in (α, ∞), then any solution of the equation (1.1) through (t_0, x_0) exists on (α, ∞). In particular, every solution of (6.6) exists on $(-\infty, \infty)$ provided that $A(t), h(t)$ are continuous.

As another illustration, we prove a simple result on stability. Suppose $f: R^{n+1} \to R^n$ and there exists a continuous function $\lambda(t)$, $-\infty < t < \infty$, such that $x \cdot f(t, x) \leq -\lambda(t)x \cdot x$ where "$\cdot$" denotes the scalar product of two vectors. Notice this implies $f(t, 0) = 0$. If we let $|x|^2 = x \cdot x$ and suppose x is a solution of $\dot{x} = f(t, x)$ on an interval I containing t_0, then

$$\frac{d}{dt}|x|^2 = \frac{d}{dt}(x \cdot x) = 2x \cdot f(t, x) \leq -2\lambda(t)|x|^2.$$

If $\omega(t, u) = -2\lambda(t)u$ and u is the solution of $\dot{u} = \omega(t, u)$, $u(t_0) = |x(t_0)|^2$, then $u(t) = \{\exp(-2\int_{t_0}^{t} \lambda(s) \, ds)\}|x(t_0)|^2$ exists for all t in $(-\infty, \infty)$. Therefore, Corollary 6.2 and the continuation theorem implies that the solution $x(t)$ not only exists on I but exists for all t in $(-\infty, \infty)$ and

$$|x(t)| \leq \exp\left(-\int_{t_0}^{t} \lambda(s) \, ds\right)|x(t_0)|, \quad t \geq t_0.$$

If $\lambda(t) \geq 0$, then the solution $x = 0$ of $\dot{x} = f(t, x)$ is stable and actually uniformly stable. If $\lambda(t) \geq 0$ and $\int_{t_0}^{\infty} \lambda(s) \, ds = +\infty$, then the solution $x = 0$ is asymptotically stable.

EXERCISE 6.1. If $\lambda(t) \geq 0$ and $\int_{t_0}^{\infty} \lambda(s) \, ds = +\infty$ for all t_0, is the solution $x = 0$ of the previous discussion uniformly asymptotically stable? Discuss the case where $\lambda(t)$ is not of fixed sign.

EXERCISE 6.2.   Suppose $f\colon R^{n+1} \to R^n$ is continuous and there exists a positive definite matrix B such that $x \cdot Bf(t, x) \leq -\lambda(t)x \cdot x$ for all t, x where $\lambda(t)$ is continuous for t in $(-\infty, \infty)$. Prove that any solution of the equation $\dot{x} = f(t, x)$, $x(t_0) = x_0$, exists on $[t_0, \infty)$ and give sufficient conditions for stability and asymptotic stability. (Hint: Find the derivative of the function $V(x) = x \cdot Bx$ along solutions and use the fact that there is a positive constant μ such that $x \cdot Bx \geq \mu x \cdot x$ for all x.)

EXERCISE 6.3.   Consider the equation $\dot{x} = f(t, x)$, $|f(t, x)| \leq \phi(t)|x|$ for all t, x in $R \times R$, $\int^{\infty} \phi(t)\, dt < \infty$.

   (a) Prove that every solution approaches a constant as $t \to \infty$.

   (b) If, in addition,

$$|f(t, x) - f(t, y)| \leq \phi(t)|x - y| \quad \text{for all } x, y,$$

prove there is a one to one correspondence between the initial values and the limit values of the solution.

   (c) Does the above result imply anything for the equation

$$\dot{x} = -x + a(t)x, \qquad \int^{\infty} |a(t)|\, dt < \infty?$$

(Hint: Consider the transformation $x = e^{-t}y$.)

   (d) Does this imply anything about the system

$$\dot{x}_1 = x_2,$$

$$\dot{x}_2 = -x_1 + a(t)x_1, \qquad \int^{\infty} |a(t)|\, dt < \infty,$$

where x_1, x_2 are scalars?

EXERCISE 6.4.   Consider the initial value problem

$$\ddot{z} + \alpha(z, \dot{z})\dot{z} + \beta(z) = u(t), \qquad z(0) = \xi, \qquad \dot{z}(0) = \eta,$$

with $\alpha(z, w)$, $\beta(z)$ continuous together with their first partial derivatives for all z, w, u continuous and bounded on $(-\infty, \infty)$, $\alpha \geq 0$, $z\beta(z) \geq 0$. Show there is one and only one solution to this problem and the solution can be defined on $[0, \infty)$. Hint: Write the equation as a system by letting $z = x$, $\dot{z} = y$, define $V(x, y) = y^2/2 + \int_0^x \beta(s)\, ds$ and study the rate of change of $V(x(t), y(t))$ along the solutions of the two dimensional system.

COROLLARY 6.5.   Let $\omega(t, u)$ satisfy the conditions of Theorem 6.1 and in addition be nondecreasing in u. If $u(t)$ is the same function as in Theorem

6.1 and $v(t)$ is continuous and satisfies

$$(6.6) \qquad v(t) \leq v_a + \int_a^t \omega(s, v(s)) \, ds, \quad a \leq t \leq b, \quad v_a \leq u(a),$$

then $v(t) \leq u(t)$, $a \leq t \leq b$.

PROOF. Let $V(t)$ be the right hand side of (6.6) so that $v(t) \leq V(t)$. Then $\dot{V}(t) = \omega(t, v(t)) \leq \omega(t, V(t))$, $V(a) = v_a \leq u(a)$. Theorem 6.1 implies $V(t) \leq u(t)$ for $a \leq t \leq b$ and this proves the corollary.

COROLLARY 6.6. (Gronwall's inequality). If α is a real constant, $\beta(t) \geq 0$, and $\phi(t)$ are continuous real functions for $a \leq t \leq b$ which satisfy

$$\phi(t) \leq \alpha + \int_a^t \beta(s)\phi(s) \, ds, \quad a \leq t \leq b,$$

then

$$\phi(t) \leq \left(\exp \int_a^t \beta(s) \, ds \right) \alpha, \quad a \leq t \leq b.$$

PROOF. We apply Corollary 6.5 with $v_a = \alpha$, $\omega(t, u) = \beta(t)u$. Then $\dot{u} = \beta(t)u$, $u(a) = \alpha$ is given by $u(t) = \left(\exp \int_a^t \beta(s) \, ds \right) \alpha$, which proves the corollary.

Actually, Gronwall's inequality is easily proved by other techniques. In the applications to follow, we actually need a generalization of this inequality so we state and prove it without using Theorem 6.1.

LEMMA 6.2. (Generalized Gronwall inequality). If ϕ, α are real valued and continuous for $a \leq t \leq b$, $\beta(t) \geq 0$ is integrable on $[a, b]$ and

$$\phi(t) \leq \alpha(t) + \int_a^t \beta(s)\phi(s) \, ds, \quad a \leq t \leq b,$$

then

$$\phi(t) \leq \alpha(t) + \int_a^t \beta(s)\alpha(s) \left(\exp \int_s^t \beta(u) \, du \right) ds, \qquad a \leq t \leq b.$$

PROOF. Let $R(t) = \int_a^t \beta(s)\phi(s) \, ds$. Then

$$\frac{dR(t)}{dt} = \beta(t)\phi(t) \leq \beta(t)\alpha(t) + \beta(t)R(t)$$

except for a set of measure zero. Thus,

$$\frac{dR(t)}{dt} - \beta(t)R(t) \leq \beta(t)\alpha(t),$$

$$\frac{d}{ds} \left(\exp - \int_a^s \beta(u) \, du \right) R(s) \leq \left(\exp - \int_a^s \beta(u) \, du \right) \beta(s)\alpha(s),$$

except for a set of measure zero. Integrating from a to t, we obtain

$$\left(\exp -\int_a^t \beta(u)\,du\right)R(t) \le \int_a^t \left(\exp -\int_a^s \beta(u)\,du\right)\beta(s)\alpha(s)\,ds.$$

and thus

$$R(t) \le \int_a^t \left(\exp \int_s^t \beta(u)\,du\right)\beta(s)\alpha(s)\,ds.$$

This estimate proves the lemma.

If α is continuous with its first derivative $\dot{\alpha} \ge 0$, then integrating by parts in Lemma 6.2 gives

$$\varphi(t) \le \alpha(a)\exp\left(\int_a^t \beta\right) + \int_a^t \dot{\alpha}(s)\exp\left(\int_s^t \beta\right)ds$$

$$\le \left[\exp \int_a^t \beta\right]\left[\alpha(a) + \int_a^t \dot{\alpha}(s)\,ds\right]$$

$$\le \alpha(t)\exp\int_a^t \beta$$

since $\beta \ge 0$. Gronwall's inequality is now a special case.

I.7. Autonomous Systems—Generalities

If $x(t)$ is a solution of (1.1) defined on an interval (a, b), we have previously introduced the concept of a *trajectory* associated with this solution as the set in R^{n+1} defined by $\bigcup_{a \le t \le b}(t, x(t))$. The *path* or *orbit* of a trajectory is the projection of the trajectory into R^n, the space of dependent variables in (1.1). The space of dependent variables is usually called the *state space* or *phase space*. The phase coordinates for a scalar nth order equation in x is the vector $(x, x^{(1)}, x^{(2)}, \ldots, x^{(n-1)})$. System (1.1) is called *autonomous* if the vector field, that is, the function f in (1.1), is independent of t. In this section we consider some general properties of autonomous systems; namely, the differential system

(7.1) $$\dot{x} = f(x)$$

where $f\colon \Omega \to R^n$ is continuous and Ω is an open set in R^n. A basic property of autonomous systems is the following: if $x(t)$ is a solution of (7.1) on an interval (a, b), then for any real number τ, the function $x(t - \tau)$ is a solution of (7.1) on the interval $(a + \tau, b + \tau)$. This is clear since the differential equation remains unchanged by a translation of the independent variable. Thus, from a single solution of (7.1) one can define a one parameter family of solutions.

We shall henceforth assume that for any p in Ω, there is a unique solution $\phi(t, p)$ of (7.1) passing through p at $t = 0$. The function $\phi(t, p)$ is defined on an open set $\Sigma \subset R^{n+1}$ and satisfies the properties:

  (i) $\phi(0, p) = p$;
  (ii) $\phi(t, p)$ is continuous in Σ;
  (iii) $\phi(t + \tau, p) = \phi(t, \phi(\tau, p))$ on Σ

In fact, it follows from Theorem 3.4 that $\phi(t, p)$ is continuous. Relation (iii) holds since both functions satisfy the equation, are equal for $t = 0$ and we have assumed uniqueness.

From the above definition, the *path* or *orbit* $\gamma = \gamma(p)$ through a fixed $p \in \Omega$ is the set in R^n defined by $\gamma(p) = \{x \in R^n \colon$ there exist $(t, \phi) \in \Sigma$ with $\phi(t, p) = x\}$. It is clear that $\phi(t, p)$ and $\phi(t + \tau, p)$ are different parametrizations of the same orbit $\gamma(p)$.

There is a unique path γ through a given p in Ω. Indeed, paths through p are projections of all solutions of (7.1) which pass through any of the points on the line (τ, p), $-\infty < \tau < \infty$. But, $\phi(t + \tau, p)$ is the unique solution of (7.1) passing through (τ, p) and we have seen above these functions are all parametrizations of the same curve. Notice this last conclusion implies that no two paths can intersect.

An *equilibrium point* or *critical point* or *singular point* of an n-dimensional vector field $f(x)$ is a point p such that $f(p) = 0$. If p is a critical point, then $x(t) = p$, $-\infty < t < \infty$, satisfies (7.1). The trajectory of the critical point p is the line in R^{n+1} given by $x = p$, $-\infty < t < \infty$ and the orbit of a critical point is the point itself. A *regular point* is a point which is not critical.

If p is a critical point of (7.1), then no trajectory other than $x(t) = p$ can reach the line $x = p$, $-\infty < t < \infty$ by the process of continuation for this would violate uniqueness. This implies: *if p is a critical point and $x(t) \neq p$ tends to p, then either $t \to +\infty$ or $t \to -\infty$.*

A *curve* λ in R^n is the range of a continuous mapping of an interval $I \subset R$ into R^n. The curve is said to be differentiable if the associated mapping is differentiable. Given a continuous $f \colon \Omega \to R^n$, Ω open in R^n, $f = (f_1, \ldots, f_n)$, we say a curve λ is a *solution of the equations*

(7.2)
$$\frac{dx_1}{f_1(x)} = \frac{dx_2}{f_2(x)} = \cdots = \frac{dx_n}{f_n(x)}$$

if λ is differentiable and the differential dx along λ is parallel to $f(x)$ when $f(x) \neq 0$, and λ is a point when $f(x) = 0$.

LEMMA 7.1. The solution of (7.2) at any point p of Ω is the orbit of (7.1) through p.

PROOF. If $p = (p_1, \ldots, p_n)$ is a critical point, then γ and λ are both equal to p. If p is not a critical point, then one of the components of f say f_1

is such that $f_1(p) \neq 0$. Therefore, $f_1(x) \neq 0$ for x in a neighborhood U of p. In U system (7.2) is therefore equivalent to the ordinary differential system

$$\frac{dx_\alpha}{dx_1} = \frac{f_\alpha(x)}{f_1(x)}, \qquad x_\alpha(p_1) = p_\alpha, \qquad \alpha = 2, 3, \ldots, n.$$

From the existence Theorem 1.1, these equations have a solution $x_\alpha(x_1)$ which exists for $|x_1 - p_1|$ sufficiently small. We parametrize λ in the following way. Consider the autonomous scalar equation

$$\frac{dx_1}{dt} = f_1(x_1, x_2(x_1), \ldots, x_n(x_1)).$$

This equation has a solution $x_1(t)$, $x_1(0) = p_1$, which exists for $|t|$ small. One now easily shows that $x_1(t)$, $x_\alpha(x_1(t))$, $\alpha = 2, \ldots, n$, is a solution of (7.1). Since the orbit of (7.1) through any point of Ω is a solution of (7.2), this proves $\lambda = \gamma$ and the lemma.

A homeomorphic image of a closed or open line segment is called an *arc*. A homeomorphic image of the circumference of a circle is called a *Jordan curve*. A path γ is said to be closed if it is a Jordan curve.

LEMMA 7.2. *A necessary and sufficient condition for a path of (7.1) to be closed is that it corresponds to a nonconstant periodic solution of (7.1).*

PROOF. If γ is a closed path of (7.1) and p is a point of γ, there is a $\tau \neq 0$ such that $\phi(\tau, p) = \phi(0, p) = p$. By uniqueness of solutions $\phi(t + \tau, p) = \phi(t, p)$ for all t which says $\phi(t, p)$ has period τ. Conversely, suppose $\phi(t, p) = \phi(t + \tau, p)$ for all t, $\phi(t, p)$ nonconstant, and τ is the least period of $\phi(t, p)$; i.e. $p \neq \phi(t, p)$, $0 < t < \tau$. As t varies in $[0, \tau)$, $\phi(t, p)$ describes a curve in R^n which is the homeomorphic image of the segment $[0, \tau)$ with $\phi(0, p) = \phi(\tau, p)$. On the other hand, the line segment $[0, \tau]$ with 0 and τ identified is homeomorphic to the unit circle. This completes the proof.

EXERCISE 7.1. Suppose the autonomous equation $\dot{x} = f(x)$ has a nonconstant periodic solution $x^0(t)$. Define stability of this solution. Can the stability ever be asymptotic? What is the strongest type of stability that you would expect in such a situation?

We now give a few examples to illustrate the above concepts.

Example 7.1. If x is a real scalar, $\dot{x} = x$, then $\phi(t, p) = e^t p$, the trajectory through p is the set $(t, e^t p)$, $-\infty < t < \infty$ and the path through p is the set $\{x > 0\}$ if $p > 0$, $\{x = 0\}$ if $p = 0$ and $\{x < 0\}$ if $p < 0$. See Fig. 7.1 where the arrow on a curve in phase space denotes the manner in which the path is described with increasing time.

Example 7.2. If x is a real scalar, $\dot{x} = -x(1 - x)$, then $\phi(t, p) = pe^{-t}/[1 - p + pe^{-t}]$. The paths are the sets $\{x > 1\}$, $\{x = 1\}$, $\{0 < x < 1\}$,

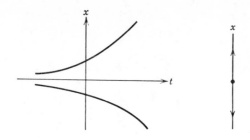

Figure I.7.1

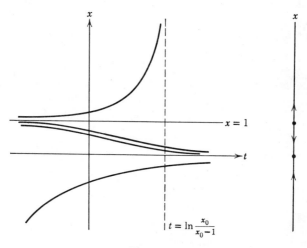

Figure I.7.2

$\{x = 0\}$, $\{x < 0\}$. See Fig. 7.2. The equilibrium point $x = 1$ is unstable and $x = 0$ is asymptotically stable.

Example 7.3. If y is a real scalar, then the equation $\ddot{y} + y = 0$ is equivalent to the system $\dot{x}_1 = x_2$, $\dot{x}_2 = -x_1$ where $x_1 = y$. The phase space is R^2. For any real constants a, b one verifies that $\phi_1(t) = a \sin(t + b)$, $\phi_2(t) = a \cos(t + b)$ is a solution of this system and any solution of the equation can be written in this form. Any trajectory lies on a circular cylinder and the path through any point is a circle passing through this point with center at the origin. See Fig. 7.3. An equilibrium point of a two dimensional autonomous system which has the property that every neighborhood of the point contains an orbit which is a closed curve (periodic solution) is called a *center*. Thus, the solution $x^1 = x^2 = 0$ of this example is a center.

In this example, we did not need to integrate the equations to obtain the parametric representation of the orbits in phase space. In fact, Lemma 7.1 implies that the orbits are the solutions of the scalar equation

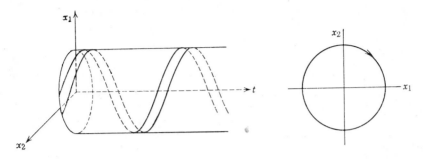

Figure I.7.3

$$\frac{dx_1}{dx_2} = -\frac{x_2}{x_1}$$

which has the solutions $x_1^2 + x_2^2 = $ constant, the constant of course being determined by the initial values. The manner in which the orbits are described with increasing time is easily obtained from the original equations.

Example 7.4. Suppose $\varepsilon > 0$ is given, x_1, x_2 are real scalars, $r^2 = x_1^2 + x_2^2$ and consider the system

$$\dot{x}_1 = -x_2 + \varepsilon x_1(1 - r^2),$$
$$\dot{x}_2 = x_1 + \varepsilon x_2(1 - r^2).$$

The phase space for this system is R^2. If $x_1 = r \cos \theta$, $x_2 = r \sin \theta$, then the system is equivalent to the system

$$\dot{\theta} = 1,$$
$$\dot{r} = \varepsilon r(1 - r^2).$$

One easily verifies that the trajectories and paths are as in Fig. 7.4. In this case, the paths are spirals outside and inside the circle of radius one and the equilibrium point $(0, 0)$ together with the circle of radius one. The equilibrium

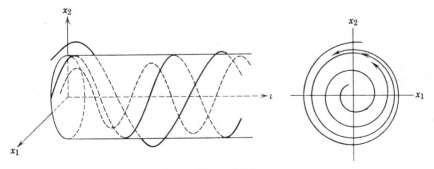

Figure I.7.4

point $(0, 0)$ is called a *focus*; that is, solutions in a neighborhood of it spiral toward it as $t \to +\infty$ (or $-\infty$). To say a solution spirals toward zero as $t \to +\infty$ (or $-\infty$) is to say that any ray emanating from zero is crossed by the orbit of the solution an infinite number of times and the solution approaches zero as $t \to +\infty$ (or $-\infty$). The orbit $(r = 1)$ which is a closed curve in this example is called a *limit cycle*, the reason for the terminology becoming clear in later discussions.

Notice that, for $\varepsilon = 0$ this example is the same as Example 7.3. However, for any $\varepsilon > 0$, no matter how small, the phase portrait of the two equations are completely different.

Example 7.5. In this example, we illustrate that the solutions of a differential equation may be much more complicated than any of the previous examples. A torus is the homeomorphic image of the cross product of two circles. Suppose θ, ϕ are the angles shown in Fig. 7.5 describing a coordinate

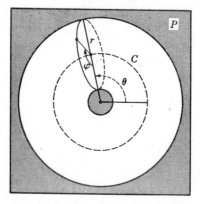

Figure I.7.5

system on the torus, If θ, ϕ satisfy the differential equation $\dot{\theta} = 1$, $\dot{\phi} = \omega$, where ω is a constant, then a solution corresponding to an orbit γ goes around the torus traversing the angle ϕ with period $2\pi/\omega$ and the angle θ with period 2π. Therefore, for ω irrational, the path is not closed, but it does have the property that the closure of γ is the whole torus!

EXERCISE 7.2. Prove the statement in Example 7.5 concerning the closure of γ for ω irrational. Can you describe the behavior when ω is rational?

Example 7.6. In this example, we show in R^3 that the complicated behavior in Example 7.5 can be shared by all solutions of an equation in a solid torus. A solid torus is the homeomorphic image of the cross product of a circle and a disk. Suppose the region is that depicted in Fig. 7.5 and the coordinate system (r, θ, ϕ) is as shown with $0 \leqq r \leqq 2$ and $0 \leqq \theta$, $\phi \leqq 2\pi$. The value $r = 0$ is a circle C which lies in a plane P and the surface $r =$ constant, $0 \leqq \phi$, $\theta \leqq 2\pi$ is a torus, which has C at its center.

For the differential equations, choose

$$\dot{\theta} = 1,$$
$$\dot{\phi} = \omega,$$
$$\dot{r} = r(1 - r),$$

where ω is a constant. The torii $r = 0$ and $r = 1$ are invariant in the sense that any solution with initial value in these surfaces remain in them for all t in $(-\infty, \infty)$. Except for the periodic solution corresponding to C, all other solutions approach the torus described by $r = 1$. Therefore, from Example 7.5, the closure of every orbit except C contains the torus $r = 1$ if ω is irrational.

EXERCISE 7.3.   Discuss the phase portrait of the solutions of the second order equation

$$\dot{x}_1 = x_2(x_2^2 - x_1^2),$$
$$\dot{x}_2 = -x_1(x_2^2 - x_1^2).$$

What are the orbits and the equilibrium points? Which equilibrium points are stable? What does the application of Lemma 7.1 yield for this system?

Suppose $\psi: S_\alpha \to R^n$ is a given continuously differentiable function on the ball $S_\alpha = \{u \text{ in } R^{n-1}, |u| < \alpha\}$ with $\psi(0) = p$ and rank $[\partial\psi(u)/\partial u] = n - 1$ for all u in S_α. The set $\{x \text{ in } R^n, x = \psi(u), u \text{ in } S_\alpha\}$ is called a differentiable $(n - 1)$-cell E^{n-1} through p. Such an E^{n-1} is said to be *transverse* to the path γ_p of (7.1) at p if for each $p' \in E_p^{n-1}$, the path $\gamma_{p'}$ through p' is not tangent to E^{n-1} at p'. This is equivalent to saying that dx along $\gamma_{p'}$ at the point p' is not a linear combination of the columns of $\partial\psi(u)/\partial u$ for $|u| < \alpha$ and this in turn is equivalent to saying that

$$D(x, u) \overset{\text{def}}{=} \det\left[\frac{\partial\psi(u)}{\partial u}, f(x)\right] \neq 0,$$

for $u < \alpha$. Suppose $D(p, 0) \neq 0$. Since $D(x, u)$ is continuous, there is an α sufficiently small so that $D(x, u) \neq 0$ for $|x - p| < \alpha$, $|u| < \alpha$. For this α, E_p^{n-1} is transverse to γ_p at p. If p is a regular point of (7.1), there always exists an E_p^{n-1} transverse to γ at p. A closed transversal through γ at p is defined in the same way using the closed ball of radius α.

An *open path cylinder* of (7.1) is a set which is homeomorphic to an open cylinder (the cross product of an open ball in R^{n-1} and an open interval) and consists only of arcs of paths of (7.1). A *closed path cylinder* of (7.1) is a set which is homeomorphic to a closed cylinder and consists only of arcs of paths of (7.1.) The *bases* of a closed path cylinder are the images under the homeomorphism of the bases of the closed cylinder. If C is a path cylinder (either open or closed) the arcs of paths of (7.1) lying in C will be called the *generators of* C.

LEMMA 7.3.   Suppose f has continuous first partial derivatives in Ω. If p is a regular point of (7.1) and E^{n-1} is a differentiable $(n-1)$-cell transverse at p to the path γ of (7.1) through p, then there is a path cylinder C containing p in its interior. In particular, every path γ' of (7.1) with a point in C must cross E^{n-1} at a p' where E^{n-1} is transverse to γ'.

PROOF.   Let E^{n-1} have a parametric representation given by $E^{n-1} = \{x \text{ in } R^n : x = \psi(u), u \text{ in } S_\alpha\}$, where $S_\alpha = \{u \text{ in } R^{n-1} : |u| < \alpha\}$, and ψ has a continuous first derivative. Let $\phi(t, p')$, $\phi(0, p') = p'$, be the solution of (7.1) which describes the path γ' through p'. There is an interval I containing zero in its interior such that each $\phi(t, p')$, $p' \in E^{n-1}$ is defined for $t \in I$. The function $\phi(t, p') = \phi(t, \psi(u))$ can therefore be considered as a mapping T of $I \times S_\alpha$ into R^n. From Theorem 3.3, this mapping is continuously differentiable in $I \times S_\alpha$. If we define

$$F(x, t, u) = -x + \psi(u) + \int_0^t f(\phi(s, \psi(u))\, ds$$

for $x \in R^n$, $(t, u) \in I \times S_\alpha$, then the fact that $\phi(t, \psi(u))$ is a solution of (7.1) implies $F(\phi(t, \psi(u)), t, u) = 0$ for $(t, u) \in I \times S_\alpha$. We now consider the relation $F(x, t, u) = 0$ as implicitly defining t, u as a function of x. Since

$$\det \left[\frac{\partial F(x, t, u)}{\partial(t, u)} \right]$$

equals $D(p, 0)$ for $(x, t, u) = (p, 0, 0)$, this determinant must be different from zero in a neighborhood of $(p, 0, 0)$. The implicit function theorem therefore implies that the inverse mapping of T exists and is continuously differentiable in a neighborhood of p. This shows there are $\alpha > 0$, $\tau > 0$ such that the mapping T is a continuously differentiable homeomorphism (or diffeomorphism) of $I_\tau \times S_\alpha$, $I_\tau = \{t : |t| < \tau\}$, into a neighborhood of p. Furthermore, $\{T(t, u), t \text{ in } I_\tau\}$ coincides with the arc of γ' described by the solution $\phi(t, p')$, $-\tau < t < \tau$. The range of T is an open path cylinder. It is clear there is also a closed path cylinder. This proves the lemma.

We now extend this local result to a global result in the sense that the time interval involved in the description of the path may be arbitrarily large as long as it is finite. First we prove a result for paths which are not closed. More precisely, we prove

LEMMA 7.4. Suppose f has continuous first partial derivatives in Ω, γ is any path, p is a regular point, pq is an arc of γ, E_p^{n-1}, E_q^{n-1} are differentiable $(n-1)$-cells transverse to γ at p, q, respectively. Then there is a closed path cylinder whose axis is pq and whose bases are in E_p^{n-1}, E_q^{n-1}.

PROOF. We can assume that $E_p^{n-1} \cap E_q^{n-1}$ is empty. Relative to E_p^{n-1}, E_q^{n-1}, we can construct local open path cylinders C_p, C_q. Let t_p be the time to traverse the path γ from p to q; i.e. $\phi(t_p, p) = q$. From continuity with respect to initial data, one can choose an open $(n-1)$-cell $\bar{E}_p^{n-1}$ in E_p^{n-1} such that $\phi(t_p, p')$ belongs to C_q for every p' in $\bar{E}_p^{n-1}$.

Lemma 7.3 implies that each point $\phi(t_p, p')$ must lie on an arc in C_q of a trajectory of (7.1) and must cross E_q^{n-1} at a point q'. Let the time to traverse the arc $\gamma_{p'}$ from p' to q' be $t_{p'}$. The mapping $p' \to t_{p'}$ is continuously differentiable and an application of the implicit function theorem similar to the above implies the mapping $p' \to \phi(t_{p'}, p')$ is a homeomorphism. Since f in (7.1) is bounded on C_q and C_p, there is a $\nu > 0$ such that the time to leave C_p along an arc of a path through $p' \in E_p^{n-1}$ as well as to leave C_q along an arc of a path through $q' \in E_q^{n-1}$ is greater than ν. Choose $\nu < t_p$.

Let us show that $p'q'$ is a closed arc if the diameter of $\bar{E}_p^{n-1}$ is sufficiently small. If $p'q'$ is not an arc, then $\gamma_{p'}$ must be a closed curve. Thus, there is a $\tau_{p'}$, $\nu < \tau_{p'} < t_{p'} - \nu$ such that $\phi(\tau_{p'}, p') = p'$. If no such $\bar{E}_p^{n-1}$ exists so that $p'q'$ is an arc, then there is a sequence of $p_k' \in \bar{E}_p^{n-1}$, $\tau_{p'_k}$, $\nu < \tau_{p'_k} < t_{p'_k} - \nu$, $p_k' \to p$ as $k \to \infty$ such that $\phi(\tau_{p'_k}, p_k') = p_k'$. The $\tau_{p'_k}$ must be bounded since $t_{p'_k} \to t_p$ as $k \to \infty$ and $\nu < \tau_{p'_k} < t_{p'_k} - \nu$. Therefore, there is a subsequence which we label the same as before such that $\tau_{p'_k} \to \tau_0$ as $k \to \infty$ and $0 < \tau_0 < t_p - \nu/2$. But this clearly implies that the path γ_p described by $\phi(t, p)$ satisfies $\phi(\tau_0, p) = p$. This is a contradiction since pq was assumed to be an arc.

The path cylinder C is obtained as the union of the arcs of the trajectories $p'q'$ with p' in $\bar{E}_p^{n-1}$. It remains only to show that this is homeomorphic to a closed cylinder. For $I = [0, 1]$, define the mapping $G: \bar{E}_p^{n-1} \times I \to R^n$ by $G(p', s) = \phi(st_{p'}, p')$, where $t_{p'}'$ is defined above. It is clear that this mapping is a homeomorphism and therefore C is a closed path cylinder. This proves the lemma.

Now suppose γ is a closed-path. Lemma 7.2 implies γ is the orbit of a nonconstant periodic solution $\phi(t, p)$ of (7.1) of least period $\bar{t} > 0$. Take a transversal E_p^{n-1} at p. There is another transversal $\bar{E}_p^{n-1}$ at p, $\bar{E}_p^{n-1} \subseteq E_p^{n-1}$ such that, for any $q \in \bar{E}_p^{n-1}$, there is a $t_q > 0$, continuously differentiable in q, $x(t_q, q)$ in E_p^{n-1}, $x(t, q)$ not in E_p^{n-1} for $0 < t < t_q$, and the mapping $F: \bar{E}_p^{n-1} \times [0, 1) \to R^n$ defined by $F(q, s) = x(st_q, q)$ is a diffeomorphism. The set $F(\bar{E}_p^{n-1} \times [0, 1))$ is called a *path ring* enclosing γ. We have proved the following result.

LEMMA 7.5.   If γ is a closed path, there is a path ring enclosing γ.

It may be that a solution of an autonomous equation is not defined for all t in R as the example $\dot{x} = x^2$ shows. In the applications, one is usually only interested in studying the behavior of the solutions in some bounded set G and it is very awkward to have to continually speak of the domain of definition of a solution. We can avoid this situation by replacing the original differential equation by another one for which all solutions are defined on $(-\infty, \infty)$ and the paths defined by the solutions of the two coincide inside G. When the paths of two autonomous differential equations coincide on a set G, we say the differential equations are *equivalent on* G.

LEMMA 7.6.   If f in (7.1) is defined on R^n and $G \subset R^n$ is open and bounded, there exists a function $g: R^n \to R^n$ such that $\dot{x} = g(x)$ is equivalent to (7.1) on G and the solutions of this latter equation are defined on $(-\infty, \infty)$.

PROOF.   If $f = (f_1, \ldots, f_n)$, we may suppose without loss of generality that $G \subset \{x: |f_j(x)| \leq 1, j = 1, 2, \ldots, n\}$. Define $g = (g_1, \ldots, g_n)$ by $g_j = f_j \phi_j$, where each ϕ_j is defined by

$$\phi_j(x) = \begin{cases} 1 & \text{if } |f_j(x)| \leq 1, \\[2mm] \dfrac{1}{f_j(x)} & \text{if } f_j(x) > 1, \\[2mm] -\dfrac{1}{f_j(x)} & \text{if } f_j(x) < -1. \end{cases}$$

Corollary 6.3 implies that g satisfies the conditions of the lemma since $|g(x)|$ is bounded in R^n.

I.8. Autonomous Systems—Limit Sets, Invariant Sets

In this section we consider system (7.1) and suppose f satisfies enough conditions on R^n to ensure that the solution $\phi(t, p)$, $\phi(0, p) = p$, is defined for all t in R and all p in R^n and satisfies the conditions (i)–(iii) listed at the beginning of Section I.7.

The orbit $\gamma(p)$ of (7.1) through p is defined by $\gamma(p) = \{x: x = \phi(t, p), -\infty < t < \infty\}$. If q belongs to $\gamma(p)$, then $\gamma(q) = \gamma(p)$ as remarked earlier. The positive semiorbit through p is $\gamma^+(p) = \{x: x = \phi(t, p), t \geq 0\}$ and the negative semiorbit through p is $\gamma^-(p) = \{x: x = \phi(t, p), t \leq 0\}$. If we do not wish to distinguish a particular point on an orbit, we will write γ, γ^+, γ^- for the orbit, positive semiorbit, negative semiorbit, respectively.

The positive or ω-limit set of an orbit γ of (7.1) is the set of points in

R^n which are approached along γ with increasing time. More precisely, a point q belongs to the ω-*limit set* or *positive limit set* $\omega(\gamma)$ of an orbit γ if there exists a sequence of real numbers $\{t_k\}$, $t_k \to \infty$ as $k \to \infty$ such that $\phi(t_k, p) \to q$ as $k \to \infty$. Similarly, a point q belongs to the α-*limit set* or *negative limit set* $\alpha(\gamma)$ if there is a sequence of real numbers $\{t_k\}$, $t_k \to -\infty$ as $k \to \infty$ such that $\phi(t_k, p) \to q$ as $k \to \infty$.

It is easy to prove that equivalent definitions of the ω-limit set and α-limit set are

$$\omega(\gamma) = \bigcap_{p \in \gamma} \overline{\gamma^+(p)} = \bigcap_{\tau \in (-\infty, \infty)} \overline{\bigcup_{t \geq \tau} \phi(t, p)}$$

$$\alpha(\gamma) = \bigcap_{p \in \gamma} \overline{\gamma^-(p)} = \bigcap_{\tau \in (-\infty, \infty)} \overline{\bigcup_{t \leq \tau} \phi(t, p)}$$

where the bar denotes closure.

A set M in R^n is called an *invariant set* of (7.1) if, for any p in M, the solution $\phi(t, p)$ of (7.1) through p belongs to M for t in $(-\infty, \infty)$. Any orbit of (7.1) is obviously an invariant set of (7.1). A set M is called *positively (negatively) invariant* if for each p in M, $\phi(t, p)$ belongs to M for $t \geq 0$ ($t \leq 0$).

THEOREM 8.1. *The α- and ω-limit sets of an orbit γ are closed and invariant. Furthermore, if $\gamma^+(\gamma^-)$ is bounded, then the ω-(α-) limit set is nonempty compact and connected*, $\mathrm{dist}(\phi(t, p), \omega(\gamma(p))) \to 0$ *as* $t \to \infty$ *and* $\mathrm{dist}(\phi(t, p), \alpha(\gamma(p))) \to 0$ *as* $t \to -\infty$.

PROOF. The closure is obvious from the definition. We now prove the positive limit sets are invariant. If q is in $\omega(\gamma)$, there is a sequence $\{t_n\}$, $t_n \to \infty$ as $n \to \infty$ such that $\phi(t_n, p) \to q$ as $n \to \infty$. Consequently, for any fixed t in $(-\infty, \infty)$, $\phi(t + t_n, p) = \phi(t, \phi(t_n, p)) \to \phi(t, q)$ as $n \to \infty$ from the continuity of ϕ. This shows that the orbit through q belongs to $\omega(\gamma)$ or $\omega(\gamma)$ is invariant. A similar proof shows that $\alpha(\gamma)$ is invariant.

If γ^+ (γ^-) is bounded, then the ω- (α-) limit set is obviously nonempty and bounded. The closure therefore implies compactness. It is easy to see that $\mathrm{dist}(\phi(t, p), \omega(\gamma(p))) \to 0$ as $t \to \infty$, $\mathrm{dist}(\phi(t, p), \alpha(\gamma(p))) \to 0$ as $t \to -\infty$. This latter property clearly implies that $\omega(\gamma)$ and $\alpha(\gamma)$ are connected and the theorem is proved.

COROLLARY 8.1. *The limit sets of an orbit must contain only complete paths.*

A set M in R^n is called a *minimal set* of (7.1) if it is nonempty, closed and invariant and has no proper subset which possesses these three properties.

LEMMA 8.1. *If A is a nonempty compact, invariant set of (7.1), there is a minimal set $M \subset A$.*

PROOF. Let F be a family of nonempty subsets of R^n defined by $F = \{B: B \subset A, B \text{ compact, invariant}\}$. For any B_1, B_2 in F, we say $B_2 < B_1$ if $B_2 \subset B_1$. For any $F_1 \subset F$ totally ordered by " $<$ ", let $C = \bigcap_{B \in F_1} B$. The family F_1 has the finite intersection property. Indeed, if B_1, B_2 are in F_1, then either $B_1 < B_2$ or $B_2 < B_1$ and, in either case, $B_1 \cap B_2$ is nonempty and invariant or thus belongs to F_1. The same holds true for any finite collection of elements in F_1. Thus, C is not empty, compact and invariant and for each B in F_1, $C < B$. Now suppose an element D of F is such that $D < B$ for each B in F_1. Then $D \subset B$ for each B in F_1 which implies $D \subset C$ or $D < C$. Therefore C is the minimum of F_1. Since each totally ordered subfamily of F admits a minimum, it follows from Zorn's lemma that there is a minimal element of F. It is easy to see that a minimal element is a minimal set of (7.1) and the proof is complete.

Let us return to the examples considered in Section I.7 to help clarify the above concepts. In example 7.1, the ω-limit set of every orbit except the orbit consisting of the critical point $\{0\}$ is empty. The α-limit set of every orbit is $\{0\}$. The only minimal set is $\{0\}$. In example 7.2, the ω-limit set of the orbits $\{0 < x < 1\}$, $\{x < 0\}$, is $\{0\}$, the α-limit set of $\{x > 1\}$, $\{0 < x < 1\}$ is $\{0\}$ and $\{0\}$ and $\{1\}$ are both minimal sets. In example 7.3, the ω- and α-limit set of any orbit is itself, every orbit is a minimal set and any circular disk about the origin is invariant. In example 7.4, the circle $\{r = 1\}$ and the point $\{r = 0\}$ are minimal sets, the circle $\{r = 1\}$ is the ω-limit set of every orbit except $\{r = 0\}$, while the point $\{r = 0\}$ is the α-limit set of every orbit inside the unit circle. In example 7.6, the torus $r = 1$ is a minimal set as well as the circle $r = 0$, the ω-limit set of every orbit except $r = 0$ is the torus $r = 1$ and the α-limit set of every orbit inside the torus $r = 1$ is the circle $r = 0$.

Let us give one other artificial example to show that the ω-limit sets do not always need to be minimal sets. Consider r and θ as polar coordinates which satisfy the equations

$$\dot{\theta} = \sin^2 \theta + (1 - r)^3,$$

$$\dot{r} = r(1 - r).$$

The ω-limit set of all orbits which do not lie on the sets $\{r = 1\}$ and $\{r = 0\}$ is the circle $r = 1$. The circle $r = 1$ is invariant but the orbits of the equation on $r = 1$ consist of the points $\{\theta = 0\}$, $\{\theta = \pi\}$ and the arcs of the circle $\{0 < \theta < \pi\}$, $\{\pi < \theta < 2\pi\}$, The minimal sets on this circle are just the two points $\{\theta = 0\}$, $\{\theta = \pi\}$.

EXERCISE 8.1. Give an example of a two dimensional system which has an orbit whose ω-limit set is not empty and disconnected.

THEOREM 8.2. If K is a positively invariant set of system (7.1) and K is homeomorphic to the closed unit ball in R^n, there is at least one equilibrium point of system (7.1) in K.

PROOF. For any $\tau_1 > 0$, consider the mapping taking p in K into $\phi(\tau_1, p)$ in K. From Brouwer's fixed point theorem, there is a p_1 in K such that $\phi(\tau_1, p_1) = p_1$ and, thus, a periodic orbit of (7.1) of period τ_1. Choose a sequence $\tau_m > 0$, $\tau_m \to 0$ as $m \to \infty$ and corresponding points p_m such that $\phi(\tau_m, p_m) = p_m$. We may assume this sequence converges to a p^* in K as $m \to \infty$ since there is always a subsequence of the p_m which converge. For any t and any integer m, there is an integer $k_m(t)$ such that $k_m(t)\tau_m \leqq t < k_m(t)\tau_m + \tau_m$ and $\phi(k_m(t)\tau_m, p_m) = p_m$ for all t since $\phi(t, p_m)$ is periodic of period τ_m in t. Furthermore,

$$
\begin{aligned}
|\phi(t, p^*) - p^*| &\leqq |\phi(t, p^*) - \phi(t, p_m)| + |\phi(t, p_m) - p_m| + |p_m - p^*| \\
&= |\phi(t, p^*) - \phi(t, p_m)| + |\phi(t - k_m(t)\tau_m, p_m) - p_m| \\
&\quad + |p_m - p^*|,
\end{aligned}
$$

and the right hand side approaches zero as $m \to \infty$ for all t. Therefore, p^* is an equilibrium point of (7.1) and the theorem is proved.

Some of the most basic problems in differential equations deal with the characterization of the minimal sets and the behavior of the solutions of the equations near minimal sets. Of course, one would also like to be able to describe the manner in which the ω-limit set of any trajectory can be built up from minimal sets and orbits connecting the various minimal sets. In the case of two dimensional systems, these questions have been satisfactorily answered. For higher dimensional systems, the minimal sets have not been completely classified and the local behavior of solutions has been thoroughly discussed only for minimal sets which are very simple. Our main goal in the following chapters is to discuss some approaches to these questions.

I.9. Remarks and Suggestions for Further Study

For a detailed proof of Peano's theorem without using the Schauder theorem, see Coddington and Levinson [1], Hartman [1]. When uniqueness of trajectories of a differential equation is not assumed, the union of all trajectories through a given point forms a type of funnel. For a discussion of the topological properties of such funnels, see Hartman [1].

There are many other ways to generalize the concept of a differential equation. For example, one could permit the vector field $f(t, x)$ to be continuous in t, but discontinuous in x. Also, $f(t, x)$ could be a set valued function. In spite of the fact that such equations are extremely important in some applications to control theory, they are not considered in this book. The interested reader may consult Flugge-Lotz [1], André and Seibert [1], Fillipov [1], Lee and Marcus [1].

The results on differential inequalities in Section 6 are valid in a much more general setting. In fact, one can use upper right hand derivatives in place of right hand derivatives, the assumption of uniqueness can be eliminated by considering maximal solutions of the majorizing equation and even some types of vector inequalities can be used. Differential inequalities are also very useful for obtaining uniqueness theorems for vector fields which are not Lipschitzian. See Coppel [1], Hartman [1], Szarski [1], Laksmikantham and Leela [1].

Sections 7 and 8 belong to the geometric theory of differential equations begun by Poincaré [1] and advanced so much by the books of Birkhoff [1], Lefschetz [1], Nemitskii and Stepanov [1], Auslander and Gottschalk [1]. The presentation in Section 7 relies heavily upon the book of Lefschetz [1]. A function $\phi: R \times R^n$ into R^n which satisfies properties (i–iii) listed at the beginning of Section 7 is called a *dynamical system*. Dynamical systems can and have been studied in great detail without any reference to differential equations (see Gottschalk and Hedlund [1], Nemitskii and Stepanov [1]). All results in Section 7 remain valid for dynamical systems. However, the proofs are more difficult since the implicit function theorem cannot be invoked. The concepts of Section 8 are essentially due to Birkhoff [1].

The definitions of stability given in Section 4 are due to Liapunov [1]. For other types of stability see Cesari [1], Yoshizawa [2].

CHAPTER II

Two Dimensional Systems

The purpose of this chapter is to discuss the global behavior of solutions of differential equations in the plane and differential equations without critical points on a torus. In particular, in Section 1, the ω-limit set of any bounded orbit in the plane is completely characterized, resulting in the famous Poincaré-Bendixson theorem. Then this theorem is applied to obtain the existence and stability of limit cycles for some special types of equations. In Section 2, all possible ω-limit sets of orbits of smooth differential equations without singular points on a torus are characterized, yielding the result that the ω-limit set of an orbit is either a periodic orbit or the torus itself.

Differential equations on the plane are by far the more important of the two types discussed since any system with one degree of freedom is described by such equations. On the other hand, in the restricted problem of three bodies in celestial mechanics, the interesting invariant sets are torii and, thus, the theory must be developed. Also, as will be seen in a later chapter, invariant torii arise in many other applications.

II.1. Planar Two Dimensional Systems—The Poincaré-Bendixson Theory

In this section, we consider the two dimensional system

$$\text{(1.1)} \qquad \dot{x} = f(x)$$

where x is in R^2, $f \colon R^2 \to R^2$ is continuous with its first partial derivatives and such that the solution $\phi(t, p)$, $\phi(0, p) = p$, of (1.1) exists for $-\infty < t < \infty$. The solution $\phi(t, p)$ is the unique solution through $(0, p)$ and, therefore, is continuous for (t, p) in R^3. For each fixed t, recall that the mapping $\phi(t, \cdot) \colon R^2 \to R^2$ is a homeomorphism.

The beautiful results for 2-dimensional planar systems are made possible because of the Jordan curve theorem which is now stated without proof. Recall that a Jordan curve is the homeomorphic image of a circle.

JORDAN CURVE THEOREM. Any Jordan curve J in R^2 separates the plane; more precisely, $R^2 \backslash J = S_e \cup S_i$ where S_e and S_i are disjoint open sets, S_e is unbounded and called the exterior of J, S_i is bounded and called the interior of J and both sets are arcwise connected.

A set B is arcwise connected if p, q in B implies there is an arc pq joining p and q which lies entirely in B.

Let p be a regular point, L be a closed transversal containing p, L_0 be its interior,

$$V = \{p \text{ in } L_0 : \text{ there is a } t_p > 0 \text{ with } \phi(t_p, p) \text{ in } L_0 \text{ and}$$
$$\phi(t, p) \text{ in } R^2 \backslash L \text{ for } 0 < t < t_p\},$$

and let $W = h^{-1}(V)$ where $h \colon [-1, 1] \to L$ is a homeomorphism. Also, let $g \colon W \to (-1, 1)$ be defined by $g(w) = h^{-1}\phi(t_{h(w)}, h(w))$. See Fig. 1.1.

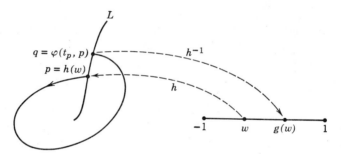

Figure II.1.1

LEMMA 1.1. The set W is open, g is continuous and increasing on W and the sequence $\{g^k(w)\}$, $k = 0, 1, \ldots, n \leq \infty$ is monotone, where $g^k(w) = g(g^{k-1}(w))$, $k = 1, 2, \ldots, g^0(w) = w$.

PROOF. For any p in $V \subset L_0$ let $q = \phi(t_p, p)$ in L_0. From Section I.7, we have proved that the arc pq of the path through p can be enclosed in an open path cylinder with pq as axis and the bases of the cylinder lying in the interior L_0 of the transversal L. This proves W is open. From continuity with respect to initial data, t_p is continuous and we get continuity of g.

To prove the last part of the lemma, consider the Jordan curve J given by $C = \{x \colon x = \phi(t, p), 0 \leq t \leq t_p\}$ and the segment of L joining p and q. If $p = q$, then $\gamma(p)$ is a periodic orbit and the sequence $\{g^k(w)\}$, $w = h^{-1}(p)$, consists of only one point. Thus, suppose for definiteness $h^{-1}(p) < h^{-1}(q)$; that is, $g(w) > w_0$, $w_0 = h^{-1}(p)$, $w = h^{-1}(q)$. Let S_i, S_e denote the interior and exterior of J, respectively. From the definition of a transversal L, paths can cross L in only one direction. Therefore, that part of L_0 given by $h[(g(w), 1)]$ must be completely in either S_i or S_e since otherwise an orbit would cross the segment pq of L_0 in a direction opposite to the direction the orbit through p crosses L. Therefore, if $g^2(w)$ is defined, it must belong to $(g(w), 1)$ which by induction shows that the sequence above is monotone. Suppose $w_1 > w$, w_1 in W. If $g(w_1)$ is defined, then $g(w_1) > g(w)$ for the same reason as before. This completes the proof of the lemma.

We used the differentiability of $f(x)$ in the above proof when we proved the existence of a path cylinder in Section I.7. As remarked at the end of Chapter I, this assumption is unnecessary and one can prove the existence of a path cylinder only under the assumption of uniqueness of solutions. In all of the proofs that follow in this section, this is the only place differentiability of $f(x)$ is used. In particular, the Poincaré-Bendixson theorem below is valid without differentiability of $f(x)$.

COROLLARY 1.1. The ω-limit set $\omega(\gamma)$ of an orbit γ can intersect the interior L_0 of a transversal L in only one point. If $\omega(\gamma) \cap L_0 = p_0$, $\gamma = \gamma(p)$, then either $\omega(\gamma) = \gamma$, and γ is a periodic orbit or there exists a sequence $\{t_k\}$, $t_k \to \infty$ as $k \to \infty$ such that $\varphi(t_k, p)$ is in L_0, $\phi(t_k, p) \to p_0$ monotonically; that is, the sequence $h^{-1}(\phi(t_k, p))$ is monotone.

PROOF. Suppose $\omega(\gamma) \cap L_0$ contains a point p_0. From the definition of ω-limit set, there is a sequence $\{t'_k\}$, $t'_k \to \infty$ as $k \to \infty$ such that $\phi(t'_k, p) \to p_0$ as $k \to \infty$. But from Section I.7, there must be a path cylinder containing p_0 such that any orbit passing sufficiently near p_0 must contain an arc which crosses the transversal L at some point. Therefore, there exist points $q_k = \phi(t_k, p)$ in L_0, $t_k \to \infty$ as $k \to \infty$ such that $q_k \to p_0$ as $k \to \infty$. But Lemma 1.1 implies that the q_k approach p_0 monotonically in the sense that $h^{-1}(q_k)$ is a monotone sequence. Suppose now p'_0 is any other point in $\omega(\gamma) \cap L_0$. Then the same argument holds to get a sequence $q'_k \to p'_0$ monotonically. Lemma 1.1 then clearly implies that $p'_0 = p_0$ and the corollary is proved.

COROLLARY 1.2. If γ^+ and $\omega(\gamma^+)$ have a regular point in common, then γ^+ is a periodic orbit.

PROOF. If p_0 in $\gamma^+ \cap \omega(\gamma^+)$ is regular, there is a transversal of (1.1) containing p_0 in its interior. From Corollary 1.1, if $\omega(\gamma^+) \neq \gamma^+$, there is a sequence $q_k = \phi(t_k, p) \to p_0$ monotonically. Since p_0 is in γ^+, this contradicts Lemma 1.1. Corollary 1.1 therefore implies the result.

THEOREM 1.1. If M is a bounded minimal set of (1.1), then M is either a critical point or a periodic orbit.

PROOF. If γ is an orbit in M, then $\alpha(\gamma)$ and $\omega(\gamma)$ are not empty and belong to M. Since $\alpha(\gamma)$ and $\omega(\gamma)$ are closed and invariant we have $\alpha(\gamma) = \omega(\gamma) = M$. If M contains a critical point, then it must be the point itself since it is equal to $\omega(\gamma)$ for some γ. If $M = \omega(\gamma)$ does not contain a critical point, then $\gamma \subset \omega(\gamma)$ implies γ and $\omega(\gamma)$ have a regular point in common which implies by Corollary 1.2 that γ is periodic. Therefore $\gamma = \omega(\gamma) = M$ and this proves Theorem 1.1.

LEMMA 1.2. If $\omega(\gamma^+)$ contains regular points and also a periodic orbit γ_0, then $\omega(\gamma^+) = \gamma_0$.

PROOF. If not, then the connectedness of $\omega(\gamma^+)$ implies the existence of a sequence $\{p_n\}$, p_n in $\omega(\gamma^+)\backslash\gamma_0$ and a p_0 in γ_0 such that $p_n \to p_0$ as $n \to \infty$. Since p_0 is regular, there is a closed transversal L such that p_0 is in the interior L_0 of L. From Corollary 1.1, $\omega(\gamma^+) \cap L_0 = \{p_0\}$. From the existence of a path cylinder in Section I.7, there is neighborhood N of p_0 such that any orbit entering N must intersect L_0. In particular, $\gamma(p_n)$ for n sufficiently large must intersect L_0. But we know this occurs at p_0. Thus p_n belongs to γ_0 for n sufficiently large which is a contradiction.

THEOREM 1.2 (Poincaré-Bendixson Theorem). If γ^+ is a bounded positive semiorbit and $\omega(\gamma^+)$ does not contain a critical point, then either

(i) $\gamma^+ = \omega(\gamma^+)$,

or

(ii) $\omega(\gamma^+) = \bar{\gamma}^+\backslash\gamma^+$.

In either case, the ω-limit set is a periodic orbit. The same result is valid for a negative semiorbit.

PROOF. By assumption and Theorem I.8.1, $\omega(\gamma^+)$ is nonempty, compact invariant and contains regular points only. Therefore, by Lemma I.8.1, there is a bounded minimal set M in $\omega(\gamma^+)$ and M contains only regular points. Theorem 1.1 implies M is a periodic orbit γ_0. Lemma 1.2 now implies the theorem.

An *invariant set* M of (1.1) is said to be *stable* if for every ε-neighborhood U_ε of M there is a δ-neighborhood U_δ of M such that p in U_δ implies $\gamma^+(p)$ in U_ε. M is said to be *asymptotically stable* if it is stable and in addition there is a $b > 0$ such that p in U_b implies $\mathrm{dist}(\phi(t, p), M) \to 0$ as $t \to \infty$. If M is a periodic orbit, one can also define stability from the inside and outside of M in an obvious manner.

COROLLARY 1.3. For a periodic orbit γ_0 to be asymptotically stable it is necessary and sufficient that there is a neighborhood G of γ_0 such that $\omega(\gamma(p)) = \gamma_0$ for any p in G.

PROOF. We first prove sufficiency. Clearly $\mathrm{dist}(\phi(t, p), \gamma_0) \to 0$ as $t \to \infty$ for every p in G. Suppose L is a transversal at p_0 in γ_0 and suppose p is in

$G \cap S_e$, q is in $G \cap S_i$, where S_e and S_i are the exterior and interior of γ_0, respectively. From Corollary 1.1, there are sequences $p_k = \phi(t_k, p)$, $q_k = \phi(t'_k, q)$ in L approaching p_0 as $k \to \infty$. Consider the neighborhood U_k of γ_0 which lies between the curves given by the arc $p_k p_{k+1}$ of $\gamma(p)$ and the segment of L between p_k and p_{k+1} and the arc $q_k q_{k+1}$ of $\gamma(p)$ and the segment of L between q_k and q_{k+1}. U_k is a neighborhood of γ_0. The sequences $\{t_k\}$, $\{t'_k\}$ satisfy $t_{k+1} - t_k \to \alpha$, $t'_{k+1} - t'_k \to \alpha$ as $k \to \infty$ where α is the period of γ_0. This follows from the existence of a path ring around γ_0. Continuity with respect to initial data then implies for any given ε-neighborhood U_ε of γ_0, there is a k sufficiently large so that p in U_k implies $\phi(t, p)$ in U_ε for $t \geqq 0$ and γ_0 is stable.

To prove the converse, suppose γ_0 is asymptotically stable. Then there must exist a neighborhood G of γ_0 which contains no equilibrium points and $G \backslash \gamma_0$ contains no periodic orbits. The Poincaré-Bendixson theorem implies the ω-limit set of every orbit is a periodic orbit. Since γ_0 is the only such orbit in G, this proves the corollary.

COROLLARY 1.4. Suppose γ_1, γ_2 are two periodic orbits with γ_2 in the interior of γ_1 and no periodic orbits or critical points lie between γ_1 and γ_2. Then both orbits cannot be asymptotically stable on the sides facing one another.

PROOF. Suppose γ_1, γ_2 are stable on the sides facing one another. Then there exist positive orbits γ'_1, γ'_2 in the region between γ_1, γ_2 such that $\gamma_1 = \bar{\gamma}'_1 \backslash \gamma'_1$, $\gamma_2 = \bar{\gamma}'_2 \backslash \gamma'_2$. For any p_1 in γ_1, p_2 in γ_2 construct transversals L_1, L_2. There exist $p'_1 \neq p''_1$ in $\gamma'_1 \cap L_1$, $p'_2 \neq p''_2$ in $\gamma'_2 \cap L_2$. Consider the region S bounded by the Jordan curve consisting of the arc $p'_1 p''_1$ of γ'_1 and the segment of the transversal L_1 between p'_1 and p''_1 and the curve consisting of the arc $p'_2 p''_2$ of γ'_2 and the segment of the transversal L_2 between p'_2 and p''_2 (see Fig. 1.2). The region S contains a negative semiorbit. Thus, the Poincaré-Bendixson Theorem implies the existence of a periodic orbit in this region. This contradiction proves the corollary.

THEOREM 1.3. Let γ^+ be a positive semiorbit in a closed bounded subset K of R^2 and suppose K has only a finite number of critical points. Then one of the following is satisfied:

    (i) $\omega(\gamma^+)$ is a critical point;

    (ii) $\omega(\gamma^+)$ is a periodic orbit;

    (iii) $\omega(\gamma^+)$ contains a finite number of critical points and a set of orbits γ_i with $\alpha(\gamma_i)$ and $\omega(\gamma_i)$ consisting of a critical point for each orbit γ_i. See Fig. 1.3.

PROOF. $\omega(\gamma^+)$ contains at most a finite number of critical points. If $\omega(\gamma^+)$ contains no regular points, then it must be just one point since it is

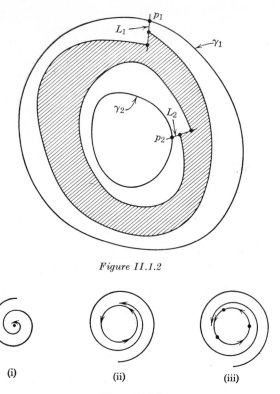

Figure II.1.2

(i)   (ii)   (iii)

Figure II.1.3

connected. This is case (i). Suppose $\omega(\gamma^+)$ has regular points and also contains a periodic orbit γ_0. Then $\omega(\gamma^+) = \gamma_0$ from Lemma 1.2.

Now suppose $\omega(\gamma^+)$ contains regular points and no periodic orbits. Let γ_0 be an orbit in $\omega(\gamma^+)$. Then $\omega(\gamma_0) \subset \omega(\gamma^+)$. If p_0 in $\omega(\gamma_0)$ is a regular point and L is a closed transversal to p_0 with interior L_0, then Corollary 1.1 implies $\omega(\gamma^+) \cap L_0 = \omega(\gamma_0) \cap L_0 = \{p_0\}$ and γ_0 must meet L_0 at some q_0. Since γ_0 belongs to $\omega(\gamma^+)$ we have $q_0 = p_0$ which implies by Corollary 1.2 that γ_0 is periodic. This contradiction implies $\omega(\gamma_0)$ has no regular points. But, $\omega(\gamma_0)$ is connected and therefore consists of exactly one point, a critical point. A similar argument applies to the α-limit sets and the theorem is proved.

COROLLARY 1.5.   If γ^+ is a positive semiorbit contained in a compact set in Ω and $\omega(\gamma^+)$ contains regular points and exactly one critical point p_0, then there is an orbit in $\omega(\gamma^+)$ whose α- and ω-limit sets are $\{p_0\}$.

We now discuss the possible behavior of orbits in a neighborhood of a periodic orbit. Let γ_0 be a periodic orbit and L_0 be a transversal at p_0 in γ_0,

h: $(-1, 1) \to L_0$ be a homeomorphism with $h(0) = p_0$. If g is the function defined in Lemma 1.1, then $g(0) = 0$ since γ_0 is periodic. Since the domain W of definition of g is open, 0 is in W, g is continuous and increasing, there is an $\varepsilon > 0$ such that g is defined and $g(w) > 0$ for w in $(0, \varepsilon)$ and $g(w) < 0$ for w in $(-\varepsilon, 0)$. We discuss in detail the case $g(w) > 0$ on $(0, \varepsilon)$ and the case $g(w) < 0$ on $(-\varepsilon, 0)$ is treated in a similar manner. Three possibilities present themselves. There is an ε_1, $0 < \varepsilon_1 < \varepsilon$, such that

  (i) $g(w) < w$ for w in $(0, \varepsilon_1)$;
  (ii) $g(w) > w$ for w in $(0, \varepsilon_1)$;
  (iii) $g(w) = w$ for a sequence $w_n > 0$, $w_n \to 0$ as $n \to \infty$.

In case (i), $g^k(w)$ is defined for each $k > 0$, is monotone decreasing and $g^k(w) \to 0$ as $k \to \infty$. In fact, it is clear that $g^k(w)$ is defined for $k > 0$. Lemma 1.1 states that $g^k(w)$ is monotone and the hypothesis implies this sequence is decreasing. Therefore, $g^k(w) \to w_0 \geqq 0$ as $k \to \infty$. But, this implies $g(w_0) = w_0$ and therefore $w_0 = 0$. Similarly, in case (ii), if we define $g^{-k}(w)$ to be the inverse of $g^k(w)$ then $g^{-k}(w)$, is defined for each $k > 0$, is decreasing and $g^{-k}(w) \to 0$ as $k \to \infty$.

If we interpret these three cases in terms of orbits and limit sets, we have

THEOREM 1.4.   If γ_0 is a periodic orbit and G is an open set containing γ_0, $G_e = G \cap S_e$, $G_i = G \cap S_i$ where S_e and S_i are the interior and exterior of γ_0, then one of the following situations occur:
  (i)   there is a G such that either $\gamma_0 = \omega(\gamma(p))$ for every p in G_e or $\gamma_0 = \alpha(\gamma(p))$ for every p in G_e;
  (ii) for each G, there is a p in G_e, p not in γ_0, such that $\alpha(\gamma(p)) = \gamma(p)$ is a periodic orbit.
Similar statements hold for G_i.

We call γ_0 a *limit cycle* if there is a neighborhood G of γ_0 such that either $\omega(\gamma(p)) = \gamma_0$ for every $p \in G$ or $\alpha(\gamma(p)) = \gamma_0$ for every $p \in G$.

The Poincaré-Bendixson theorem suggests a way to determine the existence of a nonconstant periodic solution of an autonomous differential equation in the plane. More specifically, one attempts to construct a domain D in R^2 which is equilibrium point free and is positively invariant; that is, any solution of (1.1) with initial value in D remains in D for $t \geqq 0$. In such a case, we are assured that D contains a positive semiorbit γ^+ and thus a periodic solution from the Poincaré-Bendixson theorem. Furthermore, if we can ascertain that there is only one periodic orbit in D, it will be asymptotically stable from Theorem 1.4 and Corollary 1.3.

These ideas are illustrated for the Liénard type equation

(1.2) $$\ddot{u} + g(u)\dot{u} + u = 0$$

where $g(u)$ is continuous and the following conditions are satisfied:

(1.3)   (a) $G(u) = \text{def} \int_0^u g(s)\,ds$ is odd in u,

    (b) $G(u) \to \infty$ as $|u| \to \infty$ and there is a $\beta > 0$ such that $G(u) > 0$
        for $u > \beta$ and is monotone increasing.

    (c) There is an $\alpha > 0$ such that $G(u) < 0$ for $0 < u < \alpha$, $G(\alpha) = 0$.

Equation 1.2 is equivalent to the system of equations

(1.4)
$$\dot{u} = v - G(u),$$
$$\dot{v} = -u.$$

System (1.4) is a special case of system (1.1) with $x = (u, v)$ and has a unique
orbit through any point in R^2 since G has a continuous first derivative.
System (1.4) has only one critical point; namely, $u = 0$, $v = 0$, and the orbits
of (1.4) are the solutions of the first order equation

(1.5)
$$\frac{dv}{du} = -\frac{u}{v - G(u)}.$$

From (1.4), the function $u = u(t)$ is increasing for $v > G(u)$, decreasing if
$v < G(u)$ and the function $v = v(t)$ is decreasing if $u > 0$, increasing if $u < 0$.
Also, the slopes of the paths $v = v(u)$ described by (1.5) are horizontal on the
v-axis and vertical on the curve $v = G(u)$. These facts and hypothesis (1.3b)
on $G(u)$ imply that a solution of (1.4) with initial value $A = (0, v_0)$ for v_0
sufficiently large describes an orbit with an arc of the general shape shown in
Fig. 1.4.

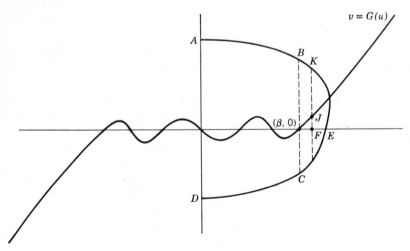

Figure II.1.4

Observe that (u, v) a solution of (1.4) implies $(-u, -v)$ is also a solution from hypothesis (1.3a). Therefore, if we know a path $ABECD$ exists as in Fig. 1.4, then the reflection of this path through the origin is another path. In particular, if $A = (0, v_0)$, $D = (0, -v_1)$, $v_1 < v_0$, then the complete positive semiorbit of the path through any point $A' = (0, v_0')$, $0 < v_0' < v_0$ must be bounded. In fact, it must lie in the region bounded by the arc $ABECD$, its reflection through the origin and segments on the v-axis connecting these arcs to form a Jordan curve. The above symmetry property also implies that (1.4) can have a periodic orbit if and only if $v_1 = v_0$.

We show there exists a $v_0 > 0$ sufficiently large so that a solution as in Fig. 1.4 exists with $A = (0, v_0)$, $D = (0, -v_1)$, $v_1 < v_0$. Consider the function $V(u, v) = (u^2 + v^2)/2$. If u, v are solutions of (1.4) and (1.5), then

(1.6)   (a) $\quad \dfrac{dV}{dt} = -uG(u),$

                       (b) $\quad \dfrac{dV}{du} = -\dfrac{uG(u)}{v - G(u)}$

                       (c) $\quad \dfrac{dV}{dv} = G(u).$

Using these expressions, we have

$$V(D) - V(A) = \int_{ABECD} dV = \left(\int_{AB} + \int_{CD} \right) \frac{-uG(u)}{v - G(u)} \, du + \int_{BEC} G(u) \, dv$$

along the orbits of (1.4). It is clear that this first expression approaches zero monotonically as $v_0 \to \infty$. If F is any point on the u-axis in Fig. 1.4 between $(\beta, 0)$ and E, and $\phi(v_0) = \int_{BEC} G(u) \, dv$, then

$$-\phi(v_0) = -\int_{BEC} G(u) \, dv = \int_{CEB} G(u) \, dv > \int_{EK} G(u) \, dv > FJ \times FK$$

where FJ, FK are the lengths of the line segments indicated in Fig. 1.4. For fixed F, $FK \to \infty$ as $v_0 \to \infty$ and this proves $\phi(v_0) \to -\infty$ as $v_0 \to \infty$. Thus, there is a v_0 such that $V(D) < V(A)$. But this implies $v_1 < v_0$ and the semiorbit through A must be bounded. On the other hand, this semiorbit must also be bounded away from the origin since (1.6a) and hypothesis (1.3c) implies that $dV/dt \geqq 0$ along solutions of (1.4) if $|u| < \alpha$. Finally, the Poincaré-Bendixson Theorem implies the existence of a periodic solution of (1.4) and we have

THEOREM 1.5. *If G satisfies the conditions (1.3), then equation (1.2) has a nonconstant periodic solution.*

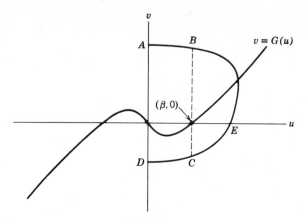

Figure II.1.5

If further hypotheses are made on G, then the above method of proof will yield the existence of exactly one nonconstant periodic solution. In fact, we can prove

THEOREM 1.6. If G satisfies the conditions (1.3) with $\alpha = \beta$, then equation (1.2) has exactly one periodic orbit and it is asymptotically stable.

PROOF. With the stronger hypotheses on G, every solution with initial value $A = (0, v_0)$, $v_0 > 0$, has an arc of an orbit as shown in Fig. 1.5. With the notations the same as in the proof of Theorem 1.5 and with $E = (u_0, 0)$, we have

$$V(D) - V(A) = \int_{ABECD} G(u) \, dv > 0,$$

if $u_0 < \alpha$. This implies no periodic orbit can have $u_0 < \alpha$.

For $u_0 > \alpha$, if we introduce new variables $x = G(u)$, $y = v$ to the right of line BC in Fig. 1.5 (this is legitimate since $G(u)$ is monotone increasing in this region), then the arc BEC goes into an arc $B^*E^*C^*$ with end points on the y-axis and the second expression $\phi(v_0) = \int_{BEC} G(u) \, dv = \int_{B^*E^*C^*} x \, dy$ is the negative of the area bounded by the curve $B^*E^*C^*$ and the y-axis. Therefore, $\phi(v_0)$ is a monotone decreasing function of v_0. It is easy to check that $\int_{AB} + \int_{BC} G(u) du$ is decreasing in v_0 and so $V(D) - V(A)$ is decreasing in v_0. Also, in the proof of Theorem 1.5, it was shown that $V(D) - V(A)$ approaches $-\infty$ as $v_0 \to \infty$. Therefore, there is a unique v_0 for which $V(D) = V(A)$ and thus a unique nonconstant periodic solution. Theorem 1.4 and Corollary 1.3 imply the stability properties of the orbit and the proof is complete.

An important special case of Theorem 1.6 is the van der Pol equation

(1.7) $$\ddot{u} - k(1 - u^2)\dot{u} + u = 0, \quad k > 0.$$

In the above crude analysis, we obtained very little information concerning the location of the unique limit cycle given in Theorem 1.6. When a differential equation contains a parameter, one can sometimes discuss the precise limiting behavior as the parameter tends to some value. This is illustrated with van der Pol's equation (1.7). Suppose k is very large; more specifically, suppose $k = \varepsilon^{-1}$ and let us determine the behavior of the periodic solution as $\varepsilon \to 0^+$. Oscillations of this type are called *relaxation oscillations*. System (1.7) is equivalent to

(1.8) $$\varepsilon \dot{u} = v - G(u),$$
$$\dot{v} = - \varepsilon u,$$

where $G(u) = u^3/3 - u$. From Theorem 1.6, equation (1.8) has a unique asymptotically stable limit cycle $\Gamma(\varepsilon)$ for every $\varepsilon > 0$. From (1.8), if ε is small and the orbit is away from the curve $v = G(u)$ in Fig. 1.6, then the u

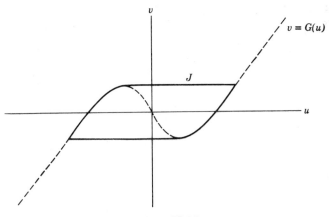

Figure II.1.6

coordinate has a large velocity and the v coordinate is moving slowly. Therefore, the orbit has a tendency to jump in horizontal directions except when it is very close to the curve $v = G(u)$. These intuitive remarks are made precise in

THEOREM 1.7. As $\varepsilon \to 0$, the limit cycle of (1.8) approaches the Jordan curve J shown in Fig. 1.6 consisting of arcs of the curve $v = G(u)$ and horizontal line segments.

To prove this, we construct a closed annular region U containing J such that dist(U, J) is any preassigned constant and yet for ε sufficiently small, all paths cross the boundary of U inward. U will thus contain (from the Poincaré-Bendixson theorem) the limit cycle $\Gamma(\varepsilon)$. The construction of

U is shown in Fig. 1.7 where h is a positive constant. The straight lines 81 and 45 are tangent to $v = G(u) + h$, $v = G(u) - h$ respectively and the lines 56, 12, 9–10, 13–14, are horizontal while 23, 67, 11–12, 15–16 are vertical. The remainder of the construction should be clear. The inner and outer

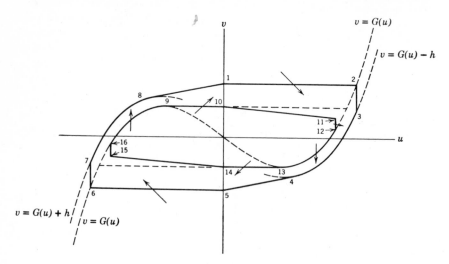

Figure II.1.7

boundaries are chosen to be symmetrical about the origin. Also marked on the figure are arrows designating the direction segments of the boundaries crossed. These are obtained directly from the differential equation and are independent of $\varepsilon > 0$. It is necessary to show that the other segments of the boundary are also crossed inward by orbits if ε is small. By symmetry, it is only necessary to discuss 34, 45 and 10–11.

At any point $(u, G(u) - h)$ on 34, along the orbits of (1.8), we have

$$\frac{dv}{du} = \frac{-\varepsilon^2 u}{v - G(u)} = \frac{\varepsilon^2 u}{h} < \frac{\varepsilon^2 u(3)}{h}$$

where $u(3)$ is the value of u at point 3. Hence for ε small enough, this is less than $g(4) < g(u)$ which is the slope of the curve $G(u) - h$. Thus, $\dot{v} < 0$ on this arc implies the orbits enter the region along this arc.

Along the arc 45, we have $|v - G(u)| > h$ and, hence, the absolute value of the slope of the path $|dv/du| = |-\varepsilon^2 u/[v - G(u)]| < \varepsilon^2 u(4)/h$ approaches zero as $\varepsilon \to 0$. For ε small enough this can be made $< g(4)$ which is the slope of the line 45. Thus, $\dot{v} < 0$ on this arc implies the orbits enter into U if ε is small enough.

Let K be the length of the arc 11–12. For K small enough, $|v - G(u)| > K$ along the arc 10–11. Hence, $|dv/du|$ along orbits of (1.8) is less than $\varepsilon^2 u/K < \varepsilon^2 u(11)/K$, which approaches zero as $\varepsilon \to 0$. Thus, for ε small, the orbits enter U since $\dot{u} > 0$ on this arc.

This shows that given a region U of the above type, one can always choose ε small enough to ensure that the orbits cross the boundary of U inward. This proves the desired result since it is clear that U can be made to approximate J as well as desired by appropriately choosing the parameters used in the construction.

EXERCISE 1.1.   Prove the following *Theorem*. Any open disk in R^2 which contains a bounded semiorbit of (1.1) must contain an equilibrium point. Hint: Use the Poincaré-Bendixson Theorem and Theorem I.8.2.

EXERCISE 1.2.   Give a generalization of Exercise 1.1 which remains valid in R^3? Give an example.

EXERCISE 1.3.   Prove the following *Theorem*. If div f has a fixed sign (excluding zero) in a closed two cell Ω, then Ω has no periodic orbits. Hint: Prove by contradiction using Green's theorem over the region bounded by a periodic orbit in Ω.

EXERCISE 1.4.   Consider the two dimensional system $\dot{x} = f(t, x)$, $f(t + 1, x) = f(t, x)$, where f has continuous first derivatives with respect to x. Suppose Ω is a subset of R^2 which is homeomorphic to the closed unit disk. Also, for any solution $x(t, x_0)$, $x(0, x_0) = x_0$, suppose there is a $T(x_0)$ such that $x(t, x_0)$ is in Ω for all $t \geq T(x_0)$. Prove by Brouwer's fixed point theorem that there is an integer m such that the equation has a periodic solution of period m. Does there exist a periodic solution of period 1?

EXERCISE 1.5.   Suppose f as in exercise (1.4) and there is a $\lambda > 0$ such that $x'f(t, x) \leq -\lambda |x|^2$ for all t, x. If $g(t) = g(t + 1)$ is a continuous function, prove the equation $\dot{x} = f(t, x) + g(t)$ has a periodic solution of period 1.

EXERCISE 1.6.   Suppose γ_0 is a periodic orbit of a two dimensional system and let G_e and G_i be the sets defined in Theorem 1.4. Is it possible for an equation to have $\alpha(\gamma(p)) = \gamma_0$ for all noncritical points p in G_i and $\omega(\gamma(q)) = \gamma_0$ for all q in G_e? Explain.

EXERCISE 1.7.   For Lienard's equation, must there always be a periodic orbit which is stable from the outside? Must there be one stable from the inside? Explain.

EXERCISE 1.8   Is it possible to have a two dimensional system such that each orbit in a bounded annulus is a periodic orbit? Can this happen for analytic systems? Explain.

II.2. Differential Systems on a Torus

In this section, we discuss the behavior of solutions of the pair of first order equations

(2.1)
$$\dot{\phi} = \Phi(\phi, \theta),$$
$$\dot{\theta} = \Theta(\phi, \theta),$$

where

(2.2)
$$\Phi(\phi + 1, \theta) = \Phi(\phi, \theta + 1) = \Phi(\phi, \theta),$$
$$\Theta(\phi + 1, \theta) = \Theta(\phi, \theta + 1) = \Theta(\phi, \theta).$$

We suppose Φ, Θ are continuous and there is a unique solution of (2.1) through any given point in the ϕ, θ plane. Since Φ, Θ are bounded, the solutions will exist on $(-\infty, \infty)$.

If opposite sides of the unit square in the (ϕ, θ)-plane are identified, then this identification yields a torus $\mathscr{T}$ and equations (2.1) can be interpreted as a differential equation on a torus. An orbit of (2.1) in the (ϕ, θ)-plane when interpreted on the torus may appear as in Fig. 2.1.

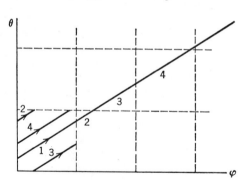

Figure II.2.1

We also suppose that (2.1) has no equilibrium points and, in particular, that $\Phi(\phi, \theta) \neq 0$ for all ϕ, θ. The phase portrait for (2.1) is then determined by

(2.3)
$$\frac{d\theta}{d\phi} = A(\phi, \theta),$$

$$A(\phi + 1, \theta) = A(\phi, \theta + 1) = A(\phi, \theta),$$

where $A(\phi, \theta)$ is continuous for all ϕ, θ. The discussion will center around the solutions of (2.3).

The torus $\mathcal{T}$ can be embedded in R^3 by the relations

(2.4) $x = (R + r \cos 2\pi\theta) \cos 2\pi\phi$.

$y = (R + r \cos 2\pi\theta) \sin 2\pi\phi,$

$z = r \sin 2\pi\theta,$

$0 \leqq \phi < 1, 0 \leqq \theta < 1, 0 < r < R.$

This embedding is convenient for the sake of terminology only. We shall refer to the circle $\phi = $ constant as a *meridian* and $\theta = $ constant as a *parallel*.

Example 2.1. In (2.1), if $\Phi = 1$, $\Theta = \omega$, a constant, then $A(\theta, \phi) = \omega$ in (2.3). The solution $\theta(\phi, \phi_0, \theta_0)$, $\theta(\phi_0, \phi_0, \theta_0) = \theta_0$ of (2.3) is $\theta(\phi, \phi_0, \theta_0) = \omega(\phi - \phi_0) + \theta_0$.

Case (i). If ω is rational, say $\omega = p/q$, p, q integers, then $\theta(\phi_0 + q, \phi_0, \theta_0) = \theta_0 + p$. But, on the torus $\mathcal{T}$, the point (ϕ_0, θ_0) is the same as $(\phi_0 + q, \theta_0 + p)$ for any integers p, q. Thus, the orbit through (ϕ_0, θ_0) on $\mathcal{T}$ is a closed curve for every initial point (ϕ_0, θ_0). This implies that the functions in (2.4) expressed as a function of t with the parametric representation of θ, ϕ being given by (2.1) are periodic in t.

Case (ii). If ω is irrational, there do not exist integers p, q such that $\theta(\phi_0 + q, \phi_0, \theta_0) = p + \theta_0$ for any θ_0. Therefore, no orbit on $\mathcal{T}$ will be closed and the functions in (2.4) will not be periodic. We show in this case that every orbit on $\mathcal{T}$ is dense in $\mathcal{T}$. It is sufficient to show that the orbit is dense in the meridian $\phi = 0$ since the trajectories in the plane all have constant slope ω. There is a constant $\delta > 0$ such that for any γ in $[0, 1)$, there is a sequence of integer pairs (q_k, p_k), $q_k \to \infty$ as $k \to \infty$ such that $|\omega - p_k/q_k - \gamma/q_k| < \delta/q_k^2$. On $\mathcal{T}$, $(q_k, \theta(q_k, 0, 0)) = (0, \omega q_k) = (0, p_k + \gamma + \eta_k) = (0, \gamma + \eta_k)$ where $\eta_k \to 0$ as $k \to \infty$. This proves the orbit through $(0, 0)$ is dense but the same is obviously true for any other orbit. The functions (2.4) will be quasiperiodic in t (for the definition, see the Appendix) with the parametric representation of θ, ϕ being given by (2.1) since for any initial point (ϕ_0, θ_0) they are obviously representable in the form $x(t) = \delta_1(t, \omega t)$, $y(t) = \delta_2(t, \omega t)$, $z(t) = \delta_3(t, \omega t)$ where each δ_j is periodic in each argument of period 1.

Example 2.2. If $A(\phi, \theta) = \sin 2\pi\theta$, there are two closed orbits $\theta = 0$, $\theta = 1/2$ on $\mathcal{T}$. It is clear that for any θ_0, $0 < \theta_0 < 1$, the corresponding orbit on $\mathcal{T}$ has ω-limit set as the closed orbit $\theta = 1/2$ and α-limit set the closed orbit $\theta = 0$ since the sets $\theta = 0$ and $\theta = 1$ on $\mathcal{T}$ are the same.

There are two striking differences in these examples. In example 2.1 and ω irrational, the α- and ω-limit set of any orbit on $\mathscr{T}$ is $\mathscr{T}$ itself. In example 2.1 with ω rational, every orbit is closed whereas in example 2.2, there are isolated closed orbits which are the limit sets of all other orbits. For smooth vector fields, it will be shown below that the ω-limit and α-limit sets of any orbit of a general equation (2.3) on $\mathscr{T}$ must either be $\mathscr{T}$ itself or a closed orbit.

Since every solution of (2.3) exists on $-\infty < \phi < \infty$, it follows that every orbit on $\mathscr{T}$ must cross the meridian C given by $\phi = 0$ and therefore it is sufficient to take the initial values as $(0, \xi)$. Let $\theta(\phi, \xi)$, $\theta(0, \xi) = \xi$, designate the solution of (2.3) that passes through $(0, \xi)$. From the assumption of uniqueness of solutions of (2.3), the function $\theta(\phi, \xi)$ is a monotone increasing function in ξ for each ϕ. Also, the mapping $\xi \to \theta(1, \xi)$ is a homeomorphism of the real line onto itself and thus induces a homeomorphism T of C onto itself. In fact,

$$TP = P_1, \; P = (0, \xi), \; P_1 = (1, \theta(1, \xi)) = (0, \theta(1, \xi)).$$

From the uniqueness of solutions of (2.3), $\theta(1, \xi)$ is a monotone increasing function and thus T preserves the orientation of C. Also, periodicity of (2.3) and uniqueness of solutions imply

(2.5) $$\theta(\phi, \xi + m) = \theta(\phi, \xi) + m$$

for any integer m.

It is easy to see that

$$T^n P \overset{\text{def}}{=} T(T^{n-1}P) = (0, \theta(n, \xi)); \; T^{m+n}(P) = T^m(T^n P), \; T^0 P = P,$$

for all $m, n = 0, \pm 1, \ldots$. First of all, the definition makes sense for negative values of the integers since $\theta(1, P)$ is a homeomorphism and $T^0 P = P$ imply that $P = T(T^{-1}P)$ uniquely defines T^{-1}. One then inductively defines T^{-2}, T^{-3}, etc. To prove the stated assertions, notice that periodicity in (2.3) and uniqueness of solutions imply that

$$\theta(m, \theta(n, \xi)) = \theta(n, \theta(m, \xi)) = \theta(m + n, \xi).$$

for all integers m, n. This relation immediately yields the above assertions.

THEOREM 2.1. The *rotation number* of (2.3),

$$\rho \overset{\text{def}}{=} \lim_{|n| \to \infty} \frac{\theta(n, \xi)}{n},$$

exists and is independent of ξ. Also, ρ is rational if and only if some power of T has a fixed point; that is, there is a closed orbit on the torus.

The rotation number can be interpreted as the average rotation about the meridian for one trip around the parallel or the average slope of the line in the (ϕ, θ)-plane which passes through the origin and the point $(n, \theta(n, \xi))$ on the trajectory through $(0, \xi)$.

PROOF OF THEOREM 2.1. The proof is in four parts.

(i) If ρ exists for a ξ, then it exists for every ξ and is independent of ξ. We need only consider $0 \leq \xi$, $\bar{\xi} < 1$ since $\theta(\phi, \xi)$ satisfies (2.5). For $0 \leq \xi$, $\bar{\xi} < 1$, relation (2.5) and the monotonicity of $\theta(\phi, \xi)$ in ξ imply that

$$\theta(\phi, \bar{\xi}) - 1 = \theta(\phi, \bar{\xi} - 1) \leq \theta(\phi, \xi) \leq \theta(\phi, \bar{\xi} + 1) = \theta(\phi, \bar{\xi}) + 1,$$

and this gives the desired result.

(ii) ρ always exists. The following proof was communicated to the author by M. Peixoto. From (i), we need only consider $\xi = 0$. For any real ξ there is an integer m such that $0 \leq \xi - m \leq 1$ and thus $\theta(\phi, 0) \leq \theta(\phi, \xi - m)$ $\leq \theta(\phi, 1)$ and (2.5) implies

$$\theta(\phi, 0) \leq \theta(\phi, \xi) - m \leq \theta(\phi, 0) + 1.$$

If $\gamma = \xi - m$, then $0 \leq \gamma \leq 1$, and this last relation implies $\theta(\phi, 0) - \gamma \leq$ $\theta(\phi, \xi) - \xi \leq \theta(\phi, 0) + 1 - \gamma$. Since $0 \leq \gamma \leq 1$, we have

$$\theta(\phi, 0) - 1 \leq \theta(\phi, \xi) - \xi \leq \theta(\phi, 0) + 1$$

for all ϕ, ξ. In particular, for any integer m,

(2.6) $$\theta(m, 0) - 1 \leq \theta(m, \xi) - \xi \leq \theta(m, 0) + 1.$$

From this relation, we have $\theta(2m, 0) = \theta(m, \theta(m, 0))$ satisfies $2\theta(m, 0) - 2 \leq$ $2\theta(m, 0) - 1 \leq \theta(2m, 0) \leq 2\theta(m, 0) + 1 \leq 2\theta(m, 0) + 2$.

By successively applying (2.6), we obtain

$$n\theta(m, 0) - n \leq \theta(nm, 0) \leq n\theta(m, 0) + n$$

for all $n \geq 0$, and

$$n\theta(-m, 0) - n \leq \theta(-nm, 0) \leq n\theta(-m, 0) + n$$

for all $n \geq 0$. Thus,

$$\left| \frac{\theta(nm, 0)}{nm} - \frac{\theta(m, 0)}{m} \right| \leq \frac{1}{|m|}$$

for all $n, m \neq 0$. Interchanging the role of n, m, we have

$$\left| \frac{\theta(nm, 0)}{nm} - \frac{\theta(n, 0)}{n} \right| \leq \frac{1}{|n|}$$

for all $n, m \neq 0$. The triangle inequality implies

$$\left| \frac{\theta(n, 0)}{n} - \frac{\theta(m, 0)}{m} \right| \leq \frac{1}{|n|} + \frac{1}{|m|},$$

and thus ρ exists with $|\rho - \theta(m, 0)/m| \leq 1/|m|$ for all m.

We now give another proof.

The reader may go immediately to part (iii) of the proof of the theorem if he so desires. The idea is best illustrated by another example. The slope of a line L through the origin of the (ϕ, θ)-plane is uniquely determined by the manner in which this line partitions the integer pairs (m, n). More specifically, if R_0 is the set of rationals n/m such that the pair (m, n) is below L and R_1 the set of rationals such that (m, n) is above L, then one shows that all rational numbers with the possible exception of one are included in R_0 or R_1, $R_0 \cap R_1 = \varnothing$, and therefore the cut defined by R_0 and R_1 defines a real number which is the slope of the line in question.

This same idea can be used to show that the rotation number ρ exists. Let $L(m, n)$ be the trajectory of (2.3) through the point (m, n) in the (ϕ, θ)-plane and let $L = L(0, 0)$. The curve $L(m, n)$ is the same as L except for a translation. We say $L(m, n)$ is above L if $\theta(0, m, n) > 0$ and below L if $\theta(0, m, n) < 0$, where $\theta(\phi, m, n)$ is the solution of (2.3) with $\theta(m, m, n) = n$. It is clear from uniqueness that this is equivalent to saying that $L(m, n)$ is above (below) L if $\theta(\phi, m, n) > \theta(\phi, 0, 0)$ $(< \theta(\phi, 0, 0))$. If $L(m, n)$ is above L, then $L(km, kn)$, k a positive integer, is also above L. In fact, $L(2m, 2n)$ is above $L(m, n)$ since L goes to $L(m, n)$ and $L(m, n)$ goes to $L(2m, 2n)$ by the translation (m, n). An induction process gives the result. Applying the translation $(-m, -n)$ to the same curves L and $L(m, n)$, we see that $L(-m, -n)$ is below L and in general $L(-km, -kn)$, $k > 0$ is below L if $L(m, n)$ is above L. These remarks show there is no ambiguity to the classification of R_0 and R_1 as before. Namely, n/m belongs to $R_0(R_1)$ if $L(m, n)$ is below (above) L. The classes R_0 and R_1 are nonempty because if $n > 0$ is sufficiently large, then $L(1, n)$ is above L and $L(1, -n)$ is below L. If n/m is not in R_0, and $s/r > n/m$, then s/r is in R_1. In fact, $L(rm, rn)$ is either above L or coincides with L and $L(rm, sm)$ is above $L(rm, rn)$ since it is obtained from it by the translation $(0, sm - rn)$ and $sm - rn > 0$. Therefore $L(rm, sm)$ is above L which implies $rm/sm = r/s$ belongs to R_1. Similarly, if n/m is not in R_1 and $s/r < m/n$, then s/r is in R_0. Thus, all rational numbers with possibly one exception are included in R_0 or in R_1 and R_0 and R_1 define a real number ρ.

It remains to show that ρ is the rotation number defined in the theorem. Suppose m is a given integer and let n be the largest integer such that n/m is in R_0. Then $n \leqq \rho m \leqq n + 1$ and every point of $L(m, n)$ is below L and every point of $L(m, n + 1)$ is not below L. Therefore, since the points on $L(m, n)$ are $(\phi + m, \theta(\phi, 0) + n)$, we have

$$\theta(\phi, 0) + n < \theta(\phi + m, 0) \leqq \theta(\phi, 0) + n + 1 .$$

In particular, for $\phi = 0$,

$$n < \theta(m, 0) \leqq n + 1,$$

and

$$\left|\frac{\theta(m, 0)}{m} - \rho\right| \leqq \left|\frac{n}{m} - \rho\right| + \frac{1}{|m|}.$$

One can approximate ρ by a sequence of rationals n/m, $|m| \to \infty$, such that $|\rho - n/m| < \kappa/|m|$ for some constant κ. Thus,

$$\left|\frac{\theta(m, 0)}{m} - \rho\right| \leqq \frac{\kappa + 1}{|m|}.$$

This clearly implies that $\rho = \lim_{|m| \to \infty} \theta(m, 0)/m$.

(iii) If some power of T has a fixed point, then ρ is rational. In fact, if there is an integer m and a ξ in $[0, 1)$ such that $T^m \xi = \xi$, there is also an integer k such that $\theta(m, \xi) = \xi + k$. Since $\rho = \lim_{|m| \to \infty} \theta(m, \xi)/m$ exists we have

$$\rho = \lim_{|n| \to \infty} \frac{\theta(nm, \xi)}{nm} = \lim_{|n| \to \infty} \frac{\xi + nk}{nm} = \frac{k}{m}$$

and ρ is rational.

(iv) If ρ is rational, some power of T has a fixed point. Suppose $\rho = k/m$, k, m integers, $m > 0$, and T^m has no fixed points. Then, for every ξ, $0 \leqq \xi \leqq 1$, $\theta(m, \xi) \neq \xi + k$. If we suppose $\theta(m, \xi) > \xi + k$ for ξ in $[0, 1)$, then there is an $a > 0$ such that $\theta(m, \xi) - \xi - k \geqq a > 0$, ξ in $[0, 1)$. For any ζ in $(-\infty, \infty)$, there are an integer p and a ξ in $[0, 1)$ such that $\zeta = p + \xi$. Relation (2.5) then implies $\theta(m, \zeta) - \zeta - k \geqq a$ for all ζ in $(-\infty, \infty)$. A repeated application of this inequality yields $\theta(rm, \xi) - \xi \geqq r(k + a)$ for any integer r. Dividing by rm and letting $r \to \infty$ we have $\rho \geqq k/m + a/m$ which is a contradiction. This completes the proof of the theorem.

COROLLARY 2.1. Among the class of functions $A(\phi, \theta)$ which are Lipschitzian in θ, the rotation number $\rho = \rho(A)$ of (2.3) varies continuously with A; that is, for any $\varepsilon > 0$ and A there is a $\delta > 0$ such that $|\rho(A) - \rho(B)| < \varepsilon$ if $\max_{0 \leqq \phi, \theta \leqq 1} |A(\phi, \theta) - B(\phi, \theta)| < \delta$.

PROOF. If $\theta_A(\phi, 0)$ and $\theta_B(\phi, 0)$ designate the solutions of (2.3) for A and B, respectively, $z(\phi) = \theta_A(\phi, 0) - \theta_B(\phi, 0)$, and L is the Lipschitz constant for A, then

$$\frac{dz}{d\phi} = [A(\phi, z(\phi) + \theta_B(\phi, 0)) - A(\phi, \theta_B(\phi, 0))]$$

$$- [B(\phi, \theta_B(\phi, 0)) - A(\phi, \theta_B(\phi, 0))],$$

and

$$D_r |z| \leqq \left|\frac{dz}{d\phi}\right| \leqq L|z| + \sup_{0 \leqq \phi, \theta \leqq 1} |B(\phi, \theta) - A(\phi, \theta)|$$

for all ϕ. Thus,

$$|\theta_A(\phi, 0) - \theta_B(\phi, 0)| \leqq L^{-1}e^{L\phi} \sup_{0 \leqq \phi,\, \theta \leqq 1} |B(\phi, \theta - A(\phi, \theta)|$$

for all $\phi \geqq 0$.

In the proof of part (ii) of Theorem 2.1, an estimate on the rate of approach of the sequence $\theta_A(m, 0)/m$ to the rotation number $\rho(A)$ was obtained; namely, $|\theta_A(m, 0)/m - \rho(A)| < 1/|m|$ for all m. Therefore,

$$
\begin{aligned}
|\rho(A) - \rho(B)| &\leqq \left| \rho(A) - \frac{\theta_A(m, 0)}{m} \right| \\
&\quad + \left| \frac{\theta_A(m, 0) - \theta_B(m, 0)}{m} \right| + \left| \frac{\theta_B(m, 0)}{m} - \rho(B) \right| \\
&\leqq \frac{2}{|m|} + \left| \frac{\theta_A(m, 0) - \theta_B(m, 0)}{m} \right|
\end{aligned}
$$

for all integers m. For any $\varepsilon > 0$, choose $|m|$ so large that $1/|m| < \varepsilon/3$. For any such given but fixed m, choose $\delta > 0$ such that $|\theta_A(m, 0) - \theta_B(m, 0)| \leqq 1$ if $\max_{0 \leqq \theta,\, \phi \leqq 1} |A(\phi, \theta) - B(\phi, \theta)| < \delta$. This fact and the preceding inequality prove the result.

The conclusion of Corollary 2.1 actually is true without assuming $A(\phi, \theta)$ is Lipschitzian in θ. The proof would use a strengthened version of Theorem I.3.4 on the continuous dependence of solutions of differential equations on the vector field when uniqueness of solutions is assumed. It is an interesting exercise to prove these assertions.

THEOREM 2.2. If the rotation number ρ is rational, then every trajectory of (2.3) on the torus is either a closed curve or approaches a closed curve.

PROOF. (Peixoto). Since ρ is rational, there exists a closed trajectory γ on $\mathscr{T}$ which intersects every meridian of $\mathscr{T}$. Therefore, $\mathscr{T} \backslash \gamma$ is topologically equivalent to an annulus Γ. The differential equation (2.3) on $\mathscr{T} \backslash \gamma$ is equivalent to a planar differential equation on Γ. Since there are no equilibrium points, the Poincaré-Bendixson Theorem, Theorem 1.2, yields the conclusion of the theorem.

The remainder of this chapter is devoted to a discussion of the behavior of the orbits of (2.3) when the rotation number ρ is irrational.

Let $T: C \rightarrow C$ be the mapping induced by (2.3) which takes the meridian C of $\mathscr{T}$ into itself. For any P in C, let

$$D(P) = \{T^n P, \ n = 0, \pm 1, \pm 2, \ldots\},$$

and $D'(P)$ be the set of limit points of $D(P)$. Also, let $\varnothing$ be the empty set.

LEMMA 2.1. Suppose ρ is irrational, m, n are given integers, P is a given point in C and α, β are the closed arcs of C with $\alpha \cap \beta = \{T^m P, T^n P\}$, $\alpha \cup \beta = C$. Then $D(Q) \cap \alpha^0 \neq \varnothing$, $D(Q) \cap \beta^0 \neq \varnothing$ for every Q in C, where α^0, β^0 are the interiors of α, β respectively.

PROOF. The set $\bigcup_k T^{k(m-n)} \alpha^0$ covers C. For, if not, the sequence $\{T^{k(m-n)}(T^n P)\}$ would approach a limit P_0 and $T^{(m-n)} P_0 = P_0$ which, from Theorem 2.1, contradicts the fact that ρ is irrational. Consequently, for any Q in C, there is an integer p such that Q is in $T^{p(m-n)} \alpha^0$; that is, $T^{p(n-m)} Q$ is in α^0 and $D(Q) \cap \alpha^0 \neq \varnothing$. The same argument applies to β.

THEOREM 2.3. If ρ is irrational, $D'(P) = F$ is the same for all P, $TF = F$ and either

(i) $F = C$ (the ergodic case)

or

(ii) F is a nowhere dense perfect set.

PROOF. If S belongs to $D'(P)$, there is a sequence $\{P_k\} \subset D(P)$ approaching S as $k \to \infty$. For any points P_k, P_{k+1} of this sequence and Q in C, it follows from Lemma 2.1 that there is an integer n_k such that $T^{n_k} Q$ belongs to the shortest of the arcs α, β on C connecting these two points. Therefore $T^{n_k} Q \to S$ as $k \to \infty$ and $D'(P) \subset D'(Q)$. The argument is clearly the same to obtain $D'(Q) \subset D'(P)$ which proves the first statement of the theorem.

If Q is in F, then there is a sequence n_k and a P such that $T^{n_k} P \to Q$ as $n \to \infty$. This clearly implies TQ belongs to F and $T^{-1} Q$ belongs to F. Therefore $TF = F$.

If R is an arbitrary element of F, then the fact that $F = D'(Q)$ for every Q implies for any $Q \in F$ there is a sequence of integers n_k such that $T^{n_k} Q \to R$. Therefore, the set of limit points of F is F itself and F is perfect.

Suppose F contains a closed arc γ of C. Then γ contains a closed subarc α with endpoints $T^n P$, $T^m P$ for some integers n, m and P in C. Therefore, by Lemma 2.1, $\bigcup_k T^k \alpha$ covers C and since $T\alpha$, $T^2 \alpha$, ... belong to F we have $F = C$. This proves the theorem.

Our next objective is to obtain sufficient conditions which will ensure that T is ergodic; that is, the limit set F of the iterates of T is C.

Let $P_n = T^n P, n = 0, \pm 1, \pm 2, \dots$. If ρ is irrational and α is any closed arc of C with P as an endpoint, Lemma 2.1 implies there is an integer n such that either P_n or P_{-n} is the only point P_k in the interior α^0 of α for $|k| \leq n$. Since no power of T has a fixed point, for any $N > 0$, α can be chosen so small that $n \geq N$. For definiteness, suppose P_{-n} is in α^0. Let $P_0 P_{-n}$ denote the arc of C with endpoints P_0, P_{-n} and which also belongs to α. We associate an orientation to this arc which is the same as the orientation of

C. Also, let $P_k P_{k-n}$, $k = 0, 1, \ldots, n-1$, designate the arc of C joining P_k, P_{k-n} which has the same orientation as C.

LEMMA 2.2.   The arcs $P_k P_{k-n}$, $k = 0, 1, \ldots, n-1$, are disjoint.

PROOF. If the assertion is not true, then there exists an ℓ from the set $\{-n, -n+1, \ldots, n-1\}$ and a k from $\{0, 1, \ldots, n-1\}$ such that P_ℓ belongs to the interior $P_k P_{k-n}^0$ of $P_k P_{k-n}$. Therefore, $P_{\ell-k}$ is in $P_0 P_{-n}^0$, from the orientation preserving nature of powers of T. This is impossible in case $-n \leq l - k < n$ from the choice of n. Suppose $-2n + 1 \leq l - k < -n$. Since P_ℓ belongs to $P_k P_{k-n}^0$, it follows that $P_{\ell+n} P_\ell$ and $\overline{\overline{P}}_k P_{k-n}$ intersect and, in particular, P_k is in $P_{\ell+n} P_\ell^0$. Thus $P_{k-n-\ell}$ is in $P_0 P_{-n}^0$ which is impossible since $0 < k - \ell - n < n$. This proves the lemma.

THEOREM 2.4.   Let $\psi(\xi) = \theta(1, \xi)$, $0 \leq \xi \leq 1$. If ρ is irrational and ψ possesses a continuous first derivative $\psi' > 0$ which is of bounded variation, then T is ergodic.

PROOF. Let $\xi_k = \psi^k(\xi)$, $k = 0, \pm 1, \ldots$, $\psi^0(\xi) = \xi$, be recursively defined by choosing $\psi^{-1}(\xi)$ as the unique solution of $\psi^0(\xi) = \psi(\psi^{-1}(\xi))$. Then $T^k P = (0, \psi^k(\xi))$, $P = (0, \xi)$. From the product rule for differentiation, we have

$$\frac{d\psi^k(\xi)}{d\xi} = \prod_{j=0}^{k-1} \psi'(\xi_j), \quad \frac{d\psi^{-k}(\xi)}{d\xi} = \left[\prod_{j=0}^{k-1} \psi'(\xi_{j-k}) \right]^{-1}$$

where $\psi'(\xi) = d\psi(\xi)/d\xi$. Suppose P and n are chosen as prior to Lemma 2.2. Since $P_k P_{k-n}$, $k = 0, 1, \ldots, n-1$, are disjoint we have

$$\left| \log\left(\frac{d\psi^n(\xi)}{d\xi} \frac{d\psi^{-n}(\xi)}{d\xi} \right) \right| = \left| \log\left(\prod_{j=0}^{n-1} \psi'(\xi_j) \right) - \log\left(\prod_{j=0}^{n-1} \psi'(\xi_{j-n}) \right) \right|$$

$$= \left| \sum_{j=0}^{n-1} [\log \psi'(\xi_j) - \log \psi'(\xi_{j-n})] \right|$$

$$\leq V$$

where V is the total variation of $\log \psi'$. This function has bounded variation from the hypothesis on ψ'. Therefore,

$$e^{-V} \leq \frac{d\psi^n(\xi)}{d\xi} \frac{d\psi^{-n}(\xi)}{d\xi} \leq e^V$$

uniformly on n and ξ.

Suppose α is any arc on C of length δ. If δ_k is the length of $T^k \alpha$, then

$$\delta_k + \delta_{-k} = \int \left(\frac{d\psi^k}{d\xi} + \frac{d\psi^{-k}}{d\xi} \right) d\xi \geq 2 \int_a \left(\frac{d\psi^k}{d\xi} \frac{d\psi^{-k}}{d\xi} \right)^{1/2} d\xi \geq 2 \, \delta e^{-V/2}$$

and, therefore, $\delta_k + \delta_{-k}$ does not approach zero as $k \to \infty$.

If $C\backslash F$ is not empty (that is, T is not ergodic), then take an open arc α in $C\backslash F$ with end points in F. This can be done since F is nowhere dense and perfect. Since $TF = F$ and T preserves orientation, all of the arcs $T^k\alpha, k = 0, \pm 1, \ldots$ are in $C\backslash F$. Also $T^k\alpha \cap T^j\alpha = \varnothing, k \neq j$, since the end points of these arcs are in F and if one coincided with another the end points would correspond to a fixed point of a power of T. Therefore, compactness of C yields $\delta_k + \delta_{-k} \to 0$ as $k \to \infty$. This contradiction implies $C\backslash F$ is empty and proves the theorem.

Remark. The smoothness assumptions on ψ in Theorem 2.4 are satisfied if $A(\phi, \theta)$ in (2.3) has continuous first and second partial derivatives with respect to θ. In fact, Theorem I.3.3 and exercise I.3.2. imply that $\psi'(\xi), \psi''(\xi)$ are continuous and, in particular, $\psi'(\xi)$ is of bounded variation for $0 \leq \xi \leq 1$. Also, this same theorem states that $\partial\theta(\phi, \xi)/\partial\xi$ is a solution of the scalar equation

$$\frac{dy}{d\phi} = \frac{\partial A(\phi, \theta)}{\partial \theta}\, y,$$

with initial value 1 at $\phi = 0$. Thus, $\psi'(\xi) = \partial\theta(1, \xi)/\partial\xi > 0, 0 \leq \xi \leq 1$.

Denjoy [1] has shown by means of an example that Theorem 2.4 is false if the smoothness conditions on ψ are relaxed.

There is no known way to determine the explicit dependence of the rotation number ρ of (2.3) on the function $A(\phi, \theta)$, and, thus, in particular, to assert whether or not ρ is irrational. However, the result of Denjoy was the first striking example of the importance of smoothness in differential equations to eliminate unwanted pathological behavior.

Suppose the notation is the same as in Theorem 2.4 and the proof of Theorem 2.4.

LEMMA 2.3. If ρ is irrational and ξ is a fixed real number, then the function $g(\xi_n + m) = n\rho + m, \xi_n = \psi^n(\xi), n, m$ integers, is an increasing function on the sequence of real numbers $\{\xi_n + m\}$.

PROOF. · Throughout this proof, n, m, r, s will denote integers. The order of the elements in $\{\xi_n + m\}$ does not depend upon ξ; that is, $\xi_n + m < \xi_r + s$ implies $\zeta_n + m < \zeta_r + s$ for any ζ. This is equivalent to saying that $\xi_n - \xi_r < s - m$ implies $\zeta_n - \zeta_r < s - m$ for any ζ. If this were not true, there would be an η such that $\eta_n - \eta_r$ is an integer which in turn implies some power of T has a fixed point, contradicting the fact that ρ is irrational. It suffices therefore to choose $\xi = 0$.

Recall that $\psi^m(0) = \theta(m, 0)$. If $p \leq \theta(m, 0) \leq r$, then a repeated application of (2.6) yields

$$\theta(m, 0) + (k - 1)p \leq \theta(km, 0) \leq \theta(m, 0) + (k - 1)r,$$

for any $k > 0$. Thus,

$$\frac{\theta(m, 0)}{k} + \left(1 - \frac{1}{k}\right)p \leq m \frac{\theta(km, 0)}{km} \leq \frac{\theta(m, 0)}{k} + \left(1 - \frac{1}{k}\right)r.$$

Taking the limit as $k \to \infty$, we obtain $p \leq m\rho \leq r$. Since ρ is irrational, $p < m\rho < r$.

Now suppose $\theta(n, 0) + m < \theta(r, 0) + s$, that is, $\theta(n, m) < \theta(r, s)$. Then $\theta(n - r, m) < \theta(0, s)$ or $\theta(n - r, 0) + m < s$. From the preceding paragraph, this implies $\rho(n - r) < s - m$, or $\rho n + m < \rho r + s$, which was to be proved.

THEOREM 2.5. If T is ergodic with (irrational) rotation number ρ, then T is topologically equivalent to a rotation of the circle C by an angle $2\pi\rho$; that is, there is a homeomorphism G of C onto C such that $GT = RG$ where R is the rotation of C through the angle $2\pi\rho$.

PROOF. Let g be the increasing function defined in Lemma 2.3, $A = \{n\rho + m\}$, $B = \{\xi_n + m\}$, n, m integers. Since T is ergodic, B is dense on the real numbers and since ρ is irrational, A is also dense on the real numbers. The function g is continuous from B to A. Since B is dense, g has a unique continuous increasing extension to all of the reals. We again designate this function by g.

If $\eta = \xi_n + m$, $g(\eta) = n\rho + m$, then

$$(2.7) \qquad g(\eta + 1) = n\rho + m + 1 = g(\eta) + 1,$$

$$g(\psi(\eta)) = g(\psi(\xi_n + m)) = g(\psi(\xi_n) + m) = g(\xi_{n+1} + m)$$
$$= (n + 1)\rho + m = g(\eta) + \rho.$$

Since B is dense and g is continuous, it follows that $g(\eta + 1) = g(\eta) + 1$, $g(\psi(\eta)) = g(\eta) + \rho$ for all real η. The homeomorphism $G: C \to C$ is defined by $G(\eta) = g(\eta)$, $0 \leq \eta < 1$. The relation $g(\psi(\eta)) = g(\eta) + \rho$ implies that $GT = RG$ and the theorem is proved.

THEOREM 2.6. (Bohl). If T is ergodic, there exists a function $\omega(y, z)$ which is continuous in y, z and

$$\omega(y + 1, z) = \omega(y, z + 1) = \omega(y, z)$$

for all y, z such that every solution θ of (2.3) satisfies

$$(2.8) \qquad \theta(\phi) = \rho\phi + c + \omega(\phi, \rho\phi + c),$$

where c is a constant and ρ is the rotation number. Conversely, for any constant c, $\theta(\phi)$ given in (2.8) satisfies (2.3). Each c in $[0, 1)$ corresponds to a unique $\theta(0) \pmod 1$.

PROOF. Let ξ be any real number and $\theta(\phi, \xi)$ be the solution of (2.3) with $\theta(0, \xi) = \xi$. We know that

$$\theta(\phi, \xi + 1) = \theta(\phi, \xi) + 1,$$
$$\theta(\phi + 1, \xi) = \theta(\phi, \psi(\xi)),$$

where as before $\psi(\xi) = \theta(1, \xi)$. Let g be the function given in the proof of Theorem 2.5 and let h denote the inverse of g. From the properties (2.7) of g, we have

$$h(c + 1) = h(c) + 1,$$
$$\psi(h(c)) = h(c + \rho).$$

If $\bar{\theta}(\phi, c) = \theta(\phi, h(c))$, then

$$\bar{\theta}(\phi, c + 1) = \theta(\phi, h(c + 1)) = \theta(\phi, h(c) + 1) = \theta(\phi, h(c)) + 1 = \bar{\theta}(\phi, c) + 1,$$

and

$$\bar{\theta}(\phi + 1, c) = \theta(\phi + 1, h(c)) = \theta(\phi, \psi(h(c))) = \theta(\phi, h(c + \rho)) = \bar{\theta}(\phi, c + \rho).$$

If $\omega(y, z) = \bar{\theta}(y, z - \rho y) - z$ for all real y, z, then one easily observes that $\omega(y, z + 1) = \omega(y + 1, z) = \omega(y, z)$. Therefore,

$$\theta(\phi, h(c)) \overset{\text{def}}{=} \bar{\theta}(\phi, c) = \rho\phi + c + \omega(\phi, \rho\phi + c),$$

which proves the theorem.

EXERCISE 2.1. All functions below are continuous, periodic in θ, ϕ of period 1 and smooth enough to ensure a unique solution of the equations for any initial values.

    (a) What is the rotation number of $d\theta/d\phi = \sin 2\pi\theta$?

    (b) If $|g(\theta)| < 1$, $0 \leq \theta \leq 1$, what is the rotation number of $d\theta/d\phi = \sin 2\pi\theta + g(\theta)$?

    (c) Using the Poincaré-Bendixson theorem, prove that the rotation number of $d\theta/d\phi = \sin 2\pi\theta + \varepsilon g(\theta, \phi)$ is rational if $|\varepsilon|$ is small enough.

    (d) Use the Brouwer fixed point theorem and the construction used in part (c) to show that the rotation number of the equation in (c) is zero if $|\varepsilon|$ is small enough.

EXERCISE 2.2. Suppose ω is irrational. For any $\varepsilon > 0$, show there exists a function $g(\theta, \phi)$ continuous in θ, ϕ of period 1 such that $\max\{|g(\theta, \phi)|, 0 \leq \theta, \phi \leq 1\} < \varepsilon$ and the rotation number of $d\theta/d\phi = \omega + g(\theta, \phi)$ is rational and not all orbits are closed on the torus. Hint: Choose sequences of integers $\{p_k\}$, $\{q_k\}$, $q_k \to \infty$ as $k \to \infty$ such that $|\omega - p_k/q_k| < \gamma/q_k^2$ for some constant γ and consider $g(\theta, \phi) = a \sin 2\pi(b\theta + c\phi)$ for appropriate constants a, b, c.

EXERCISE 2.3.   In Exercise 2.1, arbitrarily small changes in the vector field did not change the rotation number, whereas in Exercise 2.2, an appropriate small change in the vector field did change this number. Can you explain why this is so? Can you give a general result for equation (2.3) which will exhibit the same properties relative to the rotation number as Exercises 2.1 and 2.2? What would you say is the most typical behavior among all vector fields (2.3)?

EXERCISE 2.4.   (a) For any continuous function $f(\phi)$ of period 1, show that the equation $d\theta/d\phi = 2\pi\theta + f(\phi)$ has a unique periodic solution of period 1. (Hint: Look at $\int_{\infty}^{\phi} e^{2\pi(\phi-s)}f(s)\,ds$).

(b) Designate this unique solution by Kf. If P is the Banach space of continuous periodic functions of period 1 with the topology of uniform convergence, show that $K: P \to P$ is continuous and linear.

(c) Suppose $g(\theta, \phi)$ has period 1 in θ, ϕ, is continuous in θ, ϕ and lipschitzian in θ. Using the contraction principle and part (b), show there is an $\varepsilon_0 > 0$ such that the equation $d\theta/d\phi = 2\pi\theta + (\sin 2\pi\theta - 2\pi\theta) + \varepsilon g(\theta, \phi)$ has a solution which is periodic in ϕ of period 1 if $|\varepsilon| < \varepsilon_0$.

II.3. Remarks and Suggestions for Further Study

For many more details on the general theory of two dimensional systems, see Hartman [1], Lefschetz [1], Sansone and Conti [1]. The book of Sansone and Conti [1] also contains a wealth of applications of the Poincaré-Bendixson theory to the existence of periodic solutions of a second order autonomous equation. Finding periodic solutions of nonautonomous equations by extensions of the ideas in Exercises 1.4 and 1.5 can be found, for example, in Cartwright [1], Levinson [1], Lefschetz [1], Sansone and Conti [1]. An interesting discussion of the global stability of the origin in the plane is given by Olech [1].

One result obtained for two dimensional systems in the plane was that the only compact minimal sets are points and closed curves. For a singularity free "smooth" vector field on the torus, the only minimal sets on the torus are closed curves and the torus itself. An outstanding problem for many years was the characterization of the minimal sets of "smooth" vector fields on an arbitrary compact two dimensional manifold. This problem was solved by Schwartz [1] when he proved that the only possible minimal sets are points, closed curves and a torus, the latter being possible only when the manifold is itself a torus.

It is difficult to pose questions in n dimensions with answers as elegant as the ones given in this chapter. One problem that has been studied in considerable detail arises from a more careful study of Theorems 2.5 and 2.6. Consider a differential equation on an n-dimensional torus

$$\dot{x} = f(x)$$

where $x = (x_1, \ldots, x_n)$, $f(x_1, \ldots, x_n)$ is periodic in each variable of period 1. *Problem*: When does there exist a transformation of variables $x = g(y)$ such that the new equation for y is $\dot{y} = \omega$ where ω is a constant n-vector? If such a transformation exists, then the resulting flow is easy to analyze and one has a generalization of Theorem 2.6. For $f(x) = \omega + \varepsilon h(x)$ where ε is small and ω belonging to a certain class of irrationals, some results along this line are available (see Arnol'd [1]). The techniques of proof are currently the most important tools in the study of stability in celestial mechanics (see Kolmogorov [1], Arnol'd [2], Moser [1]).

In Section 1, the van der Pol equation (1.7) was discussed when the parameter k was large. In the coordinates described by (1.8), it becomes clear that the asymptotically stable periodic orbit has the property that the system moves along its orbit at a reasonable pace as long as it remains close to a given curve and then moves at a very rapid pace away from this curve. Such relaxation oscillations are very important in the applications and the reader may consult van der Pol [1, 2], LaSalle [1], Minorsky [1], Pontryagin and Mishchenko [1].

CHAPTER III

Linear Systems and Linearization

In the previous chapter, much information concerning the qualitative behavior of solutions of two dimensional autonomous systems was obtained without using any particular properties of the vector field. In higher dimensions and even in two dimensions when more specific information is desired, one must resort to techniques which are more analytical in nature and yield information only in a neighborhood of some solution or an invariant set of solutions.

If $\phi(t)$ is a solution of

$$\dot{x} = f(x), \tag{1}$$

where f has continuous first derivatives, then $x = \phi(t) + y$ in (1) implies

$$\dot{y} = A(t)y + g(t, y) \tag{2}$$

where $A(t) = \partial f(\phi(t))/\partial x$ and $g(t, 0) = 0$, $\partial g(t, 0)/\partial y = 0$. It is quite natural to study the linear equation

$$\dot{z} = A(t)z, \tag{3}$$

and to inquire about the relationship between the solution z of (3) and the solution y of (2) in a neighborhood of $y = 0$. More specifically, is the linear equation (3) a good "approximation" to equation (2) near $y = 0$? Some questions of this type are considered in this chapter.

More precisely, the theory of general linear systems is given in Section 1 with Section 2 being devoted to characterizations of the concepts of stability for linear systems and the preservation of these stability properties for special perturbations of a linear system. Section 3 is a specialization of the general theory to n^{th} order equations. In Section 4, the fundamental solution of linear autonomous systems is characterized in terms of the elementary functions with the phase portrait of the two dimensional systems discussed in Section 5. Section 6 is devoted to an autonomous equation (2) with no eigenvalue of A on the imaginary axis. It is shown in this section that many of the qualitative properties of the solutions of the linear and nonlinear equation are the same near $y = 0$. Section 7 gives the theory of linear equations with periodic coefficients with the Floquet representation included. Sections 8, 9 and 10 are devoted to stability theory for special classes of linear equations with periodic coefficients.

III.1. General Linear Systems

Consider the linear system of n first order equations

(1.1) $$\dot{x}_j = \sum_{k=1}^{n} a_{jk}(t)x_k + h_j(t), \quad j = 1, 2, \ldots, n,$$

where the a_{jk} and h_j for j, $k = 1, 2, \ldots, n$ are continuous real or complex valued functions on the interval $(-\infty, +\infty)$. In matrix notation, equation (1.1) can be written in more compact form as

(1.2) $$\dot{x} = A(t)x + h(t)$$

where $A = (a_{jk})$, j, $k = 1, 2, \ldots, n$; $h = \text{col}(h_1, \ldots, h_n)$. The basic characteristic property of linear systems of the form (1.2) is the *Principle of Superposition*: If x is a solution of (1.2) corresponding to the "forcing function" or "input" h and y is a solution corresponding to g then $cx + dy$ is a solution corresponding to the forcing or input function $ch + dg$ for any real or complex numbers c and d.

This principle is verified by direct computation.

In particular, if $c = -d = 1$, $h = g$, then x and y being solutions of (1.2) corresponding to the forcing function h implies that $x - y$ is a solution of the homogeneous equation

(1.3) $$\dot{x} = A(t)x.$$

If $\phi(t, c)$ is a general solution of the homogeneous equation (1.3) and $x_p(t)$ is any particular solution of (1.2), then this latter property states that any solution of (1.2) is of the form $\phi(t, c) + x_p(t)$ for some c. Below, we describe theoretically how to obtain the general solution of (1.3) and then show that a particular solution of (1.2) can be found from the knowledge of the general solution of (1.3) and a quadrature.

From the properties of the coefficient matrix A and the forcing function h in (1.2), the basic existence and uniqueness theorem in Chapter I implies that the initial value problem for (1.2) has a unique solution which exists on an interval containing the initial time. Theorem I.6.2 also asserts that every solution of (1.2) exists on the infinite interval $(-\infty, +\infty)$.

A set of vectors $x^1, \ldots, x^n$ are said to be *linearly independent* if $\sum_{j=1}^{n} c_j x^j = 0$ for any complex constants c_j implies $c_j = 0$ for $j = 1, \ldots, n$.

The vectors $x^1, \ldots, x^n$ are said to be *linearly dependent* if they are not linearly independent. The proof of the following lemma is left as an exercise.

LEMMA 1.1.   The vectors $x^1, \ldots, x^n$ are linearly independent if and only if $\det[x^1, \ldots, x^n] \neq 0$.

An $n \times n$ matrix $X(t)$, $t > 0$, is said to be an $n \times n$ *matrix solution* of (1.3) if each column of $X(t)$ satisfies (1.3). A *fundamental matrix solution* of (1.3) is an $n \times n$ matrix solution $X(t)$ of (1.3) such that $\det X(t) \neq 0$. A *principal matrix solution* of (1.3) at initial time t_0 is a fundamental matrix solution such that $X(t_0) = I$, the identity matrix. The principal matrix solution at t_0 will be designated by $X(t, t_0)$.

From the above definition of a fundamental matrix solution it is clear that a fundamental matrix solution is simply a matrix solution of (1.3) such that the n columns of $X(t)$ are linearly independent. An elementary but useful property of a matrix solution of (1.3) is

LEMMA 1.2.   If $X(t)$ is an $n \times n$ matrix solution of (1.3), then either $\det X(t) \neq 0$ for all t or $\det X(t) = 0$ for all t.

PROOF.   If there exists a real number τ such that the matrix $X(\tau)$ is singular, then there exists a nonzero vector c such that $X(\tau)c = 0$. For this vector c, linearity of (1.3) implies the function $x(t) = X(t)c$ is a solution of (1.3). Since the function $x = 0$ is obviously a solution of the equation, uniqueness implies $X(t)c = 0$ for all t in $(-\infty, \infty)$ and thus $\det X(t) = 0$ for all t in $(-\infty, \infty)$. This proves the lemma.

COROLLARY 1.1.   If X_0 is an $n \times n$ nonsingular matrix and if $X(t)$ is the matrix solution of (1.3) with $X(0) = X_0$, then $X(t)$ is a fundamental matrix solution of (1.3).

Corollary 1.1 shows that the determination of a fundamental matrix solution consists only of selecting n linearly independent initial vectors and finding the corresponding n linearly independent solutions of the system (1.3). The fact that the matrix $X(t)$ is nonsingular for all t yields the following important

LEMMA 1.3.   If $X(t)$ is any fundamental matrix solution of (1.3), then a general solution of (1.3) is $X(t)c$ where c is an arbitrary n vector.

PROOF.   It is clear that the function $x(t) = X(t)c$ is a solution of (1.3) for any constant n-vector c. Furthermore, if t_0 and x_0 are given, then $X(t)c$, $c = X^{-1}(t_0)x_0$, is the solution of (1.3) which passes through the point (t_0, x_0).

LEMMA 1.4.   If $X(t)$ is a fundamental matrix solution of (1.3), then the matrix $X^{-1}(t)$ is a fundamental matrix solution of the *adjoint equation*

$$(1.4) \qquad\qquad \dot{y} = -yA(t)$$

where y is a row n-vector; that is, each row of X^{-1} satisfies (1.4) and $\det X^{-1}(t) \neq 0$.

PROOF. Let $Y(t) = X^{-1}(t)$. Since $YX = I$, the identity, it follows that $\dot{Y}X + YAX = 0$. Since X is non-singular this implies that $Y = X^{-1}$ is a matrix solution of the adjoint equation (1.4). This is equivalent to saying that each row of the matrix X^{-1} is a solution of (1.4). X^{-1} is obviously nonsingular.

THEOREM 1.1. If X is a fundamental matrix solution of (1.3) then every solution of (1.2) is given by formula

(1.5)
$$x(t) = X(t)\left[X^{-1}(\tau)x(\tau) + \int_{\tau}^{t} X^{-1}(s)h(s)\, ds\right]$$

for any real number τ in $(-\infty, +\infty)$.

Formula (1.5) is referred to as the *variation of constants formula* for (1.2). The reason for this is clear since it implies that every solution has the form $X(t)c(t)$ where $c(t)$ is the function given in the brackets in (1.5). This is the same form as the general solution of (1.3) given by Lemma 1.3 except that the vector c now depends upon the independent variable t.

PROOF OF THEOREM 1.1. If we rewrite equation (1.2) as $\dot{x} - Ax = h$ and use the fact that X^{-1} is a matrix solution of the adjoint equation (1.4), then equation (1.2) is equivalent to

$$\frac{d}{ds}[X^{-1}(s)x(s)] = X^{-1}(s)h(s).$$

Integrating this expression from τ to t for any real numbers τ and t, we obtain

$$X^{-1}(t)x(t) - X^{-1}(\tau)x(\tau) = \int_{\tau}^{t} X^{-1}(s)h(s)\, ds.$$

A rearrangement of the terms in this expression yields equation (1.5) and the theorem is proved.

We give another proof of this theorem. Since $X(t)$ is nonsingular for each t, it follows that the transformation of variables $x = X(t)y$ defines a homeomorphism of R^n into R^n for each t. If this transformation is applied to (1.2), then by Lemma 1.4,

$$\dot{y} = \frac{d}{dt}[X^{-1}x] = X^{-1}(Ax + h) - X^{-1}Ax = X^{-1}h.$$

This implies that

$$y(t) = X^{-1}(\tau)x(\tau) + \int_{\tau}^{t} X^{-1}(s)h(s)$$

since $y(\tau) = X^{-1}(\tau)x(\tau)$. The formula $x(t) = X(t)y(t)$ yields relation (1.5).

If for any given τ in $(-\infty, \infty)$, we let $X(t, \tau)$, $X(\tau, \tau) = I$, designate the principal matrix solution of (1.3) at τ, then

$$(1.6) \qquad X(t, \tau) = X(t, s)X(s, \tau)$$

for all t, τ, s. In fact, considered as functions of t, both sides of this relation satisfy (1.3) and coincide for $t = s$. Uniqueness yields the result. This formula leads to a simplification of the variation of constants formula; namely

$$(1.7) \qquad x(t) = X(t, \tau)x(\tau) + \int_{\tau}^{t} X(t, s)h(s)\, ds.$$

LEMMA 1.5. If $X(t)$ is an $n \times n$ matrix solution of (1.3), then

$$(1.8) \qquad \det X(t) = [\det X(t_0)] \exp\left(\int_{t_0}^{t} \mathrm{tr}\, A(s)\, ds\right)$$

for all t, t_0 in $(-\infty, \infty)$, where $\mathrm{tr}\, A$ is the sum of the diagonal elements of A.

The proof is easily supplied by showing that $\det X(t)$ satisfies the scalar equation $\dot{z} = [\mathrm{tr}\, A(t)]z$ and is left as an exercise for the reader.

The above theory is valid for linear systems with $A(t)$, $h(t)$ integrable in t. If the elements of the matrix $A(t)$ and the components of $h(t)$ are only Lebesgue integrable on every compact subset of R, then the results of Section I.5 imply that the initial value problem for (1.2) has a unique solution. If $x(t) = x(t, t_0, x_0)$, $x(t_0, t_0, x_0) = x_0$, is a solution of (1.2) then

$$|x(t)| \leq |x_0| + \int_{t_0}^{t} |h(s)|\, ds + \int_{t_0}^{t} |A(s)|\, |x(s)|\, ds, \qquad t \geq t_0,$$

as long as $x(t)$ is defined. The generalized Gronwall inequality (Lemma I.6.2 and an integration by parts implies

$$|x(t)| \leq \left(\exp\int_{t_0}^{t} |A(s)|\, ds\right)\left[|x_0| + \int_{t_0}^{t} |h(s)|\, ds\right]$$

for all $t \geq t_0$ for which $x(t)$ is defined. The continuation theorem implies $x(t)$ is defined for $t \geq t_0$. Replacing t by $t_0 - u$ and using similar estimates one obtains existence of $x(t, t_0, x_0)$ for $t \leq t_0$. The general structure of the solutions of the homogeneous equation (1.3) and the variation of constants formula are proved exactly as before.

We also need some estimates on the dependence of the solutions of (1.2) on the right hand side of the differential equation. Suppose A, B are $n \times n$ integrable matrix functions and h, g are n-vector integrable functions on compact subsets of R and consider the equations

$$(1.9) \qquad \begin{aligned} &\text{(a)} \quad \dot{x} = A(t)x + h(t), \\ &\text{(b)} \quad \dot{y} = B(t)y + g(t). \end{aligned}$$

If x, y are solutions of (1.9a), (1.9b) respectively, then $x - y$ satisfies the equation

$$\dot{z} = Az + (A - B)y + h - g.$$

Thus,

$$|x(t) - y(t)| \leq |x(t_0) - y(t_0)| + \int_{t_0}^{t} [|A(s)| \, |x(s) - y(s)| + |A(s) - B(s)| \, |y(s)|$$

$$+ |h(s) - g(s)|] \, ds, \qquad t \geq t_0.$$

Using the generalized Gronwall inequality (Lemma I.6.2), we have

$$(1.10) \qquad |x(t) - y(t)| \leq \left(\exp \int_{t_0}^{t} |A(s)| \, ds \right) |x(t_0) - y(t_0)|$$

$$+ \int_{t_0}^{t} \left(\exp \int_{s}^{t} |A(u)| \, du \right) [|A(s) - B(s)| \, |y(s)|$$

$$+ |h(s) - g(s)|] \, ds.$$

In particular, if $X_A(t, t_0)$, $X_A(t_0, t_0) = I$, $X_B(t, t_0)$, $X_B(t_0, t_0) = I$, are fundamental matrix solutions of $\dot{x} = A(t)x$, $\dot{y} = B(t)y$, respectively, then

$$(1.11) \qquad |X_A(t, t_0) - X_B(t, t_0)| \leq \sup_{t_0 \leq u \leq t} |X_B(u, t_0)| \left(\exp \int_{t_0}^{t} |A(s)| \, ds \right)$$

$$\times \int_{t_0}^{t} |A(s) - B(s)| \, ds,$$

for all $t \geq t_0$. Relation (1.11) implies that the principal matrix solution of $\dot{x} = Ax$ is a continuous function of the integrable matrix functions A defined on $[0, t]$ with $|A| = \int_{0}^{t} |A(s)| \, ds$.

When a linear differential equation contains general time varying coefficients, the remarks in this section essentially comprise the theory concerning the specific structure of the solutions. For special equations, much more detailed information is available. For equations with constant or periodic coefficients, the general structure of the solutions is known and discussed in Sections 4 and 7.

III.2. Stability of Linear and Perturbed Linear Systems

In this section, characterizations of stability for linear systems are given and the results applied to stability of perturbed linear systems. We first remark that the stability of any solution of homogeneous linear equation is determined by the stability of the zero solution. Therefore, for linear systems, it is permissible to say system (1.3) is stable, uniformly stable, etc.

THEOREM 2.1.   Let $X(t)$ be a fundamental matrix solution of (1.3) and let β be any number in $(-\infty, \infty)$. The system (1.3) is

(i)   stable for any t_0 in $(-\infty, \infty)$ if and only if there is a $K = K(t_0) > 0$ such that

(2.1) $|X(t)| \leqq K, \qquad t_0 \leqq t < \infty;$

(ii)   uniformly stable for $t_0 \geqq \beta$ if and only if there is a $K = K(\beta) > 0$ such that

(2.2) $|X(t)X^{-1}(s)| \leqq K, \qquad t_0 \leqq s \leqq t < \infty;$

(iii)   asymptotically stable for any t_0 in $(-\infty, \infty)$ if and only if

(2.3) $|X(t)| \to 0 \qquad \text{as } t \to \infty;$

(iv)   uniformly asymptotically stable for $t_0 \geqq \beta$ if and only if it is *exponentially asymptotically stable*; that is, there are $K = K(\beta) > 0$, $\alpha = \alpha(\beta) > 0$ such that

(2.4) $|X(t)X^{-1}(s)| \leqq Ke^{-\alpha(t-s)}, \qquad \beta \leqq s \leqq t < \infty.$

PROOF.   (i)   Suppose t_0 is any given real number in $(-\infty, \infty)$ and (2.1) holds. Any solution $x(t)$ satisfies $x(t) = X(t)X^{-1}(t_0)x(t_0)$. Let $|X^{-1}(t_0)| = L$. Then $|x(t)| \leqq KL|x(t_0)| < \varepsilon$ if $|x(t_0)| < \varepsilon/KL$ and the zero solution of (1.3) is stable. Conversely, if for any $\varepsilon > 0$ there is a $\delta = \delta(\varepsilon, t_0) > 0$ such that $|X(t)X^{-1}(t_0)x(t_0)| < \varepsilon$ for $|x(t_0)| < \delta$, then

$$|X(t)X^{-1}(t_0)| = \sup_{|\xi| \leqslant 1} |X(t)X^{-1}(t_0)\xi|$$

$$= \sup_{|x(t_0)| \leqslant \delta} |X(t)X^{-1}(t_0)\delta^{-1}x(t_0)| \leqq \varepsilon\delta^{-1}$$

for $t \geqq t_0$. This proves (i).

(ii)   If (2.2) is satisfied, then $x(t) = X(t)X^{-1}(t_0)x(t_0)$ for any $t_0 \geqq \beta$ and $|x(t)| \leqq K|x(t_0)| < \varepsilon$, $t \geqq \tau$, if $|x(t_0)| < \varepsilon/K$ which is uniform stability. The converse follows as in (i) with the observation that δ is independent of t_0.

(iii)   If (2.3) is satisfied, then for any t_0 in $(-\infty, \infty)$ there is a $K = K(t_0)$ such that $|X(t)| \leqq K$, $t \geqq t_0$ and (i) implies stability. Since $x(t) = X(t)X^{-1}(t_0)x(t_0)$ we have $|x(t)| \to 0$ as $t \to \infty$. The converse is trivial.

(iv)   If (2.4) is satisfied, then (1.3) is uniformly stable from (ii). Suppose $|x(t_0)| \leqq 1$. For any $\eta > 0$, $0 < \eta < K$, let $T = -\alpha^{-1}\log(\eta/K)$. Then

$$|x(t)| = |X(t)X^{-1}(t_0)x(t_0)| \leqq Ke^{-\alpha(t-t_0)}|x(t_0)| \leqq \eta,$$

if $t \geq t_0 + T$, $t_0 \geq \beta$; that is, uniform asymptotic stability of the zero solution of (1.3).

Conversely, suppose the solution $x = 0$ is uniformly asymptotically stable for $t_0 \geq \beta$. There is a $b > 0$ such that for any η, $0 < \eta < b$, there is a $T = T(\eta) > 0$ such that

(2.5) $|X(t)X^{-1}(t_0)x(t_0)| < \eta$ for $t \geq t_0 + T$,

and all $t_0 \geq \beta$, $|x(t_0)| \leq b$. Thus, $|X(t)X^{-1}(t_0)| < \eta b^{-1}$ for $t \geq t_0 + T$, $t_0 \geq \beta$. In particular,

(2.6) $|X(t + T)X^{-1}(t)| < \eta b^{-1} < 1$, $t \geq \beta$.

Since the solution $x = 0$ is uniformly stable, we have from (ii) that there is an $M = M(\beta)$ such that $|X(t)X^{-1}(s)| \leq M$, $\beta \leq s \leq t < \infty$. Suppose

$$\alpha = -T^{-1} \log(\eta b^{-1}),$$
$$K = M e^{\alpha T}.$$

For any $t \geq t_0$, there is an integer $k \geq 0$ such that $kT \leq t - t_0 < (k + 1)T$. Thus, using (2.6),

$$\begin{aligned}
|X(t)X^{-1}(t_0)| &\leq |X(t)X^{-1}(t_0 + kT)| \cdot |X(t_0 + kT)X^{-1}(t_0)| \\
&\leq M \cdot |X(t_0 + kT)X^{-1}(t_0)| \\
&\leq M(\eta b^{-1})|X(t_0 + (k-1)T)X^{-1}(t_0)| \\
&\leq M(\eta b^{-1})^k = M e^{-\alpha kT} \\
&= K e^{-\alpha(k+1)T} \leq K e^{-\alpha(t-t_0)}.
\end{aligned}$$

This proves (iv) and Theorem 2.1.

Using the variation of constants formula and Gronwall's inequality (Corollary I.6.6), it is very easy to obtain the following stability results for perturbed linear systems.

THEOREM 2.2. Suppose β is given in $(-\infty, \infty)$ and the system (1.3) is uniformly stable for $t_0 \geq \beta$. If the $n \times n$ continuous matrix function $B(t)$ satisfies $\int_\beta^\infty |B(t)| \, dt < \infty$, then the system

(2.7) $\dot{x} = [A(t) + B(t)]x$

is uniformly stable.

PROOF. If $X(t)$ is a fundamental matrix solution of (1.3), then the variation of constants formula implies that any solution of (2.7) has the form

(2.8) $x(t) = X(t)X^{-1}(t_0)x(t_0) + \int_{t_0}^t X(t)X^{-1}(s)B(s)x(s) \, ds,$

for all t, t_0 in $(-\infty, \infty)$. If $t_0 \geq \beta$, then Theorem 2.1 implies there is a constant $K = K(\beta)$ such that $|X(t)X^{-1}(t_0)| \leq K$ for all $t \geq t_0$. Consequently,

$$|x(t)| \leq K|x(t_0)| + \int_{t_0}^{t} K|B(s)| \cdot |x(s)|\, ds, \qquad t \geq t_0 \,.$$

Gronwall's inequality implies

$$|x(t)| \leq K\left(\exp K \int_{t_0}^{t} |B(s)|\, ds\right)|x(t_0)|, \qquad t \geq t_0 \,.$$

This clearly implies uniform stability of (2.7) and the theorem.

THEOREM 2.3.   Suppose β is given in $(-\infty, \infty)$ and system (1.3) is uniformly asymptotically stable for $t_0 \geq \beta$. If the $n \times n$ continuous matrix function $B(t)$ satisfies

$$(2.9) \qquad \int_{t_0}^{t} |B(s)|\, ds \leq \gamma(t - t_0) + \tau, \qquad t \geq t_0 \geq \beta$$

for some constants $\tau = \tau(\beta)$, $\gamma = \gamma(\beta)$, then there is an $r > 0$ such that system (2.7) is uniformly asymptotically stable if $\gamma < r$.

PROOF.   If $X(t)$ is a fundamental matrix solution of (1.3) and (1.3) is uniformly asymptotically stable for $t_0 \geq \beta$, then Theorem 2.1 implies there are constants $K = K(\beta) > 0$, $\alpha = \alpha(\beta) > 0$ such that (2.4) is satisfied. Using (2.8), one observes that the solution $x(t)$ of (2.7) satisfies

$$|x(t)| \leq Ke^{-\alpha(t-t_0)}|x(t_0)| + K\int_{t_0}^{t} e^{-\alpha(t-s)}|B(s)| \cdot |x(s)|\, ds, \qquad t \geq t_0 \,.$$

If $z(t) = e^{\alpha t}|x(t)|$, this inequality implies

$$z(t) \leq Kz(t_0) + \int_{t_0}^{t} K|B(s)|z(s)\, ds, \qquad t \geq t_0 \,.$$

An application of Gronwall's inequality and (2.9) yield

$$z(t) \leq K\left(\exp K \int_{t_0}^{t} |B(s)|\, ds\right)z(t_0) \leq K_1 e^{K\gamma(t-t_0)}z(t_0), \qquad K_1 = Ke^{K\tau},$$

which in turn implies $|x(t)| \leq K_1|x(t_0)| \exp[-(\alpha - K\gamma)(t - t_0)]$, $t \geq t_0$. If $r = \alpha K^{-1}$ and $\gamma < r$, then system (2.7) is uniformly asymptotically stable and the theorem is proved.

THEOREM 2.4.   Suppose β is given in $(-\infty, \infty)$ and system (1.3) is uniformly asymptotically stable for $t_0 \geq \beta$. If $f(t, x)$ is continuous for (t, x) in $R \times R^n$ and for any $\varepsilon > 0$, there is a $\sigma > 0$ such that

$$(2.10) \qquad |f(t, x)| \leq \varepsilon|x| \quad \text{for} \quad |x| < \sigma, \quad t \text{ in } R,$$

then the solution $x = 0$ of

(2.11) $\dot{x} = A(t)x + f(t, x)$

is uniformly asymptotically stable for $t_0 \geq \beta$.

PROOF. The hypothesis of uniform asymptotic stability for $t_0 \geq \beta$ of the system (1.3) implies there are constants $K = K(\beta) > 0$, $\alpha = \alpha(\beta) > 0$ such that (2.4) is satisfied. Any solution x of (2.11) satisfies

(2.12) $x(t) = X(t)X^{-1}(t_0)x(t_0) + \int_{t_0}^{t} X(t)X^{-1}(s)f(s, x(s))\, ds.$

Choose ε so that $\varepsilon K < \alpha$ and let σ be chosen so that (2.10) is satisfied. For those values of $t \geq t_0$ for which $x(t)$ satisfies $|x(t)| < \sigma$, we have

$$|x(t)| \leq K e^{-\alpha(t-t_0)}|x(t_0)| + \int_{t_0}^{t} \varepsilon K e^{-\alpha(t-s)}|x(s)|\, ds.$$

Proceeding exactly as in the proof of the previous theorem, we obtain

(2.13) $|x(t)| \leq K\,|x(t_0)|\, e^{-(\alpha - \varepsilon K)(t-t_0)}$

for all values of $t \geq t_0$ for which $|x(t)| < \sigma$. Since $\alpha - \varepsilon K > 0$, this inequality implies $|x(t)| < \sigma$ for all $t \geq t_0$ as long as $|x(t_0)| < \sigma/K$. Consequently (2.13) holds for all $t \geq t_0$ provided $|x(t_0)| < \sigma/K$. Relation (2.13) clearly implies uniform asymptotic stability of the solution $x = 0$ and the theorem is proved.

Theorem 2.4 is a generalization of the famous theorem of Lyapunov on stability with respect to the first approximation. The reason for the terminology is the following: Suppose $g\colon R^{n+1} \to R^n$ is continuous, $g(t, x)$ has continuous first partial derivatives with respect to x and $g(t, 0) = 0$ for all t. If $A(t)x = [\partial g(t, 0)/\partial x]x$ and $f(t, x) = g(t, x) - A(t)x$, then $f(t, 0) = 0$, $\partial f(t,x)/\partial x$ approaches zero as x approaches zero uniformly in t and the conditions of Theorem 2.4 are satisfied. The function $A(t)x$ is clearly the linear approximation to the right hand side of the eqatuion $\dot{x} = g(t,x)$.

There are many more results available on stability of perturbed linear systems and clearly many variants that could be obtained by abstracting the essential elements of the proofs given above. The exercises at the end of this section indicate some of the possibilities.

In spite of the fact that the proofs of the above theorems are extremely simple and the results are not surprising to the intuition, extreme care must be exercised in treating arbitrary perturbed linear systems.

The example $\ddot{u} - 2t^{-1}\dot{u} + u = 0$ has a fundamental system of solutions $\sin t - t \cos t$, $\cos t + t \sin t$ and is therefore unstable. This equation can be considered as a perturbation of the uniformly stable system $\ddot{u} + u = 0$.

The following example is an equation (1.3) which is asymptotically stable, but not uniformly, and system (2.7) is unstable for an appropriate $B(t)$ such that $\int^{\infty} |B(s)|\, ds < \infty$ and $|B(t)| \to 0$ as $t \to \infty$. Consider the system (1.3) with

$$(2.14) \qquad A(t) = \begin{bmatrix} -a & 0 \\ 0 & \sin \log t + \cos \log t - 2a \end{bmatrix},$$

and $1 < 2a < 1 + e^{-\pi}$. The solutions of (1.3) are

$$x_1(t) = c_1 \exp(-at),$$
$$x_2(t) = c_2 \exp(t \sin \log t - 2at),$$

where c_1, c_2 are arbitrary constants. For any c_1, c_2, $x_1(t)$, $x_2(t) \to 0$ as $t \to \infty$ exponentially. If $B(t)$ is defined by

$$B(t) = \begin{bmatrix} 0 & 0 \\ e^{-at} & 0 \end{bmatrix},$$

then $\int^{\infty} |B(s)|\, ds < \infty$ and $|B(t)| \to 0$ as $t \to \infty$. The solutions of the perturbed equation (2.7) with initial time $t_0 = 0$ are

$$x_1(t) = c_1 \exp(-at),$$
$$x_2(t) = \left[c_2 + c_1 \int_0^t \exp(-s \sin \log s)\, ds \right] \exp(t \sin \log t - 2at).$$

If α is chosen so that $0 < \alpha < \pi/2$, $t_n = \exp[(2n - 1/2)\pi]$, $n = 1, 2, \ldots$, then $\sin \log s \leqq -\cos \alpha$ for $t_n \leqq s \leqq t_n e^{\alpha}$. Hence,

$$\int_0^{t_n e^{\pi}} \exp(-s \sin \log s)\, ds > \int_{t_n}^{t_n e^{\alpha}} \exp(-s \sin \log s)\, ds$$

$$\geqq \int_{t_n}^{t_n e^{\alpha}} \exp(s \cos \alpha)\, ds$$

$$> t_n(e^{\alpha} - 1)\exp(t_n \cos \alpha).$$

Choose $c_2 = 0$, $c_1 = 1$. Since $\sin \log(t_n e^{\pi}) = 1$, we have

$$|x_2(t_n e^{\pi})| \geqq t_n(e^{\alpha} - 1)\exp(bt_n),$$

where $b = (1 - 2a)e^{\pi} + \cos \alpha$. If we choose α so that $b > 0$, then $|x_2(t_n e^{\pi})| \to \infty$ as $n \to \infty$ and the system is unstable.

EXERCISE 2.1.   Suppose there is a constant K such that a fundamental matrix solution X of the real system (1.3) satisfies $|X(t)| \leqq K$, $t \geqq \beta$ and

$$\liminf_{t \to \infty} \int_\beta^t tr\, A(s)\, ds > -\infty.$$

Prove that X^{-1} is bounded on $[\beta, \infty)$ and no nontrivial solution of (1.3) approaches zero as $t \to \infty$.

EXERCISE 2.2. Suppose A satisfies the conditions in Exercise 2.1 and $B(t)$ is a continuous real $n \times n$ matrix for $t \geq \beta$ with $\int_\beta^\infty |A(t) - B(t)| < \infty$. Prove that every solution of $\dot{y} = B(t)y$ is bounded on $[\beta, \infty)$. For any solution x of (1.3), prove there is a unique solution y of $\dot{y} = B(t)y$ such that $y(t) - x(t) \to 0$ as $t \to \infty$.

EXERCISE 2.3. Suppose system (1.3) is uniformly asymptotically stable, f satisfies the conditions of Theorem 2.4 and $b(t) \to 0$ as $t \to \infty$. Prove there is a $T > \beta$ such that any solution $x(t)$ of

$$\dot{x} = A(t)x + f(t, x) + b(t)$$

approaches zero as $t \to \infty$ if $|x(T)|$ is small enough.

EXERCISE 2.4. Generalize the result of Exercise 2.3 with $b(t)$ replaced by $g(t, x)$ where $g(t, x) \to 0$ as $t \to \infty$ uniformly for x in compact sets.

EXERCISE 2.5. Suppose there exists a continuous function $c(t)$ such that $\int_t^{t+1} c(s)\, ds \leq \gamma$, $t \geq \beta$, for some constant $\gamma = \gamma(\beta)$ and $f\colon R^{n+1} \to R^n$ is continuous with $|f(t, x)| \leq c(t)|x|$. Prove there is a constant $r > 0$ such that the solution $x = 0$ of (2.11) is uniformly asymptotically stable if $\gamma < r$.

EXERCISE 2.6. Generalize Exercises 2.3 and 2.4 with f satisfying the conditions of Exercise 2.5.

III.3. n^{th} Order Scalar Equations

Due to the frequency of occurrence of nth order scalar equations in the applications, it is worthwhile to transform the information obtained in Section 1 to equations of this type. Suppose y is a scalar, $a_1, \ldots, a_n$ and g are continuous real or complex valued functions on $(-\infty, +\infty)$ and consider the equation

$$(3.1) \qquad D^n y + a_1(t)D^{n-1}y + \cdots + a_n(t)y = g(t),$$

where D represents the operation of differentiation with respect to t. The function $D^2 y$ is the second derivative of y with respect to t, and so forth.

Equation (3.1) is equivalent to

$$(3.2) \quad \begin{cases} \dot{x} = Ax + h \\ x = \begin{bmatrix} y \\ Dy \\ \vdots \\ D^{n-2}y \\ D^{n-1}y \end{bmatrix}, \quad A = \begin{bmatrix} 0 & 1 & 0 & \cdots & 0 \\ 0 & 0 & 1 & \cdots & 0 \\ \vdots & & & & \\ 0 & 0 & 0 & \cdots & 1 \\ -a_n & -a_{n-1} & -a_{n-2} & \cdots & -a_1 \end{bmatrix} \quad h = \begin{bmatrix} 0 \\ 0 \\ \vdots \\ 0 \\ g \end{bmatrix}. \end{cases}$$

From this representation of (3.1), a solution of (3.2) is a column vector of dimension n, but the $(j + 1)^{\text{th}}$ component of the solution vector is obtained by differentiation of the first component j times with respect to t and this first component must be a solution of (3.1). Consequently, any $n \times n$ matrix solution $[\xi^1, \ldots, \xi^n]$, ξ^j an n-vector, of (3.2) must satisfy $\xi^j = \text{col}(\phi_j, D\phi_j, \ldots, D^{n-1}\phi_j)$, where ϕ^j, $j = 1, 2, \ldots, n$, is a solution of (3.1).

If $\phi_1, \ldots, \phi_n$ are n-scalar functions which are $(n - 1)$-times continuously differentiable, the *Wronskian* $\Delta(\phi_1, \ldots, \phi_n)$ of $\phi_1, \ldots, \phi_n$ is defined by

$$(3.3) \qquad \Delta(\phi_1, \ldots, \phi_n) = \det \begin{bmatrix} \phi_1 & \phi_2 & \cdots & \phi_n \\ D\phi_1 & D\phi_2 & \cdots & D\phi_n \\ \vdots & \vdots & & \vdots \\ D^{n-1}\phi_1 & D^{n-1}\phi_2 & \cdots & D^{n-1}\phi_n \end{bmatrix}.$$

A set of scalar functions $\phi_1, \ldots, \phi_n$ defined on $a \leq t \leq b$ are said to be *linearly dependent* on $[a, b]$ if there exist constants $c_1, \ldots, c_n$ not all zero such that $c_1\phi_1(t) + \cdots + c_n\phi_n(t) = 0$ for all t in $[a, b]$. Otherwise, the functions are *linearly independent* on $[a, b]$.

LEMMA 3.1. *If* $\phi_1, \ldots, \phi_n$ *are* $n - 1$ *times continuously differentiable scalar functions on an interval* I, *then* $\phi_1, \ldots, \phi_n$ *are linearly independent on* I *if the Wronskian* Δ *defined in (3.3) is different from zero on* I.

PROOF. Suppose there is a linear combination of the ϕ_i which is zero on I. More precisely, suppose that

$$\sum_{j=0}^{n} c_j \phi_j(t) = 0, \qquad t \text{ in } I,$$

where the c_j are constant. By repeated differentiation of this relation, the following system of equations for the constants c_j is obtained:

$$\sum_{j=1}^{n} c_j D^k\phi_j(t) = 0, \qquad k = 0, 1, 2, \ldots, n - 1.$$

Or, in matrix form,

$$\begin{bmatrix} \phi_1 & \phi_2 & \cdots & \phi_n \\ D\phi_1 & D\phi_2 & \cdots & D\phi_n \\ \vdots & \vdots & & \\ D^{n-1}\phi_1 & D^{n-1}\phi_2 & \cdots & D^{n-1}\phi_n \end{bmatrix} \begin{bmatrix} c_1 \\ \vdots \\ \vdots \\ c_n \end{bmatrix} = 0.$$

This latter relation must be satisfied for all t in I. On the other hand, by hypothesis, the determinant of the coefficient matrix is different from zero for all t in I which implies that the only solution of this system of equations is $c_1 = \cdots = c_n = 0$; that is, the functions ϕ_j are linearly independent.

The converse of this statement is not true. In fact, there can be n-functions with $n-1$ continuous derivatives on an interval I such that the Wronskian is identically zero for all t in I and yet the functions are linearly independent. For $n = 2$, one merely has to choose two differentiable functions ϕ_1, ϕ_2 on $[0, 3]$ such that $\phi_1(t) = 0$, t in $[0, 2]$, $\phi_2(t) = 0$, t in $[1, 3]$ and the functions are $\neq 0$ otherwise.

On the other hand, it is clear from the proof of Lemma 3.1 that linear dependence of $\phi_1, \ldots, \phi_n$ on I implies $\Delta \equiv 0$ on I.

If $\phi_1, \ldots, \phi_n$ are solutions of the homogeneous equation

(3.4) $$D^n y + a_1(t) D^{n-1} y + \cdots + a_n(t) y = 0,$$

then $\Delta(\phi_1, \ldots, \phi_n)$ is the determinant of an $n \times n$ matrix solution of the homogeneous system (3.2); that is, system (3.2) with $h = 0$. Lemmas 1.2, 1.5, formula (3.2) and the remarks above imply

THEOREM 3.1. If $\phi_1, \ldots, \phi_n$ are n solutions of (3.4), then $\Delta(\phi_1, \ldots, \phi_n)(t)$ is $\neq 0$ for all t in $(-\infty, \infty)$ or identically zero. More specifically,

$$\Delta(\phi_1, \ldots, \phi_n)(t) = \Delta(\phi_1, \ldots, \phi_n)(0) \exp\left(-\int_0^t a_1(s)\, ds\right).$$

Thus, the n solutions $\phi_1, \ldots, \phi_n$ of (3.4) are linearly independent if and only if $\Delta(\phi_1, \ldots, \phi_n)(0) \neq 0$.

Let us determine the variation of constants formula for the solutions of equation (3.1). Since (3.1) is equivalent to (3.2), it is sufficient to determine only the variation of constants formula for the first component of the vector solution of (3.2). If $X(t, \tau)$, $X(\tau, \tau) = I$, is the principal matrix solution of $\dot{x} = Ax$, then any solution of (3.2) satisfies relation (1.7). If the first row of the matrix $X(t, \tau)$ is designated by $(\phi_1(t, \tau), \ldots, \phi_n(t, \tau))$, then the first component y of the vector x in (1.7) is given by

(3.5) $$y(t) = \sum_{j=1}^{n} [D^{j-1} y(\tau)] \phi_j(t, \tau) + \int_\tau^t \phi_n(t, s) g(s)\, ds,$$

where g is the function given in (3.1). The reason for such a simple expression is due to the fact that the vector h in (2.2) has all components zero except the last which is g. Since the function $\phi_n(t, s)$ together with its derivatives up through order $n - 1$ form the last column of the matrix $X(t, s)$ it follows that $\phi_n(t, s)$ is the solution of the homogeneous equation (3.4) which for $t = s$ vanishes together with all its derivatives up through order $n - 2$, and the derivative of order $n - 1$ with respect to t is equal to one. This is summarized in

THEOREM 3.2.   For any real number s in $(-\infty, +\infty)$, let $\phi(t, s)$ be the unique solution of the homogeneous equation (3.4) which satisfies the initial data

$$(3.6) \qquad y(s) = Dy(s) = \cdots = D^{n-2}y(s) = 0, \qquad D^{n-1}y(s) = 1.$$

Then the unique solution of equation (3.1) which vanishes together with all derivatives up through order $n - 1$ at $t = \tau$ is given by

$$(3.7) \qquad y(t) = \int_\tau^t \phi(t, s)g(s)\, ds.$$

Another relation which is easily obtained from the general theory of Section 1 is the equation adjoint to equation (3.4). This is derived from the general definition of the adjoint equation given in Section 1. The adjoint to equation (3.2) with $h = 0$ is given by $\dot{w} = -wA$, $w = (w_1, \ldots, w_n)$, or, equivalently,

$$(3.8) \qquad \begin{cases} Dw_1 = a_n w_n \\ Dw_2 = -w_1 + a_{n-1} w_n \\ \cdots \\ Dw_n = -w_{n-1} + a_1 w_n. \end{cases}$$

It is not obvious that equation (3.8) is equivalent to an nth order scalar equation of any type. However, if each of the functions a_k, $k = 1, \ldots, n$, has a sufficient number of continuous derivatives, this is the case. In fact, differentiation of the $(k + 1)^{\text{th}}$ equation in (3.8) k-times with respect to t yields the equivalent set of equations

$$(3.9) \qquad Dw_1 = a_n w_n,$$
$$D^{k+1}w_{k+1} = -D^k w_k + D^k(a_{n-k} w_n), \qquad k = 1, 2, \ldots, n - 1.$$

If $z = w_n$, then one easily concludes from (3.9) that

$$(3.10) \qquad D^n z - D^{n-1}(a_1 z) + \cdots + (-1)^n a_n z = 0.$$

Equation (3.10) is referred to as the adjoint to equation (3.4). If the a's are constant this is the same differential equation as the original one except for some changes in sign in the coefficients.

III.4. Linear Systems with Constant Coefficients

In this section, we consider the homogeneous equation

(4.1) $$\dot{x} = Ax$$

and the nonhomogeneous equation

(4.2) $$\dot{x} = Ax + h(t)$$

where A is an $n \times n$ real or complex constant matrix and h is a continuous real or complex n-vector function on $(-\infty, \infty)$.

The principal matrix solution $P(t)$ of (4.1) at $t = 0$ is such that each column of $P(t)$ satisfies (4.1) and $P(0) = I$, the identity. The matrix $P(t)$ satisfies the following property: $P(t + s) = P(t)P(s)$ for all t, s in $(-\infty, \infty)$. In fact, for each fixed s in $(-\infty, \infty)$, both $P(t + s)$ and $P(t)P(s)$ satisfy (4.1) and for $t = 0$ these matrices coincide. Thus, the uniqueness theorem implies the result. This relation suggests that $P(t)$ behaves as an exponential function. Therefore, we make the following

Definition 4.1.   The $n \times n$ matrix e^{At}, $-\infty < t < \infty$, $e^0 = I$, the identity, is defined as the principal matrix solution of (4.1) at $t = 0$.

The matrix e^{At} satisfies the following properties:

  (i) $e^{A(t+s)} = e^{At}e^{As}$

  (ii) $(e^{At})^{-1} = e^{-At}$

  (iii) $\dfrac{d}{dt} e^{At} = Ae^{At} = e^{At}A$

  (iv) $e^{At} = I + At + \dfrac{1}{2!} A^2 t^2 + \cdots + \dfrac{1}{n!} A^n t^n + \cdots$

  (v)   a general solution of (4.1) is $e^{At}c$ where c is an arbitrary constant n-vector.

  (vi) If $X(t)$, $\det X(0) \neq 0$ is an $n \times n$ matrix solution of (4.1), then $e^{At} = X(t)X^{-1}(0)$.

Property (i) was proved above and (ii) is a consequence of (i). Property (v) is Lemma 1.3 for this special case. (iii) follows from uniqueness and the observation that both Ae^{At} and $e^{At}A$ are matrix solutions of (4.1). This same argument proves (vi).

Before proving (iv), it is necessary to make a few remarks about the meaning of a matrix power series. If $f(z)$ is a function of the complex variable z which is analytic in a neighborhood of $z = 0$, the power series $f(A)$ is defined as the formal power series obtained by substituting for each term in the power series of $f(z)$ the matrix A for the complex variable z. Of course, such an expression will be meaningful if and only if for the particular matrix A each element of the power series matrix converges. One can actually use such a series as a definition for matrix functions $f(A)$ when $f(z)$ is an analytic function of the complex variable z. On the other hand, it should be noted that property (iv) above is not a definition of e^{At} but is going to be derived from the fact that e^{At} is the principal matrix solution of (3.1) at $t = 0$.

PROOF OF (IV).   Since $P(t) \overset{\text{def}}{=} e^{At}$ is the principal matrix solution of (4.1) at $t = 0$ it follows that $P(t)$ must satisfy the integral equation

$$(4.3) \qquad P(t) = I + \int_0^t A P(s)\, ds.$$

If one attempts to solve equation (4.3) by the obvious method of successive approximations

$$(4.4) \qquad P^{(0)} = I,$$

$$P^{(k+1)}(t) = I + \int_0^t A P^{(k)}(s)\, ds, \qquad k = 0, 1, 2, \ldots,$$

then one observes that the expression for $P^{(k)}$ is given by the formula

$$P^{(k)}(t) = I + At + \frac{1}{2!} A^2 t^2 + \cdots + \frac{1}{k!} A^k t^k.$$

In the proof of the Picard-Lindelöf Theorem in Chapter I, it was shown that this sequence converged for $|t| \leq \alpha$ and α small. We now wish to show that the sequence of matrices $P^{(k)}(t)$ converge uniformly for all t in a compact set in $(-\infty, \infty)$. If we let $P_{ij}^{(k)}(t)$ be the $(ij)^{\text{th}}$ element of the matrix $P^{(k)}(t)$, then there is a constant $\beta > 0$ independent of k such that $|P_{ij}^{(k)}(t)| \leq \beta^{-1} |P^{(k)}(t)|$. Thus,

$$\beta |P_{ij}^{(k)}(t)| \leq 1 + \alpha t + \frac{1}{2!} \alpha^2 t^2 + \cdots + \frac{1}{k!} \alpha^k t^k \leq e^{\alpha t},$$

where for simplicity we have put $\alpha = |A|$. Furthermore,

$$\beta |P_{ij}^{(k+1)}(t) - P_{ij}^{(k)}(t)| \leq \frac{1}{(k+1)!} \alpha^{k+1} t^{k+1},$$

and the sequence $(P_{ij}^{(k)}(t))$ converges uniformly for all t in a compact set of $(-\infty, \infty)$. This shows that the power series in property (iv) is well defined and thus the sequence of matrices $\{P^{(k)}(t)\}$ converges uniformly on all compact subsets of $(-\infty, \infty)$. From (4.4), it follows that the limit is the principal matrix solution $P(t) = e^{At}$ of (4.1). This proves property (iv).

In spite of the above apparent simplicity, the matrix e^{At} is a rather complicated individual. Many of the operations that are valid for the scalar exponential function are also true for the matrix function e^{At}. However, other operations do not behave in the same way as for scalars since in general matrix multiplication is not commutative. The following two exercises indicate that caution must be exercised in operating with e^{At}.

EXERCISE 4.1. Prove that $Be^{At} = e^{At}B$ for all t if and only if $BA = AB$.

EXERCISE 4.2. Prove that $e^{At}e^{Bt} = e^{(A+B)t}$ for all t if and only if $BA = AB$.

With a general solution of equation (4.1) being given by $e^{At}c$, where c is a constant n-vector and the matrix e^{At} having the power series representation given in property (iv), one might ask if there is anything left to discuss about linear equations with constant coefficients. Unfortunately, the previous notation is very misleading and we have not answered the following questions: In what precise sense does the function $x(t) = e^{At}c$ where c is a constant n-vector behave as the scalar exponential function? What is an effective means for computing e^{At}?

Both of the questions are very closely related and reduce to a detailed discussion of the eigenvalues and eigenvectors of a matrix. The general structure of the solutions of the differential equation (3.1) is determined completely from the solution of an algebraic problem. This fact expresses the simplicity of linear systems of differential equations with constant coefficients and also explains why there is always a definite attempt in the applications to simulate models of physical systems by using equations with constant coefficients.

A complex number λ is called an *eigenvalue (proper value, characteristic value)* of an $n \times n$ matrix A if there exists a non-zero vector v such that

(4.5) $$ Av = \lambda v \qquad \text{or} \qquad (A - \lambda I)v = 0. $$

If λ is an eigenvalue of the matrix A and v is any non-zero solution of equation (4.5), then v is called an *eigenvector (proper value, characteristic vector)* associated with the eigenvalue λ. Hence, λ is an eigenvalue of the matrix A if and only if λ is a solution of the characteristic equation

(4.6) $$ \det(A - \lambda I) = 0. $$

The characteristic equation is a polynomial of degree n in λ and, therefore, has n solutions, not all of which may be distinct. On the other hand, if $\lambda_1, \ldots, \lambda_k$ are distinct eigenvalues of the matrix A and $v^1, \ldots, v^k$ are corresponding eigenvectors, then $v^1, \ldots, v^k$ are linearly independent. The following lemma is obvious.

LEMMA 4.1. If λ is an eigenvalue of the matrix A and v is an eigenvector associated with λ, then the function $x(t) = e^{\lambda t}v$ is a solution of the differential equation $\dot{x} = Ax$.

LEMMA 4.2.   If A is an $n \times n$ matrix with n-distinct eigenvalues $\lambda_1, \ldots, \lambda_n$ and $v^1, \ldots, v^n$ are the corresponding eigenvectors, then a general solution of equation (4.1) is given by

$$V(t)c = c_1 v^1 e^{\lambda_1 t} + \cdots + c_n v^n e^{\lambda_n t},$$

where $V(t) = (v^1 e^{\lambda_1 t}, \ldots, v^n e^{\lambda_n t})$ and $c = (c_1, \ldots, c_n)$ is an arbitrary complex n-vector. Furthermore, $e^{At} = V(t)V(0)^{-1}$.

PROOF.   If we let $V(t)$ be defined as above, then Lemma 1.3 implies it is sufficient to show that $V(t)$ is a fundamental matrix solution of equation (4.1). Since $V(0) = (v^1, \ldots, v^n)$ and the vectors $v^1, \ldots, v^n$ are linearly independent, $V(0)$ is non-singular. This fact, Lemma 4.1 and Corollary 1.1 imply that $V(t)$ is a fundamental matrix solution of (4.1). A general solution is therefore $V(t)c$ where c is an arbitrary n-vector. The last assertion of the lemma is precisely property (vi) above of e^{At}.

If the matrix A has real elements then the eigenvalues will occur in complex conjugate pairs. This will imply that the corresponding eigenvectors can be chosen as complex conjugates. Therefore, if only real solutions of (4.1) are desired, the constants $c_1, \ldots, c_n$ given in the general solution will need to satisfy some restriction of complex conjugacy. In fact, if λ_1 and λ_2 are complex conjugates, then the eigenvectors v^1, v^2 can be chosen as complex conjugates. In the general solution the constants c_1 and c_2 then may be chosen as complex conjugates to obtain a real solution of equation (4.1).

We now describe the general procedure for obtaining e^{At}. To do this we need some elementary concepts from linear algebra. Let C^n be complex n-dimensional space and let S_1, S_2 be linear subspaces of C^n. The subspaces S_1 and S_2 are *linearly independent* if $c_1 y_1 + c_2 y_2 = 0$, y_1 in S_1, y_2 in S_2, implies that $c_1 = c_2 = 0$. The *direct sum*, $S_1 \oplus S_2$, of independent subspaces S_1, S_2 is a subspace S of C^n whose elements y are given by $y = y_1 + y_2$, y_1 in S_1, y_2 in S_2. If A is an $n \times n$ matrix and S is a subspace of C^n, then S is *invariant under* A if for any y in S the vector Ay is in S. If A is an $n \times n$ matrix the *null space of* A, denoted by $N(A)$, is the set of all y in C^n such that $Ay = 0$. If λ is an eigenvalue of the $n \times n$ matrix A, $r(\lambda)$ is the least integer k such that $N(A - \lambda I)^{k+1} = N(A - \lambda I)^k$. The set $N(A - \lambda I)^{r(\lambda)}$ is a subspace of C^n and is called the *generalized eigenspace* of A corresponding to the eigenvalue λ. This generalized eigenspace is denoted by the symbol $M_\lambda(A)$. The dimension of $M_\lambda(A)$ is equal to the algebraic multiplicity of the eigenvalue A; that is, the multiplicity of the eigenvalue as a zero of the characteristic polynomial $\det(A - \lambda I)$. We say an eigenvalue λ of A has *simple elementary divisors* if $M_\lambda(A) = N(A - \lambda I)$. The following result is basic and a proof can be found in any book on linear algebra.

LEMMA 4.3.   If A is an $n \times n$ matrix and $\lambda_1, \ldots, \lambda_s$ are the distinct eigenvalues of A, then the corresponding generalized eigenspaces $M_{\lambda_1}(A)$, $\ldots, M_{\lambda_s}(A)$ are linearly independent, are invariant under the matrix A and $C^n = M_{\lambda_1}(A) \oplus \cdots \oplus M_{\lambda_s}(A)$.

Since the generalized eigenspaces $M_{\lambda_1}(A), \ldots, M_{\lambda_s}(A)$ are linearly independent if $\lambda_1, \ldots, \lambda_s$ are the distinct eigenvalues of A, it follows that any vector x^0 in C^n can be represented uniquely as

$$(4.7) \qquad x^0 = \sum_{j=1}^{s} x^{0,j}, \qquad x^{0,j} \text{ in } M_{\lambda_j}(A).$$

Since the generalized eigenspaces $M_{\lambda_j}(A)$ are invariant under the matrix A, any solution of the differential equation (4.1) with initial value in $M_{\lambda_j}(A)$ will remain in $M_{\lambda_j}(A)$ for all values of t. This is immediately obvious since the tangent vector to the curve described by the solution of the differential equation in the phase space lies in the subspace $M_{\lambda_j}(A)$.

For any x^0 in C^n and any complex number λ,

$$e^{At}x^0 = e^{(A-\lambda I)t}e^{\lambda t}x^0$$

$$= e^{(A-\lambda I)t}x^0 e^{\lambda t}$$

$$= \left(\sum_{k=0}^{\infty} (A - \lambda I)^k \frac{t^k}{k!} \right) x^0 e^{\lambda t}.$$

If λ is an eigenvalue of A and $M_{\lambda}(A)$ is the corresponding generalized eigenspace, then x^0 in $M_{\lambda}(A)$ implies that the above infinite series is actually only a polynomial since $(A - \lambda I)^k x^0 = 0$ for all $k \geq r(\lambda)$. Therefore, for any x^0 in $M_{\lambda}(A)$, $e^{At}x^0$ is given by

$$(4.8) \qquad e^{At}x^0 = \left[\sum_{k=0}^{r(\lambda)-1} (A - \lambda I)^k \frac{t^k}{k!} \right] x^0 e^{\lambda t}.$$

In summary, if λ is an eigenvalue of the matrix A and $M_{\lambda}(A)$ is the corresponding generalized eigenspace, then any solution of the differential equation with its initial value in $M_{\lambda}(A)$ must lie in $M_{\lambda}(A)$ for all values of t and the solution of the differential equation is a polynomial in t with vector coefficients times $e^{\lambda t}$. Furthermore, this polynomial in t can be obtained by a direct evaluation of the infinite series for $e^{(A-\lambda I)t}$.

These remarks together with Lemma 4.3 and the decomposition (4.7) yield

THEOREM 4.1. If $\lambda_1, \ldots, \lambda_s$ are the distinct eigenvalues of the $n \times n$ matrix A and $M_{\lambda_1}(A), \ldots, M_{\lambda_s}(A)$ are the corresponding generalized eigenspaces, then the solution of the initial value problem

$$(4.9) \qquad \dot{x} = Ax, \qquad x(0) = x^0,$$

is given by

$$(4.10) \qquad x(t) = \sum_{j=1}^{s} \left[\sum_{k=0}^{r(\lambda_j)-1} (A - \lambda_j I)^k \frac{t^k}{k!} \right] x^{0,j} e^{\lambda_j t},$$

where the vector $x^{0,j}$ belongs to the generalized eigenspace $M_{\lambda_j}(A)$ and is determined by the unique decomposition of the initial vector x^0 according to equation (4.7).

The specific manner in which one applies this result is as follows. The dimension of the generalized eigenspace $M_\lambda(A)$ of an eigenvalue of A is equal to the algebraic multiplicity of the eigenvalue A. For a given eigenvalue λ of the matrix A one first obtains the null space of the matrix $A - \lambda I$. If the dimension of this null space is not equal to the algebraic multiplicity of the eigenvalue λ, one proceeds to calculate the null space of $(A - \lambda I)^2$. If the dimension of this subspace is equal to the algebraic multiplicity of the eigenvalue λ, the subspace is $M_\lambda(A)$. This process is continued until a subspace is obtained which has the dimension of the algebraic multiplicity of the eigenvalue λ. In this process, one also has obtained a basis for the generalized eigenspace $M_\lambda(A)$. Having done this for each eigenvalue λ of the matrix A, one has a basis for C^n. In terms of this basis the element x^0 can be expanded uniquely, which determines the $x^{0,j}$ and therefore the solution $x(t)$ by formula (4.10). The general solution of equation (4.1) is obtained by replacing $x^{0,j}$ in (4.10) by an arbitrary linear combination of the basis vectors for the generalized eigenspaces $M_{\lambda_j}(A)$, $j = 1, 2, \ldots, s$.

EXERCISE 4.1.   Using Theorem 4.1, find the general real solution of the equation $\dot{x} = Ax$ with

$$(a) \quad A = \begin{bmatrix} 0 & 1 \\ -1 & 0 \end{bmatrix};$$

$$(b) \quad A = \begin{bmatrix} 0 & 1 \\ 0 & 0 \end{bmatrix};$$

$$(c) \quad A = \begin{bmatrix} 1 & 1 \\ 1 & -1 \end{bmatrix};$$

$$(d) \quad A = \begin{bmatrix} 0 & -1 & 2 \\ 0 & 1 & 0 \\ 1 & 1 & -1 \end{bmatrix};$$

$$(e) \quad A = \begin{bmatrix} 0 & 1 & 0 \\ 4 & 3 & -4 \\ 1 & 2 & -1 \end{bmatrix}.$$

Theorem 4.1 gives a complete description of the general form of the solution of a linear equation with constant coefficients; namely, any solution of (4.1) is a sum of exponential terms with coefficients which are polynomials in t. The exponents of the exponential terms involve the eigenvalues of A. The degree of the corresponding polynomial in t is no more than one less than the dimension of the generalized eigenspace of the eigenvalue. A more complete description of the number of linearly independent solutions of (4.1) of the form $p(t)e^{\lambda t}$, where $p(t)$ is a polynomial of a given degree is given by the *Jordan canonical form*: For an $n \times n$ matrix A, there is a nonsingular matrix Q such that

(4.11) $\qquad Q^{-1}AQ = \operatorname{diag}(C_1, \ldots, C_p);$

$$C_j = \lambda_j I + R_j;$$

$$R_j = \begin{bmatrix} 0 & 1 & 0 & \cdots & 0 & 0 \\ 0 & 0 & 1 & \cdots & 0 & 0 \\ \cdot & \cdot & \cdot & \cdots & \cdot & \cdot \\ 0 & 0 & 0 & \cdots & 0 & 1 \\ 0 & 0 & 0 & \cdots & 0 & 0 \end{bmatrix}.$$

and $\lambda_j, j = 1, 2, \ldots, p$, is an eigenvalue of A. The $\lambda_1, \ldots, \lambda_p$ are not necessarily distinct and the dimension n_j of C_j is not necessarily equal to the $r(\lambda_j)$ mentioned above.

If Q, A are related by (4.11), then

(4.12) $\qquad Q^{-1}e^{At}Q = e^{[\operatorname{diag}(C_1, \ldots, C_p)]t} = \operatorname{diag}(e^{C_1 t}, \ldots, e^{C_p t}),$

$$e^{C_j t} = e^{\lambda_j t}e^{R_j t}, \; e^{R_j t} = \begin{bmatrix} 1 & t & \dfrac{t^2}{2!} & \cdots & \dfrac{t^{n_j-1}}{(n_j-1)!} \\ 0 & 1 & t & \cdots & \dfrac{t^{n_j-2}}{(n_j-2)!} \\ \cdot & \cdot & \cdot & \cdots & \\ 0 & 0 & 0 & \cdots & 1 \end{bmatrix}.$$

The representation (4.12) gives more specific information about e^{At}, but depicts the same general structure of the solutions of (4.1) as Theorem 4.1.

A similarity transformation is a transformation which takes a matrix A into a matrix $Q^{-1}AQ$, Q nonsingular. Similarity transformations are very important in differential equations. In fact, if Q is nonsingular and $x = Qy$ in (4.1), then $\dot{y} = Q^{-1}AQy$. The Jordan canonical form implies that such transformations can be used to reduce a differential equation to a very simple form.

THEOREM 4.2. (i) A necessary and sufficient condition that the system (4.1) be stable is that the eigenvalues of A have real parts $\leqq 0$ and those with zero real parts have simple elementary divisors.

(ii) A necessary and sufficient condition that the system (4.1) be asymptotically stable is that all eigenvalues of A have real parts < 0. If this is the case, there exist positive constants K, α such that

(4.13) $$|e^{At}| \leq Ke^{-\alpha t}, \quad t \geq 0.$$

PROOF. Since (4.1) is autonomous, stability implies uniform stability, asymptotic stability implies uniform asymptotic stability. Theorem. 2.1 implies that stability and boundedness of (4.1) are equivalent and asymptotic stability and exponential asymptotic stability are equivalent.

(i) If all solutions of (4.1) are bounded, then there can be no eigenvalue of A with positive real parts from Lemma 4.1. Furthermore, if there are eigenvalues with zero real part and $M_\lambda(A) \neq N(A - \lambda I)$, then Theorem 4.1 implies there is a solution with a term $te^{\lambda t}$ times a nonzero constant vector. Such a solution would not be bounded. Conversely, if all eigenvalues have nonpositive real parts and $M_\lambda(A) = N(A - \lambda I)$ for those eigenvalues with zero real parts, then the general representation theorem, Theorem 4.1, implies that no powers of t occur except when multiplied by an $e^{\lambda t}$ with $\mathrm{Re}\lambda < 0$. Therefore, the solutions are bounded.

(ii) This part is proved in the same manner using Theorem 4.1 and Theorem 2.1.

Since $(e^{At})^{-1} = e^{-At}$, formula (1.7) implies the variation of constants formula for (4.2) is

(4.14) $$x(t) = e^{A(t-\tau)}x(\tau) + \int_\tau^t e^{A(t-s)}h(s)\,ds.$$

As a particular illustration of this relation, consider the scalar equation

$$\ddot{u} + u = g(t).$$

This equation is equivalent to the system

$$\dot{u} = v,$$
$$\dot{v} = -u + g(t).$$

If $x = (u, v)$,

$$A = \begin{bmatrix} 0 & 1 \\ -1 & 0 \end{bmatrix}, \qquad f = \begin{bmatrix} 0 \\ g \end{bmatrix},$$

then

$$e^{At} = \begin{bmatrix} \cos t & \sin t \\ -\sin t & \cos t \end{bmatrix},$$

and the first component of (4.14) is

$$u(t) = u(\tau)\cos(t - \tau) + \dot{u}(\tau)\sin(t - \tau) + \int_\tau^t \sin(t - s)g(s)\,ds,$$

a well known formula in elementary differential equations.

III.5. Two Dimensional Linear Autonomous Systems

As an application of Theorem 4.1, we classify the different behaviors of the solutions of

(5.1) $$\dot{x} = Ax, \qquad A = \begin{bmatrix} a & b \\ c & d \end{bmatrix}, \qquad \det A \neq 0.$$

where a, b, c, d are real constants. If λ_1, λ_2 designate the eigenvalues of A, there are many cases to consider.

Case 1. λ_1, λ_2 real, $\lambda_2 < \lambda_1$. Let v^1, v^2 designate unit eigenvectors of A associated with the eigenvalues λ_1, λ_2, respectively. A general real solution of (5.1) is

(5.2) $$x(t) = c_1 e^{\lambda_1 t} v^1 + c_2 e^{\lambda_2 t} v^2,$$

where c_1, c_2 are arbitrary real constants. For a given $c_1, c_2, c_1^2 + c_2^2 > 0$, the unit tangent vector to the orbit γ described by (5.2) approaches a vector parallel to v^1 as $t \to \infty$ if $c_1 \neq 0$ and $\pm v^2$ as $t \to -\infty$ if $c_2 \neq 0$.

Case 1a. Negative roots, $\lambda_2 < \lambda_1 < 0$ (*Stable node*). All solutions approach zero as $t \to \infty$ and the above information on asymptotic directions allows one to sketch the orbits as in Fig. 5.1. The straight lines L_1 and L_2 are the lines that contain the eigenvector v^1 and v^2, respectively. The origin is stable and called a stable node.

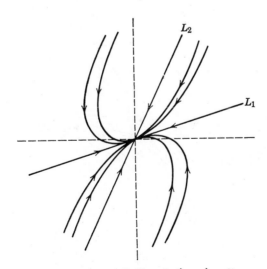

Figure III.5.1   Stable node $(\lambda_2 < \lambda_j < 0)$.

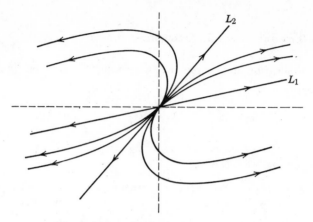

Figure III.5.2   Unstable node $(0 < \lambda_2 < \lambda)$.

Case 1b.   Positive roots, $0 < \lambda_2 < \lambda_1$ *(Unstable node).* This is similar to Case 1a except for reversing the arrows and changing the lines of tangency of the curves. See Fig. 5.2. The origin is unstable and called an unstable nope.

Case 1c.   One positive and one negative root, $\lambda_2 < 0 < \lambda_1$ *(Saddle point).* From (5.2), those orbits which lie on L_2 approach zero as $t \to +\infty$ and those which lie on L_1 approach zero as $t \to -\infty$. All other orbits are unbounded. The zero solution is unstable and the orbits are depicted in Fig. 5.3. The origin is called a saddle point.

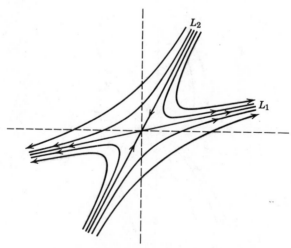

Figure III.5.3   Saddle point $\lambda_2 < 0 < \lambda_1$.

Case 2. Complex roots. Since A is real, we have $\lambda_1 = \alpha + i\beta$, $\lambda_2 = \alpha - i\beta$, α, β real, $\beta > 0$. If v^1 and v^2 are chosen as complex conjugate, a general solution for the real solutions of (5.1) is

$$x(t) = c_1 e^{(\alpha+i\beta)t} v^1 + \bar{c}_1 e^{(\alpha-i\beta)t} \bar{v}^1 = 2\,Re(c_1 e^{(\alpha+i\beta)t} v^1),$$

where c^1 is an arbitrary complex number and $\bar{b}$ designates the complex conjugate of a vector b.

If $v^1 = u + iv$, u, v real, then u, v are linearly independent and if $c_1 = ae^{i\delta}$ where a, δ are real, then a general real solution of (5.1) is

(5.3) $x(t) = 2ae^{\alpha t}[u \cos(\beta t + \delta) - v \sin(\beta t + \delta)].$

The expression (5.3) gives all of the essential properties of the solutions. If $\beta t + \delta = k\pi$, k an integer, then the orbit of the solution crosses the line U generated by u and if $\beta t + \delta = (2k + 1)\pi/2$, k an integer, it crosses the line V generated by v. The components of the solution curve in the direction u and v oscillate and are $\pi/2$ radians out of phase. Therefore, the orbit must resemble a spiral.

Case 2a. Purely imaginary roots *(Center)*. If the eigenvalues of A are $\pm i\beta$, a general real solution of (5.1) is

$$x(t) = a[u \cos(\beta t + \delta) - v \sin(\beta t + \delta)],$$

where a, δ are abitrary real constants and u, v are as above. The orbits are closed curves and every solution is periodic of period $2\pi/\beta$. These curves are ellipses with center at $(0,0)$ (see Fig. 5.4). The origin is stable and called a center.

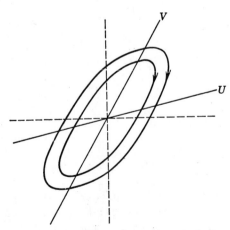

Figure III.5.4   Center $\lambda_1 = \lambda_2$, $Re\ \lambda_1 = 0$, $Im\ \lambda \neq 0$.

Case 2b. Complex roots with negative real parts (*Stable focus*). If $\alpha < 0$, it follows from (5.3) that all solutions approach zero as $t \to \infty$ and the orbits are spirals. The equilibrium point is called a stable focus (see Fig. 5.5).

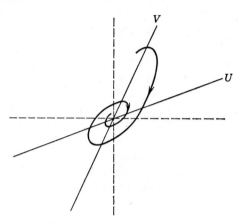

Figure III.5.5   Stable focus.

Case 2c. Complex roots with positive real parts (*Unstable focus*). If $\alpha > 0$, then solutions approach zero as $t \to -\infty$ and the equilibrium point is called an unstable focus (see Fig. 5.6).

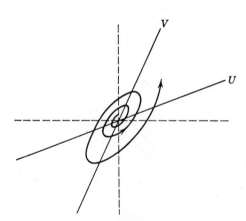

Figure III.5.6   Unstable focus.

Case 3. Both eigenvalues equal (*Improper node*). The eigenvalues are real. If there are two real linearly independent eigenvectors v^1, v^2 associated with the eigenvalue λ (i.e., λ has simple elementary divisors), then a general solution is

$$x(t) = (c_1 v^1 + c_2 v^2)e^{\lambda t},$$

where c_1, c_2 are arbitrary real constants. The unit tangent vector to the orbit of x is constant for any c_1, c_2. Therefore, all orbits are on straight lines passing through the origin (see Fig. 5.7).

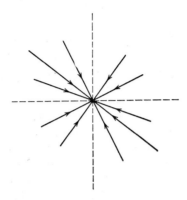

Figure III. 5.7   Stable improper node.

If there is only one linearly independent eigenvector v^1 of λ, then the general solution given by Theorem 4.1 is

$$x(t) = (c_1 + c_2 t)e^{\lambda t}v^1 + c_2 e^{\lambda t}v^2,$$

where v^2 is any vector independent of v^1. By direct computation, one shows that the tangent to the orbit becomes parallel to v^1 as $t \to +\infty$ and $t \to -\infty$. A typical situation is shown in Fig. 5.8.

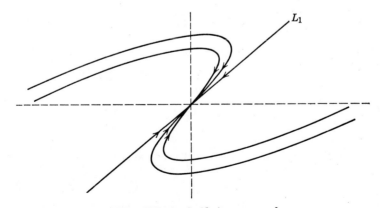

Figure III.5.8   Stable improper node.

III.6. The Saddle Point Property

In the previous section, we have given a rather complete characterization of the behavior of the solutions of a two dimensional linear system with constant coefficients. It is of interest to study the conditions under which the behavior of the solutions near an equilibrium point of a particular type is not "changed" by "small" perturbations of the right hand side of the differential equation. By considering only linear perturbations Bx with $|B|$ small, it is easy to see that the only types of equilibrium points for two dimensional linear systems which are preserved are the focus proper node and saddle point. On the other hand, the limiting behavior at ∞ is insensitive to small perturbations for all types except the center. In this section, we discuss for linear systems of arbitrary order a special type of equilibrium point which is insensitive to perturbations in the differential equation. The classification of the equilibrium point is crude in the sense that it does not take into account the "fine" structure of the trajectories but only general asymptotic properties of the trajectories. The questions discussed here are treated in a much more general context in a later chapter, but, in spite of the duplication of effort, it seems appropriate to consider the special case at this time.

The equilibrium point $x = 0$ of (4.1) is called a *saddle point of type (k)* of (4.1) if the eigenvalues of the matrix A have nonzero real parts with $k \geqq 0$ eigenvalues with positive real parts. For $n = 2$, this definition of a saddle point does not coincide with the one in Section 5. In fact, an equilibrium point which is completely stable $(k = 0)$ as well as one which is completely unstable $(k = 2)$ is referred to in this section as a saddle point.

If $x = 0$ is a saddle point of (4.1) of type (k), the space C^n can be decomposed as

$$(6.1) \qquad\qquad C^n = C^n_+ \oplus C^n_-,$$
$$C^n_+ = \pi_+ C^n, \; C^n_- = \pi_- C^n,$$

where π_+, π_- are projection operators, $\pi_+ + \pi_- = I$, C^n_+ has dimension k, C^n_- has dimension $n - k$, C^n_+, C^n_- are invariant under A and there are constants $K > 0$, $\alpha > 0$ such that

$$(6.2) \qquad\quad \text{(a)} \quad |e^{At}\pi_+ x| \leqq K e^{\alpha t}|\pi_+ x|, \qquad t \leqq 0,$$
$$\text{(b)} \quad |e^{At}\pi_- x| \leqq K e^{-\alpha t}|\pi_- x|, \qquad t \geqq 0,$$

for all x in C^n. These relations are immediate from the observation that there exists a nonsingular matrix U such that $U^{-1}AU = \text{diag}(A_+, A_-)$ where A_+ is a $k \times k$ matrix whose eigenvalues have positive real parts and A_- is an $(n - k) \times (n - k)$ matrix whose eigenvalues have negative real parts. From

Theorem 4.2, there are constants $K_1 > 0$, $\alpha > 0$ such that $|e^{A_+t}| \leq K_1 e^{\alpha t}$, $t \leq 0$ and $|e^{A_-t}| \leq K_1 e^{-\alpha t}$, $t \geq 0$. The first relation is obtained by replacing t by $-t$ in the equation $\dot{u} = A_+ u$. Let $C_+^n = \{x \text{ in } C^n \text{ such that } x = Uy$, $y = (u, 0)$, u a k-vector$\}$ and $C_-^n = \{x \text{ in } C^n \text{ such that } x = Uy$, $y = (0, v)$, v an $(n - k)$-vector$\}$. Since U is nonsingular $C_+^n \oplus C_-^n = C^n$ and this defines the projection operators π_+, π_- above. It is clear that (6.1) and (6.2) are satisfied.

The subspace C_+^n defined in (6.1) is called the *unstable manifold* passing through zero and C_-^n is called the *stable manifold* passing through zero. The orbits of solutions on the unstable manifold approach zero as $t \to -\infty$ and the orbits of solutions on the stable manifold approach zero as $t \to \infty$. Furthermore, the construction of C_+^n, C_-^n clearly shows that the only solutions of (4.1) whose orbits remain in a given neighborhood of the origin for all $t \geq 0$ ($t \leq 0$) must have their initial values on C_-^n (C_+^n).

What type of behavior is expected when a linear system with $x = 0$ as a saddle point is subjected to perturbations which are small near $x = 0$? The following example is instructive. Consider the second order system

$$\dot{x}_1 = x_1,$$
$$\dot{x}_2 = -x_2 + x_1^3,$$

whose general solution is

$$x_1(t) = e^t a,$$

$$x_2(t) = e^{-t}\left(b - \frac{a^3}{4}\right) + \frac{a^3}{4} e^{3t},$$

where a, b are arbitrary constants. It is easily seen that $x_1(t)$, $x_2(t) \to 0$ as $t \to \infty$ if and only if $a = 0$ and b is arbitrary. Also, $x_1(t)$, $x_2(t) \to 0$ as $t \to -\infty$ if and only if $b = a^3/4$ and a is arbitrary. If $b = a^3/4$, then the corresponding orbit in the phase plane is $x_2 = x_1^3/4$. The phase plane portrait is shown in Fig. 6.1. Notice that the stable and unstable manifolds near $x = 0$ are essentially the same as the ones for the linear system. This example motivates the following definition of a saddle point for nonlinear equations.

Suppose x_0 is an equilibrium point of the equation

(6.3) $$\dot{x} = Ax + f(x),$$

where f is continuous on C^n. We say x_0 is a *saddle point of type* (k) of (6.3) if there is a bounded, open neighborhood V of x_0 and two sets U_k, S_{n-k}, in C^n, $U_k \cap S_{n-k} = \{x_0\}$, of dimensions k and $n - k$, respectively, U_k negatively invariant, S_{n-k} positively invariant with respect to (6.3), such that the orbit of any solution of (6.3) which remains in V for $t \leq 0$ ($t \geq 0$) must have initial value in $U_k(S_{n-k})$ and the orbit of any solution with initial value in

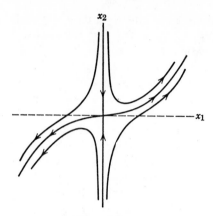

Figure III. 6.1

$U_k(S_{n-k})$ approaches x_0 as $t \to -\infty$ $(+\infty)$. The set U_k is called the *unstable manifold* and S_{n-k} the *stable manifold*.

If $x = 0$ is a saddle point of (4.1) of type (k), we say that the *saddle point property is preserved relative to a given class* $\mathscr{F}$ of functions if, for each f in $\mathscr{F}$, the system (6.3) has an equilibrium point $x_0 = x_0(f)$ which is a saddle point of type (k).

Our main interest centers around the preservation of the saddle point property when the family $\mathscr{F}$ is a family of "small" perturbations and the equilibrium point $x_0 = x_0(f)$ satisfies $x_0(0) = 0$. If f has continuous first derivatives and we measure f by specifying that the norm of f in a bounded neighborhood V of $x = 0$ as

$$|f| = \sup_{x \in V}|f(x)| + \sup_{x \in V}\left|\frac{\partial f(x)}{\partial x}\right|.$$

then there is no loss in generality in assuming that $f(x) = o(|x|)$ as $|x| \to 0$. In fact, there is a $\delta > 0$ such that the matrix $A + \partial f(x)/\partial x$ as a function of x has k eigenvalues with positive real parts, $n - k$ with negative real parts for $|x| < \delta$. From the implicit function theorem, the equation $Ax + f(x) = 0$ has a unique solution x_0 in the region $|x| < \delta$. The transformation $x = x_0 + y$ yields the equation

$$\dot{y} = \left[A + \frac{\partial f(x_0)}{\partial x}\right] y + f(x_0 + y) - f(x_0) - \frac{\partial f(x_0)}{\partial x} y$$

$$\overset{\text{def}}{=} By + g(y),$$

where B has k eigenvalues with positive real parts, $n - k$ with negative real parts and $g(y) = o(|y|)$ as $|y| \to 0$.

On the strength of this remark, we consider the preservation of the saddle point property for equation (6.3) for families of continuous functions f which at least satisfy $f(x) = o(|x|)$ as $|x| \to 0$.

LEMMA 6.1.   If $f: C^n \to C^n$ is continuous, $x = 0$ is a saddle point of type (k) of (4.1), π_+, π_- are the projection operators defined in (6.1), then, for any solution $x(t)$ of (6.3) which exists and is bounded on $[0, \infty)$, there is an x_- in $\pi_- C^n$ such that $x(t)$ satisfies

$$(6.4) \quad x(t) = e^{At}x_- + \int_0^t e^{A(t-s)}\pi_- f(x(s))\, ds + \int_\infty^0 e^{-As}\pi_+ f(x(t+s))\, ds,$$

for $t \geqq 0$. For any solution $x(t)$ of (6.3) which exists and is bounded on $(-\infty, 0]$, there is an x_+ in $\pi_+ C^n$ such that

$$(6.5) \quad x(t) = e^{At}x_+ + \int_0^t e^{A(t-s)}\pi_+ f(x(s))\, ds + \int_{-\infty}^0 e^{-As}\pi_- f(x(t+s))\, ds$$

for $t \leqq 0$. Conversely, any solution of (6.4) bounded on $[0, \infty)$ and any solution of (6.5) bounded on $(-\infty, 0]$ is a solution of (6.3).

PROOF.   Suppose $x(t)$ is a solution of (6.3) which exists for $t \geqq 0$ and $|x(t)| \leqq M$ for $t \geqq 0$. There is a constant L such that $|\pi_+ x| \leqq L|x|$ for all x in C^n and, thus, $|\pi_+ x(t)| \leqq ML$ for all $t \geqq 0$. Since f is continuous there is a constant N such that $|f(x(t))| \leqq N$, $t \geqq 0$. For any σ in $[0, \infty)$, the solution $x(t)$ satisfies

$$\pi_+ x(t) = e^{A(t-\sigma)}\pi_+ x(\sigma) + \int_\sigma^t e^{A(t-s)}\pi_+ f(x(s))\, ds, \qquad t \text{ in } [0, \infty),$$

since $A\pi_+ = \pi_+ A$, $A\pi_- = \pi_- A$.

Since the matrix A satisfies (6.2),

$$\left| e^{A(t-\sigma)}\pi_+ x(\sigma) \right| \leqq K e^{\alpha(t-\sigma)}|\pi_+ x(\sigma)| \leqq KLM e^{\alpha(t-\sigma)},$$

for $t \leqq \sigma$ and, therefore, approaches zero as $\sigma \to \infty$. Also, for $t \leqq \sigma$,

$$\left| \int_\sigma^t e^{A(t-s)}\pi_+ f(x(s))\, ds \right| \leqq K \int_t^\sigma e^{\alpha(t-s)}|\pi_+ f(x(s))|\, ds$$

$$\leqq KLN \int_t^\sigma e^{\alpha(t-s)}\, ds$$

$$= \frac{KLN}{\alpha}[1 - e^{\alpha(t-\sigma)}],$$

$$\leqq \frac{KLN}{\alpha}.$$

Therefore, the integral $\int_{\infty}^{t} e^{A(t-s)}\pi_{+}f(x(s))\, ds$ exists. From the above integral equation for $\pi_{+}x(t)$, this implies

$$\pi_{+}x(t) = \int_{\infty}^{0} e^{-As}\pi_{+}f(x(t+s))\, ds.$$

Since $x(t) = \pi_{+}x(t) + \pi_{-}x(t)$, this relation and the variation of constants formula yield (6.4). Relation (6.5) is proved in a completely analogous manner. The last statement of the lemma is verified by direct computation to complete the proof.

Notice that x_{-}, x_{+} in (6.4), (6.5) are not the initial values of the solutions at $t = 0$, but are determined only after the solution is known.

LEMMA 6.2. Suppose $\alpha > 0$, $\gamma > 0$, K, L, M are nonnegative constants and u is a nonnegative bounded continuous solution of either the inequality

$$(6.6) \quad u(t) \leq Ke^{-\alpha t} + L \int_{0}^{t} e^{-\alpha(t-s)}u(s)\, ds + M \int_{0}^{\infty} e^{-\gamma s}u(t+s)\, ds, \quad t \geq 0,$$

or the inequality

$$(6.7) \quad u(t) \leq Ke^{\alpha t} + L \int_{t}^{0} e^{\alpha(t-s)}u(s)\, ds + M \int_{-\infty}^{0} e^{\gamma s}u(t+s)\, ds, \quad t \leq 0.$$

If

$$(6.8) \qquad \beta \overset{\text{def}}{=} \frac{L}{\alpha} + \frac{M}{\gamma} < 1,$$

then, in either case,

$$(6.9) \qquad u(t) \leq (1-\beta)^{-1}Ke^{-[\alpha-(1-\beta)^{-1}L]|t|}.$$

PROOF. We only need to prove the lemma for u satisfying (6.6) since the transformation $t \to -t$, $s \to -s$ reduces the discussion of (6.7) to (6.6). We first show that $u(t) \to 0$ as $t \to \infty$. If $\delta = \overline{\lim}_{t\to\infty} u(t)$, then u bounded implies δ is finite. If θ satisfies $\beta < \theta < 1$, then $\delta > 0$ implies there is a $t_1 \geq 0$ such that $u(t) \leq \theta^{-1}\delta$ for $t \geq t_1$. From (6.6), for $t \geq t_1$, we have

$$(6.10) \qquad u(t) \leq Ke^{-\alpha t} + Le^{-\alpha t}\int_{0}^{t_1} e^{\alpha s}u(s)\, ds + \left(\frac{L}{\alpha} + \frac{M}{\gamma}\right)\theta^{-1}\delta.$$

Since the lim sup of the right hand side of (6.10) as $t \to \infty$ is $< \delta$, this is a contradiction. Therefore, $\delta = 0$ and $u(t) \to 0$ as $t \to \infty$.

If $v(t) = \sup_{s \geq t} u(s)$, then $u(t) \to 0$ as $t \to \infty$ implies for any t in $[0, \infty)$, there is a $t_1 \geq t$ such that $v(t) = v(s) = u(t_1)$ for $t \leq s \leq t_1$, $v(s) < v(t_1)$ for

$s > t_1$. From (6.6), this implies

$$v(t) = u(t_1) \leqq Ke^{-\alpha t_1} + L \int_0^t e^{-\alpha(t_1-s)}v(s)\,ds$$

$$+ L \int_t^{t_1} e^{-\alpha(t_1-s)}v(s)\,ds + M \int_0^\infty e^{-\gamma s}v(t'+s)\,ds$$

$$\leqq Ke^{-\alpha t_1} + L \int_0^t e^{-\alpha(t_1-s)}v(s)\,ds + \beta v(t),$$

where $\beta = L/\alpha + M/\gamma < 1$. If $z(t) = e^{\alpha t}v(t)$, then $t_1 \geqq t$ implies

$$z(t) \leqq (1-\beta)^{-1}K + (1-\beta)^{-1}L \int_0^t z(s)\,ds.$$

From Gronwall's inequality, we obtain $z(t) \leqq (1-\beta)^{-1}K \exp(1-\beta)^{-1}Lt$ and, thus, the estimate (6.9) in the lemma for $u(t)$.

EXERCISE 6.1.   Suppose a, b, c are nonnegative continuous functions on $[0, \infty)$, u is a nonnegative bounded continuous solution of the inequality

$$u(t) \leqq a(t) + \int_0^t b(t-s)u(s)\,ds + \int_0^\infty c(s)u(t+s)\,ds, \qquad t \geqq 0,$$

and $a(t) \to 0$, $b(t) \to 0$ as $t \to \infty$, $\int_0^\infty b(s)\,ds < \infty$, $\int_0^\infty c(s)\,ds < \infty$. Prove that $u(t) \to 0$ as $t \to \infty$ if

$$\int_0^\infty b(s)\,ds + \int_0^\infty c(s)\,ds < 1.$$

If Γ is any subset in C^n which contains zero, and $C^n = \pi_+C^n \oplus \pi_-C^n$, π_+, π_- projection operators with $\pi_+\pi_- = \pi_-\pi_+ = 0$, we say Γ *is tangent to* π_-C^n at zero if $|\pi_+x|/|\pi_-x| \to 0$ as $x \to 0$ in Γ. Similarly, we say Γ *is tangent to* π_+C^n if $|\pi_-x|/|\pi_+x| \to 0$ as $x \to 0$ in Γ.

THEOREM 6.1.   Suppose η is a continuous, nondecreasing, nonnegative function on $[0, \infty)$ with $\eta(0) = 0$ and let $\mathscr{L}i\mathit{p}(\eta)$ designate the family of continuous functions $f: C^n \to C^n$ such that

(6.11)   (a) $f(0) = 0$,

              (b) $|f(x) - f(y)| \leqq \eta(\sigma)|x - y|$, $|x|, |y| \leqq \sigma$.

If $x = 0$ is a saddle point of type (k) of (4.1), then the saddle point property is preserved relative to the family $\mathscr{L}i\mathit{p}(\eta)$. If, for any f in $\mathscr{L}i\mathit{p}(\eta)$, $U_k = U_k(f)$, $S_{n-k} = S_{n-k}(f)$ are the unstable and stable manifolds of the equilibrium point $x = 0$ of (6.3), then U_k is tangent to π_+C^n at $x = 0$ and S_{n-k} is tangent to π_-C^n at $x = 0$, where π_+C^n and π_-C^n are the unstable and stable manifolds of the saddle point $x = 0$ of (4.1). Furthermore, there are positive constants M, γ such that

(6.12)   (a) $|x(t)| \leq Me^{-\gamma t}|x(0)|$, $t \geq 0$, $x(0)$ in S_{n-k},

         (b) $|x(t)| \leq Me^{\gamma t}|x(0)|$, $t \leq 0$, $x(0)$ in U_k.

It is worthwhile to consider the following schematic representation Fig. 6.2 of Theorem 6.1. The picture for $f = 0$ is global whereas for $f \neq 0$, it is local. In the diagram, we have indicated orbits of (6.3) other than the ones on U_k and S_{n-k}. Actually, we do not prove that the orbits which do not intersect U_k or S_{n-k} behave as shown but only assert that these orbits must leave a certain neighborhood of zero with increasing t.

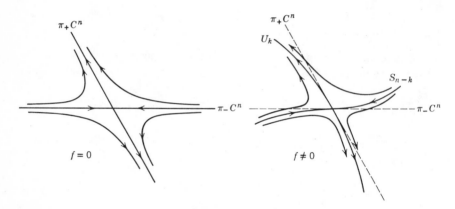

Figure. III.6.2

PROOF OF THEOREM 6.1.   From Lemma 6.1, for any solution x of (6.3) which is bounded on $[0, \infty)$, there is an x_- in $\pi_- C^n$ such that $x(t)$ satisfies (6.4). We first discuss the existence of solutions of (6.4) on $[0, \infty)$ for any x_- in $\pi_- C^n$ sufficiently small. Since π_-, π_+ are projections, there is a constant K_1 such that $|\pi_+ x| \leq K_1|x|$, $|\pi_- x| \leq K_1|x|$ for all x in C^n. Suppose K, α are the constants given in (6.2) and $\eta(\sigma)$, $\sigma \geq 0$, is the function given in (6.11). Choose δ so that

(6.13) $4KK_1\eta(\delta) < \alpha$, $8K^2K_1\eta(\delta) < 3\alpha$.

With this choice of δ and for any x_- in $\pi_- C^n$ with $|x_-| \leq \delta/2K$, define $\mathscr{G}(x_-, \delta)$ as the set of continuous functions $x \colon [0, \infty) \to C^n$ such that $|x| = \sup_{0 \leq t < \infty} |x(t)| \leq \delta$ and $\pi_- x(0) = x_-$. $\mathscr{G}(x_-, \delta)$ is a closed bounded subset of the Banach space of all bounded continuous functions taking $[0, \infty)$ into C^n with the uniform topology. For any x in $\mathscr{G}(x_-, \delta)$, define Tx by

$$(6.14) \quad (Tx)(t) = e^{At}x_- + \int_0^t e^{A(t-s)}\pi_- f(x(s))\,ds + \int_\infty^0 e^{-As}\pi_+ f\left(x(t+s)\right)ds,$$

for $t \geqq 0$. Since x is in $\mathscr{G}(x_-, \delta)$, it is easy to see that Tx is defined and continuous for $t \geqq 0$ with $[\pi_- Tx](0) = x_-$. From (6.2), (6.11), (6.13), we obtain

$$|(Tx)(t)| \leqq Ke^{-\alpha t}|x_-| + \int_0^t Ke^{-\alpha(t-s)}|\pi_- f(x(s))|\,ds + \int_0^\infty Ke^{-\alpha s}|\pi_+ f\left(x(t+s)\right)|\,ds$$

$$\leqq Ke^{-\alpha t}|x_-| + \frac{KK_1}{\alpha}\,\eta(\delta)|x|[2 - e^{-\alpha t}]$$

$$\leqq K|x_-| + \frac{2KK_1}{\alpha}\,\eta(\delta)\delta$$

$$< \frac{\delta}{2} + \frac{\delta}{2} = \delta,$$

for $t \geqq 0$. Thus, $|Tx| < \delta$ and $T\colon \mathscr{G}(x_-, \delta) \to \mathscr{G}(x_-, \delta)$.
Furthermore, the same type of estimates yields

$$|(Tx)(t) - (Ty)(t)| \leqq \frac{2KK_1}{\alpha}\,\eta(\delta)|x - y| \leqq \frac{1}{2}\,|x - y|,$$

for $t \geqq 0$. Thus T is a contraction on $\mathscr{G}(x_-, \delta)$ and there is a unique fixed point $x^*(\cdot, x_-)$ in $\mathscr{G}(x_-, \delta)$ and this fixed point satisfies (6.4).

Using the same estimates as above, one shows that the function $x^*(\cdot, x_-)$ is continuous in x_- and $x^*(\cdot, 0) = 0$. However, more precise estimates of the dependence of $x^*(\cdot, x_-)$ on x_- are needed. If we let $x^* = x^*(\cdot, x_-)$, $\tilde{x}^* = x^*(\cdot, \tilde{x}_-)$, then, from (6.4),

$$|x^*(t) - \tilde{x}^*(t)| \leqq Ke^{-\alpha t}|x_- - \tilde{x}_-| + KK_1\eta(\delta)\int_0^t e^{-\alpha(t-s)}|x^*(s) - \tilde{x}^*(s)|\,ds$$

$$+ KK_1\eta(\delta)\int_0^\infty e^{-\alpha s}|x^*(t+s) - \tilde{x}^*(t+s))|\,ds$$

for $t \geqq 0$. We may now apply Lemma 6.2 to this relation. In Lemma 6.2, let $\gamma = \alpha$, $M = L = KK_1\eta(\delta)$. If $u(t) = |x^*(t) - \tilde{x}^*(t)|$, δ satisfies (6.13) and appropriate identification of constants are made in Lemma 6.2, then

$$(6.15) \quad |x^*(t, x_-) - x^*(t, \tilde{x}_-)| \leqq 2K\left(\exp -\frac{\alpha t}{2}\right)|x_- - \tilde{x}_-|, \qquad t \geqq 0.$$

Since $x^*(\cdot, 0) = 0$, relation (6.15) implies these solutions satisfy a relation of the form (6.12a) and approach zero exponentially at $t \to \infty$.

Let $B_{\delta/2K}$ denote the open ball of radius $\delta/2K$ in C^n with center at the origin. Let S^*_{-k} designate the initial values of all those solutions of (6.3) which remain inside B_δ for $t \geqq 0$ and have $\pi_- x(0)$ in $B_{\delta/2K}$. From the above proof, $S_{n-k}^* = \{x \colon x = x^*(0, x_-),\ x_-$ in $(\pi_- C^n) \cap B_{\delta/2K}\}$. Let $g(x_-) = x^*(0, x_-)$, x_- in $(\pi_- C^n) \cap B_{\delta/2K}$. The function g is a continuous map of $(\pi_- C^n) \cap B_{\delta/2K}$ onto S_{n-k}^* and is given by

$$(6.16) \qquad g(x_-) = x_- + \int_\infty^0 e^{-As} \pi_+ f(x^*(s, x_-))\, ds.$$

From (6.2), (6.11), (6.13), (6.15), we have

$$|g(x_-) - g(\tilde{x}_-)| \geqq |x_- - \tilde{x}_-| - \int_0^\infty K e^{-\alpha s} K_1 \eta(\delta) |x^*(s, x_-) - x^*(s, \tilde{x}_-))|\, ds$$

$$\geqq |x_- - \tilde{x}_-| \left[1 - \frac{4K^2 K_1 \eta(\delta)}{3\alpha} \right]$$

$$\geqq \frac{1}{2} |x_- - \tilde{x}_-|,$$

for all x_-, $\tilde{x}_-$ in $(\pi_- C^n) \cap B_{\delta/2K}$. Therefore g is one-to-one. Since $g^{-1} = \pi_-$ is continuous it follows that g is a homeomorphism. This shows that S_{n-k}^* is homeomorphic to the open unit ball in C^{n-k} and, in particular, has dimension $n - k$. However, S_{n-k}^* may not be positively invariant. If we extend S_{n-k}^* to a set S_{n-k} by adding to it all of the positive orbits of solutions with initial values in S_{n-k}^*, then S_{n-k} is positively invariant and also homeomorphic to the open unit ball in C^{n-k} from the uniqueness of solutions of the equation. The set S_{n-k} coincides with S_{n-k}^* when x in S_{n-k} implies $|\pi_- x| < \delta/2K$.

From (6.14), (6.15) and the fact that $x^*(\cdot, 0) = 0$, we also obtain

$$|\pi_+ x^*(0, x_-)| \leqq K K_1 \int_0^\infty e^{-\alpha s} \eta(|x^*(s, x_-)|) |x^*(s, x_-)|\, ds$$

$$\leqq K K_1 \int_0^\infty e^{-\alpha s} \eta(2K|x_-|) 2K|x_-|\, ds$$

$$= \frac{2K^2 K_1}{\alpha} \eta(2K|x_-|) |x_-|.$$

Consequently $|\pi_+ x^*(0, x_-)| / |x_-| \to 0$ as $|x_-| \to 0$ in S_{n-k} which shows that S_{n-k} is tangent to $\pi_- C^n$ at $x = 0$.

Using relation (6.5), one constructs the set U_k in a completely analogous manner. This completes the proof of the theorem.

In the proof of Theorem 6.1, it was actually shown that the mapping g taking $\pi_- C^n \cap B_{\delta/2K}$ into S_{n-k}^* is Lipschitz continuous [see relations (6.15) and (6.16)]. Since the solutions of (6.3) also depend Lipschitz continuously on the initial data if (6.11) is satisfied, it follows that the *stable manifold*

S_{n-k} and also the *unstable manifold* U_k are *Lipschitz continuous*; that is, $S_{n-k}(U_k)$ is homeomorphic to the unit ball in $C^{n-k}(C^k)$ by a mapping which is Lipschitz continuous. It is also clear from the proof of Theorem 6.1 that the Lipschitz condition of the type specified in (6.11b) was unnecessary. One could have assumed only that

$$|f(x) - f(y)| \leqq \gamma |x - y|,$$

where γ satisfies (6.13) with $\eta(\delta)$ replaced by γ. Of course, the assertion of tangency in the theorem may not hold in this more general situation.

An even weaker version of Theorem 6.1 can be proved for functions which may not be lipschitzian but satisfy

(6.17) $$|f(x)| \leqq \mu |x|$$

for x in some neighborhood V of $x = 0$ and μ satisfies (6.13) with $\eta(\sigma)$ replaced by μ. In fact, let δ be chosen such that $|x| < \delta$ implies x in V and let $\mathscr{G}(x_-, \delta)$ be defined as before as all continuous functions taking $[0, \infty)$ into C^n which are bounded by δ, but use the topology of uniform convergence on compact subsets of $[0, \infty)$. If the mapping T is defined as in (6.14), one can show that $T \colon \mathscr{G}(x_-, \delta) \to \mathscr{G}(x_-, \delta)$ in the same manner as before. For any $\varepsilon > 0$, choose τ so large that

$$|(Tx)(t) - (Ty)(t)| \leqq KK_1 \int_0^t e^{-\alpha(t-s)} |f(x(s)) - f(y(s))|\, ds$$

$$+ KK_1 \int_0^\tau e^{-\alpha s} |f(x(t+s)) - f(y(t+s))|\, ds$$

$$+ \varepsilon/2$$

for any x, y in $\mathscr{G}(x_-, \delta)$. Therefore, given any compact set G in $[0, \infty)$, one can always choose a $\delta > 0$ such that $|x(t) - y(t)| < \delta$ on G implies $|(Tx)(t) - (Ty)(t)| < \varepsilon$ on G. This implies T is continuous on $\mathscr{G}(x_-, \delta)$ in the topology of uniform convergence on compact sets. Since $(Tx)(t)$ is a solution of the equation $\dot{z} = Az + f(x(t))$, and $|Tx| \leqq \delta$ for all x in $\mathscr{G}(x_-, \delta)$, it follows that $|d(Tx)(t)/dt|$ is uniformly bounded and T is a completely continuous map of $\mathscr{G}(x_-, \delta)$ into $\mathscr{G}(x_-, \delta)$. Using the Schauder-Tychonov theorem, one can assert the existence of a fixed point $x^*(\,\cdot\,, x_-)$ of T in $\mathscr{G}(x_-, \delta)$. Furthermore, since $x^*(\,\cdot\,, x_-)$ satisfies (6.4), it follows that

$$|x^*(t, x_-)| \leqq Ke^{-\alpha t}|x_-| + KK_1\mu \int_0^t e^{-\alpha(t-s)} |x^*(s, x_-)|\, ds$$

$$+ KK_1\mu \int_0^\infty e^{-\alpha s} |x^*(t+s, x_-)|\, ds.$$

Again, using Lemma 6.2, one obtains

(6.18) $|x^*(t, x_-)| \leqq 2K\left(\exp - \dfrac{\alpha t}{2}\right)|x_-|, \qquad t \geqq 0.$

Also, exactly as in the proof of Theorem 6.1, one obtains

(6.19) $|\pi_+ x^*(0, x_-)| \leqq \dfrac{2K^2 K_1}{\alpha} \mu |x_-|.$

Thus, all solutions $x(t)$ of (6.3) such that $|x(t)| \leqq \delta$, $t \geqq 0$ and $|\pi_- x(0)| \leqq \delta/2K$ approach zero as $t \to \infty$ exponentially and in fact satisfy (6.19) for an appropriate x_-. If we designate S^* as the set of initial values of such solutions and extend S^* to a set S by adding to it all of the positive orbits of solutions with initial values in S^*, then it is reasonable to call S a stable manifold of (6.3).

If f satisfies the stronger condition

(6.20) $f(x) = o(|x|) \quad \text{as} \quad |x| \to 0,$

then there is an $\eta(\sigma)$ continuous for $\sigma \geqq 0$, $\eta(0) = 0$, such that $|f(x)| \leqq \eta(\sigma)|x|$ for $|x| \leqq \sigma$. The estimate (6.19) can then be improved to

(6.21) $|\pi_+ x^*(0, x_-)| \leqq \dfrac{2K^2 K_1}{\alpha} \eta(2K|x_-|)|x_-|.$

Relation (6.21) shows that S is tangent to $\pi_- C^n$ at $x = 0$. In the same manner, one obtains an unstable manifold U which is tangent to $\pi_+ C^n$ at $x = 0$. One will still have the property $S \cap U = \{0\}$, but cannot assert that S has dimension $n - k$ and U has dimension k.

If f satisfies the condition (6.17), the estimate (6.19) shows that the stable manifold must lie in a region containing $\pi_- C$, the region being bounded by two planes which approach $\pi_- C$ as $\mu \to 0$. The same remark applies to the unstable manifold.

If f satisfies (6.20), then the tangency of the stable and unstable manifolds of (6.3) to the stable and unstable manifolds of (4.1) at $x = 0$ implies the following: For any neighborhood N of $(\pi_- C^n)$ intersected with the ball of radius one with center at the origin, there is a neighborhood V of $x = 0$ such that the stable manifold of (6.3) relative to the neighborhood V lies in N. The same remark applies to the unstable manifold. An important consequence of this remark is the following corollary, the first part of which is a special case of Theorem 2.4.

COROLLARY 6.1. Suppose $f: C^n \to C^n$ is continuous and $f(x) = o(|x|)$ as $|x| \to 0$. If the eigenvalues of A have negative real parts, then the solution $x = 0$ of (6.3) is uniformly asymptotically stable. If one eigenvalue of A has a positive real part, then the solution $x = 0$ is unstable.

The next example due to C. Olech shows that the dimension of the stable and unstable manifolds may increase under perturbations which are continuous and $o(|x|)$ as $|x| \to 0$, but which are not differentiable at $x = 0$. Let $\xi(x)$, $-\infty < x < \infty$, be any function with continuous second derivatives in a neighborhood of $x = 0$, $\xi(0) = \xi'(0) = 0$, $\xi(x) \neq 0$ for $x \neq 0$ and $\xi(x) + x\xi'(x) = o(|x|)$ as $x \to 0$. For any a, let $\psi(a, y)$ be any continuously differentiable function, $0 \leq \psi(a, y) \leq 1$, $-\infty < y < \infty$, $\psi(a, 0) = 1$, $\psi(a, -a) = 0$. Consider the second order equation

(6.22) $\dot{x}_1 = -x_1,$

$\dot{x}_2 = x_2 - \psi(\xi(x_1), x_2 - \xi(x_1))[\xi(x_1) + x_1\xi'(x_1)].$

Under the above hypotheses on ξ, ψ, we see that the perturbation is $o(|x|)$ as $|x| \to 0$, $x = (x_1, x_2)$. Also, $x_2 = 0$, $x_1 = e^{-t}a$, and $x_2 = \xi(x_1)$, $x_1 = e^{-t}a$, where a is arbitrary, are solutions of (6.22). These solutions approach zero as $t \to \infty$, the corresponding orbits belong to the stable manifold of $x = 0$, and obviously these orbits are distinct. The fan near $x = 0$ consisting of the orbits of the solutions whose initial values are between the curve $x_2 = 0$ and $x_2 = \xi(x_1)$ in Fig. 6.3 must also belong to the stable manifold of the solution $x = 0$ of (6.22). In fact, one easily shows that the perturbation being $o(|x|)$ as $|x| \to 0$ implies there is a cone enclosing the positive x_1-axis so that near $x = 0$ the tangent vector to any orbit cannot be perpendicular to the x_1-axis. This immediately yields the result.

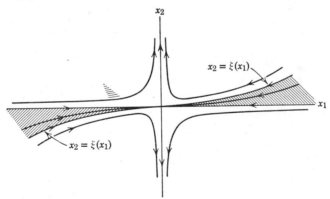

Figure III.6.3

III.7. Linear Periodic Systems

Consider the homogeneous linear periodic system

(7.1) $\dot{x} = A(t)x, \qquad A(t + T) = A(t), \qquad T > 0$

where $A(t)$ is a continuous[1] $n \times n$ real or complex matrix function of t. Our first objective in this section is to give a complete characterization of the general structure of the solutions of (7.1).

LEMMA 7.1.   If C is an $n \times n$ matrix with det $C \neq 0$, then there is a matrix B such that $C = e^B$.

PROOF.   If P is nonsingular and there is a matrix B such that $C = e^B$, then $P^{-1}CP = e^{P^{-1}BP}$. Therefore, we may assume that C is in Jordan canonical form; that is,

$$C = \text{diag}(C_1, \ldots, C_p),$$

$$C_j = \lambda_j I + R_j,$$

$$R_j = \begin{bmatrix} 0 & 1 & 0 & \cdots & 0 \\ 0 & 0 & 1 & \cdots & 0 \\ \cdot & \cdot & \cdot & & \\ 0 & 0 & 0 & \cdots & 1 \\ 0 & 0 & 0 & \cdots & 0 \end{bmatrix}.$$

By hypothesis, each $\lambda_j \neq 0$, $j = 1, 2, \ldots, p$. To prove the lemma, it is sufficient to show that each C_j can be written as $C_j = e^{B_j}$. Therefore, we drop the subscripts and suppose $C = \lambda I + R$ where $\lambda \neq 0$ and R is a matrix of the same type as the R_j above; in particular, $R^k = 0$ for all $k \geq m$, for some integer m. Since $\lambda \neq 0$, $C = \lambda(I + R/\lambda)$. Let

$$B = (\log \lambda)I + S,$$

$$S = -\sum_{j=1}^{m-1} \frac{(-R)^j}{j\lambda^j}.$$

The matrix S is the matrix power series obtained by taking the power series for $\log(1 + t)$ near $t = 0$ and replacing t by R/λ. Since $R^k = 0$ for $k \geq m$, there is no problem of convergence. On the other hand, one can show directly by substitution in the power series for e^B that $C = e^B$. The lemma is proved.

EXERCISE 7.1.   For any real matrix D, det $D \neq 0$, show there is a real matrix B such that $e^B = D^2$. If C is a real matrix in Lemma 7.1 and there is a real matrix B such that $e^B = C$, must there exist a real matrix D such that $C = D^2$?

THEOREM 7.1.   (Floquet).   Every fundamental matrix solution $X(t)$ of (7.1) has the form

(7.2) $X(t) = P(t)e^{Bt}$

where $P(t)$, B are $n \times n$ matrices, $P(t + T) = P(t)$ for all t, and B is a constant.

---

[1] This assumption is for simplicity only. The theory is valid for $A(t)$ which are periodic and Lebesgue integrable if (7.1) is required to be satisfied except for a set of Lebesgue measure zero. No changes in proofs are required.

PROOF. Suppose $X(t)$ is a fundamental matrix solution of (7.1). Then $X(t + T)$ is also a fundamental matrix solution since $A(t)$ is periodic of period T. Therefore, there is a nonsingular matrix C such that

$$X(t + T) = X(t)C.$$

From Lemma 7.1, there is a matrix B such that $C = e^{BT}$. For this matrix B, let $P(t) = X(t)e^{-Bt}$. Then

$$P(t + T) = X(t + T)e^{-B(t+T)} = X(t)e^{BT}e^{-B(t+T)} = P(t),$$

and the theorem is proved.

COROLLARY 7.1. There exists a nonsingular periodic transformation of variables which transforms (7.1) into an equation with constant coefficients.

PROOF. Suppose $P(t)$, B are defined by (7.2) and let $x = P(t)y$ in (7.1). The equation for y is

$$\dot{y} = P^{-1}(AP - \dot{P})y.$$

Since $P = Xe^{-Bt}$, it follows that $\dot{P} = AP - PB$ and this proves the result.

EXERCISE 7.2. Prove that B in the representation (7.2) can always be chosen to be a real matrix if $A(t)$ in (7.1) is real and it is only required that $P(t + 2T) = P(t)$.

Theorem 7.1 states that any solution of (7.1) is a linear combination of functions of the form $e^{\lambda t}p(t)$ where $p(t)$ is a polynomial in t with coefficients which are periodic in t of the same period as the period of the coefficients in the differential equation.

A *monodromy matrix* of (7.1) is a nonsingular matrix C associated with a fundamental matrix solution $X(t)$ of (7.1) through the relation $X(t + T) = X(t)C$. The eigenvalues ρ of a monodromy matrix are called the *characteristic multipliers of* (7.1) and any λ such that $\rho = e^{\lambda T}$ is called a *characteristic exponent of* (7.1). Notice that the characteristic exponents are not uniquely defined, but the multipliers are. The real parts of the characteristic exponents are uniquely defined and we can always choose the exponents λ as the eigenvalues of B, where B is any matrix so that $C = e^{BT}$. The characteristic multipliers do not depend upon the particular monodromy matrix chosen; that is, the particular fundamental solution used to define the monodromy matrix. In fact, if $X(t)$ is a fundamental matrix solution, $X(t + T) = X(t)C$ and $Y(t)$ is any other fundamental matrix solution, then there is a nonsingular matrix D such that $Y(t) = X(t)D$. Therefore, $Y(t + T) = X(t + T)D = X(t)CD = Y(t)D^{-1}CD$ and the monodromy matrix for $Y(t)$ is $D^{-1}CD$. On the other hand, matrices which are similar have the same eigenvalues. We shall usually use the term monodromy matrix for $X(T)$ where $X(t)$, $X(0) = I$ is a fundamental matrix solution of (7.1).

120 ORDINARY DIFFERENTIAL EQUATIONS

LEMMA 7.2. A complex number λ is a characteristic exponent of (7.1) if and only if there is a nontrivial solution of (7.1) of the form $e^{\lambda t}p(t)$ where $p(t+T) = p(t)$. In particular, there is a periodic solution of (7.1) of period T (or $2T$ but not T) if and only if there is a multiplier $= +1$ (or -1)

PROOF. If $e^{\lambda t}p(t)$, $p(t+T) = p(t) \neq 0$, satisfies (7.1), Theorem 7.1 implies there is an $x_0 \neq 0$ such that $e^{\lambda t}p(t) = P(t)e^{Bt}x_0$. Thus, $P(t)e^{Bt}[e^{BT} - e^{\lambda T}I]x_0 = 0$ and, thus, $\det(e^{BT} - e^{\lambda T}I) = 0$. Conversely, if there is a λ such that $\det(e^{BT} - e^{\lambda T}I) = 0$, then choose $x_0 \neq 0$ such that $(e^{BT} - e^{\lambda T}I)x_0 = 0$. One can choose the representation (7.2) so that λ is actually an eigenvalue of B. Then $e^{Bt}x_0 = e^{\lambda t}x_0$ for all t and $P(t)e^{Bt}x_0 = P(t)x_0 e^{\lambda t}$ is the desired solution. The last assertion is obvious.

LEMMA 7.3. If $\rho_j = e^{\lambda_j T}$, $j = 1, 2, \ldots, n$, are the characteristic multipliers of (7.1), then

(7.3)
$$\prod_{j=1}^{n} \rho_j = \exp\left(\int_0^T tr\, A(s)\, ds\right),$$

$$\sum_{j=1}^{n} \lambda_j \equiv \frac{1}{T} \int_0^T tr\, A(s)\, ds \left(\mod \frac{2\pi i}{T}\right).$$

PROOF. Suppose C is the monodromy matrix for the matrix solution $X(t)$, $X(0) = I$, of (7.1). Then Lemma 1.5 implies

$$\det C = \det X(T) = \exp\left(\int_0^T tr\, A(s)\, ds\right).$$

The statements of the lemma now follow immediately from the definitions of characteristic multipliers and exponents.

THEOREM 7.2. (i) A necessary and sufficient condition that the system (7.1) be uniformly stable is that the characteristic multipliers of (7.1) have modulii ≤ 1 (the characteristic exponents have real parts ≤ 0) and the ones with modulii $= 1$ (the characteristic exponents with real parts $= 0$) have simple elementary divisors.

(ii) A necessary and sufficient condition that the system (7.1) be uniformly asymptotically stable is that all characteristic multipliers of (7.1) have modulii < 1 (all characteristic exponents have real parts < 0). If this is the case and $X(t)$ is a matrix solution of (7.1), there exist $K > 0$, $\alpha > 0$ such that

$$|X(t)X^{-1}(s)| \leq Ke^{-\alpha(t-s)}, \qquad t \geq s.$$

PROOF. The proof is essentially the same as the proof of Theorem 4.2 if one uses Corollary 7.1.

At first glance, it might appear that linear periodic equations share the same simplicity as linear equations with constant coefficients. However, there is a very important difference—the characteristic exponents are defined only after the solutions of (7.1) are known and there is no obvious relation between the characteristic exponents and the matrix $A(t)$. The following example illustrates that the eigenvalues of the matrix $A(t)$ cannot be used to determine the asymptotic behavior of the solutions.

Example 7.1.   (Markus and Yamabe [1]).   If

$$(7.4) \qquad A(t) = \begin{bmatrix} -1 + \dfrac{3}{2}\cos^2 t & 1 - \dfrac{3}{2}\cos t \sin t \\[2ex] -1 - \dfrac{3}{2}\sin t \cos t & -1 + \dfrac{3}{2}\sin^2 t \end{bmatrix},$$

then the eigenvalues $\lambda_1(t)$, $\lambda_2(t)$ of $A(t)$ are $\lambda_1(t) = [-1 + i\sqrt{7}]/4$, $\lambda_2(t) = \bar{\lambda}_1(t)$ and, in particular, the real parts of the eigenvalues have negative real parts. On the other hand, one can verify directly that the vector

$$(-\cos t, \sin t)\exp\left(\dfrac{t}{2}\right)$$

is a solution of (7.1) with $A(t)$ given in (7.4) and this solution is unbounded as $t \to \infty$. One of the characteristic multipliers is e^π. The other multiplier is $e^{-2\pi}$ since (7.3) implies that the product of the multipliers is $e^{-\pi}$.

The problem of determining the characteristic multipliers or exponents of linear periodic systems is an extremely difficult one. Except for scalar second order equations and, more generally, Hamiltonian and canonical systems, very little is known at all. Even the equations of the form $\dot{x} = Ax + \varepsilon\Phi(t)x$ where ε is a small parameter, A is a constant matrix and x is a vector of dimension >2 are extremely difficult and exhibit very striking behavior. This shall be illustrated in a later chapter by means of examples.

III.8. Hill's Equation

In this section, we give a rather detailed discussion of the stability properties of the Hill equation,

$$(8.1) \qquad \ddot{y} + (a + \phi(t))y = 0, \qquad \phi(t + \pi) = \phi(t),$$

where a is constant and ϕ is assumed real and continuous. Actually, there is no change in the theory if ϕ is integrable and bounded, but in such a situation, we have to say the equation is satisfied almost everywhere.

The ultimate goal is to characterize the values of the parameter a for which there is stability of the equation. The previous section implies this is equivalent to determining the qualitative structure of the characteristic multipliers of (8.1) as a function of a.

Equation (8.1) is equivalent to the system

$$(8.2) \qquad \dot{x} = [C(a) + A(t)]x, \qquad A(t) = \begin{bmatrix} 0 & 0 \\ -\phi(t) & 0 \end{bmatrix}, \qquad x = \begin{bmatrix} y \\ \dot{y} \end{bmatrix},$$

$$C(a) = \begin{bmatrix} 0 & 1 \\ -a & 0 \end{bmatrix}.$$

Suppose

$$(8.3) \qquad X(t) \overset{\text{def}}{=} \begin{bmatrix} y_1(t) & y_2(t) \\ \dot{y}_1(t) & \dot{y}_2(t) \end{bmatrix}, \qquad X(0) = I,$$

is the principal matrix solution of (8.2) at $t = 0$. The characteristic multipliers of (8.2) are the eigenvalues of the matrix $X(\pi)$; that is, the roots of the equation

$$\det(X(\pi) - \rho I) = 0.$$

Since $tr \, [C(a) + A(t)] = 0$, (7.3) implies that the characteristic multipliers are the roots of the equation

$$(8.4) \qquad \rho^2 - 2B(a)\rho + 1 = 0,$$

$$2B(a) = tr \, X(\pi) = y_1(\pi) + \dot{y}_2(\pi)$$

where y_1, y_2 are the solutions of (8.1) defined above. From Chapter I, the function $B(a)$ is entire in the parameter a.

In view of Lemma 7.2 and the fact that the multipliers ρ_1, ρ_2 of (8.4) satisfy $\rho_1 \rho_2 = 1$, the equation (8.1) can be stable only if $|\rho_1| = |\rho_2| = 1$. The next lemma shows this can never be the case if a is complex.

LEMMA 8.1. If a is complex, $Im \, a \neq 0$, equation (8.1) is unstable and no characteristic multiplier of (8.4) has modulus 1.

PROOF. As remarked above, equation (8.1) can be stable only if both characteristic multipliers have modulii one. Therefore, the lemma is proved if we show no characteristic multiplier has modulus one if a is complex. If a characteristic multiplier has modulus one, then Lemma 7.2 implies there must be a solution of (8.1) of the form $e^{i\lambda t}p(t)$ where λ is real and $p(t + \pi) = p(t) \not\equiv 0$. If $e^{i\lambda t}p(t) = u + iv$, $a = \alpha + i\beta$, u, v, α, β, real, then

$$\ddot{u} + [\alpha + \phi(t)]u = \beta v,$$

$$\ddot{v} + [\alpha + \phi(t)]v = -\beta u.$$

This implies

$$\ddot{u}v - \ddot{v}u = \beta(u^2 + v^2)$$

and, thus, upon integrating,

$$\beta \int_0^t |p(s)|^2 \, ds \stackrel{\text{def}}{=} \beta \int_0^t [u^2(s) + v^2(s)] \, ds = \dot{u}(t)v(t) - \dot{v}(t)u(t) + c,$$

where c is a constant. Since the right hand side is bounded for all t and p is periodic of period π this gives a contradiction unless either $\beta = 0$ or $p = 0$. This proves the lemma.

LEMMA 8.2. The equation $B^2(a) = 1$ can have only real solutions. $B(a) = 1$ (or -1) is equivalent to the statement that there is a periodic solution of (8.1) of period π (or 2π). If a is real, then $B^2(a) < 1$ implies all solutions of (8.1) are bounded and quasiperiodic on $(-\infty, \infty)$. If a is real and $B^2(a) > 1$, there is an unbounded solution of (8.1).

PROOF. The roots ρ_1, ρ_2 of (8.4) are $\rho_1 = B(a) + \sqrt{B^2(a) - 1}$, $\rho_2 = B(a) - \sqrt{B^2(a) - 1}$. If $B^2(a) = 1$, then $\rho_1 = \rho_2 = B(a) = \pm 1$ and Lemma 7.2 yields the statement concerning periodic solutions. This implies $B^2(a) = 1$ can have only real solutions from Lemma 8.1. If a is real, then $B(a)$ is real and $B^2(a) < 1$ is equivalent to $\rho_1 = \bar{\rho}_2$, $\rho_1 \neq \rho_2$, $|\rho_1| = 1$. Lemma 7.2 implies the existence of two linearly independent solutions which are quasiperiodic. Thus, every solution is quasiperiodic. If $B^2(a) > 1$, then one characteristic multiplier is >1 and one is <1. Theorem 7.2 completes the proof of the lemma.

Lemmas 8.1 and 8.2 imply that system (8.1) can never be asymptotically stable. In the remainder of the discussion, the parameter a is taken to be real. An interval I will be called an *a-interval of stability (instability)* of *(8.1)* if for each a in I, equation (8.1) is stable (unstable). Lemma 8.2 implies that the transition from an a-interval of stability to an a-interval of instability can only occur at those values of a for which $B^2(a) = 1$. Therefore, the basic problem is to find those values of a for which $B^2(a) = 1$ and discuss the behavior of the function $B(a)$ in a neighborhood of such values. Theorem 8.1 below gives a qualitative description of the manner in which the a-stability and a-instability intervals of (8.1) are situated on the real line. The following lemmas lead to a proof of this theorem.

LEMMA 8.3. If $a + \phi(t) \leq 0$ for all t in $[0, \pi]$, then there is an unbounded solution of (8.1). If, in addition, $a + \phi(t)$ is not identically zero, then $B(a) > 1$.

PROOF. As before, let $y_1(t)$ be the solution of (8.1) with $y_1(0) = 1$, $\dot{y}_1(0) = 0$, and let $\psi(t) = -(a + \phi(t)) \geq 0$. From (8.1),

(8.5) $$(\dot{y}_1(t))^2 = 2 \int_0^t \psi(s) y_1(s) \dot{y}_1(s) \, ds$$

for all $t \geq 0$. Since $y_1(0) = 1$, $\ddot{y}_1(0) = \psi(0) y_1(0) \geq 0$. Thus, $\dot{y}_1(t) \geq 0$ on an interval $0 \leq t \leq \eta$. If $\dot{y}_1(t) = 0$ for all $t \geq 0$, then $y_1(t) = 1$ is a solution of

(8.1) which implies $a + \phi(t) \equiv 0$ for all t and conversely. In this case, there are unbounded solutions since $y(t) = t$ is a solution. Suppose $a + \phi(t)$ is not identically zero and let η be such that $\dot{y}_1(t) = 0$ for $0 \leq t \leq \eta$ and, for any $\tau > 0$, there is a t in $(\eta, \eta + \tau)$ such that $\dot{y}_1(t) \neq 0$. Such an η always exists. One can choose τ so small that $y_1(s) > 0$ for $0 \leq s \leq \eta + \tau$. Since $\psi(s) \geq 0$, it follows from (8.5) that $\dot{y}_1(t) > 0$ on $(\eta, \eta + \tau)$. The right hand side of (8.5) is a nondecreasing function of t and, therefore, $\dot{y}_1(t) > 0$ for all $t > \eta$ and $\dot{y}_1(t)$ is monotone increasing for $t \geq \eta$. Finally

$$y_1(t) = 1 + \int_0^t \dot{y}_1(s) \, ds \geq 1 + \int_{\eta+\tau}^t \dot{y}_1(s) \, ds \geq 1 + \dot{y}_1(\eta + \tau)(t - \eta - \tau)$$

for $t \geq \eta + \tau$, $\tau > 0$. Since $\dot{y}_1(\eta + \tau) > 0$, this shows $y_1(t)$ is unbounded and proves the first part of the lemma.

To prove the second part of the lemma, we first recall the fact that we have just proved that $\dot{y}_1(t) \geq 0$ for all t and, also, $y_1(\pi) > 1$. Now, consider the solution $y_2(t)$ of (8.1) with $y_2(0) = 0$, $\dot{y}_2(0) = 1$. The function $y_2(t)$ will satisfy

$$(\dot{y}_2(t))^2 = 1 + 2 \int_0^t \psi(s) y_2(s) \dot{y}_2(s) \, ds.$$

Since $\psi(s) \geq 0$ and is not identically zero, it follows that $\dot{y}_2(\pi) > 1$. Therefore, $2B(a) = y_1(\pi) + \dot{y}_2(\pi) > 2$ and the lemma is proved.

Lemma 8.3 shows that $a + \phi(t)$ must take on some positive values if the solutions of (8.1) are to remain bounded. Since ϕ is bounded, there is an a^* such that $a^* + \phi(t) < 0$. Thus $B(a) > 1$ and the equation (8.1) is unstable for $-\infty < a < a^*$. We show below that this a-instability interval is bounded above.

LEMMA 8.4.   The functions $B(a) - 1$, $B(a) + 1$ have an infinite number of real zeros $\{a_0 < a_1 < a_2 \cdots\}$, $\{a_1^* < a_2^* < \cdots\}$, respectively, and a_k, a_k^* approach $+\infty$ as $k \to \infty$.

PROOF.   We actually show that $B(a)$ is an entire function of order $1/2$, that is, $B(a)$ is an entire function and, for any $\varepsilon > 0$, $|B(a)| \exp[|a|^{\varepsilon+1/2}] \to 0$ as $|a| \to \infty$ and, for $\varepsilon < 0$, this function is unbounded. Since any entire function of fractional order must have an infinite number of zeros, it follows that the functions $B(a) - 1$, $B(a) + 1$ must have an infinite number of zeros. Lemma 8.2 implies these zeros are real. The only possible accumulation point of the zeros is $\pm\infty$, but the remark after Lemma 8.3 implies the accumulation point must be $+\infty$.

If $X(t)$ is defined as in (8.3), then

$$X(t) = I + \int_0^t [C(a) + A(s)] X(s) \, ds.$$

Denote the right hand side of this integral equation by $(TX)(t)$ and define the sequence $\{X^{(k)}\}$ of functions by $X^{(0)} = I$, $X^{(k+1)} = TX^{(k)}$, $k = 0, 1, \ldots$. Each function $X^{(k)}$ is an entire function of a. Suppose a belongs to a compact set V and t is in a compact set U. Exactly as in the proof of the power series representation of e^{At} (see Section 4), one shows the sequence $X^{(k)}(t) = X^{(k)}(t, a)$ converges uniformly to $X(t) = X(t, a)$ for t in U, a in V. Therefore, $X(t, a)$ is an entire function of a and $B(a)$ is an entire function of a.

To show the order relation, let $\omega^2 = a$ and consider the variation of constants formula for (8.2) treating $A(t)x$ as the forcing function. If y_1, y_2 are defined as above, then

$$(8.6) \qquad y_1(t, a) = \cos \omega t - \int_0^t \omega^{-1} \sin \omega(t - s)\phi(s)y_1(s, a)\,ds,$$

$$y_2(t, a) = \omega^{-1} \sin \omega t - \int_0^t \omega^{-1} \sin \omega(t-s)\phi(s)y_2(s, a)\,ds.$$

Since $|\omega^{-1} \sin \omega t| \leqq e^{|\omega|t}$, $|\cos \omega t| \leqq e^{|\omega|t}$, $t \geqq 0$, if $|\omega| \geqq 1$, it follows that

$$|y_j(t, a)| \leqq e^{|\omega|t} + K \int_0^t e^{|\omega|(t-s)}|y_j(s, a)|\,ds, \qquad j = 1, 2,$$

for $0 \leqq t \leqq \pi$, where K is a bound on ϕ. If $z(t) = e^{-|\omega|t}|y_j(t, a)|$, then

$$z(t) \leqq 1 + K \int_0^t z(s)\,ds.$$

Gronwall's inequality implies $z(t) \leqq e^{Kt} \leqq e^{K\pi}$, $0 \leqq t \leqq \pi$, and, therefore, $|y_j(t, a)| \leqq e^{K\pi}e^{|\omega|t}$, $0 \leqq t \leqq \pi$, $j = 1, 2$. Since $\dot{y}_2(t, a)$ satisfies

$$\dot{y}_2(t, a) = \cos \omega t - \int_0^t \cos \omega(t - s)\phi(s)y_2(s, a)\,ds,$$

we have the existence of a constant L such that $|\dot{y}_2(t, a)| \leqq Le^{|\omega|t}$, $0 \leqq t \leqq \pi$. Therefore, the order of $B(a)$ is $\leqq 1/2$.

To prove the order is $\geqq 1/2$, it is no loss in generality to assume that $\phi(t) \leqq -1$ for all t. In fact, we can always replace ϕ by $\phi - M - 1$, a by $a + M + 1$, where M is a bound on $|\phi(t)|$. Also, choose $a < 0$ so that $\sqrt{a} = i\mu$ with $\mu > 0$. For $a < 0$, $\phi(t) \leqq -1$, it was shown in the proof of Lemma 8.3 that $y_1(t, a) > 1$, $\dot{y}_2(t, a) > 1$ for $t > 0$. This fact, $-\phi(t) \geqq 1$ and (8.6) imply that

$$\dot{y}_1(t, a) \geqq \int_0^t \cosh \mu(t - s)y_1(s, a)\,ds$$

$$\geqq \int_0^t \cosh \mu(t - s)\,ds$$

$$= \frac{\sinh \mu t}{\mu}$$

Thus, $y_1(t, a) \geqq (e^{\mu t} - 1)/2\mu^2 + 1$ for all t in $[0, \pi]$. Since $\dot{y}_2(\pi, a) \geqq 0$, it follows that $B(a)$ is of order at least $1/2$. This completes the proof of the lemma.

LEMMA 8.5. If b is a root of $B(a) = 1$ such that $B'(b) \overset{\text{def}}{=} dB(b)/db \leqq 0$ then $B'(a) < 0$ for $b < a < c^*$ provided $B(a) > -1$ on this interval. If b^* is a root of $B(a) = -1$ such that $B'(b^*) \geqq 0$, then $B'(a) > 0$ for $b^* < a < c$ provided $B(a) < 1$ on this interval.

PROOF. Let $X(t, a) = X(t)$ be defined as in (8.3). From Lemma 1.5, $\det X(t) = 1$ for all t, and, thus,

$$X^{-1}(t) = \begin{bmatrix} \dot{y}_2(t) & -y_2(t) \\ -\dot{y}_1(t) & y_1(t) \end{bmatrix}.$$

Also, from Theorem I.3.3 and the variation of constants formula for a linear system,

$$\frac{\partial X(t, a)}{\partial a} = X(t, a) \int_0^t X^{-1}(s, a) \frac{\partial C(a)}{\partial a} X(s, a)\, ds,$$

where $C(a)$ is the matrix in (8.2). Thus,

$$\frac{\partial C(a)}{\partial a} = \begin{bmatrix} 0 & 0 \\ -1 & 0 \end{bmatrix}.$$

In the same way,

$$\frac{\partial^2 X(t, a)}{\partial a^2} = 2X(t, a) \int_0^t X^{-1}(t, a) \frac{\partial C(a)}{\partial a} \frac{\partial X(s, a)}{\partial a}\, ds.$$

From the definition of $B(a)$, these relations imply

$$(8.7) \qquad 2B'(a) = (\alpha - \dot{\beta}) \int_0^\pi y_1 y_2\, ds - \beta \int_0^\pi y_1^2\, ds + \dot{\alpha} \int_0^\pi y_2^2\, ds,$$

$$B''(a) = \alpha \int_0^\pi y_2 \frac{\partial y_1}{\partial a}\, ds - \beta \int_0^\pi y_1 \frac{\partial y_1}{\partial a}\, ds + \dot{\alpha} \int_0^\pi y_2 \frac{\partial y_2}{\partial a}\, ds - \dot{\beta} \int_0^\pi y_1 \frac{\partial y_2}{\partial a}\, ds,$$

where for simplicity in notation $\alpha = \alpha(a) = y_1(\pi, a)$, $\beta = \beta(a) = y_2(\pi, a)$, $\dot{\alpha} = \dot{\alpha}(a) = \dot{y}_1(\pi, a)$, $\dot{\beta} = \dot{\beta}(a) = \dot{y}_2(\pi, a)$. Using the fact that $\det X(t) = 1$ for all t, we have

$$(8.8) \qquad 4(B^2 - 1) = (\alpha + \dot{\beta})^2 - 4(\alpha\dot{\beta} - \beta\dot{\alpha}) = (\alpha - \dot{\beta})^2 + 4\dot{\alpha}\beta.$$

Suppose b is such that $B(b) = 1$ and $B'(b) \leqq 0$. We wish to prove there is a $\delta > 0$ such that $B'(a) < 0$ for $b < a < b + \delta$. If $B'(b) < 0$, it is clear that such a δ exists. Suppose $B'(b) = 0, B(b) = 1$. For this value of b, Relation (8.8) implies $4\dot{\alpha}\beta = -(\alpha - \dot{\beta})^2$. From (8.7), we have

$$2\dot{\alpha}B'(b) = \int_0^\pi \left[\dot{\alpha}y_2 + \frac{1}{2}(\alpha - \dot{\beta})y_1\right]^2 = 0$$

and thus, $\dot{\alpha}y_2(s) + (\alpha - \dot{\beta})y_1(s)/2 = 0$ for $0 \le s \le \pi$. Since y_1, y_2 are linearly independent, we have $\dot{\alpha} = 0$, $\alpha = \dot{\beta}$. Since $\overline{\overline{B}}'(\overline{\overline{b}}) = 0$, relation (8.7) implies $\beta = 0$. The fact that $\alpha + \dot{\beta} = 2B(b) = 2$ implies $\alpha = \dot{\beta} = 1$. Also, direct evaluation in (8.7) and an integration by parts yields

$$B''(b) = \left(\int_0^\pi y_1 y_2 \, ds\right)^2 - \int_0^\pi y_1^2 \, ds \int_0^\pi y_2^2 \, ds.$$

The Schwarz inequality implies $B''(b) < 0$ since the functions y_1, y_2 are linearly independent. Therefore, there must be a δ such that $B'(a) < 0$ for $b < a < b + \delta$.

Suppose now there is a c^* such that $B'(a) < 0$ for $b < a < c^*$ and $B'(c^*) = 0$ with $B(c^*) > -1$. Then $B^2(c^*) - 1 < 0$ and (8.8) implies $\dot{\alpha}(c^*)\beta(c^*) < 0$ and, in particular, $\dot{\alpha}(c^*) \ne 0$. One easily shows from (8.7) and (8.8) that

$$2\dot{\alpha}B'(a) = \int_0^\pi \left(\dot{\alpha}y_2 + \frac{1}{2}(\alpha - \dot{\beta})y_1\right)^2 ds - \frac{1}{4}(B^2 - 1)\int_0^\pi y_1^2 \, ds$$

for any a such that $\dot{\alpha}(a) \ne 0$. Since $B^2(c^*) < 1$, $\dot{\alpha}(c^*) \ne 0$, it is clear that $B'(c^*) \ne 0$, which is a contradiction. This proves the lemma for the case $B(b) = 1$. The case $B(b) = -1$ is treated in essentially the same manner to complete the proof of the lemma.

In the middle part of the proof of this lemma, the following relationship was proved.

LEMMA 8.6 If for a particular b, $B^2(b) = 1$, $B'(b) = 0$, then $B''(b) < 0$ if $B(b) = 1$, and $B''(b) > 0$ if $B(b) = -1$. In particular, a root of $B^2(b) = 1$ can be at most double. A necessary and sufficient condition for a double root at b is

$$y_1(\pi, b) = \dot{y}_2(\pi, b) = 1 \text{ (or } -1), \qquad \dot{y}_1(\pi, b) = y_2(\pi, b) = 0.$$

LEMMA 8.7. If a_0 is the smallest root of the equation $B^2(a) = 1$, then a_0 is simple and $B'(a_0) < 0$.

PROOF. From Lemma 8.3, $B(a) > 1$ for $a < a_0$. If a_0 were a double root of $B(a) = 1$, then Lemma 8.6 would imply it is a maximum, which is impossible. This proves the lemma.

By combining the information in the above lemmas we obtain

THEOREM 8.1. There exist two sequences $\{a_0 < a_1 \leq a_2 \leq \cdots\}$, $\{a_1^* \leq a_2^* \leq a_3^* \leq \cdots\}$ of real numbers, $a_k, a_k^* \to \infty$ as $k \to \infty$,

$$a_0 < a_1^* \leq a_2^* < a_1 \leq a_2 < a_3^* \leq a_4^* < a_3 \leq a_4 < \cdots,$$

such that equation (8.1) has a periodic solution of least period π (or 2π) if and only if $a = a_k$ for some $k = 0,1,2,\ldots$ (or a_k^* for some $k = 1,2,\ldots$). The equation (8.1) is stable in the intervals

$$(a_0, a_1^*), \qquad (a_2^*, a_1), \qquad (a_2, a_3^*), \qquad (a_4^*, a_3), \ldots.$$

and unstable in the intervals

$$(-\infty, a_0], \qquad (a_1^*, a_2^*), \qquad (a_1, a_2), \qquad (a_3^*, a_4^*), \qquad (a_3, a_4), \ldots.$$

The equation (8.1) is stable at a_{2k+1} or a_{2k+2} (or a_{2k+1}^* or a_{2k+2}) if and only if $a_{2k+1} = a_{2k+2}$ (or $a_{2k+1}^* = a_{2k+2}^*$), $k \geq 0$. Equation (8.1) is always unstable if a is complex.

PROOF. We remark first of all that (8.1) is unstable if a is complex (Lemma 8.1). Lemma 8.4 implies the existence of the two infinite sequences $\{a_k\}$, $\{a_k^*\}$. Lemma 8.3 implies $a_0 \neq -\infty$. Lemma 8.3 and 8.7 imply that the first zero of $B^2(a) = 1$ is a_0, it is simple and $(-\infty, a_0]$ is an interval of instability. Lemma 8.5 implies $a_0 < a_1^*$ and Lemmas 8.5 and 8.2 imply (a_0, a_1^*) is an interval of stability. If the equation (8.1) is stable at a_1^*, then $B(a) = -1$ would have a double root at a_1^*. Lemma 8.6 would imply $B(a)$ has a minimum at a_1^* and then Lemma 8.5 implies $a_1^* = a_2^*$. If $a_1^* < a_2^*$ then $B(a) < -1$ for $a_1^* < a < a_2^*$ and Lemma 8.2 implies (a_1^*, a_2^*) is an interval of instability. Lemma 8.5 implies $a_2^* < a_1$ and the argument proceeds inductively to yield the theorem.

The accompanying Fig. 8.1 is a possible graph of the function $B(a)$.

The previous analysis of the general equation (8.1) has shown that it is extremely difficult to decide whether or not the solutions of (8.1) are bounded for a given a and ϕ. On the other hand, the general theory has pinpointed the computations that are necessary to decide this question; namely, the determination of those special values of a for which there exist solutions of period π or 2π and the number of such linearly independent solutions for these values.

One very important special case of (8.1) for which we can give some explicit conditions on the coefficients for which there is stability is the Mathieu equation

(8.9)
$$\frac{d^2y}{d\tau^2} + (\sigma^2 + 2q \cos \omega\tau)y = 0,$$

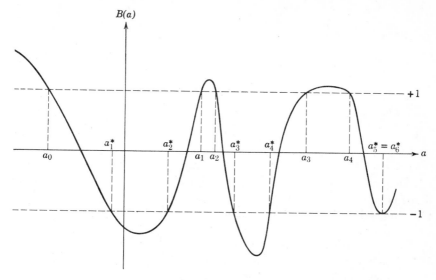

Figure III.8.1

where $\sigma \geqq 0$, q, ω are real constants. If we let $\omega\tau = 2t$, this equation is equivalent to

$$(8.10) \qquad \ddot{y} + \left(\left(\frac{2\sigma}{\omega}\right)^2 + \frac{8q}{\omega^2}\cos 2t\right)y = 0,$$

which is a special case of (8.1) with $a = (2\sigma/\omega)^2$, $\phi = (8q/\omega^2)\cos 2t$. Let us investigate this equation for q near 0. The equation for the characteristic multipliers is

$$\rho^2 - 2B\rho + 1 = 0,$$

where $B = B(\sigma, q, \omega)$ is a continuous function of σ, q, ω^{-1}. For $q = 0$, a principal matrix solution of (8.10) is

$$X = \begin{bmatrix} \cos\dfrac{2\sigma}{\omega}t & \dfrac{\omega}{2\sigma}\sin\dfrac{2\sigma}{\omega}t \\[2ex] -\dfrac{2\sigma}{\omega}\sin\dfrac{2\sigma}{\omega}t & \cos\dfrac{2\sigma}{\omega}t \end{bmatrix}$$

Therefore, $B(\sigma, 0, \omega) = \cos 2\pi\sigma/\omega$ and $B^2(\sigma, 0, \omega) < 1$ if $2\sigma \neq k\omega$ [or $a \neq k^2$], $k = 0, 1, 2, \dots$. Thus, for any σ and ω for which $2\sigma \neq k\omega$, there is a $q = q(\sigma, \omega)$ such that $B^2(\sigma, q, \omega) < 1$ and the equation (8.9) is stable (Lemma 8.2). These stability regions are shown in the (σ^2, q)-plane in Fig. 8.2.

One can give a very geometric proof of this stability result in the

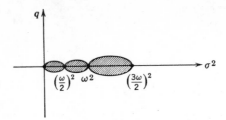

Figure III.8.2

following way. If ρ_1, ρ_2 are characteristic multipliers of (8.9), then ρ_1^{-1}, ρ_2^{-1} are also characteristic multipliers since $\rho_1\rho_2 = 1$. Also, since σ, q, ω are real $\bar\rho_1$, $\bar\rho_2$ are multipliers. Therefore, if $\rho_1 \neq \rho_2$, $|\rho_1| = |\rho_2| = 1$ for $q = 0$, then $|\rho_1| = |\rho_2| = 1$, $\rho_1 \neq \rho_2$ for q sufficiently small since the multipliers are continuous functions of q.

To determine whether or not (8.9) is stable or unstable for $\sigma^2 = (k\omega/2)^2$, k an integer, is extremely difficult. Analytical methods applicable to this problem will be discussed in a later chapter. For equation (8.9), in any neighborhood of any of the points $(0, [k\omega/2]^2)$, k an integer, it can actually be shown there are values of (σ^2, q) such that (8.9) is *unstable*.

There are a large number of results in the literature which are concerned with the estimation of the stability regions in Theorem 8.1 in terms of the function $a + \phi(t)$ in (8.1). We give a theorem of Borg [1] on the first stability region which generalizes a result of Liapunov [1].

THEOREM 8.2. If $p(t + \pi) = p(t) \not\equiv 0$ for all t, $\int_0^\pi p(t)\, dt \geqq 0$, $p(t)$ is continuous, real and

$$\pi \int_0^\pi |p(t)|\, dt \leq 4,$$

then all solutions of the equation

(8.11) $\ddot u + p(t)u = 0,$

are bounded on $(-\infty, \infty)$.

PROOF. It is only necessary to show that no characteristic multiplier of (8.11) is real. If a characteristic multiplier ρ of (8.11) is real, then there is a real solution $u(t) \not\equiv 0$, $u(t + \pi) = \rho u(t)$ for all t. Either $u(t) \neq 0$ for all t or it has infinitely many zeros with two consecutive zeros $a, b, 0 \leqq b - a \leqq \pi$. In the first case $u(\pi) = \rho u(0)$, $\dot u(\pi) = \rho \dot u(0)$ and $\dot u(\pi)/u(\pi) = \dot u(0)/u(0)$. Since $\ddot u/u + p = 0$, an integration by parts yields

$$\int_0^\pi \frac{\dot u^2(t)}{u^2(t)}\, dt + \int_0^\pi p(t)\, dt = 0,$$

which is impossible from the hypothesis on p. In the second case, we may assume that $u(t) > 0$ on $a < t < b$. Let $u(c) = \max_{a < t < b} |u(t)|$. The hypothesis on p implies

$$\frac{4}{\pi} \geq \int_0^\pi |p(t)|\, dt \geq \int_a^b \frac{|\ddot{u}(t)|}{|u(t)|}\, dt$$

$$> \frac{1}{u(c)} \int_a^b |\ddot{u}(t)|\, dt \geq \frac{1}{u(c)} |\dot{u}(\alpha) - \dot{u}(\beta)|,$$

for any α, β in (a, b). From the mean value theorem, there exist α, β such that $\dot{u}(\alpha) = u(c)/(c - a)$, $-\dot{u}(\beta) = u(c)/(b - c)$. Therefore,

$$\frac{4}{\pi} > \frac{1}{c - a} + \frac{1}{b - c} = \frac{b - a}{(c - a)(b - c)} \geq \frac{4}{b - a} \geq \frac{4}{\pi},$$

since $4xy \leq (x + y)^2$ for all x, y. This contradiction shows that the characteristic multipliers are complex and proves the result.

EXERCISE 8.1. Consider the equation

(8.12) $$\ddot{u} + [a - 2qS(t)]u = 0,$$

where $S(t) = +1$ if $-\pi/2 \leq t < 0$, $= -1$ if $0 \leq t < \pi/2$, $S(t + \pi) = S(t)$ for all t. Show that every neighborhood of a point $(a_0, 0)$, $a_0 > 0$, in the (a, q) plane where $\cos \pi \sqrt{a_0} = \pm 1$ contains points for which (8.12) has unbounded solutions. Hint: If $a > |2q|$, $r^2 = a + 2q$, $s^2 = a - 2q$, one can show that $A = (\rho_1 + \rho_2)/2$ is given by

$$A = \frac{1}{4rs}\left[(s + r)^2 \cos \frac{\pi}{2}(s + r) - (s - r)^2 \cos \frac{\pi}{2}(s - r) \right].$$

III.9. Reciprocal Systems

Consider the system

(9.1) $$\dot{x} = A(t)x, \quad A(t + T) = A(t), \quad -\infty < t < \infty,$$

where $A(t)$ is a continuous real or complex $n \times n$ matrix and $T > 0$ is a constant. Following Lyapunov, system (9.1) is called *reciprocal* if for every characteristic multiplier ρ of (9.1) the number $\bar{\rho}^{-1}$ is also a characteristic multiplier. If A is real, the characteristic multipliers occur in complex conjugate pairs and (9.1) will be reciprocal if ρ a characteristic multiplier implies ρ^{-1} is a characteristic multiplier. Let $\mathscr{A}$ designate the Banach space of continuous real or complex valued $n \times n$ matrix functions $A(t)$, $-\infty < t < \infty$, $A(t+T) = A(t)$ with $|A| = \sup_{0 \le t \le T} |A(t)|$.[2]

If A in $\mathscr{A}$ is reciprocal, then Theorem 7.2 implies that (9.1) is stable if and only if all characteristic multipliers of (9.1) have simple elementary divisors and modulii equal to 1. Therefore, stability of reciprocal systems (9.1) is equal to boundedness of the solutions on $(-\infty, \infty)$. We shall say (9.1) is *stable on* $(-\infty, \infty)$ when all solutions are bounded on $(-\infty, \infty)$. An element A of $\mathscr{A}$ is said to be *strongly stable on* $(-\infty, \infty)$ relative to a set $\mathscr{B}$ in $\mathscr{A}$ if there is a $\delta > 0$ such that the system

$$(9.2) \qquad \dot{x} = B(t)x,$$

is stable on $(-\infty, \infty)$ for all B in $\mathscr{B}$ with $|A - B| < \delta$. We let $\mathscr{R}\mathscr{A}$ designate the set of A in $\mathscr{A}$ for which (9.1) is reciprocal.

LEMMA 9.1.   If A is in $\mathscr{R}\mathscr{A}$ and $\rho_0 = \rho_0(A)$, $|\rho_0| = 1$, is a simple characteristic multiplier of (9.1), then there is a $\delta_0 > 0$ such that the system (9.2) has a simple characteristic multiplier $\rho_0(B)$, $|\rho_0(B)| = 1$, for every B in $\mathscr{R}\mathscr{A}$ with $|A - B| < \delta_0$.

PROOF.   Suppose A in $\mathscr{R}\mathscr{A}$ and $\rho_0 = \rho_0(A)$ is a simple characteristic multiplier of (9.1), $|\rho_0| = 1$. Formula (1.11) implies that the matrix solution $X_A(t)$; $X_A(0) = I$, of (9.1) is a continuous function of A in $\mathscr{A}$. In particular, the characteristic multipliers of (9.1) are continuous functions of A in $\mathscr{A}$. Therefore, there is a disk $D_\varepsilon(\rho_0)$, $\varepsilon > 0$, in the complex plane of radius ε and center ρ_0 and a $\delta_1 > 0$ such that (9.2) has exactly one characteristic multiplier $\rho_0(B)$ in $D_\varepsilon(\rho_0)$ for all B in $\mathscr{R}\mathscr{A}$, $|A - B| < \delta_1$. Since (9.2) is reciprocal, $\bar{\rho}^{-1}(B)$, is also a characteristic multiplier. But, $\bar{\rho}_0^{-1}(B) = \rho_0(B)/|\rho_0(B)|^2 \ne \rho_0(B)$ unless $|\rho_0(B)| = 1$. On the other hand, the hypothesis $|\rho_0(A)| = 1$ implies $\bar{\rho}_0^{-1}(A) = \rho_0(A)$ and by continuity of $\rho_0(A)$ in A, we can find a $\delta_0 < \delta_1$ such that $\bar{\rho}_0^{-1}(B)$, $\rho_0(B)$ belong to $D_\varepsilon(\rho_0)$ if $|A - B| < \delta_0$, B in $\mathscr{R}\mathscr{A}$. This implies $|\rho_0(B)| = 1$ for $|A - B| < \delta_0$, B in $\mathscr{R}\mathscr{A}$, and proves the lemma.

THEOREM 9.1.   If A is in $\mathscr{R}\mathscr{A}$ and all of the characteristic multipliers of (9.1) are distinct and have unit modulii, then A is strongly stable relative to $\mathscr{R}\mathscr{A}$.

PROOF.   This is immediate from Lemma 9.1 and the Floquet representation of the solutions of a periodic system.

LEMMA 9.2.   If A in $\mathscr{A}$ is real and there exists an $n \times n$ nonsingular

---

[2] If A is not continuous but only Lebesgue integrable, then the results below are valid with $|A| = \int_0^T |A(s)| \, ds$.

matrix D such that either

    (i) $DA(t) = -A(-t)D$

or

    (ii) $DA(t) = -A'(t)D,$ (A' is the transpose of A)

then A is in $\mathscr{RA}$. In (i) the principal matrix solution $X(t)$ satisfies $X^{-1}(-t)DX(t) = D$ and in (ii) it satisfies $X'(t)DX(t) = D$ for all t.

PROOF. Let $X(t)$, $X(0) = I$, be a matrix solution of (9.1). If $Y(t)$, $Y(0) = Y_0$ is an $n \times n$ matrix solution of the adjoint equation $\dot{y} = -yA(t)$, then $Y(t)X(t) = Y_0$ for all t.

Case i. If $DA(t) = -A(-t)D$, then $Y(t) = X^{-1}(-t)D$ satisfies the adjoint equation. In fact, $\dot{Y}(t) = -\dot{X}^{-1}(-t)D = X^{-1}(-t)A(-t)D = -X^{-1}(-t)DA(t) = -Y(t)A(t)$. Therefore, $X^{-1}(-t)DX(t) = D$ which implies $X(t)$ is similar to $X(-t)$ for all t. If $X(t) = P(t)e^{Bt}$, then $P(0) = I$ and $X(t)$ similar to $X(-t)$ implies the roots of $\det(e^{BT} - \rho I) = 0$ and $\det(e^{-BT} - \mu I) = 0$ are the same. Obviously, these roots are the reciprocals of each other and this proves case (i).

Case ii. If $DA(t) = -A'(t)D$, then $Y(t) = X'(t)D$ is a solution of the adjoint equation. In fact, $\dot{Y} = \dot{X}'D = X'A'D = -X'DA = -YA$. Thus, $X'(t)DX(t) = D$ for all t and $X'(t)$ is similar to $X^{-1}(t)$ for all t. The remainder of the argument proceeds as in case (i).

By far the most important reciprocal systems are Hamiltonian systems; namely, the systems

(9.3) $E\dot{x} = H(t)x,$

where $H' = H$ is a real $2k \times 2k$ matrix of period T,

$$E = \begin{bmatrix} 0 & I_k \\ -I_k & 0 \end{bmatrix},$$

and I_k is the $k \times k$ unit matrix. Since $E^2 = -I_{2k}$, $E' = -E$ system (9.3) is a special case of system (9.1) with $A = -EH$ and $EA = H = H' = A'E' = -A'E$. Thus, (9.3) is reciprocal since this is a special case of case (ii) of Theorem 9.2 with $D = E$. Furthermore, the matrix solution $X(t)$, $X(0) = I$ of (9.3) satisfies

(9.4) $X'(t)EX(t) = E.$

The set of all matrices which satisfy (9.4) is called the *real symplectic group* or sometimes such a matrix is called *E-orthogonal*.

A general class of complex reciprocal systems consists of the *canonical systems*,

(9.5) $J\dot{x} = H(t)x,$

where H is Hermitian (i.e. $H^* = H$, where H^* is the conjugate transpose of H) and

$$(9.6) \qquad J = i \begin{bmatrix} I_p & 0 \\ 0 & -I_q \end{bmatrix}, \qquad i = \sqrt{-1},$$

To prove this system is reciprocal, let $X(t)$, $X(0) = I$, be a principal matrix solution of (9.5). Then $A = -JH$ is the coefficient matrix in (9.5) and

$$\frac{d}{dt} X^*(t) J X(t) = -X^*(t)H^*(t)J^*JX(t) - X^*(t)J^2H(t)X(t) = 0,$$

and $X^*(t)JX(t)$ is a constant. Since $X(0) = I$, we have

$$(9.7) \qquad X^*(t)JX(t) = J,$$

for all t. Thus $X(T)$ is similar to $X^{*-1}(T)$ and the result follows immediately. Matrices which satisfy (9.7) are called J-*unitary*. Notice that J-unitary matrices are nonsingular.

System (9.5) includes as a special case system (9.3) for the following reason. Any two skew Hermitian matrices A, B with the same eigenvalues with the same multiplicity are unitarily equivalent; that is, if $A^* = -A$, $B^* = -B$, then there is a matrix U such that $U^*U = UU^* = I$ and $U^*AU = B$. If, in addition, A and B are real, there is a real unitary (orthogonal) matrix U such that $U'AU = B$. Since E is skew Hermitian with the eigenvalue i of multiplicity k there is a unitary U such that $UEU^* = J$, where J is given in (9.6) with $p = q = k$. If we let $x = U^*y$ then (9.3) is transformed into (9.5) with H replaced by UHU^*. A matrix U which accomplishes this is

$$(9.8) \qquad U = \frac{1}{\sqrt{2}} \begin{bmatrix} I_k & -iI_k \\ -iI_k & I_k \end{bmatrix}.$$

A special case of (9.5) is the second order system

$$(9.9) \qquad \ddot{u} + Q\dot{u} + P(t)u = 0,$$

where u is a k-vector, Q is a constant matrix, $Q^* = -Q$, $P^*(t) = P(t)$. In fact, system (9.9) can be written in the form

$$K\dot{x} = H(t)x,$$

where

$$x = \begin{bmatrix} u \\ \dot{u} \end{bmatrix}, \qquad K = \begin{bmatrix} -Q & -I_k \\ I_k & 0 \end{bmatrix}, \qquad H(t) = \begin{bmatrix} P(t) & 0 \\ 0 & I_k \end{bmatrix}.$$

There is a nonsingular matrix P such that $PKP^* = J$ and, therefore, the transformation $x = P^*y$ reduces (9.9) to a special case of (9.5).

Case (i) in Lemma 9.2 expresses some even and oddness properties of the coefficient matrix $A(t)$. To illustrate, consider the second order matrix system

$$(9.10) \qquad \ddot{u} + P(t)u = 0,$$

where $P(t + T) = P(t)$ is a $k \times k$ continuous real matrix. This is equivalent to the system of order $2k$,

$$(9.11) \qquad \dot{x} = A(t)x, \qquad A(t) = \begin{bmatrix} 0 & I_k \\ -P(t) & 0 \end{bmatrix}.$$

If $P = (P_{jk})$, j, $k = 1, 2$ is partitioned so that P_{11} is an $r \times r$ matrix and P_{22} is an $s \times s$ matrix with $P_{jk}(t) = (-1)^{j+k} P_{jk}(-t)$, then (9.11) is reciprocal. In fact, case (i) of Lemma 9.2 is satisfied with $D = \mathrm{diag}(I_r, -I_s, -I_r, I_s)$.

An even more special reciprocal system is the system

$$(9.12) \qquad \ddot{u} + Fu = 0,$$
$$F = \mathrm{diag}(\sigma_1^2, \ldots, \sigma_n^2), \qquad \sigma_j > 0.$$

The matrix F is periodic of any period. For any period $T > 0$, the characteristic multipliers of (9.12) are $\rho_{2j-1} = \bar{\rho}_{2j}$, $\rho_{2j-1} = e^{i\sigma_j T}$, $j = 1, 2, \ldots, n$. These multipliers are distinct if and only if

$$(9.13) \qquad 2\sigma_j \not\equiv 0 \ (\mathrm{mod} \ \omega), \qquad \sigma_j \pm \sigma_k \not\equiv 0 \ (\mathrm{mod} \ \omega), \qquad j \neq k,$$

where $T = 2\pi/\omega$. Consequently, if (9.13) is satisfied, Theorem 9.1 implies there is a $\delta > 0$ such that all solutions of (9.2) are bounded in $(-\infty, \infty)$ provided that B is in $\mathscr{R}\mathscr{A}$ and

$$|B - A| < \delta, \qquad A = \begin{bmatrix} 0 & I \\ -F & 0 \end{bmatrix},$$

and F is defined in (9.12). In particular, if $\Phi(t) = \Phi(t + T)$ is real, symmetric or satisfies the even and oddness conditions above and (9.13) is satisfied, there is an $\varepsilon_0 > 0$ such that all solutions of the system

$$(9.14) \qquad \ddot{u} + (F + \varepsilon\Phi(t))u = 0$$

are bounded in $(-\infty, \infty)$ for all $|\varepsilon| \leq \varepsilon_0$. Compare this result with the one at the end of Section 8 for a single second order equation.

If some of the conditions (9.13) are not satisfied, it is very difficult to determine whether there is an $\varepsilon_0 > 0$ such that solutions of (9.14) are bounded for $|\varepsilon| \leq \varepsilon_0$. For Hamiltonian systems, some general results are available (see Section 10) but, for the other cases, only special equations have been discussed. Iterative schemes for reaching a decision are available and will be discussed in a subsequent chapter.

If system (9.14) is not a reciprocal system, it can be shown by example (see Chapter VIII) that even when (9.13) is satisfied, solutions of (9.14) may be unbounded for any $\varepsilon \neq 0$.

EXERCISE 9.1. Suppose $B(t)$ is an integrable T-periodic matrix such that the characteristic multipliers of $\dot{x} = B(t)x$ are distinct and have unit modulii. If $A(t)$ is an integrable T-periodic matrix such that $\dot{x} = A(t)x$ is reciprocal, then there is a $\delta > 0$ such that $\int_0^T |A(s) - B(s)| \, ds < \delta$ implies the equation $\dot{x} = A(t)x$ is stable on $(-\infty, \infty)$. Hint: Use the continuity of the fundamental matrix solution of $\dot{x} = A(t)x$ in A which is implied by formula (1.11).

III.10. Canonical Systems

As in Section 9, let $\mathscr{A}$ be the Banach space of $n \times n$ complex integrable matrix functions of period T with $|A| = \int_0^T |A(t)| dt$. Let $\mathscr{C}\mathscr{A}$ be the subspace consisting of those matrices of the form $-JH$, $H^* = H$ and

$$(10.1) \qquad J = i \begin{bmatrix} I_p & 0 \\ 0 & -I_q \end{bmatrix}.$$

If A belongs to $\mathscr{C}\mathscr{A}$, then the associated periodic differential system is a canonical system

$$(10.2) \qquad J\dot{x} = H(t)x, \qquad H^* = H,$$

Our main objective in this section is to give necessary and sufficient conditions in order that system (10.2) be strongly stable on $(-\infty, \infty)$ relative to $\mathscr{C}\mathscr{A}$.

We have seen in Section 9 that a canonical system (10.2) is reciprocal and, therefore, stable on $(-\infty, \infty)$ if and only if all characteristic multipliers of (10.2) are on the unit circle and have simple elementary divisors; or, equivalently, that all eigenvalues of the monodromy matrix S have simple elementary divisors and modulii 1. This latter statement is equivalent to saying there is a nonsingular matrix U such that

$$U^{-1}SU = \operatorname{diag}(e^{i\nu_1}, \ldots, e^{i\nu_n}), \qquad \nu_j \text{ real.}$$

It was also shown in Section 9 that the monodromy matrix S of a principal matrix solution of (10.2) is J-unitary; that is,

$$(10.3) \qquad S^*JS = J.$$

Since the stability properties of (10.2) depend only upon the eigenvalues of S, we use the following terminology: A *J-unitary matrix S is stable* if all eigenvalues have modulii 1 and simple elementary divisors. A *J-unitary matrix S is strongly stable* if there is a $\delta > 0$ such that every J-unitary matrix R for which $|R - S| < \delta$ is stable.

Preliminary to the discussion of stability, we introduce the following terminology. For any x, y in C^n, define the bilinear form

$$\langle x, y \rangle = i^{-1} y^* J x.$$

For Hamiltonian systems, the expression $\langle x, y \rangle$ is related to the Lagrange bracket. In fact, if U is given in (9.8) and $x = U^* u$, $y^* = v^* U$, then $\langle u, v \rangle = i^{-1} v^* E u$. For real vectors v, u, $v^* E u$ is the Lagrange bracket.

Since $i^{-1} J$ is Hermitian, it is clear that $\langle x, y \rangle = \overline{\langle y, x \rangle}$ and $\langle x, x \rangle$ is real. The *J-norm of x* is $\langle x, x \rangle$. A subspace V of C^n is called *non-negative* if $\langle x, x \rangle \geq 0$ for all x in V and *positive* if $\langle x, x \rangle > 0$ for all $x \neq 0$ in V. Two vectors x, y are called *J-orthogonal* if $\langle x, y \rangle = 0$. If $\langle x, y \rangle = 0$ for all y, then $Jx = 0$ which implies $x = 0$. This immediately implies that S is J-unitary if and only if $\langle Sx, Sy \rangle = \langle x, y \rangle$ for all x, y. It is immediate from the definition of J that the vectors $e^1 = (1, 0, \ldots, 0)$, $e^2 = (0, 1, \ldots, 0)$, $\ldots$, $e^n = (0, 0, \ldots, 1)$ satisfy

(10.4)
$$\langle e^j, e^k \rangle = 0, \qquad j \neq k,$$
$$\langle e^j, e^j \rangle = 1, \qquad 1 \leq j \leq p,$$
$$\langle e^j, e^j \rangle = -1, \qquad p < j \leq n.$$

LEMMA 10.1. *Eigenvectors associated with distinct eigenvalues of a stable J-unitary matrix are J-orthogonal. The eigenvectors of a stable J-unitary matrix span C^n.*

PROOF. If x, y are eigenvectors associated with λ, μ, respectively, $\lambda \neq \mu$ and $|\lambda| = 1$, then $\langle x, y \rangle = \langle Sx, Sy \rangle = \lambda \bar{\mu} \langle x, y \rangle$. If $\lambda \bar{\mu} \neq 1$, then $\langle x, y \rangle = 0$. Since the system is assumed stable $\bar{\mu}^{-1} = \mu$ and $\lambda \neq \mu$ imply $\lambda \bar{\mu} = \lambda/\mu \neq 1$. This proves $\langle x, y \rangle = 0$. The proof that the eigenvectors of a stable J-unitary matrix span the space is now supplied in exactly the same way that one proves the corresponding result for unitary matrices in linear algebra.

LEMMA 10.2. *A nonnegative eigenspace V of a stable J-unitary matrix is positive. A nonpositive eigenspace of a stable J-unitary matrix is negative.*

PROOF. Suppose V is nonnegative. If x is in V and $\langle x, x \rangle = 0$, then for any y in V and any complex number λ,

$$0 \leq \langle y + \lambda x, y + \lambda x \rangle = \langle y, y \rangle + 2 Re\lambda \langle x, y \rangle.$$

We must have $\langle x, y \rangle = 0$ for otherwise λ could be chosen in such a manner as to make this expression negative. Therefore, $\langle x, y \rangle = 0$ for all y in V. This fact together with Lemma 10.1 imply $\langle x,y \rangle = 0$ for all y. Thus, $x = 0$. A similar argument applies when V is nonpositive and proves the lemma.

THEOREM 10.1.   A matrix S is a stable J-unitary matrix if and only if there is a J-unitary matrix U such that

$$(10.5) \qquad U^{-1}SU = \mathrm{diag}(e^{i\nu_1}, \ldots, e^{i\nu_n}),$$

where each ν_j is real.

PROOF.   Let $G = \mathrm{diag}(e^{i\nu_1}, \ldots, e^{i\nu_n})$. If $S = UGU^{-1}$ and U is J-unitary, it is clear that S is stable and J-unitary.

Conversely, suppose S is a stable J-unitary matrix and $\lambda = e^{i\nu}$ is an eigenvalue of multiplicity r. One can choose a J-orthonormal basis $v^1, \ldots, v^r$ for the eigenspace V_λ so that $\langle v^j, v^k \rangle = 0$, $j \neq k$, $= \pm 1$ if $j = k$; $j, k = 1, 2, \ldots, r$. For if not, there would be an eigenvector v such that v is J-orthogonal to V_λ. This is impossible from Lemma 10.1 since it would imply v is J-orthogonal to the whole space C^n. Some of the v^j may have $\langle v^j, v^j \rangle = +1$ and some may have this expression equal -1. From Lemma 10.1, we can choose a J-orthonormal basis $u^1, \ldots, u^n$ of eigenvectors for the whole space and we can order these vectors in such a way that $\langle u^j, u^k \rangle = 0$, $j \neq k$, $\langle u^j, u^j \rangle = 1$, $j \leq p'$ and $= -1$ for $p' < j \leq n$. But the law of inertia for Hermitian forms and (10.4) imply that $p' = p$. If $U = (u^1, \ldots, u^n)$, then (10.5) is satisfied. Furthermore, for the vectors e^j in (10.4), $\langle Ue^j, Ue^k \rangle = \langle u^j, u^k \rangle = \langle e^j, e^k \rangle$ for all j, k. Thus, $\langle Ux, Uy \rangle = \langle x, y \rangle$ for all x, y and U is J-unitary. This proves the theorem.

It was shown in the proof of the above theorem that for any stable J-unitary matrix, a complete set of J-orthonormal eigenvectors u^j and corresponding eigenvalues $\lambda_j = e^{i\nu_j}$ can be obtained. We shall say that the eigenvalue λ_j is of *positive type* (or *negative type*) if $\langle u^j, u^j \rangle = 1$ (or -1). In the Russian literature, the terminology is *first kind* (or *second kind*). There are p eigenvalues of positive type and $n - p$ of negative type. The following theorem asserts that a stable J-unitary matrix is strongly stable if and only if a multiple eigenvalue is not of both positive and negative type.

THEOREM 10.2.   A stable J-unitary matrix is strongly stable if and only if each of its eigenspaces is definite (that is, either positive or negative).

PROOF.   Suppose S is a stable J-unitary matrix and has an eigenvalue λ, $|\lambda| = 1$, whose eigenspace is not definite. Lemma 10.2 implies there are eigenvectors v^1, v^2 such that $\langle v^1, v^2 \rangle = 0$, $\langle v^1, v^1 \rangle = 1$, $\langle v^2, v^2 \rangle = -1$.

Choose $v^3, \ldots, v^n$, J-orthogonal to v^1, v^2 so that $v^1, \ldots, v^n$ form a basis for C^n. For any $\alpha \geqq 0$, define the linear transformation $R: C^n \to C^n$ by

$$Rv^1 = S[(\cosh \alpha)v^1 + (\sinh \alpha)v^2],$$
$$Rv^2 = S[(\sinh \alpha)v^1 + (\cosh \alpha)v^2],$$
$$Rv^k = Sv^k, \qquad k = 3, \ldots, n.$$

One easily verifies that $\langle Rv^j, Rv^k \rangle = \langle v^j, v_k \rangle$ for all j, k and, thus, the matrix associated with the transformation R is J-unitary. Also, $v^1 + v^2$ is an eigenvector of R associated with the eigenvalue λe^α, and R has an eigenvalue with modulus >1 for any $\alpha > 0$. Thus, R is not stable on $(-\infty, \infty)$. Since R approaches S as $\alpha \to 0$, this implies S is not strongly stable.

Conversely, suppose S is a stable J-unitary matrix whose eigenspaces are definite. Let λ_k, $k = 1, 2, \ldots, r$, be the distinct eigenvalues of S with multiplicity n_k and let V_k of dimension n_k be the corresponding eigenspaces. Then

$$C^n = V_1 \oplus V_2 \oplus \cdots \oplus V_r,$$

and we let P_k denote the corresponding projection operators of C^n onto V_k; that is, for any x in C^n, $P_k x$ is in V_k and $P_k x = x$ if x is in V_k. These projection operators satisfy $P_k P_j = 0$, $j \neq k$, $P_k^2 = P_k$, and can be defined by the formula

(10.6) $$P_k = \frac{1}{2\pi i} \int_{\gamma_k} (\zeta I - S)^{-1} \, d\zeta,$$

where γ_k is a positively oriented circumference of a circle with center λ_k and radius so small that it contains no other eigenvalue of S. If R is any $n \times n$ matrix with $|R - S|$ sufficiently small, then each of circumferences γ_k encloses exactly n_k eigenvalues (counted with their multiplicities) of R. Therefore, the integrals

$$Q_k = \frac{1}{2\pi i} \int_{\gamma_k} (\zeta I - R)^{-1} \, d\zeta,$$

define projection operators Q_k of C^n onto subspaces W_k of dimension n_k such that the W_k are invariant under R and $C^n = W_1 \oplus W_2 \oplus \cdots \oplus W_r$. The subspace W_k is the algebraic sum of the generalized eigenspaces of R associated with the eigenvalues of R enclosed by γ_k. Also, this formula shows that Q_k is a continuous function of R and $|Q_k - P_k| \to 0$ as $R \to S$.

Our next objective is to show that $\langle x, x \rangle$ is definite on each W^k. For x in W_k,

$$\langle x, x \rangle = \langle Q_k x, Q_k x \rangle$$
$$= \langle P_k x, P_k x \rangle + \langle (Q_k - P_k)x, (Q_k - P_k)x \rangle$$
$$+ 2Re\langle P_k x, (Q_k - P_k)x \rangle.$$

From the definition of $\langle x, y \rangle$, we have $|\langle x, y \rangle| \leq |J| \cdot |x| \cdot |y|$ for all x, y in C^n. Also, since S is definite on eigenspaces, there is an $\alpha > 0$ such that $|\langle x, x \rangle| \geq \alpha |x|^2$ for all x in V_k and each $k = 1, 2, \ldots, r$. Therefore, for x in W_k,

$$|\langle x, x \rangle| \geq \alpha |P_k x|^2 - |J| \cdot |Q_k - P_k|^2 \cdot |x|^2 - 2|J| \cdot |P_k| \cdot |Q_k - P_k|^2 |x|^2$$
$$\geq [\alpha(1 - |Q_k - P_k|)^2 - |J| \cdot |Q_k - P_k|^2 - 2|J| \cdot |P_k| \cdot |Q_k - P_k|] |x|^2.$$

Consequently, if $|S - R|$ is small enough, the continuity of the projections Q_k in R implies that

$$|\langle x, x \rangle| \geq \frac{1}{2} \alpha |x|^2, \qquad x \text{ in } W_k, \qquad k = 1, 2, \ldots, r.$$

Since each W^k is invariant under R, for any integer m, it follows that

$$|R^m x|^2 \leq 2\alpha^{-1} |\langle R^m x, R^m x \rangle| = 2\alpha^{-1} |\langle x, x \rangle| \leq 2\alpha^{-1} |J| \cdot |x|^2,$$

for all x in W_k, $k = 1, 2, \ldots, r$, if R is J-unitary. For any x in C^n we have $x = Q_1 x + \cdots + Q_r x$ and, thus,

$$|R^m x|^2 \leq 2\alpha^{-1} |J| [|Q_1| + \cdots + |Q_r|]^2 |x|^2.$$

It is easy to see this implies all eigenvalues must have simple elementary divisors and modulus 1. Thus, R is stable which in turn implies S is strongly stable. This proves the theorem.

THEOREM 10.3. System (10.2) is strongly stable on $(-\infty, \infty)$ if and only if the monodromy matrix is strongly stable.

PROOF. It was remarked at the beginning of this section that (10.2) is stable on $(-\infty, \infty)$ if and only if the monodromy matrix is stable. Since the solutions of (10.2) depend continuously upon A in $\mathscr{A}$, then (10.2) is strongly stable on $(-\infty, \infty)$ if the monodromy matrix is strongly stable.

Suppose now that the monodromy matrix S of the solution $X(t)$, $X(0) = I$, of (10.2) is stable and not strongly stable. In the proof of Theorem 10.2, it was shown that any neighborhood of S contains a J-unitary matrix R which has an eigenvalue with modulus > 1. Since $S^{-1}R$ is nonsingular, Lemma 7.1 implies there is a matrix F such that $S^{-1}R = e^F$. For $|R - S|$ sufficiently small, one can show directly from the power series representation of e^F or from the implicit function theorem that F can be chosen to be a continuous function of R which vanishes for $R = S$. Furthermore, $S^{-1}R$ being J-unitary implies $e^{-F} = J^{-1}e^{F^*}J = e^{J^{-1}F^*J}$ and we can choose F so that $-F = J^{-1}F^*J$. Therefore, F can be written as $F = TJ^{-1}G$ where G is Hermitian and T is the period.

If we define $Y(t) = X(t)e^{tJ^{-1}G}$, then $Y(0) = I$, $Y(T) = R$, and $Y(t)$ is a fundamental matrix solution of the canonical system

$$J\dot{x} = L(t)x,$$

where $L = H + X^{*-1}GX^{-1}$. However, $L(t)$ may not be periodic of period T. On the other hand, we can alter $L(t)$ by a symmetric perturbation $L_1(t)$ such that $L(t) + L_1(t)$ is periodic of period T, $\int_0^T |L_1(t)|\, dt < \delta$ for any pre-assigned δ. If we let $Y_1(t)$, $Y_1(0) = I$, be the fundamental matrix solution with L replaced by $L + L_1$, then formula (1.11) implies there is a constant K such that

$$|Y_1(T) - Y(T)| \leq \delta K.$$

Therefore, for δ sufficiently small the monodromy matrix $Y_1(T)$ will have an eigenvalue with modulus >1 and (10.2) is not strongly stable. This proves the theorem.

If the eigenvalues $e^{i\nu_j}$ of S in the representation (10.5) are ordered so that the first p are of positive and the remaining $n - p$ are of negative type, then Theorem 10.2 states that a stable J-unitary matrix S is strongly stable if and only if

(10.7) $\nu_j \not\equiv \nu_k \pmod{2\pi}, \qquad 1 \leq j \leq p < k \leq n.$

If S is the monodromy matrix of a stable canonical system (10.2) and the eigenvalues of S are denoted by $e^{i\theta_j T}$, $j = 1, 2, \ldots, n$, and ordered in the same way as above, then Theorem 10.3 implies that the canonical system is strongly stable on $(-\infty, \infty)$ if and only if

(10.8) $\theta_j \not\equiv \theta_k \pmod{2\pi/T}, \qquad 1 \leq j \leq p < k \leq n.$

EXERCISE 10.1. Show that inequalities (10.8) are equivalent to the inequalities

(10.9) $\sigma_j + \sigma_k \not\equiv 0 \pmod{\omega}, \qquad j, k = 1, 2, \ldots, n,$

for system (9.12), and, thus, (9.12) is strongly stable if and only if (10.9) is satisfied. Compare these inequalities with (9.13).

Theorem 10.3 answers many of the questions concerned with the qualitative properties of the stability of canonical systems. It states very clearly the critical positions of the characteristic multipliers for which one can always find a canonical system that is unstable and yet arbitrarily near to the original one. One of the basic remaining problems is the determination of an efficient procedure to obtain these critical positions of the multipliers. A method for doing this for equations which contain a small parameter is given in Chapter VIII.

III.11. Remarks and Suggestions for Further Study

The reader interested in more results for the stability of perturbed linear systems may consult Bellman [1], Cesari [1], Coppel [1].

In the case of linear systems with constant or periodic coefficients, the stability properties of the linear system were completely determined by the characteristic exponents of the equation. Furthermore, if all characteristic exponents have negative real parts, then the linear equation is uniformly asymptotically stable. Consequently, the linear system can be subjected to a perturbation of the type given by relation (2.9) and the property of being uniformly asymptotically stable is preserved for the perturbed system. If the coefficients of the matrix $A(t)$ in (1.3) are bounded, then every solution of (1.3) is bounded by an exponential function. Therefore, it is possible to associate with each solution x of (1.3) a number $\lambda = \lambda_x$ defined by

$$\lambda = \lim_{t \to \infty} \sup \frac{1}{t} \log |x(t)|.$$

This number λ was called by Lyapunov [1] the *characteristic number* of the solution x. If $\lambda < 0$, then $|x(t)| \to 0$ as $t \to \infty$. If $\lambda_x < 0$ for all solutions x of (1.3), then all solutions approach zero exponentially. On the other hand, the example of Perron for $A(t)$ as in (2.14) has the property that perturbations satisfying (2.9) can lead to unbounded solutions in contrast to the above remarks for periodic systems. There is an extensive theory of the characteristic numbers of Lyapunov and their application to stability (see Cesari [1], Malkin [1], Nemitskii and Stepanov [1], Lillo [1]).

In the discussion of the preservation of the saddle point property in Theorem 6.1, we were only concerned with the orbits on the stable and unstable manifolds. Of course, one could consider the following problem: Suppose $x = 0$ is a saddle point of (4.1) and $f: R^n \to R^n$ has continuous first derivatives such that $f(0) = 0$, $\partial f(0)/\partial x = 0$. Does there exist a neighborhood V of $x = 0$ such that for any f of the above class, there is a transformation h which takes the trajectories of (4.1) onto the trajectories of the perturbed equation (6.3) in the neighborhood V? This problem has a long and interesting history. The reader may consult Poincaré [1] and Lyapunov [1] for the analytic case, Sternberg [1], Chen [1] for the case where h may have a finite number of derivatives and Hartman [1] for the general case.

The problem posed in the previous paragraph in a local neighborhood of a critical point can be made much more general. In fact, one can say two n-dimensional differential equations

(11.1) $\dot{x} = f(x),$

(11.2) $\dot{y} = g(y)$,

are *equivalent* in a region G of R^n if there is a homeomorphism h which takes the trajectories of (11.1) onto the trajectories of (11.2) in G. Suppose the vector fields f, g belong to some topological space $\mathscr{B}$. A system (11.1) is said to be *structurally stable* if there is a neighborhood $N(f)$ of f in $\mathscr{B}$ such that (11.1) and (11.2) are equivalent for every g in $N(f)$. The study of structural stability and equivalent systems of differential equations is at present one of the most exciting topics in differential equations. All of the concepts are equally meaningful for vector fields on arbitrary n-dimensional manifolds. The reader is referred to the basic paper of Peixoto [1] for two-dimensional systems and the paper of Smale [1] for the general problem. See also the book of Netecki [1].

The presentation of the stability theory of Hill's equation given in Section 8 relies very heavily upon the book of Magnus and Winkler [1], but does not begin to indicate the tremendous number of results that are available for this equation. The precise determination of the a-intervals of stability and instability for a given function $\phi(t)$ is extremely important in the applications. Each function $\phi(t)$ defines a class of functions depending upon a and the particular a_j, a_j^* of Theorem 8.1 yield special periodic functions of period π or 2π. It is important to discuss the expansion of arbitrary functions in terms of these special periodic functions in the same way that one develops a function in a Fourier series. For a more complete discussion as well as many references, see Arscott [1], Cesari [1], Magnus and Winkler [1], McLachlan [1].

The presentation in Section 10 follows the thesis of Howe [1] and should be considered only as introduction to the theory of stability of Hamiltonian and canonical systems. The topological characteristics of the class of strongly stable systems have been discussed as well as the regions of stability and instability for given equations. Computational methods also have been devised for equations with a small parameter. The reader is referred to Gelfand and Lidskii [1], Yakubovich [1, 2], Krein [1], Diliberto [2], Coppel and Howe [1] for results as well as further references.

CHAPTER IV

Perturbations of Noncritical Linear Systems

It is convenient to have

Definition 1. If $A(t)$ is an $n \times n$ continuous matrix function on $(-\infty, \infty)$ and $\mathscr{D}$ is a given class of functions which contains the zero function, the homogeneous system

$$(1) \qquad \qquad \dot{x} = A(t)x,$$

is said to be *noncritical with respect to* $\mathscr{D}$ if the only solution of (1) which belongs to $\mathscr{D}$ is the solution $x = 0$. Otherwise, system (1) is said to be *critical with respect to* $\mathscr{D}$.

Throughout the following, $\mathscr{B}(-\infty, \infty) = \{f : (-\infty, \infty) \to C^n,\ f$ continuous and bounded$\}$ and for any f in $\mathscr{B}(-\infty, \infty)$, $|f| = \sup_{-\infty < t < \infty} |f(t)|$. The subset $\mathscr{A}\mathscr{P}$ of $\mathscr{B}(-\infty, \infty)$ denotes the set of almost periodic functions (for definition, see Appendix) and for any f in $\mathscr{A}\mathscr{P}$, $m[f]$ denotes the module of f. If f, g are in $\mathscr{A}\mathscr{P}$, then $m[f, g]$ denotes the smallest module containing $m[f] \cup m[g]$. The subset $\mathscr{P}_T$ of $\mathscr{A}\mathscr{P}$ denotes the set of periodic functions of period T. The spaces $\mathscr{B}(-\infty, \infty)$, $\mathscr{A}\mathscr{P}$ and $\mathscr{P}_T$ with $|\cdot|$ defined as above are Banach spaces. An $n \times n$ matrix function on $(-\infty, \infty)$ is said to belong to one of these spaces if each column belongs to the space.

This chapter centers around the study of the existence and stability properties of solutions of

$$(2) \qquad \qquad \dot{x} = A(t)x + f(t, x),$$

which belong to one of the classes $\mathscr{B}(-\infty, \infty)$, $\mathscr{A}\mathscr{P}$ or $\mathscr{P}_T$ for the case in which the matrix A is in $\mathscr{P}_T$, system (1) is noncritical with respect to the class under discussion and f is "small" in a sense to be made precise later.

The presentation will proceed as follows. The nonhomogeneous *linear* system is studied first in great detail. It is shown that system (1) being noncritical with respect to $\mathscr{D}$, one of the classes $\mathscr{B}(-\infty, \infty)$, $\mathscr{A}\mathscr{P}$ or $\mathscr{P}_T$, implies that the nonhomogeneous linear equation has a unique solution in $\mathscr{D}$ which depends linearly and continuously upon the forcing function in $\mathscr{D}$.

For f in (2) "small," the contraction principle yields the existence of a solution of (2). By extending the saddle point property of Chapter III, Section 6, to nonautonomous equations, one obtains the stability properties of the solution of (2). Some generalizations of the results are given in Section 4 together with some elementary properties of Duffing's equation with large forcing and large damping in Section 5. At the end of Section 1, there is also a stability result for the case when A is not in $\mathscr{P}_T$.

LEMMA 1. (a) System (1) with A in $\mathscr{P}_T$ is noncritical with respect to $\mathscr{B}(-\infty, \infty)$ [or $\mathscr{AP}$] if and only if the characteristic exponents of (1) have nonzero real parts. (b) System (1) with A in $\mathscr{P}_T$ is noncritical with respect to $\mathscr{P}_T$ if and only if $I - X(T)$ is nonsingular, when $X(t)$, $X(0) = I$, is a fundamental matrix solution of (1).

PROOF. (a) If the characteristic exponents of (1) are assumed to have nonzero real parts, then the Floquet representation of the solutions implies that the only solution of (1) in $\mathscr{B}(-\infty, \infty)$ is $x = 0$ and (1) is noncritical with respect to $\mathscr{B}(-\infty, \infty)$ and, therefore, also with respect to $\mathscr{AP}$. Conversely, if there is no solution $x \neq 0$ of (1) in $\mathscr{B}(-\infty, \infty)$, then the Floquet representation implies there cannot be a characteristic exponent λ of (1) with $\lambda = i\theta$ since equation (1) would then have a nonzero solution $e^{i\theta t}p(t)$, $p(t + T) = p(t)$.

(b) With $X(t)$ defined as in the lemma, the general solution of (1) is $X(t)x_0$, where x_0 is an arbitrary constant vector. System (1) has a nonzero periodic solution of period T if and only if there is an $x_0 \neq 0$ such that $[X(t + T) - X(t)]x_0 = 0$ for all t. Since A is assumed to belong to $\mathscr{P}_T$, this is equivalent to the existence of an $x_0 \neq 0$ such that $[X(T) - I]x_0 = 0$; that is, $X(T) - I$ is singular. This proves the lemma.

Remark 1. If A in equation (1) is a constant, then (1) is noncritical with respect to $\mathscr{B}(-\infty, \infty)$ or $\mathscr{AP}$ if and only if all eigenvalues of A have nonzero real parts. The equation (1) is noncritical with respect to $\mathscr{P}_T$ if and only if all eigenvalues λ of A satisfy $\lambda T \not\equiv 0 \pmod{2\pi i}$; or equivalently, the purely imaginary eigenvalues $i\omega$ of A satisfy $\omega \not\equiv 0 \pmod{2\pi/T}$.

IV.1. Nonhomogeneous Linear Systems

Basic to any discussion of problems concerned with perturbed linear systems is a complete understanding of the nonhomogeneous linear system

$$(1.1) \qquad \dot{x} = A(t)x + f(t),$$

where f is a given function in some specified class.

Let us recall that the equation adjoint to (1) is $\dot{y} = -yA(t)$ and $X^{-1}(t)$ is a fundamental matrix solution of this equation if $X(t)$ is a fundamental matrix solution of (1). Since A belongs to $\mathscr{P}_T$, it follows that the adjoint equation has a nontrivial solution y with y' in $\mathscr{P}_T$ (' is the transpose of y) if and only if $y(0)[X^{-1}(T) - I] = 0$; that is, if and only if $X^{-1}(T) - I$ is singular. Equation (1) has a solution x in $\mathscr{P}_T$ if and only if $[X(T) - I]x(0) = 0$. Since the matrices $X^{-1}(T) - I$ and $X(T) - I$ are the same except for multiplication by a nonsingular matrix, it follows that the dimensions of the set of y_0 such that $y_0[X^{-1}(T) - I] = 0$ and the set of x_0 such that $[X(T) - I]x_0 = 0$ are the same. Therefore, the adjoint equation and equation (1) always have the same number of linearly independent T-periodic solutions.

LEMMA 1.1.   (Fredholm's alternative).   If A is in $\mathscr{P}_T$ and f is a given element of $\mathscr{P}_T$, then equation (1.1) has a solution in $\mathscr{P}_T$ if and only if

$$(1.2) \qquad \int_0^T y(t)f(t)\, dt = 0,$$

for all solutions y of the adjoint equation

$$(1.3) \qquad \dot{y} = -yA(t),$$

such that y' is in $\mathscr{P}_T$. If (1.2) is satisfied, then system (1.1) has an r-parameter family of solutions in $\mathscr{P}_T$, where r is the number of linearly independent solutions of (1) in $\mathscr{P}_T$.

PROOF.   Since A and f belong to $\mathscr{P}_T$, $x(t)$ is a solution of (1.1) in $\mathscr{P}_T$ if and only if $x(0) = x(T)$. If $X(t, \tau)$, $X(\tau, \tau) = I$, is a matrix solution of (1), and $x(0) = x_0$, then

$$x(t) = X(t, 0)x_0 + \int_0^t X(t, s)f(s)\, ds.$$

Since $X(t, \tau)X(\tau, s) = X(t, s)$ for all t, τ, s, the condition $x(T) = x_0$ is equivalent to

$$(1.4) \qquad [X^{-1}(T, 0) - I]x_0 = \int_0^T X^{-1}(s, 0)f(s)\, ds.$$

If $B = X^{-1}(T, 0) - I$, $b = \int_0^T X^{-1}(s, 0)f(s)\, ds$, then (1.4) is equivalent to $Bx_0 = b$. From elementary linear algebra, this matrix equation has a solution if and only if $ab = 0$ for all row vectors a such that $aB = 0$. Since $X^{-1}(t, 0)$ is a principal matrix of (1.3), the set of vectors a for which $aB = 0$ coincides with the set of initial values of those solutions y of (1.3) which are T-periodic; that is, the solution $y(t) = aX^{-1}(t, 0)$ is a T-periodic solution of (1.3) if and only if $aB = 0$. Using the definition of b, we obtain the first part of the lemma.

If (1.2) is satisfied and $\phi(t)$ is a solution of (1.1) in $\mathscr{P}_T$, then any other solution in $\mathscr{P}_T$ must be given by $x = z + \phi$, where z is a solution of (1) in $\mathscr{P}_T$. This proves the lemma.

EXERCISE 1.1. What is the analogue of Lemma 1.1 for the case in which A is in $\mathscr{P}_T$ and f is in $\mathscr{A}\mathscr{P}$?

Example 1.1. Consider the second order scalar equation

$$(1.5) \qquad \ddot{u} + u = \cos \omega t, \qquad \omega > 0.$$

If $x_1 = u$, an equivalent system is

$$(1.5)' \qquad \begin{aligned} \dot{x}_1 &= x_2, \\ \dot{x}_2 &= -x_1 + \cos \omega t, \qquad \omega > 0. \end{aligned}$$

The adjoint equation is $\dot{y}_1 = y_2$, $\dot{y}_2 = -y_1$ which has the general solution $y_1 = a \cos t + b \sin t$, $y_2 = -a \sin t + b \cos t$ where a, b are arbitrary constants. If $\omega \neq 1$, there are no nontrivial solutions of the adjoint equation of least period $T = 2\pi/\omega$. Thus, (1.2) is satisfied. Since the homogeneous equation has no nontrivial periodic solutions, the equation (1.5) has a unique periodic solution of period $2\pi/\omega$. If $\omega = 1$, then every solution of the adjoint equation has period 2π. On the other hand relation (1.2) is not satisfied since

$$\int_0^{2\pi} y(t) f(t)\, dt = \int_0^{2\pi} (-a \sin t \cos t + b \cos^2 t)\, dt = \pi b$$

and this is not zero unless $b = 0$. Therefore, the equation (1.5) has no solution of period 2π and, in fact, all solutions are unbounded since the general solution is $u = a \sin t + b \cos t + (t \sin t)/2$.

Example 1.2. Consider the system (1.1) with

$$(1.6) \qquad A = \frac{1}{2}\begin{bmatrix} -1 & 2 & 1 \\ -1 & 0 & -1 \\ -1 & 2 & 1 \end{bmatrix}, \qquad f(t) = \sin t \begin{bmatrix} -1 \\ -1 \\ 1 \end{bmatrix}.$$

One can easily show that

$$(1.7) \qquad e^{At} = \frac{1}{2}\begin{bmatrix} 1 + \cos t - \sin t & 2 \sin t & -1 + \cos t + \sin t \\ 1 - \cos t - \sin t & 2 \cos t & -1 + \cos t - \sin t \\ -1 + \cos t - \sin t & 2 \sin t & 1 + \cos t + \sin t \end{bmatrix}.$$

Since the rows of e^{-At} are solutions of the adjoint equation, it follows that all solutions of the adjoint equation have period 2π, the same period as the forcing function f, and the solutions are linear combinations of the rows of e^{-At}. To check the orthogonality condition (1.2), we need to verify that $\int_0^{2\pi} e^{-At} f(t)\, dt = 0$. One easily finds that $e^{-At} f(t) = z \sin t$, $z = (-1, -1, 1)$

and the orthogonality condition is satisfied. Therefore, system (1.1) with A and f given in (1.6) has a periodic solution of period 2π, and, in fact, every solution has period 2π.

EXERCISE 1.2.   In example 1.2, is there a unique solution of period 2π which is orthogonal over $[0, 2\pi]$ to all of the 2π-periodic solutions of the homogeneous equation—which is orthogonal to all of the 2π-periodic solutions of the adjoint equation? How does one obtain such a solution?

EXERCISE 1.3.   Show that every solution of (1.1) is unbounded if relation (1.2) is not satisfied and A, f are in $\mathscr{P}_T$.

THEOREM 1.1.   Suppose A is in $\mathscr{P}_T$ and $\mathscr{D}$ is one of the classes $\mathscr{B}(-\infty, \infty)$, $\mathscr{A}\mathscr{P}$ or $\mathscr{P}_T$. The nonhomogeneous equation (1.1) has a solution $\mathscr{K}f$ in $\mathscr{D}$ for every f in $\mathscr{D}$ if and only if system (1) is noncritical with respect to $\mathscr{D}$. Furthermore, if system (1) is noncritical with respect to $\mathscr{D}$, then $\mathscr{K}f$ is the only solution of (1.1) in $\mathscr{D}$ and is linear and continuous in f; that is, $\mathscr{K}(af + bg) = a\mathscr{K}f + b\mathscr{K}g$ for all a, b in C, f, g in $\mathscr{D}$ and there is a constant K such that

(1.8) $|\mathscr{K}f| \leq K|f|,$

for all f in $\mathscr{D}$. Finally, if $\mathscr{D} = \mathscr{A}\mathscr{P}$, then $m[\mathscr{K}f] \subset m[f, A]$.

PROOF.   *Case 1.* $\mathscr{D} = \mathscr{P}_T$.   If f belongs to $\mathscr{P}_T$, then Lemma 1.1 implies that equation (1.1) has a solution in $\mathscr{P}_T$ if and only if f is orthogonal in the sense of (1.2) to all T-periodic solutions of (1.3). But $\dot{x} = A(t)x$ and $\dot{y} = -yA(t)$ have the same number of linearly independent T-periodic solutions. Therefore, equation (1.1) has a solution in $\mathscr{P}_T$ for every f in $\mathscr{P}_T$ if and only if equation (1) has no nontrivial solutions in $\mathscr{P}_T$; that is, (1) is noncritical with respect to $\mathscr{P}_T$. If system (1) is noncritical with respect to $\mathscr{P}_T$, then Lemma 1.1 implies system (1.1) has a unique solution $\mathscr{K}f$ in $\mathscr{P}_T$ for every f in $\mathscr{P}_T$. It is clear from the uniqueness that $\mathscr{K}$ is a linear mapping of $\mathscr{P}_T$ into $\mathscr{P}_T$.
    If $X(t, \tau)$, $X(\tau, \tau) = I$, is the principal matrix solution of (1), then the function $\mathscr{K}f$ can be written explicitly as

(1.9) $(\mathscr{K}f)(t) = \displaystyle\int_0^T [X^{-1}(t + T, t) - I]^{-1}X(t, t + s)f(t + s)\, ds.$

The kernel function in this expression is known as the *Green's function* for the boundary value problem $x(0) = x(T)$ for (1.1). The explicit computations for obtaining (1.9) proceed as follows.
    If we let $(\mathscr{K}f)(0) = x_0$, then

$$(\mathscr{K}f)(t) = X(t, 0)x_0 + \int_0^t X(t, s)f(s)\, ds.$$

Since $(\mathscr{K}f)(t + T) = (\mathscr{K}f)(t)$, and $X(t,s) = X(t,\tau)X(\tau,s)$ for any t,τ,s, we have

$$[I - X(t + T,t)]X(t, 0)x_0 = -[I - X(t + T,t)]\int_0^t X(t, s)f(s)\, ds$$

$$+ X(t + T,t)\int_t^{t+T} X(t, s)f(s)\, ds.$$

Multiplication by $[I - X(t + T,t)]^{-1}$ and a substitution in the formula for $(\mathscr{K}f)(t)$ yields (1.9).

Formula (1.9) obviously implies there is a constant K such that (1.8) is satisfied. In fact, K can be chosen as

$$T^{-1}K = \sup_{0 \le s,\, t \le T} |[X^{-1}(t + T,t) - I]^{-1}X(t, t + s)|.$$

This proves Case 1.

Case 2. $\mathscr{D} = \mathscr{B}(-\infty, \infty)$. Let $X(t)$ be a fundamental matrix solution of (1). The Floquet representation implies $X(t) = P(t)e^{Bt}$ where $P(t + T) = P(t)$ and B is a constant. Furthermore, the transformation $x = P(t)y$ applied to (1.1) yields

$$(1.10) \qquad \dot{y} = By + P^{-1}(t)f(t) \overset{\text{def}}{=} By + g(t),$$

where g is in $\mathscr{B}(-\infty, \infty)$ (or $\mathscr{A}\mathscr{P}$) if f is in $\mathscr{B}(-\infty, \infty)$ (or $\mathscr{A}\mathscr{P}$). Since $P(t)$ is nonsingular, it is therefore sufficient for the first part of the theorem to show that (1.10) has a solution in $\mathscr{B}(-\infty, \infty)$ for every g in $\mathscr{B}(-\infty, \infty)$ if and only if $\dot{y} = By$ is noncritical with respect to $\mathscr{B}(-\infty, \infty)$; that is, no eigenvalues of B have zero real parts.

By a similarity transformation, we may assume that

$$B = \text{diag}(B_+, B_0, B_-)$$

where all eigenvalues of $B_+(B_0)(B_-)$ have positive (zero) (negative) real parts. If $y = (u, v, w)$, $g = (g_+, g_0, g_-)$ are partitioned so that block matrix multiplication will be compatible with the partitioning of B, then (1.10) is equivalent to the system

$$(1.11) \qquad \text{(a)} \quad \dot{u} = B_+ u + g_+,$$
$$\text{(b)} \quad \dot{v} = B_0 v + g_0,$$
$$\text{(c)} \quad \dot{w} = B_- w + g_-.$$

There are positive constants K, α so that

(1.12)   (a) $\ |e^{B_+ t}| \leq K e^{\alpha t}, \qquad t \leq 0,$

(b) $\ |e^{B_- t}| \leq K e^{-\alpha t}, \qquad t \geq 0.$

Equations (1.11a), (1.11c) have unique bounded solutions on $(-\infty, \infty)$ given, respectively, by

(1.13)   (a) $\ (\mathcal{K}_+ g_+)(t) = \displaystyle\int_\infty^0 e^{-B_+ s} g_+(t+s)\, ds,$

(b) $\ (\mathcal{K}_- g_-)(t) = \displaystyle\int_{-\infty}^0 e^{-B_- s} g_-(t+s)\, ds.$

One can either verify this directly or apply Lemma III.6.1. These remarks show that if $\dot{y} = By$ is noncritical with respect to $\mathscr{B}(-\infty, \infty)$ then there is a unique solution $\mathcal{K}f$ of (1.10) in $\mathscr{B}(-\infty, \infty)$. Furthermore, using (1.12) we see that $|\mathcal{K}_+ g_+| \leq (K/\alpha)|g_+|$, $|\mathcal{K}_- g_-| \leq (K/\alpha)|g_-|$. Therefore, $\mathcal{K}f$ satisfies (1.8).

If the matrix B has any eigenvalues with zero real parts; that is, the vector v in (1.11b) is not zero dimensional, we show there is a g_0 in $\mathscr{B}(-\infty, \infty)$ such that all solutions of (1.10) are unbounded in $(-\infty, \infty)$. This will complete the proof of Case 2. Without loss in generality, we may assume that $B_0 = \mathrm{diag}(B_{01}, \ldots, B_{0s})$, where $B_{0j} = i\omega_j I + R_j$ and R_j has only zero as an eigenvalue. It is enough to consider only one of the matrices B_{0j} since $i\omega_j$ may be eliminated by the multiplicative transformation $\exp(i\omega_j t)$. Thus, we consider the equation

$$\dot{x} = Rx + g,$$

where R has only zero as an eigenvalue and g is in $\mathscr{B}(-\infty, \infty)$. For any row vector a,

$$a\dot{x} = aRx + ag,$$

for all t. If $a \neq 0$ is chosen so that $aR = 0$ and $g = a^*$, where a^* is the conjugate transpose of a, then $a\dot{x} = |a|^2 > 0$ which implies $ax(t) \to \infty$ for every solution $x(t)$. Therefore, every solution is unbounded as $t \to \infty$. This completes the proof of case 2. Notice that the g chosen in (1.10) to make all solutions unbounded if B has eigenvalues with zero real parts was actually periodic and, therefore, $f = P(t)g$ is an almost periodic (quasiperiodic) function which makes all solutions of (1.1) unbounded if (1) has purely imaginary characteristic exponents.

Case 3. $\mathscr{D} = \mathscr{AP}$. The previous remark implies that (1.1) has a solution in $\mathscr{B}(-\infty, \infty)$ for all f in $\mathscr{AP}$ if and only if no characteristic exponents of (1) have zero real parts or, equivalently, (1) is noncritical with respect to $\mathscr{A}\mathscr{P}$. If (1) is noncritical with respect to $\mathscr{A}\mathscr{P}$, then the unique solution $\mathscr{K}f$ in $\mathscr{B}(-\infty, \infty)$ is a periodic transformation of the functions given in (1.13) where g_+, g_- are almost periodic with their modules contained in $m[f, A]$ Therefore, it remains only to show that the functions in (1.13) are in $\mathscr{A}\mathscr{P}$ and their modules are in $m[f, A]$. We make use of Definition 1 of the Appendix.

Suppose $\alpha' = \{\alpha'_n\}$ is a sequence in R. Since g is almost periodic there is a subsequence $\alpha = \{\alpha_n\}$ of α' such that $\{g(t + \alpha_n)\}$ converges uniformly on $(-\infty, \infty)$. Since $\mathscr{K}_+$ is a continuous linear operator on $\mathscr{B}(-\infty, \infty)$, this implies $\{(\mathscr{K}_+g_+)(t + \alpha_n)\}$, converges uniformly on $(-\infty, \infty)$. Thus, Definition 1 of the Appendix implies $\mathscr{K}_+g_+$ is almost periodic. Theorem 8 of the Appendix implies $m[\mathscr{K}_+g_+] \subset m[g_+] \subset m[f,A]$. The same argument applies to $\mathscr{K}_-g_-$ to complete the proof of Case 3 and the theorem.

EXERCISE 1.4.   Let $\mathscr{K}$ be the operator defined in Theorem 1.1. Prove or disprove the relation $m[\mathscr{K}f] = m[f, A]$ for every f in $\mathscr{A}\mathscr{P}$.

Theorem 1.1 clearly illustrates that requiring a certain behavior for some solutions of the nonhomogeneous equation (1.1) for forcing functions f in a large class of solutions imposes strong conditions on the homogeneous equation (1). For the case in which A does not belong to $\mathscr{P}_T$, similar conclusions can be drawn but the analysis becomes more difficult. Preliminary to the statement of the simplest result of this type are some lemmas of independent interest.

LEMMA 1.1.   If $A(t)$ is a continuous $n \times n$ matrix, $|A(t)| < M$, $0 \leq t < \infty$, and $X(t, \tau)$, $X(\tau, \tau) = I$, is the principal matrix solution of (1), then there is a $\delta > 0$ such that

$$|X(t, s) - X(t, \tau)| \leq \frac{1}{2} |X(t, \tau)|,$$

for all t, τ, $s \geq 0$ and $|s - \tau| \leq \delta$.

PROOF.   Since $X(t, s) = X(t, \tau)X(\tau, s)$ for all t, τ, s in $[0, \infty)$, it is sufficient to show there is a $\delta > 0$ such that $|X(\tau, s) - I| \leq 1/2$ for $\tau, s \geq 0$, $|\tau - s| \leq \delta$. Since

$$X(\tau, s) - I = \int_s^\tau A(u)[X(u, s) - I] \, du + \int_s^\tau A(u) \, du,$$

an application of Gronwall's inequality yields $|X(\tau, s) - I| \leq M |\tau - s| e^{M|\tau - s|}$ for all τ, s. If δ is such that $M\delta e^{M\delta} \leq 1/2$, the lemma is proved.

LEMMA 1.2. With A and $X(t, \tau)$ as in Lemma 1.1, the condition $\int_0^t |X(t, s)| \, ds < c$, a constant, for all $t \geq 0$ implies $|X(t, s)|$ uniformly bounded for $0 \leq s \leq t < \infty$; that is, the equation (1) is uniformly stable for $t_0 \geq 0$.

PROOF. Since $X(t, s)X(s, t) = I$ for all s, t, it follows that $X(t, s)$ as a function of s is a fundamental matrix solution of the adjoint equation. Therefore,

$$X(t, s) = I + \int_s^t X(t, \xi)A(\xi)d\xi$$

for all t, s. In particular, for $o \leq s \leq t$, the hypotheses of the theorem imply $|X(t, s)| \leq 1 + Mc$. The uniform stability follows from Theorem III.2.1. This proves the lemma.

THEOREM 1.2. If $A(t)$ is a continuous $n \times n$ matrix, $|A(t)| < M$, $0 \leq t < \infty$, then every solution of (1.1) is bounded on $[0, \infty)$ for every continuous f bounded on $[0, \infty)$ if and only if system (1) is uniformly asymptotically stable.

PROOF. We first prove that if every solution of (1.1) is bounded on $[0, \infty)$ for every continuous f bounded on $[0, \infty)$, then $\int_0^t |X(t, s)| \, ds < c$, a constant, for $t \geq 0$. The solution of (1.1) with $x(0) = 0$ is given by

$$x(t) = \int_0^t X(t, s)f(s) \, ds.$$

Let $\mathscr{B}[0, \infty)$ be the class of functions $f : [0, \infty) \to C^n$, f continuous and bounded and let $|f| = \sup_{0 \leq t < \infty} |f(t)|$. For any fixed t in $[0, \infty)$, consider the mapping $T_t : \mathscr{B}[0, \infty) \to \mathscr{B}[0, \infty)$ given by

$$(T_t f)(\alpha) = \begin{cases} \int_0^\alpha X(\alpha, s)f(s) \, ds, & 0 \leq \alpha \leq t \\ \int_0^t X(t, s)f(s) \, ds, & t \leq \alpha < \infty. \end{cases}$$

For each fixed t, T_t is a continuous linear map. Furthermore, by hypothesis, for each f in $\mathscr{B}[0, \infty)$, there is a constant N such that $|T_t f| \leq N$, $0 \leq t < \infty$. Consequently, the principle of uniform boundedness implies there is a constant K such that $|T_t f| \leq K |f|$ for all t in $[0, \infty)$, f in $\mathscr{B}[0, \infty)$. In particular,

$$\left| \int_0^t X(t, s)f(s)ds \right| \leq K |f|, \qquad 0 \leq t < \infty.$$

If $X = (x_{jk})$, let $f_t^{j,k}(s)$ be the function which is the sign of $x_{jk}(t, s)$, $j, k = 1, 2, \ldots, n$, for a fixed t. Choose a sequence of uniformly bounded continuous functions $f_{t,r}^{j,k}(s)$ which approach $f_t^{j,k}(s)$ pointwise almost everywhere on $[0, t]$. Choose the norm of a vector $x = (x_1, \ldots, x_n)$ to be $|x| = \max_j |x_j|$. For any given j, k, let $f_t(s), f_{t,r}(s)$ be the n-vectors with all components zero except the kth which is $f_t^{j,k}(s), f_{t,r}^{j,k}(s)$, respectively. Then

$$\lim_{r \to \infty} \int_0^t X(t, s) f_{t,r}(s) \, ds = \int_0^t X(t, s) f_t(s) \, ds.$$

From the above definition of the norm of a vector and the fact that

$$\int_0^t X(t, s) f_{t,r}(s) \, ds \Big| \leq K \, |f_{t,r}| \leq K_1, \qquad 0 \leq t < \infty,$$

it follows that

$$\int_0^t |x_{jk}(t, s)| \, ds \leq \left| \int_0^t X(t, s) f_t(s) \, ds \right| \leq K_1,$$

for all $t \geq 0$ and $j, k = 1, 2, \ldots, n$. Since all norms in C^n are equivalent, this clearly implies there is a constant c such that $\int_0^t |X(t, s)| \, ds < c, 0 \leq t < \infty$.

From Lemma 1.2, this relation implies $X(s, \tau)$ is uniformly bounded by a constant N for $0 \leq \tau \leq s < \infty$ and, thus, equation (1) is uniformly stable for $t_0 \geq 0$. Therefore,

$$\left| \int_\tau^t X(t, s) X(s, \tau) \, ds \right| \leq \int_\tau^t |X(t, s)| \cdot |X(s, \tau)| \, ds \leq Nc,$$

for all $t \geq \tau$. But the first expression in this inequality is equal to $(t - \tau) |X(t, \tau)|$. Therefore, $|X(t, \tau)|$ approaches zero as $t \to \infty$ uniformly and equation (1) is uniformly asymptotically stable.

Conversely, suppose system (1) is uniformly asymptotically stable for $t_0 \geq 0$. From Theorem III.2.1, there are positive constants K, α such that

$$|X(t, \tau)| \leq K e^{-\alpha(t-\tau)}, \qquad 0 \leq \tau \leq t < \infty.$$

The general solution of (1.1) is

$$x(t) = X(t, 0) x(0) + \int_0^t X(t, s) f(s) \, ds, \qquad t \geq 0.$$

If f is in $\mathscr{B}[0, \infty)$, then $|x(t)| \leq K |x(0)| + |f| \cdot (K/\alpha)$ for all $t \geq 0$ and thus every solution of (1.1) is in $\mathscr{B}[0, \infty)$. This proves the theorem.

Theorem 1.2 is due to Perron [1] and was the first general statement dealing with the determination of the behavior of the solutions of a homogeneous equation by observing the behavior of the solutions of a nonhomogeneous equation for forcing functions in a certain given class of functions. Investigations along this line have continued to this day with the most significant recent contributions being made by Massera and Schäffer [1]. For a better appreciation of the problem, we rephrase it in another way. Let $(\mathscr{B}, \mathscr{D})$ be two Banach spaces of functions mapping $[\sigma, \infty)$ into C^n where σ may be finite or infinite. The pair $(\mathscr{B}, \mathscr{D})$ is said to be *admissible for equation* *(1.1)* if for every f in $\mathscr{D}$, there is at least one solution x of (1.1) in $\mathscr{B}$. Theorem 1.1 states that $(\mathscr{B}(-\infty, \infty), \mathscr{B}(-\infty, \infty))$, $(\mathscr{A}\mathscr{P}, \mathscr{A}\mathscr{P})$, $(\mathscr{P}_T, \mathscr{P}_T)$ are admissible for equation (1.1) if and only if equation (1) is noncritical with respect to $\mathscr{B}(-\infty. \infty)$, $\mathscr{A}\mathscr{P}$, $\mathscr{P}_T$, respectively. Theorem 1.2 shows that $(\mathscr{B}[0, \infty)$, $\mathscr{B}[0, \infty))$ is admissible for (1.1) if equation (1) is uniformly asymptotically stable. The further investigation of such admissible pairs is extremely interesting and the reader is referred to Coppel [1, Chap. V], Hartman [1, Chap. 13] and Massera and Schäffer [1], Antosiewicz [2].

IV.2. Weakly Nonlinear Equations—Noncritical Case

Throughout this section, it will be assumed that A is a continuous $n \times n$ matrix in $\mathscr{P}_T$, $\Omega(\rho, \sigma) = \{x \text{ in } C^n, \varepsilon \text{ in } C^r \colon |x| \leq \rho, |\varepsilon| \leq \sigma\}$, $\eta(\rho, \sigma)$, $M(\sigma)$, $\rho \geq 0$, $\sigma \geq 0$, are continuous functions which are nondecreasing in both variables, $\eta(0, 0) = 0$, $M(0) = 0$, and $\mathscr{L}i\not{h}(\eta, M) = \{q \colon R \times \Omega(\rho_0, \varepsilon_0) \to C^n \colon q \text{ continuous}, |q(t, 0, \varepsilon)| \leq M(|\varepsilon|), |q(t, x, \varepsilon) - q(t, y, \varepsilon)| \leq \eta(\rho, \sigma)|x - y|$, for all (t, x, ε), (t, y, ε) in $R \times \Omega(\rho, \sigma)$, $0 \leq \rho \leq \rho_0$, $0 < \sigma < \epsilon_0$ and $q(t, x, \varepsilon)$ is continuous in x, ϵ uniformly for t in $R\}$.

If q is in $\mathscr{L}i\not{h}(\eta, M)$, then automatically $q(\cdot, x, \varepsilon)$ is in $\mathscr{B}(-\infty, \infty)$. In fact, $|q(t, x, \varepsilon)| \leq \eta(\rho, \sigma)|x| + M(|\varepsilon|)$ for (t, x, ε) in $R \times \Omega(\rho, \sigma)$. A function q will be said to be in $\mathscr{A}\mathscr{P} \cap \mathscr{L}i\not{h}(\eta, M)$ if q is in $\mathscr{L}i\not{h}(\eta, M)$ and, for each fixed ε, $q(t, x, \varepsilon)$ is almost periodic in t uniformly with respect to x for x in compact sets. A function q will be said to be in $\mathscr{P}_T \cap \mathscr{L}i\not{h}(\eta, M)$ if q is in $\mathscr{L}i\not{h}(\eta, M)$ and $q(t + T, x, \varepsilon) = q(t, x, \varepsilon)$ for all (t, x, ε) in $R \times \Omega(\rho_0, \varepsilon_0)$.

A function q will clearly be in $\mathscr{L}i\not{h}(\eta, M)$ for some η, M if $q(t, x, \varepsilon) \to 0$ and $\partial q(t, x, \varepsilon)/\partial x \to 0$ as $x \to 0$, $\varepsilon \to 0$ uniformly in t.

In this section, results are given concerning the existence of bounded, almost periodic and periodic solutions of the nonlinear equation

$$(2.1) \qquad \dot{x} = A(t)x + q(t, x, \varepsilon),$$

where q is in $\mathscr{L}i\not{h}(\eta, M)$ and the homogeneous equation (1) is noncritical. More specifically, we prove

THEOREM 2.1. Suppose $\mathscr{D}$ is one of the classes $\mathscr{B}(-\infty, \infty)$, $\mathscr{A}\mathscr{P}$ or $\mathscr{P}_T$. If q is in $\mathscr{D} \cap \mathscr{L}i\not{h}(\eta, M)$ and system (1) is noncritical with respect to $\mathscr{D}$, then there are constants $\rho_1 > 0$, $\varepsilon_1 > 0$ and a function $x^*(t, \varepsilon)$ continuous in t, ε for $-\infty < t < \infty$, $0 \leq |\varepsilon| \leq \varepsilon_1$, $x^*(t, 0) = 0$, $x^*(\cdot, \varepsilon)$ in $\mathscr{D}$, $|x^*(\cdot, \varepsilon)| \leq \rho_1$, $0 \leq |\varepsilon| \leq \varepsilon_1$, such that $x^*(t, \varepsilon)$ is a solution of (2.1) and is the only solution of (2.1) in $\mathscr{D}$ which has norm $\leq \rho_1$. If $\mathscr{D} = \mathscr{A}\mathscr{P}$ then $m[x^*(\cdot, \varepsilon)] \subset m[q, A]$, $0 \leq |\varepsilon| \leq \varepsilon_1$.

PROOF.   For a given ρ_1, $0 < \rho_1 \leq \rho_0$, let $\mathscr{D}_{\rho_1} = \{x \text{ in } \mathscr{D}: |x| \leq \rho_1\}$. Then $\mathscr{D}_{\rho_1}$ is a closed, bounded subset of the Banach space $\mathscr{D}$. For any x in $\mathscr{D}_{\rho_1}$, the function $q(\cdot, x(\cdot), \varepsilon)$ belongs to $\mathscr{D}$. Since system (1) is assumed to be non-critical with respect to $\mathscr{D}$, we may consider the transformation $w = \mathscr{T}x$, x in $\mathscr{D}$, defined by

(2.2) $$w = \mathscr{T}x = \mathscr{K}q(\cdot, x(\cdot), \varepsilon),$$

where $\mathscr{K}$ is the operator uniquely defined by Theorem 1.1. From Theorem 1.1, $\mathscr{T} : \mathscr{D}_{\rho_1} \to \mathscr{D}$. Furthermore, the fixed points of $\mathscr{T}$ in $\mathscr{D}_{\rho_1}$ coincide with the solutions of (2.1) which are in $\mathscr{D}_{\rho_1}$. We now use the contraction principle to show that $\mathscr{T}$ has a unique fixed point in $\mathscr{D}_{\rho_1}$ for ρ_1 and $|\varepsilon|$ sufficiently small.

Since q is in $\mathscr{Lip}(\eta, M)$,

(2.3) $$|q(t, x, \varepsilon)| \leq |q(t, x, \varepsilon) - q(t, 0, \varepsilon)| + |q(t, 0, \varepsilon)|$$
$$\leq \eta(\rho_1, \varepsilon_1)|x| + M(\varepsilon_1),$$

for (t, x, ε) in $R \times \Omega(\rho_1, \varepsilon_1)$. Let K be the constant defined in (1.8) and choose $\rho_1 \leq \rho_0$, $\varepsilon_1 \leq \varepsilon_0$ positive and so small that

$$K[\eta(\rho_1, \varepsilon_1)\rho_1 + M(\varepsilon_1)] < \rho_1.$$

For this choice of ρ_1, ε_1, it follows from relations (1.8), (2.2), (2.3) and the fact that q is in $\mathscr{Lip}(\eta, M)$ that

$$|\mathscr{T}x| \leq K|q(\cdot, x(\cdot), \varepsilon)| \leq K[\eta(\rho_1, \varepsilon_1)|x| + M(\varepsilon_1)]$$
$$< \rho_1,$$

$$|\mathscr{T}x - \mathscr{T}y| \leq K|q(\cdot, x(\cdot), \varepsilon) - q(\cdot, y(\cdot), \varepsilon)|$$
$$\leq K\eta(\rho_1, \varepsilon_1)|x - y|$$
$$< \theta|x - y|,$$

for all x, y in $\mathscr{D}_{\rho_1}$, $0 \leq |\varepsilon| \leq \varepsilon_1$ and $\theta < 1$ is a positive constant. Therefore, $\mathscr{T}$ is a uniform contraction on $\mathscr{D}_{\rho_1}$ for $|\varepsilon| \leq \varepsilon_1$ and has a unique fixed point $x^*(t, \epsilon)$ in $\mathscr{D}_{\rho_1}$. Since $q(t, x, \epsilon)$ is continuous in x, ϵ uniformly in t, $x^*(t, \epsilon)$ is continuous in t, ϵ, $-\infty < t < \infty$, $0 \leq |\varepsilon| \leq \varepsilon_1$. For $\varepsilon = 0$, $x = 0$ is obviously a solution of (2.1) and, therefore, $x^*(\cdot, 0) = 0$. To prove the last statement of the theorem, suppose $\alpha' = \{\alpha'_n\}$ is a sequence in $(-\infty, \infty)$. Let M be a compact set containing $\{x^*(t, \epsilon), t\epsilon(-\infty, \infty), |\epsilon| \leq \epsilon_1\}$. Theorem 11 of the Appendix implies there is a subsequence $\alpha = \{\alpha_n\}$ of α' such that $\{q(t + \alpha_n, x, \epsilon)\}$ converges uniformly for $t\epsilon(-\infty, \infty)$, $x\epsilon M$ and each fixed ϵ. Since $x^*(t, \epsilon) = \mathscr{K}q(\cdot, x^*(\cdot, \epsilon), \epsilon)$

and $\mathcal{K}$ is continuous and linear on $\mathcal{B}(-\infty, \infty)$, this implies $\{x^*(t + \alpha_n, \epsilon)\}$ converges uniformly for t in $(-\infty, \infty)$ for each fixed ϵ. The fact that

$$m[x^*(\cdot, \epsilon)] \subset m[q, A]$$

follows from Theorem 8 of the Appendix.

Theorem 2.1 has interesting implications for the equation

(2.4) $$\dot{x} = A(t)x + b(t) + \epsilon h(t, x, \epsilon),$$

where ϵ is a scalar, h belongs to $\mathcal{DL}i_p(\eta, M)$, b is in $\mathcal{D}$ and system (1) is noncritical with respect to $\mathcal{D}$. Of course, as before $\mathcal{D}$ is one of the classes $\mathcal{B}(-\infty, \infty)$, $\mathcal{AP}$ or $\mathcal{P}_T$. Under these hypotheses, Theorem 1.1 implies that the equation

$$\dot{x} = A(t)x + b(t),$$

has a unique solution $\mathcal{K}b$ in $\mathcal{D}$. If $|\mathcal{K}b| < \rho_0$ and $x = y + \mathcal{K}b$ in (2.4), then

$$\dot{y} = A(t)y + \epsilon h(t, y + \mathcal{K}b, \epsilon) \overset{\text{def}}{=} A(t)y + q(t, y, \epsilon).$$

The function q is in $\mathcal{L}i_p(\eta, M)$ with $\eta(\rho, \sigma) = \sigma \eta_1(\rho, \sigma)$, where $\eta_1(\rho, \sigma)$ is a continuous function of ρ, σ, since $q(t, y, \epsilon)$ approaches zero as $\epsilon \to 0$ at least as fast as a linear function in ϵ. Therefore, Theorem 2.1 implies the existence of a solution $x^*(t, \epsilon, b)$ of (2.4), $x^*(\cdot, \epsilon, b)$ in $\mathcal{D}$, $x^*(\cdot, 0, b) = \mathcal{K}b$ and this is the only solution of (2.4) which is in $\mathcal{D}$ and at the same time in a ρ_1-neighborhood of $\mathcal{K}b$.

IV.3. The General Saddle Point Property

Theorem 2.1 of the previous section asserts the existence of a distinguished solution $x^*(\cdot, \epsilon)$ of (2.1) in a class $\mathcal{D}$ of functions. What are the stability properties of the solution $x^*(\cdot, \epsilon)$? If the real parts of the characteristic exponents of the linear system (1) have nonzero real parts, one would expect the stability properties of $x^*(\cdot, \epsilon)$ to be the same as those of the solution $x = 0$ of (1). The purpose of this section is to prove this is actually the case and also to discuss some of the geometrical properties of the solutions of (2.1) near $x^*(\cdot, \epsilon)$ in the same manner as the saddle point property was treated in Section III.6. The proofs will follow the ones in Section III.6. very closely but are slightly more complicated due to the fact that the differential equations depend explicitly upon t.

To reduce the equations to a simpler form, let

(3.1) $$x = x^*(\cdot, \epsilon) + y,$$

where $x^*(\cdot, \epsilon)$ is the function given in Theorem 2.1. If x is a solution of equation (2.1), then y is a solution of

(3.2) $$\dot{y} = A(t)y + p(t, y, \varepsilon),$$

where $p(t, y, \varepsilon) = q(t, x^*(t, \varepsilon) + y, \varepsilon) - q(t, x^*(t, \varepsilon), \varepsilon)$. Consequently, if q is in $\mathscr{L}i\not{p}(\eta, M)$, then p is in $\mathscr{L}i\not{p}(\eta, 0)$. To simplify the equations even further let $X(t) = P(t)e^{Bt}$, $P(t + T) = P(t)$, B a constant matrix, be a fundamental matrix solution of (1) and let $y = P(t)z$. If y is a solution of (3.2), then z is a solution of

(3.3) $$\dot{z} = Bz + f(t, z, \varepsilon),$$

where $f(t, z, \varepsilon) = P^{-1}(t)p(t, P(t)z, \varepsilon)$. Finally, we can assert that if $\mathscr{D}$ is one of the classes $\mathscr{B}(-\infty, \infty)$, $\mathscr{AP}$ or $\mathscr{P}_T$, and q is in $\mathscr{D} \cap \mathscr{L}i\not{p}(\eta, M)$, then f is in $\mathscr{D} \cap \mathscr{L}i\not{p}(\eta, 0)$. Any assertions made about system (3.3) yield implications for system (2.1) which are easily traced through the above transformations.

The remainder of the discussion centers around (3.3) under the hypothesis that f is in $\mathscr{L}i\not{p}(\eta, 0)$ and the eigenvalues of the matrix B have nonzero real parts, k with positive real parts and $n - k$ with negative real parts. As in Sect on III.6, the space C^n can be decomposed as

(3.4) $$C^n = C^n_+ \oplus C^n_-,$$
$$C^n_+ = \pi_+ C^n, \; C^n_- = \pi_- C^n,$$

where π_+, π_- are projection operators, C^n_+, C^n_- have dimensions k, $n - k$, respectively, are invariant under B and there are positive constants K, α such that

(3.5) (a) $|e^{Bt}\pi_+ z| \leq Ke^{\alpha t}|\pi_+ z|$, $t \leq 0$,

(b) $|e^{Bt}\pi_- z| \leq Ke^{-\alpha t}|\pi_- z|$, $t \geq 0$.

For any σ in $(-\infty, \infty)$, let $z(t, \sigma, z^\sigma, \varepsilon)$ designate the solution of (3.3) satisfying $z(\sigma, \sigma, z^\sigma, \varepsilon) = z^\sigma$, let K designate the constant in (3.5) and, for any $\delta > 0$, define

(3.6) (a) $$S(\sigma, \delta, \varepsilon) = \left\{ z^\sigma \text{ in } C^n \colon |\pi_- z^\sigma| < \frac{\delta}{2K}, |z(t, \sigma, z^\sigma, \varepsilon)| < \delta, t \geq \sigma \right\},$$

(b) $$U(\sigma, \delta, \varepsilon) = \left\{ z^\sigma \text{ in } C^n \colon |\pi_+ z^\sigma| < \frac{\delta}{2K}, |z(t, \sigma, z^\sigma, \varepsilon)| < \delta, t \leq \sigma \right\}.$$

Also, let B^n_ρ designate $\{z \text{ in } C^n \colon |z| < \rho\}$.

THEOREM 3.1.   If f is in $\mathscr{L}i\not{p}(\eta, 0)$ and the eigenvalues of B have nonzero real parts, then there are $\delta > 0$, $\varepsilon_1 > 0$, $\beta > 0$ such that for any σ in $(-\infty, \infty)$, $0 \leq |\varepsilon| \leq \varepsilon_1$, the mapping π_- is a homeomorphism of $S(\sigma, \delta, \varepsilon)$ onto $(\pi_- C^n) \cap B^n_{\delta/2K}$, $S(\sigma, \delta, 0)$ is tangent to $\pi_- C^n$ at zero and

(3.7) $$|z(t, \sigma, z^\sigma, \varepsilon)| \leq 2K|\pi_- z^\sigma|e^{-\beta(t-\sigma)}, \qquad t \geq \sigma,$$

for any z^σ in $S(\sigma, \delta, \varepsilon)$. The mapping π_+ is a homeomorphism of $U(\sigma, \delta, \varepsilon)$ onto $(\pi_+ C^n) \cap B^n_{\delta/2K}$, $U(\sigma, \delta, 0)$ is tangent to $\pi_+ C^n$ at zero and

$$(3.8) \qquad |z(t, \sigma, z^\sigma, \varepsilon)| \leq 2K |\pi_+ z^\sigma| e^{\beta(t-\sigma)}, \qquad t \leq \sigma,$$

for any z^σ in $U(\sigma, \delta, \varepsilon)$.

Furthermore, if $g(\cdot, \sigma, \varepsilon)$: $(\pi_- C^n) \cap B^n_{\delta/2K} \to S(\sigma, \delta, \varepsilon)$ is the inverse of the homeomorphism π_-, then $g(z_-, \sigma, \varepsilon)$ is lipschitzian in z_- with lipschitz constant $2K$. If f is in $\mathscr{AP} \cap \mathscr{Lip}(\eta, 0)$, then $g(z_-, \sigma, \varepsilon)$ is almost periodic in σ with module contained in $m[f]$. If f is in $\mathscr{P}_T \cap \mathscr{Lip}(\eta, 0)$, then $g(z_-, \sigma, \varepsilon)$ is periodic in σ of period T. The same conclusions hold for the inverse of the homeomorphism π_+ of $U(\sigma, \delta, \varepsilon)$ onto $(\pi_+ C^n) \cap B^n_{\delta/2K}$.

Before proving this theorem, let us make some remarks about its geometric meaning and some of its implications. The accompanying Fig. 3.1

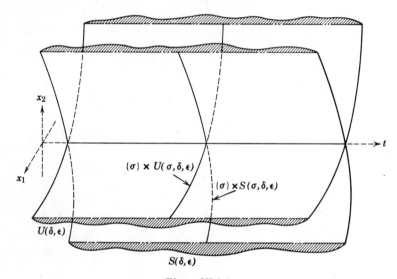

Figure IV.3.1

may be useful in visualizing the following remarks. For any $\sigma \in (-\infty, \infty)$, the set $\{\sigma\} \times S(\sigma, \delta, \varepsilon)$ is a subset of R^{n+1}. For any τ in $[\sigma, t]$, $z(t, \sigma, z^\sigma, \varepsilon) = z(t, \tau, z(\tau, \sigma, z^\sigma, \varepsilon), \varepsilon)$, and, therefore, any solution of (3.3) with initial value in $\{\sigma\} \times S(\sigma, \delta, \varepsilon)$ must cross $\{\tau\} \times S(\tau, \delta, \varepsilon)$ for any $\tau \geq 0$ for which $\pi_- z(\tau, \sigma, z^\sigma, \varepsilon)$ has norm less than $\delta/2K$. Since this may not occur for all $\tau \geq \sigma$, the set consisting of the union of the $\{\sigma\} \times S(\sigma, \delta, \varepsilon)$ for all σ in $(-\infty, \infty)$ may not be an integral manifold in the sense that the trajectory of any solution with initial value (τ, z^τ) on the manifold remains on the manifold for all $t \geq \tau$. On the other hand, this set can be extended to an integral manifold by extending the set $\{\sigma\} \times S(\sigma, \delta, \varepsilon)$ to a set $\{\sigma\} \times S^*(\sigma, \delta, \varepsilon)$ where $S^*(\sigma, \delta, \varepsilon) = S(\sigma, \delta, \varepsilon) \cup \{z: z = z(\sigma, \tau, z^\tau, \varepsilon), (\tau, z^\tau) \text{ in } \{\tau\} \times S(\tau, \delta, \varepsilon) \text{ for}$ some $\tau \leq \sigma\}$. The set $S(\delta, \varepsilon)$ consisting of the union of the $\{\sigma\} \times S^*(\sigma, \delta, \varepsilon)$ is then an integral manifold of (3.3) and is rightfully termed a *stable integral manifold* since all solutions on this manifold approach zero as $t \to \infty$ and these are the only solutions which lie in a certain neighborhood of zero for increasing time. In the same way one defines the corresponding sets $U^*(\sigma, \delta, \varepsilon) = U(\sigma, \delta, \varepsilon) \cup \{z: z = z(\sigma, \tau, z^\tau, \varepsilon), (\tau, z^\tau) \text{ in } U(\tau, \delta, \varepsilon) \text{ for some}$ $\tau \geq \sigma\}$ and an *unstable integral manifold* $U(\delta, \varepsilon)$. The sets $S(\delta, \varepsilon)$ and $U(\delta, \varepsilon)$ are hypersurfaces homeomorphic to $R \times B_1^{n-k}$ and $R \times B_1^k$, respectively, and $S(\delta, \varepsilon) \cap U(\delta, \varepsilon)$ is the t-axis.

If $k \geq 1$, the above remarks imply the solution $z = 0$ of (3.3) is *unstable* for $|\varepsilon| \leq \varepsilon_1$ and any σ in $(-\infty, \infty)$. From (3.7), if $k = 0$, the solution $z = 0$ of (3.3) is *uniformly asymptotically stable* for $|\varepsilon| \leq \varepsilon_1$ and $t_0 \geq \sigma$ for any σ in $(-\infty, \infty)$.

If B and f are real in (3.3), then the sets $S(\sigma, \delta, \varepsilon)$, $U(\sigma, \delta, \varepsilon)$ defined by taking only real initial values are in R^n. If system (3.2) is real, then the decomposition $X(t) = P(t)e^{Bt}$ of a fundamental matrix solution of (1) may not have $P(t)$, B real if it is required that $P(t + T) = P(t)$ for all T (see Section III.7). On the other hand, this decomposition can be chosen to be real if it is only required that $P(t + 2T) = P(t)$ for all t. In such a case, system (3.3) will be real, but f belongs to $\mathscr{P}_{2T} \cap \mathscr{L}i p(\eta, 0)$ if p is in $\mathscr{P}_T \cap \mathscr{L}i p(\eta, 0)$. This implies that the sets $S(\sigma, \delta, \varepsilon)$, $U(\sigma, \delta, \varepsilon)$ will be periodic in σ of period $2T$ rather than T.

It is not asserted in the statement of the theorem that $S(\sigma, \delta, \varepsilon)$ is tangent to $\pi_- C^n$ at zero. This may not be true since $f(t, y, \varepsilon)$ could contain a term which is linear in y and yet approaches zero when $\varepsilon \to 0$. Theorem 3.1 is clearly a strong generalization of Theorem III.6.1 due to the fact that the perturbation term $f(t, z, \varepsilon)$ may depend explicitly upon t.

PROOF OF THEOREM 3.1. Since this proof is so similar to the proof of Theorem III.6.1, it is not necessary to give the details but only indicate the differences. In the same way as in the proof of Lemma 6.1, one easily shows that, for any solution $z(t)$ of (3.3) which is bounded on $[\sigma, \infty)$, there must exist a z_- in C_-^m such that

$$(3.9) \qquad z(t) = e^{B(t-\sigma)}z_- + \int_\sigma^t e^{B(t-s)}\pi_- f(s, z(s), \varepsilon) \, ds$$

$$+ \int_\infty^0 e^{-Bs}\pi_+ f(t + s, z(t + s), \varepsilon) \, ds, \qquad t \geq \sigma,$$

and for any solution $z(t)$ of (3.3) which is bounded on $(-\infty, \sigma]$, there is a z_+ in C_+^m such that $z(t)$ satisfies the relation

$$(3.10) \qquad z(t) = e^{B(t-\sigma)}z_+ + \int_\sigma^t e^{B(t-s)}\pi_+ f(s, z(s), \varepsilon)\, ds$$

$$+ \int_{-\infty}^0 e^{-Bs}\pi_- f(t+s, z(t+s), \varepsilon)\, ds, \qquad t \leqq \sigma.$$

We first discuss the existence of solutions of (3.9) on $[\sigma, \infty)$ for any z_- in $\pi_- C^n$. There is a constant K_1 such that $|\pi_+ z| \leq K_1|z|$, $|\pi_- z| \leq K_1|z|$ for any z in C^n. If K, α are the constants given in (3.5), and η is the lipschitz constant of $f(t, z, \varepsilon)$ with respect to z, choose δ, ε_1, so that

$$(3.11) \qquad 4KK_1\eta(\delta, \varepsilon_1) \leqq \alpha, \qquad 8K^2 K_1\eta(\delta, \varepsilon_1) < 3\alpha.$$

With this choice of δ, ε_1 and for any z_- in $\pi_- C^n$ with $|z_-| \leq \delta/2K$, define $\mathscr{G}(\sigma, z_-, \delta)$ as the set of continuous functions $z\colon [\sigma, \infty) \to C^n$ such that $|z| = \sup_{\sigma \leqq t < \infty}|z(t)| \leq \delta$ and $\pi_- z(\sigma) = z_-$. $\mathscr{G}(\sigma, z_-, \delta)$ is a closed bounded subset of the Banach space of all bounded continuous functions taking $[\sigma, \infty)$ into C^n with the uniform topology. For any z in $\mathscr{G}(\sigma, z_-, \delta)$ define $\mathscr{T}z$ by

$$(3.12) \qquad (\mathscr{T}z)(t) = e^{B(t-\sigma)}z_- + \int_\sigma^t e^{B(t-s)}\pi_- f(s, z(s), \varepsilon)\, ds$$

$$+ \int_\infty^0 e^{-Bs}\pi_+ f(t+s, z(t+s), \varepsilon)\, ds, \qquad t \geqq \sigma.$$

Exactly as in the proof of Theorem III.6.1, one uses the contraction principle to show that $\mathscr{T}$ has a unique fixed point $z^*(\cdot, \sigma, z_-, \varepsilon)$ for $|\varepsilon| \leqq \varepsilon_1$, $z^*(\cdot, \sigma, 0, \varepsilon) = 0$, $z^*(t, \sigma, z_-, \varepsilon)$ depends continuously upon t, σ, z_-, ε and

$$(3.13) \quad |z^*(t, \sigma, z_-, \varepsilon) - z^*(t, \sigma, \tilde{z}_-, \varepsilon)| \leq 2K\left[\exp\frac{-\alpha(t-\sigma)}{2}\right]|z_- - \tilde{z}_-|$$

for $t \geqq \sigma$.

From the definition of $S(\sigma, \delta, \varepsilon)$ and the above construction of $z^*(\cdot, \sigma, z_-, \varepsilon)$, it follows that

$$(3.14) \qquad S(\sigma, \delta, \varepsilon) = \{z\colon z = z^*(\sigma, \sigma, z_-, \varepsilon),\ z_-\ \text{in}\ (\pi_- C^n) \cap B_{\delta/2k}\}$$

for $|\varepsilon| \leq \varepsilon_1$. Relation (3.14), (3.13) and the fact that $z^*(\cdot, \sigma, 0, \varepsilon) = 0$ implies relation (3.7) with $\beta = \alpha/2$. If we let $g(z_-, \sigma, \varepsilon) = z^*(\sigma, \sigma, z_-, \varepsilon)$, then

$$(3.15) \qquad g(z_-, \sigma, \varepsilon) = z_- + \int_\infty^0 e^{-Bs}\pi_+ f(\sigma+s, z^*(\sigma+s, \sigma, z_-, \varepsilon), \varepsilon)\, ds.$$

As in the proof of Theorem III.6.1,

$$|g(z_-, \sigma, \varepsilon)| - g(\tilde{z}_-, \sigma, \varepsilon)| \geqq \frac{|z_- - \tilde{z}_-|}{2}$$

for any σ in $(-\infty, \infty)$ and $|\varepsilon| \leq \varepsilon_1$. This shows that $g(\cdot, \sigma, \varepsilon)$ is a one-to-one map of $(\pi_- C^n) \cap B_{\delta/2K}$ into $S(\sigma, \delta, \varepsilon)$. Since the inverse of this map is π_- and therefore is continuous, it follows that it is a homeomorphism. The following estimate is also easy to obtain:

$$|\pi_+ z^*(\sigma, \sigma, z_-, \varepsilon)| \leq \frac{4K^2 K_1}{\alpha} \eta(2K |z_-|, |\varepsilon|) |z_-|.$$

Since $\pi_- z^*(\sigma, \sigma, z_-, \varepsilon) = z_-$, it follows from the properties of η that $S(\sigma, \delta, 0)$ is tangent to $\pi_- C^n$ at zero. Relation (3.13) implies that the set $S(\sigma, \delta, 0)$ is a lipschitzian manifold.

Now suppose f is in $\mathscr{A}\mathscr{P} \cap \mathscr{L}i \textit{p}(\eta, 0)$. We now prove the representation $g(z_-, \sigma, \varepsilon)$ of $S(\sigma, \delta \varepsilon)$ is almost periodic in σ with module contained in $m[f]$. From (3.15), it will be necessary to estimate $z^*(\sigma + s, \sigma, z_-, \varepsilon)$ as a function of σ and s. To simply the notation, let $z(t, \sigma) = z^*(t, \sigma, z_-, \varepsilon), f(t, z) = f(t, z, \varepsilon)$. The equation for $z(\sigma + t, \sigma)$ becomes

$$(3.16) \quad z(\sigma + t, \sigma) = e^{Bt} z_- + \int_0^t e^{B(t-s)} \pi_- f(\sigma + s, z(\sigma + s, \sigma)) \, ds$$

$$+ \int_\infty^0 e^{-Bs} \pi_+ f(\sigma + t + s, z(\sigma + t + s, \sigma)) \, ds.$$

The objective is to show that $z(\sigma + t, \sigma)$ is almost periodic in σ if f is in $\mathscr{A}\mathscr{P} \cap \mathscr{L}i \textit{p}(\eta, 0)$. Suppose $\alpha' = \{\alpha'_n\}$ is a sequence in $(-\infty, \infty)$. There is a subsequence $\alpha = \{\alpha_n\}$ of α' such that $\{f(t + \alpha_n, z)\}$ converges uniformly for $t\epsilon(-\infty, \infty), |z| \leq \delta$. For any n, m, let

$$\gamma_{n,m} = \sup_{-\infty < s < \infty, |z| \leq \delta} |f(s + \alpha_n, z) - f(s + \alpha_m, z)|$$

$$u_{n,m}(t, \sigma) = |z(\sigma + t + \alpha_n, \sigma + \alpha_n) - z(\sigma + t + \alpha_m, \sigma + \alpha_m)|$$

Then (3.16) implies that

$$u_{n,m}(t, \sigma) \leq KK_1 \int_0^t e^{-\alpha(t-s)} \eta(\delta, \epsilon_1) u_{n,m}(s, \sigma)$$

$$+ KK_1 \int_0^\infty e^{-\alpha s} \eta(\delta, \epsilon_1) u_{n,m}(t + s, \sigma) + \frac{2KK_1}{\alpha} \gamma_m,$$

Using the inequalities (3.11), one easily observes that $u_{n,m}(t, \sigma) \leq 4KK_1 \gamma_m/\alpha$ for all m. Since $\gamma_m \to 0$ as $m \to \infty$, it follows that $u_{n,m}(t, \sigma) \to 0$ as $m \to \infty$ and, therefore, $z(\sigma + t, \sigma)$ is almost periodic in σ with module contained in the module of f. In particular, $g(z_-, \sigma, \varepsilon) = z^*(\sigma, \sigma, z_-, \varepsilon)$ is almost periodic with module contained in $m[f]$.

If f is in $\mathscr{P}_T \cap \mathscr{L}i \textit{p}(\eta, 0)$, then $g(z_-, \sigma + T, \varepsilon) = g(z_-, \sigma, \varepsilon)$ for all σ, since one sees directly from the uniqueness of the solution of (3.16) that $z(\sigma + t + T, \sigma + T) = z(\sigma + t, \sigma)$ for all σ, t.

In the same manner as above, one uses (3.10) to prove the assertions on $U(\sigma, \delta, \varepsilon)$ to complete the proof of the theorem.

A simple example illustrating the above results is the forced van der Pol equation

$$(3.17) \qquad \begin{aligned} \dot{x}_1 &= x_2, \\ \dot{x}_2 &= -x_1 + k(1 - x_1^2)x_2 + \varepsilon g(t), \end{aligned}$$

where $k \neq 0$, ε are real parameters and g is in $\mathscr{AP}$. If $x = (x_1, x_2)$, this system is of the form (2.1) with

$$A = \begin{bmatrix} 0 & 1 \\ -1 & k \end{bmatrix}, \qquad q(t, x, \varepsilon) = \begin{bmatrix} 0 \\ -kx_1^2 x_2 + \varepsilon g(t) \end{bmatrix}.$$

Since $k \neq 0$, the eigenvalues of A have nonzero real parts and q is in $\mathscr{AP} \cap \mathscr{Lip}(\eta, M)$ with $\eta(\rho) = K\rho^2$ for some constant K and $M = \varepsilon \sup_t |g(t)|$. Therefore, Theorem 2.1 implies there are $\rho_1 > 0$, $\varepsilon_1 > 0$ such that system (3.17) has an almost periodic solution $x^*(t, \varepsilon)$ with $m[x^*(\cdot, \varepsilon)] \subset m[g]$, $x^*(\cdot, 0) = 0$ and this solution is unique in the ρ_1-neighborhood of $x_1 = x_2 = 0$. Since the eigenvalues of A have negative real parts if $k < 0$ and positive real parts if $k > 0$, Theorem 3.1 asserts that the solution $x^*(t, \varepsilon)$ is asymptotically stable if $k < 0$ and unstable if $k > 0$.

For $\varepsilon = 0$, we have seen in Theorem II.1.6 that the van der Pol has a unique asymptotically orbitally stable limit cycle for any $k > 0$. This implies geometrically that the cylinder generated by the limit cycle in (x_1, x_2, t)-space is asymptotically stable. Therefore it is intuitively clear (and could be made very precise) that for ε small there must be a neighborhood of this cylinder in which solutions of (3.17) enter and never leave. This "stable" neighborhood is certainly more interesting than the almost periodic $x^*(\cdot, \varepsilon)$ determined above since $x^*(\cdot, \varepsilon)$ is unstable for $k > 0$. A discussion of what happens in this neighborhood is much more difficult than the above analysis for almost periodic solutions and will be treated in Chapter VII.

IV.4. More General Systems

The first interesting modifications of the results of the previous sections are obtained by considering the parameter ε as appearing in the system in a different manner than in (3.1).

Consider the system

$$(4.1) \qquad \varepsilon \dot{x} = A(t)x,$$

where $\varepsilon > 0$ is a real parameter and A belongs to $\mathscr{P}_T$. In general, it is almost impossible to determine necessary and sufficient conditions on the matrix A

which insure that system (4.1) is noncritical with respect to one of the classes $\mathscr{B}(-\infty, \infty)$, $\mathscr{A}\mathscr{P}$ or $\mathscr{P}_T$ for all ε in some interval $0 < \varepsilon \leqq \varepsilon_0$. On the other hand, if A is a constant matrix, the following result is true.

LEMMA 4.1.   If A is a constant matrix and $\varepsilon_0 > 0$ is given, then system (4.1) is noncritical with respect to $\mathscr{B}(-\infty, \infty)$ (or $\mathscr{A}\mathscr{P}$) (or $\mathscr{P}_T$) for $0 < \varepsilon \leqq \varepsilon_0$ if and only if the eigenvalues of A have nonzero real parts.

PROOF.   The assertions concerning $\mathscr{B}(-\infty, \infty)$ or $\mathscr{A}\mathscr{P}$ are obvious from Lemma 1 since the eigenvalues of A/ε are $1/\varepsilon$ times the eigenvalues of A. Also, from Lemma 1, (4.1) is noncritical with respect to $\mathscr{P}_T$ if and only if $\det[I - \exp(A/\varepsilon)T] \neq 0$; that is, if and only if $\lambda T/\varepsilon \neq 2k\pi i$, $k = \pm 1, \pm 2, \ldots$ for all eigenvalues λ of A. If $Re \, \lambda \neq 0$ for all λ, this relation is satisfied. If there is a $\lambda = i\omega$, ω real, then these relations become $\varepsilon \neq \omega T/2k\pi$, $k = 0, \pm 1$, $\ldots$. It is clear these relations cannot be satisfied for all ε in an interval $(0, \varepsilon_0]$. This completes the proof of the lemma.

LEMMA 4.2.   If A is a constant matrix, $\mathscr{D}$ is one of the classes $\mathscr{B}(-\infty, \infty)$, $\mathscr{A}\mathscr{P}$ or $\mathscr{P}_T$, and system (4.1) is noncritical with respect to $\mathscr{D}$ for $0 < \varepsilon \leqq \varepsilon_0$, then the system

(4.2) $$\varepsilon \dot{x} = Ax + f(t), \qquad f \text{ in } \mathscr{D},$$

has a unique solution $\mathscr{K}_\varepsilon f$ in $\mathscr{D}$, $0 < \varepsilon \leqq \varepsilon_0$, $\mathscr{K}_\varepsilon : \mathscr{D} \to \mathscr{D}$ is a continuous linear map and there is a $K > 0$ (independent of ε) such that $|\mathscr{K}_\varepsilon f| \leqq K |f|$ for $0 < \varepsilon \leqq \varepsilon_0$.

PROOF.   The hypothesis and Lemma 4.1 imply the eigenvalues of A have nonzero real parts. If $t = \varepsilon\tau$, $y(\tau) = x(\varepsilon\tau)$, $g(\tau) = f(\varepsilon\tau)$, and x is a solution of (4.2), then

$$\frac{dy}{d\tau} = Ay + g(\tau),$$

where $g \in \mathscr{B}(-\infty, \infty)$. Theorem 1.1 implies the existence of a unique $\mathscr{K}g \in \mathscr{B}(-\infty, \infty)$ satisfying this equation, with $|\mathscr{K}g| \leqq \mathscr{K} |g|$. Since $|g| = |f|$, it follows that $\mathscr{K}_\varepsilon f \overset{\text{def}}{=} \mathscr{K}g$ satisfies the properties of the lemma.

EXERCISE 4.1.   Discuss the manner in which the solutions of (4.2) approach the solution of the equation $Ax + f(t) = 0$ as $\varepsilon \to 0$ under the hypothesis that the eigenvalues of A have negative real parts (positive real parts). What happens if A has eigenvalues of both positive and negative real parts?

Lemma 4.2 and the same proofs as used in Theorems 2.1 and 3.1 yield an immediate extension of those results to the system

(4.3) $$\varepsilon \dot{x} = Ax + f(t, x, \varepsilon),$$

where f satisfies the same conditions as stated in those theorems.

This result is stated in detail for further reference. Suppose the eigenvalues of A have nonzero real parts, let π_+, π_- be the projection operators taking C^n onto the invariant subspaces of A corresponding to the eigenvalues with positive, negative real parts, respectively, and let K, α be positive constants so that

$$(4.4) \qquad |e^{At}\pi_- x| \le K e^{-\alpha t}, \qquad t \ge 0,$$

$$|e^{At}\pi_+ x| \le K e^{\alpha t}, \qquad t \le 0.$$

For any σ in $(-\infty, \infty)$, let $x(t, \sigma, x^\sigma, \varepsilon)$ designate the solution of (4.3) satisfying $x(\sigma, \sigma, x^\sigma, \varepsilon) = x^\sigma$ and, for any $\delta > 0$ and any function $\phi : R \to C^n$, let

$$(4.5) \quad (a) \quad S(\phi, \sigma, \delta, \varepsilon) = \left\{ x = x^\sigma - \phi(\sigma) \text{ in } C^n : |\pi_-[x^\sigma - \phi(\sigma)]| < \frac{\delta}{2K} \right.$$

$$\left. |x(t, \sigma, x^\sigma, \varepsilon) - \phi(t)| < \delta, \qquad t \ge \sigma \right\},$$

$$(b) \quad U(\phi, \sigma, \delta, \varepsilon) = \left\{ x = x^\sigma - \phi(\sigma) \text{ in } C^n : |\pi_+[x^\sigma - \phi(\sigma)]| < \frac{\delta}{2K} \right.$$

$$\left. |x(t, \sigma, x^\sigma, \varepsilon) - \phi(t)| < \delta, \qquad t \le \sigma \right\},$$

In words, the set $S(\phi, \sigma, \delta, \varepsilon)$ is the set of initial values at σ of those solutions of (4.3) which have their π_- projections in a $\delta/2K$ neighborhood of $\pi_- \phi(\sigma)$ and remain in a δ-neighborhood of the curve $\phi(t)$ for all $t \ge \sigma$. A similar statement concerns $U(\phi, \sigma, \delta, \varepsilon)$ for π_+ and $t \le \sigma$. If B_ρ^n designates the set $\{x \text{ in } C^n : |x| < \rho\}$, then the following proposition holds.

THEOREM 4.1. Suppose $\mathscr{D}$ is one of the classes $\mathscr{B}(-\infty, \infty)$, $\mathscr{AP}$ or $\mathscr{P}_T$. If f is in $\mathscr{D} \cap \mathscr{Lip}(\eta, M)$ and the eigenvalues of A have nonzero real parts, then there are $\delta > 0$, $\varepsilon_1 > 0$, $\beta > 0$ and a function $x^*(t, \varepsilon)$ continuous in t, ε for $-\infty < t < \infty$, $0 \le \varepsilon \le \varepsilon_1$, $x^*(t, 0) = 0$, $x^*(\cdot, \varepsilon)$ in $\mathscr{D}$, $|x^*(\cdot, \varepsilon)| < \delta$, $0 \le \varepsilon \le \varepsilon_1$, such that $x^*(t, \varepsilon)$ is a solution of (4.3) and is the only solution of (4.3) in $\mathscr{D}$ which has norm $< \delta$. If $\mathscr{D} = \mathscr{AP}$, then $m[x^*(\cdot, \varepsilon)] \subset m[f]$, $0 \le \varepsilon \le \varepsilon_1$.

Furthermore, the mapping π_- is a homeomorphism of $S(x^*(\cdot, \varepsilon), \sigma, \delta, \varepsilon)$ onto $(\pi_- C^n) \cap B_{\delta/2K}^n$ and

$$(4.6) \quad |x(t, \sigma, x^\sigma, \varepsilon) - x^*(y, \varepsilon)| \le 2K |\pi_-[x^\sigma - x^*(\sigma, \varepsilon)]| e^{-\varepsilon^{-1}\beta(t-\sigma)}, \qquad t \ge \sigma,$$

for any $x^\sigma - x^*(\sigma, \varepsilon)$ in $S(x^*(\cdot, \varepsilon), \sigma, \delta, \varepsilon)$, $0 < \varepsilon \le \varepsilon_1$. The mapping π_+ is a homeomorphism of $U(x^*(\cdot, \varepsilon), \sigma, \delta, \varepsilon)$ onto $(\pi_+ C^n) \cap B_{\delta/2K}^n$,

$$(4.7) \quad |x(t, \sigma, x^\sigma, \varepsilon) - x^*(t, \varepsilon)| \le 2K |\pi_+[x^\sigma - x^*(\sigma, \varepsilon)]| e^{\varepsilon^{-1}\beta(t-\sigma)}, \qquad t \le \sigma,$$

for any $x^\sigma - x^*(\sigma, \varepsilon)$ in $U(x^*(\cdot, \varepsilon), \sigma, \delta, \varepsilon)$, $0 < \varepsilon \le \varepsilon_1$.

The dependence of the stable and unstable manifolds of $u^*(\cdot, \varepsilon)$ upon σ is exactly the same as in the statement of Theorem 3.1.

The only part of the proof of the above theorem which is not exactly the same as the proofs of Theorem 2.1 and 3.1 is the fact that the statements are asserted to be true on the closed interval $0 \leqq \varepsilon \leqq \varepsilon_1$ rather than the interval $0 < \varepsilon \leqq \varepsilon_1$. The proof for the interval $0 < \varepsilon \leqq \varepsilon_1$ is the same as before and from the proof itself one observes that $x^*(\cdot, \varepsilon) \to 0$ as $\varepsilon \to 0$. If one defines $x^*(\cdot, 0) = 0$, it will obviously be a continuous function on $0 \leqq \varepsilon \leqq \varepsilon_1$.

LEMMA 4.3.   If A is constant matrix and $\varepsilon_0 > 0$ is given, then the system

(4.8) $\dot{x} = \varepsilon A x,$

is noncritical with respect to $\mathscr{B}(-\infty, \infty)$ or $\mathscr{A}\mathscr{P}$ for $0 < \varepsilon \leqq \varepsilon_0$ if and only if the eigenvalues of A have nonzero real parts. There is an $\varepsilon_0 > 0$ such that system (4.8) is noncritical with respect to $\mathscr{P}_T$ for $0 < \varepsilon \leqq \varepsilon_0$ if and only if $\det A \neq 0$.

PROOF.   The first part is a restatement of Lemma 1. Also, Lemma 1 implies (4.8) is noncritical with respect to $\mathscr{P}_T$ if and only if $\varepsilon \omega T \neq 2k\pi$ for all $k = 0, \pm 1, \ldots$ and all real ω such that $\lambda = i\omega$ is an eigenvalue of A. If $\omega = 0$ is not an eigenvalue of A, there is always an $\varepsilon_0 > 0$ such that these inequalities are satisfied for $0 < \varepsilon \leqq \varepsilon_0$. If $\omega = 0$ is an eigenvalue, there is never such an ε_0 and the lemma is proved.

EXERCISE 4.2.   For what values of ε in $[0, \infty)$ is system (4.8) noncritical with respect to $\mathscr{P}_T$? For what complex values of ε is system (4.8) noncritical with respect to $\mathscr{P}_T$?

LEMMA 4.4.   If A is a constant matrix, $\mathscr{D}$ is one of the classes $\mathscr{B}(-\infty, \infty)$, $\mathscr{A}\mathscr{P}$ or $\mathscr{P}_T$, and system (4.8) is noncritical with respect to $\mathscr{D}$ for $0 < \varepsilon \leqq \varepsilon_0$, then the system

(4.9) $\dot{x} = \varepsilon(Ax + f(t)),$ f in $\mathscr{D},$

has a unique solution $\mathscr{K}_\varepsilon f$ in $\mathscr{D}$, $0 < \varepsilon \leqq \varepsilon_0$, $\mathscr{K}_\varepsilon : \mathscr{D} \to \mathscr{D}$ is a continuous linear map and there is a $K > 0$ (independent of ε) such that $|\mathscr{K}_\varepsilon f| \leqq K|f|$ for $0 < \varepsilon \leqq \varepsilon_0$.

If, in addition, f is in $\mathscr{A}\mathscr{P}$ (or $\mathscr{P}_T$) and $\int^t f$ is in $\mathscr{A}\mathscr{P}$ (or $\mathscr{P}_T$) then $|\mathscr{K}_\varepsilon f| \to 0$ as $\varepsilon \to 0$.

PROOF.   Since (4.8) is assumed to be noncritical with respect to $\mathscr{D}$, Theorem 1.1 implies the existence of the continuous linear operators $\mathscr{K}_\varepsilon$, $0 < \varepsilon \leqq \varepsilon_0$. The existence of a K as specified in the lemma for the case when $\mathscr{D}$ is $\mathscr{B}(-\infty, \infty)$ or $\mathscr{A}\mathscr{P}$ is verified exactly as in the proof of Lemma 4.2. If

$\mathscr{D} = \mathscr{P}_T$, then formula (1.9) yields $\mathscr{K}_\varepsilon$ for the special case (4.9) as

$$(4.10) \qquad (\mathscr{K}_\varepsilon f)(t) = \int_0^T \varepsilon [e^{-\varepsilon A T} - I]^{-1} e^{-\varepsilon A s} f(t+s)\, ds, \qquad 0 < \varepsilon \le \varepsilon_0.$$

Since $[e^{-\varepsilon A T} - I]^{-1}$ is a continuous function of ε on $0 < \varepsilon \le \varepsilon_0$, $\mathscr{K}_\varepsilon f$ is continuous for $0 < \varepsilon \le \varepsilon_0$. We now show that $\mathscr{K}_\varepsilon f$ defined by (4.10) has a limit as $\varepsilon \to 0$. Since system (4.8) is noncritical with respect to $\mathscr{P}_T$, Lemma 4.3 implies A is nonsingular. Also, $\lim_{\varepsilon \to 0^+} \varepsilon [e^{-\varepsilon A T} - I]^{-1} = -(AT)^{-1}$. This shows $\mathscr{K}_\varepsilon$ has a uniform bound on $(0, \varepsilon_0]$ and completes the first part of the lemma.

To prove the last part of the lemma, let $x = y + \varepsilon \int^t f$ and y will satisfy the equation

$$\dot{y} = A\varepsilon \left(y + \varepsilon \int^t f \right).$$

An application of the first part of the lemma to this equation shows that the unique solution of $\mathscr{AP}$ (or $\mathscr{P}_T$) approaches zero as $\varepsilon \to 0$ and the lemma is proved.

The condition that $\int^t f$ be in $\mathscr{AP}$ for f in $\mathscr{AP}$ is equivalent to saying that the integral of f is bounded and in particular, implies

$$(4.11) \qquad M[f] \overset{\text{def}}{=} \lim_{t \to \infty} \frac{1}{t} \int_0^t f(s)\, ds = 0.$$

In Chapter V, we show that (4.11) is sufficient to draw the same conclusion as in the last part of Lemma 4.4.

EXERCISE 4.3. Discuss the manner in which the solutions of (4.9) approach the solutions of the equation $\dot{x} = 0$. Can you give any reason in the almost periodic or periodic case besides the one discussed in Lemma 4.4 for why all solutions approach the zero solution of $\dot{x} = 0$ if $\int^t f$ is bounded ?

Lemma 4.4 and the same proofs as used in Theorems 2.1 and 3.1 yield an immediate extension of those results to the system

$$(4.12) \qquad \dot{x} = \varepsilon(Ax + f(t, x, \varepsilon)),$$

where f satisfies the same conditions as stated in those theorems.

This result is the basis for the method of averaging given in the next chapter and is therefore stated in detail for further reference. The operators π_+, π_-, constants K, α and sets $S(\phi, \sigma, \delta, \varepsilon)$, $U(\phi, \sigma, \delta, \varepsilon)$ are assumed to be the same as the ones given in (4.4) and (4.5).

THEOREM 4.2. Suppose $\mathscr{D}$ is one of the classes $\mathscr{B}(-\infty, \infty)$, $\mathscr{AP}$ or $\mathscr{P}_T$ and system (4.8) is noncritical with respect to $\mathscr{D}$ for $0 < \varepsilon \le \varepsilon_0$. If f is in

$\mathscr{D} \cap \mathscr{L}i\not{h}(\eta, M)$, then there are $\delta > 0$, $\varepsilon_1 > 0$, $\beta > 0$ and a function $x^*(t, \varepsilon)$ continuous in t, ε for $-\infty < t < \infty$, $0 \leqq \varepsilon \leqq \varepsilon_1$, $x^*(t, 0) = 0$, $x^*(\cdot, \varepsilon)$ in $\mathscr{D}$, $|x^*(\cdot, \varepsilon)| < \delta$, $0 \leqq \varepsilon \leqq \varepsilon_1$, such that $x^*(t, \varepsilon)$ is a solution of (4.12) and is the only solution of (4.12) in $\mathscr{D}$ with norm $< \delta$. If $\mathscr{D} = \mathscr{A}\mathscr{P}$, then $m[x^*(\cdot, \varepsilon)] \subset m[f], 0 \leqq \varepsilon \leqq \varepsilon_1$.

Furthermore, if the eigenvalues of A have nonzero real parts, then the mapping π_- is a homeomorphism of $S(x^*(\cdot, \varepsilon), \sigma, \delta, \varepsilon)$ onto $(\pi_- C^n) \cap B^n_{\delta/2K}$ and

(4.13) $\quad |x(t, \sigma, x^\sigma, \varepsilon) - x^*(t, \varepsilon)| \leqq 2K \left| \pi_-[x^\sigma - x^*(\sigma, \varepsilon)] \right| e^{-\epsilon\beta(t-\sigma)}, \qquad t \geqq \sigma,$

for all $x^\sigma - x^*(\sigma, \varepsilon)$ in $S(x^*(\cdot, \varepsilon), \sigma, \delta, \varepsilon)$, $0 < \varepsilon \leqq \varepsilon_1$. The mapping π_+ is a homeomorphism of $U(x^*(\cdot, \varepsilon), \sigma, \delta, \varepsilon)$ onto $(\pi_+ C^n) \cap B^n_{\delta/2K}$ and

(4.14) $\quad |x(t, \sigma, x^\sigma, \varepsilon) - x^*(t, \varepsilon)| \leqq 2K \left| \pi_+[x^\sigma - x^*(\sigma, \varepsilon)] \right| e^{\epsilon\beta(t-\sigma)}, \qquad t \leqq \sigma,$

for any $x^\sigma - x^*(\sigma, \varepsilon)$ in $U(x^*(\cdot, \varepsilon), \sigma, \delta, \varepsilon)$, $0 < \varepsilon \leqq \varepsilon_1$.

The dependence of the stable and unstable manifolds of $x^*(\cdot, \varepsilon)$ upon σ is exactly the same as in the statement of Theorem 3.1.

Using the remark following Lemma 4.4, one easily sees that the conclusions of Theorem 4.2 remain valid in the periodic and almost periodic case for the system

(4.15) $\qquad\qquad \dot{x} = \varepsilon(Ax + h(t) + f(t, x, \varepsilon)),$

provided that $\int^t h$ is bounded. Extensions of Theorem 4.2 to the case where $f = f(t, x, \epsilon, \mu)$ and the vector $\epsilon' = (\epsilon, \mu)$ is small are also easily given.

EXERCISE 4.4. Consider the equation

(4.16) $\qquad\qquad \dot{x} = Ax + h(\omega t) + f(\omega t, x, \varepsilon),$

where ω is a large parameter, the eigenvalues of A have negative real parts, h is in $\mathscr{P}_T$, $\int_0^T h(s)\, ds = 0$, and f is in $\mathscr{P}_T \cap \mathscr{L}i\not{h}(\eta, M)$. Assuming Theorem 4.2 is valid when f depends upon more than one small parameter, prove there is an $\omega_1 > 0$, $\varepsilon_1 > 0$ such that (4.16) has an asymptotically stable periodic solution $x^*(\cdot, \omega, \varepsilon)$ in $\mathscr{P}_{T/\omega}$ for $\omega \geqq \omega_1$, $0 < |\varepsilon| \leqq \varepsilon_1$ and $x^*(\cdot, \omega, \varepsilon) \to 0$ as $\omega \to \infty$, $\varepsilon \to 0$. In particular, consider the forced van der Pol equation

$$\dot{x}_1 = x_2,$$
$$\dot{x}_2 = -x_1 + k(1 - x_1^2)x_2 + \varepsilon g(\omega t),$$

where $k \neq 0$, g is in $\mathscr{P}_T$ and $\int^t g(s)$ is bounded.

Using Lemmas 1, 4.2 and 4.4 and the proof of Theorem 2.1, one can prove the following general result for systems which are coupled versions of the above types.

THEOREM 4.3. Suppose $\mathscr{D}$ is one of the classes $\mathscr{B}(-\infty, \infty)$, $\mathscr{AP}$ or $\mathscr{P}_T$, u is an n-vector, $u = (x, y, z)$ where x, y, z are n_1-, n_2-, n_3-vectors respectively, $f = f(t, u, \varepsilon) = (X, Y, Z)$ is in $\mathscr{D} \cap \mathscr{L}i\!f\!f(\eta, M)$, $\varepsilon = (\mu, \nu)$, μ a real scalar, B is an $n_2 \times n_2$ matrix in $\mathscr{P}_T$, and A, C are constant $n_1 \times n_1$, $n_3 \times n_3$ matrices, respectively, such that the system

(4.17) $\dot{x} = \mu A x, \qquad \dot{y} = B(t) y, \qquad \mu \dot{z} = C z,$

is noncritical with respect to $\mathscr{D}$ for $0 < \mu \leq \mu_0$. Then there are constants $\rho_1 > 0$, $\mu_1 > 0$, $\nu_1 > 0$ and a function $u^*(t, \varepsilon)$ continuous in t, ε for $-\infty < t < \infty$, $0 < \mu \leq \mu_1$, $0 \leq |\nu| \leq \nu_1$, $u^*(t, 0) = 0$, $u^*(\cdot, \varepsilon)$ in $\mathscr{D}$, $|u^*(\cdot, \varepsilon)| < \rho_1$, such that $u^*(t, \varepsilon)$ is a solution of the equations

(4.18) $\dot{x} = \mu[A x + X(t, x, y, z, \varepsilon)]$

                  $\dot{y} = B(t) y + Y(t, x, y, z, \varepsilon),$

                  $\mu \dot{z} = C z + Z(t, x, y, z, \varepsilon),$

and is the only solution of (4.18) in $\mathscr{D}$ which has norm $< \rho_1$. If $\mathscr{D} = \mathscr{AP}$, then $m[u^*(\cdot, \varepsilon)] \subset m[f, B]$, $0 < \mu \leq \mu_1$, $0 \leq |\nu| \leq \nu_1$.

The stability properties of the solution $u^*(\cdot, \varepsilon)$ of (4.18) are discussed exactly in the same manner as in Section 3. The stable and unstable manifolds of the solution $u^*(\cdot, \varepsilon)$ can be characterized as in Theorems 4.1 and 4.2. In particular, if all eigenvalues of A, C have negative real parts and all character-istic exponents of $\dot{y} = B(t) y$ have negative real parts, then the solution $u^*(\cdot, \varepsilon)$ is exponentially asymptotically stable. If at least one of the eigen-values of A or C or the characteristic exponents of $\dot{y} = B(t) y$ have a positive real part, then the solution $u^*(\cdot, \varepsilon)$ is unstable.

As mentioned earlier, it is difficult to remove the restriction in (4.18) that A be constant. On the other hand, it is possible to allow C to be a func-tion of t, say $C = C(t)$, provided either that the eigenvalues $\lambda(t)$ of $C(t)$ have real parts bounded away from zero or, more generally, that

$$C(t) = \mathrm{diag}(D(t), E(t))$$

where the eigenvalues of both $D(t)$ and $E(t)$ have real parts (not necessarily of the same sign) bounded away from zero. For references concerning problems in the spirit of this section, see Hale [6].

IV.5. The Duffing Equation with Large Damping and Large Forcing

Consider the equation

(5.1) $\ddot{y} + c \dot{y} + y + \varepsilon b y^3 = B \cos \nu t,$

where $c > 0$, b, ε, B and $\nu > 0$ are constants. The equivalent second order system is

(5.2)
$$\dot{y} = z,$$
$$\dot{z} = -y - cz - \varepsilon by^3 + B \cos \nu t.$$

The methods of this chapter and some elementary facts about quadratic forms will be used to prove there is an $r_0 > 0$ such that for any $r \geqq r_0$, there is an $\epsilon_0 = \epsilon_0(r) > 0$ such that, for $|\epsilon| < \epsilon_0(r)$, system (5.2) has a unique periodic solution of period $2\pi/\nu$ in the disk B_r^2 with center zero and radius r. This solution is uniformly asymptotically stable and any solution of (5.2) with initial value in B_r^2, must approach this periodic solution as $t \to \infty$.

Consider the linear nonhomogeneous system

(5.3)
$$\dot{y} = z,$$
$$\dot{z} = -y - cz + B \cos \nu t.$$

Since the eigenvalues of the coefficient matrix of the homogeneous equation

(5.4)
$$\dot{y} = z,$$
$$\dot{z} = -y - cz,$$

have negative real parts, Theorem 1.1 implies there is a unique periodic solution of (5.3) of period $2\pi/\nu$. If this solution is designated by $y^0(t)$, $z^0(t)$, then the transformation of variables $y = y^0(t) + u$, $z = z^0(t) + v$ applied to (5.2) yields the equivalent system

(5.5)
$$\dot{u} = v + \varepsilon f_1(t, u, v),$$
$$\dot{v} = -u - cv + \varepsilon f_2(t, u, v),$$

where f_1, f_2 are periodic in t of period $2\pi/\nu$, continuous in t, u, v and continuously differentiable in u, v. Actually, $f_1 \equiv 0$, $f_2 = -b(y_0(t) + u)^3$, but it is convenient for notational purposes to consider the more general system (5.5).

Since $c > 0$, Theorem 2.1 implies there is a $\rho_1 > 0$ and $\varepsilon_1 > 0$ such that system (5.5) has a unique periodic solution $(u^*(t, \varepsilon), v^*(t, \varepsilon))$, $|\varepsilon| \leqq \varepsilon_1$ of period $2\pi/\nu$ in the disk $B_{\rho_1}^2$, this periodic solution is uniformly asymptotically stable, and $u^*(t, 0) = 0 = v^*(t, 0)$. To prove the above mentioned result for (5.2), it is sufficient to show that, for any two disks B_r^2 and $B_{r_1}^2$, $r_1 < r$, there exists an $\epsilon_2 > 0$ such that, for $|\epsilon| < \epsilon_2$, any solution of (5.5) with initial value in B_r^2 must eventually enter and remain in the ball $B_{r_1}^2$. In fact, suppose r_0 is such that the periodic solution $(y^*(t, \varepsilon), z^*(t, \varepsilon))$ of (5.2) given by $y^*(t, \varepsilon) = y^0(t) + u^*(t, \varepsilon)$, $z^*(t, \varepsilon) = z^0(t) + v^*(t, \varepsilon)$, lies in $B_{r_0}^2$ for $0 \leqq t \leqq 2\pi/\nu$ and $|\varepsilon| \leqq \varepsilon_1$. Choose $r_1 < \rho_1$. For any $r \geqq r_0$, the choice $\varepsilon_0(r) = \min(\varepsilon_2, \varepsilon_1)$ gives the desired result.

We now prove the assertion about (5.5). Since the linear system (5.4) is asymptotically stable, it is intuitively clear that there should be a family of ellipses encircling the origin so that the solutions of (5.4) cross the boundaries of these ellipses from the outside to the inside with increasing time. If this is the case, then for any given ellipse one can choose ε small so that the solutions of (5.5) also cross the boundary of this ellipse in the same direction as for the linear system (5.4). These ideas are now made precise by actually constructing such a family of ellipses.

Consider the quadratic form

$$(5.6) \qquad V(u, v) = \frac{c^2 + 2}{c} u^2 + 2uv + \frac{2}{c} v^2.$$

This quadratic form is positive definite and therefore the level curves of this function are ellipses with center at the origin. The derivative of V along the solutions of (5.5) is

$$(5.7) \qquad \dot{V}(u, v) = -2(u^2 + v^2) + 2\varepsilon \left[\frac{c^2 + 2}{c} uf_1 + uf_2 + vf_1 + \frac{2}{c} vf_2 \right].$$

For any positive r, r_1, $r_1 < r$, choose positive constants $c_1 > c_2$ so that the region U contained between the curves $V(u, v) = c_1$ and $V(u, v) = c_2$ contains the region between the boundaries of the disks B_r^2, $B_{r_1}^2$ (see Fig. 5.1). Suppose

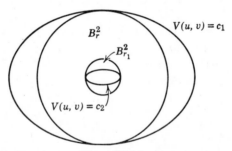

Figure IV.5.1

$\min(u^2 + v^2) = \alpha$ for u, v ranging over the set $V(u, v) = c_2$. Then $\alpha > 0$ and one can find an $\varepsilon_2 > 0$ such that the right hand side of (5.7) is less than $-\alpha/2$ for $0 \leq |\varepsilon| \leq \varepsilon_2$, $-\infty < t < \infty$ and all (u, v) in U. For any (u^0, v^0) in U, the solution $u(t), v(t)$, $u(0) = u^0$, $v(0) = v^0$, of (5.5) remains in the interior of the ellipse $V(u, v) = c_1$ and satisfies

$$V(u(t), v(t)) \leq V(u(0), v(0)) + \int_0^t \dot{V}(u(s), v(s))\, ds$$

$$\leq V(u(0), v(0)) - \frac{\alpha}{2} t.$$

Since V is positive in U, and the solution cannot reach the ellipse $V(u, v) = c_1$, there must be a $t_0 > 0$ such that $(u(t), v(t))$ remains in the interior of the ellipse $V(u, v) = c_2$ for all $t \geq t_0$. This proves the result.

IV.6. Remarks and Extensions

The technique presented in this chapter is applicable to many other types of problems. To illustrate this, let us give a brief abstract summary of the basic ideas.

Suppose I is an interval in R^n, $A(t)$ is an $n \times n$ matrix function continuous on I and $\mathscr{B}, \mathscr{D}$ are given Banach spaces of continuous n-vector functions on I. For every $f \in \mathscr{D}$, suppose the equation

(6.1) $\dot{x} = A(t)x + f(t)$

has a unique solution $\mathscr{K} f$ in $\mathscr{B}$ and $\mathscr{K} : \mathscr{D} \to \mathscr{B}$ is a continuous linear operator. At the end of section 1, we said $(\mathscr{B}, \mathscr{D})$ was *admissible* if, for every $f \in \mathscr{D}$, there is at least one solution of Eq. (6.1) in $\mathscr{B}$. Here, we are requiring uniqueness of the solution in $\mathscr{B}$. In this case, we say $(\mathscr{B}, \mathscr{D})$ is *strongly admissible*.

Let $\mathscr{C}^1(I \times R^n, R^n) = \{f : I \times R^n \to R^n, f(t, x) \text{ continuous and continuously differentiable in } x\}$. Let

$$|f|_1 = \sup\{|f(t,x)| + |\partial f(t,x)/\partial x|, (t,x) \text{ in } I \times R^n\}.$$

If ϕ is in $\mathscr{B}$, let us suppose the function $f(\cdot, \phi(\cdot))$ is in $\mathscr{D}$ and consider the problem of the existence of solutions in $\mathscr{B}$ of the equation

(6.2) $\dot{x} = A(t)x + f(t,x)$

If there exists a solution x of Eq. (6.2) in $\mathscr{B}$, then x must satisfy the equation

(6.3) $G(x,f) \overset{def}{=} x - \mathscr{K} F(x,f) = 0$

where $\mathscr{K} : \mathscr{D} \to \mathscr{B}$ is the continuous linear operator defined above and

(6.4)
$$F : \mathscr{B} \times \mathscr{C}^1(I \times R^n, R^n) \to \mathscr{D}$$
$$F(x,f)(t) = f(t, x(t)), \qquad t \in I.$$

Let us suppose the function $F(x,f)$ is continuous together with its Frechet derivative.

To solve Eq. (6.3) for $|f|_1$ small, one can use the contraction mapping principle to obtain a generalization of Theorem 2.1. Or, one can use the Implicit Function Theorem in Banach spaces observing that $G(0,0) = 0$, $\partial G(0,0)/\partial x = I$, where G is defined in Eq. (6.3). The results may then be summarized as

THEOREM 6.1. Suppose $(\mathscr{B}, \mathscr{D})$ is strongly admissible for system (6.1) and the function F defined in Relation (6.4) is continuous together with its Frechet derivative. Then there is a $\delta > 0$, $\eta > 0$, and a unique function

$$x^*:\{f \in \mathscr{C}^1(I \times R^n, R^n): |f|_1 < \delta\} \to \mathscr{B}$$

such that $x^*(f)$ is continuous together with its derivative in f, $x^*(0) = 0$, $x^*(f)$ satisfies Eq. (6.2) and is the only solution of Eq. (6.2) in $\mathscr{B}$ with norm less than η.

Theorem 6.1 includes Theorem 2.1 in the case where $f(t,x)$ has a continuous first derivative in x. We give a few examples of other applications of Theorem 6.1.

Suppose $I = [0,1]$, M,N are $n \times n$ constant matrices, $A(t)$ is an $n \times n$ continuous matrix on I, f is a continuous n-vector function on I and consider the boundary value problem

(6.5)
$$\dot{x} = A(t)x + f(t), \qquad t \text{ in } I,$$
$$Mx(0) + Nx(1) = 0$$

If $X(t)$ is the fundamental matrix solution of the homogeneous equation, $X(0) = I$, then

$$x(t) = X(t)x(0) + \int_0^t X(t)X^{-1}(s)f(s)\,ds$$

satisfies the boundary conditions if and only if

$$[M + NX(1)]\,x(0) = -X(1)\int_0^1 X^{-1}(s)f(s)\,ds$$

If $f(s) = -X(s)X^{-1}(1)b$ for a given vector b in R^n, then

$$-X(1)\int_0^1 X^{-1}(s)f(s)\,ds = b$$

This implies that Eq. (6.7) has a unique solution for every continuous function f on I if and only if $[M + NX(1)]^{-1}$ exists. If this inverse exists, then the boundary value problem (6.5) has a unique solution $\mathscr{K}f$ given by

(6.6)
$$(\mathscr{K}f)(t) = -X(t)[M + NX(1)]^{-1}X(1)\int_0^1 X^{-1}(s)f(s)\,ds$$
$$+ \int_0^t X(t)X^{-1}(s)f(s).$$

These results are summarized in the following:

LEMMA 6.1.   Let

$$\mathscr{D} = \{f:[0,1] \to R^n, f \text{ cont.}\}, |f| = \sup_t |f(t)|,$$
$$\mathscr{B} = \{x \in \mathscr{D} : Mx(0) + Nx(1) = 0\}.$$

If $X(t), X(0) = I$, is a fundamental matrix solution of $\dot{x} = A(t)x$, then $(\mathcal{B}, \mathcal{D})$ is strongly admissible for the equation

$$\dot{x} = A(t)x + f(t)$$

if and only if $[M + NX(1)]^{-1}$ exists.

   With this lemma, one can obtain the existence of a solution of the boundary value problem

$$\dot{x} = A(t)x + f(t, x)$$

$$Mx(0) + Nx(1) = 0$$

if the matrix $[M + NX(1)]^{-1}$ exists and $|f(t, x)|, |\partial f(t, x)/\partial x|$ are small.

   As another illustration, let us briefly indicate how the saddle point property may be obtained in this way. Consider the equation

(6.7) $\dot{x} = A x + f(t)$

where the eigenvalues of A have nonzero real parts and f is in $\mathcal{D} \overset{def}{=} \mathcal{B}([0, \infty))$, the space of continuous bounded n-vector functions on $[0, \infty)$. We know Eq. (6.7) has at least one solution in $\mathcal{D}$ for every f in $\mathcal{D}$. However, the equation

(6.8) $\dot{x} = A x$

has a finite dimensional subspace $\mathcal{P}_s$ of solutions which are in $\mathcal{D}$ and so the solutions in $\mathcal{D}$ of Eq. (6.7) are not unique unless $\mathcal{P}_s = \{0\}$. The pair $(\mathcal{D}, \mathcal{D})$ is admissible but generally not strongly admissible. On the other hand, we can define $\mathcal{B} \subsetneq \mathcal{D}$ so that $(\mathcal{B}, \mathcal{D})$ is strongly admissible. In fact, let $\pi_s : \mathcal{D} \to \mathcal{P}_s$ be a continuous projection operator and define $\mathcal{B} = (I - \pi_s)\mathcal{D}$. The pair $(\mathcal{B}, \mathcal{D})$ is then strongly admissible and there is a continuous linear operator $\mathcal{K} : \mathcal{D} \to \mathcal{B}$ such that $\mathcal{K}f$ is the unique solution of Eq. (6.7) with $\pi_s \mathcal{K}f = 0$. The operator $\mathcal{K}$, of sourse, depends on the projection π_s, but one can clearly choose π_s so that $\mathcal{K}f$ consists of the sum of the two integrals in Formula (6.4) of Chapter III. With this definition of $\mathcal{K}$, every solution of Eq. (6.7) in $\mathcal{D}$ is given by

(6.9) $x = \pi_s e^{A \cdot} x_0 + \mathcal{K}f.$

   Let us now consider the nonlinear problem

(6.10) $\dot{x} = A x + f(x)$

where $f(0) = 0$, $\partial f(0)/\partial x = 0$ and try to prove the existence of a saddle point at $x = 0$. In particular, let us obtain the stable manifold. We know from Eq. (6.9) that every solution in $\mathcal{D}$ must satisfy

$$x = \pi_s e^{A \cdot} x_0 + \mathcal{K}f(x)$$

If $y = x - \pi_s e^{A \cdot} x_0$, then $y \in \mathcal{B}$ and we have

(6.11) $G(y,f,x_0) \overset{def}{=} y - \mathscr{K}f(y + \pi_s e^{A \cdot} x_0) = 0.$

One can now use the Implicit Function Theorem to determine $y^*(f,x_0)$ in $\mathscr{B}$ as a continuously differentiable function of $f,x_0, y^*(0,0) = 0$, $G(y^*(f,x_0),f,x_0) = 0$. This shows there is a family of solutions for each f, which are bounded and remain in a neighborhood of zero. The dimension of the family is the dimension of the family $\pi_s e^{A \cdot} x_0$; that is, the dimension of the stable manifold of $\dot{x} = Ax$. To show these solutions approach zero, one can put more restrictions on the space $\mathscr{D}$ and repeat the same argument.

This idea can be generalized to the abstract case where $(\mathscr{B},\mathscr{D})$ is admissible and not strongly admissible (see, for example, Antosiewicz [2], Hartman [1]).

CHAPTER V

Simple Oscillatory Phenomena and the

Method of Averaging

In Chapter II, it was shown how the Poincaré-Bendixson theory could be used to determine the existence and stability properties of periodic orbits of autonomous two dimensional systems. For systems of higher dimension and even for nonautonomous two dimensional systems, the methods of Chapter II are of no assistance in the discussion of the existence of periodic or almost periodic solutions.

In Chapter IV, the existence of periodic and almost periodic solutions were discussed for systems of differential equations which were perturbations of linear systems of arbitrary order. On the other hand, it was assumed that the trivial solution was the only solution of the linear system which belonged to the class of desired solutions; that is, the system was noncritical. Applications are very common for which the linear part of the system contains nontrivial periodic solutions and the system is critical with respect to some class of forcing functions. The methods of Chapter IV are not applicable directly to such problems. However, there may be appropriate transformations of variables which bring a critical system into the framework of Chapter IV. This is precisely the basis for the *method of averaging* discussed in Section 3 below. The *method of averaging* is a general method for determining sufficient conditions for the existence and stability of periodic and almost periodic solutions of a class of nonlinear vector differential equations which contain a small parameter. It is possible to discuss the existence of periodic solutions without invoking the results of Chapter IV. In fact, a much more general theory is given in Chapter VIII for the periodic case.

Before discussing the method of averaging, general properties of conservative systems and simple nonconservative systems are treated in Sections 1 and 2. The remaining sections of the chapter are devoted to specific applications.

175

V.1. Conservative Systems

Suppose $f: R^n \to R^n$ is continuous and for any x_0 in R^n the system

(1.1) $\dot{x} = f(x)$,

has a unique solution $x(t) = x(t, x_0)$ with $x(0, x_0) = x_0$. A function $E: D \subset R^n \to R$ is said to be an *integral of* (1.1) on a region $D \subset R^n$ if E is continuous together with its first partial derivatives, E is not constant on any open set in D, and $E(x(t)) = $ constant along the solutions of (1.1). Since E is assumed to have continuous first derivatives, the last property is equivalent to $[\partial E(x(t))/\partial x] f(x(t)) = 0$. System (1.1) is said to be *conservative* if it has an integral E on R^n. The orbits in R^n of a conservative system must therefore lie on level curves of the integral E.

Suppose $x = (x_1, \ldots, x_n)$, E is an integral of (1.1) on D and x^0 in D is such that $\partial E(x^0)/\partial x_n \neq 0$. Let $c = E(x^0)$. From the implicit function theorem, it follows that the equation $E(x) = c$ can be solved for x_n as a function x_n^* of the x_k, $k < n$, x^0 and x in a sufficiently small neighborhood U of x^0. Since E is assumed to be a first integral, $E(x(t)) = c$ if $x(t)$ is the solution of (1.1.) with $x(0) = x^0$. Substituting x_n^* in (1.1) results in a system of $(n - 1)$ equations for the determination of the solution of (1.1) through x^0. The dimension of (1.1) is therefore decreased by one on U. In particular, if $n = 2$, the existence of an integral reduces the solution of the equation to a quadrature.

In the following, the notation introduced in Chapter I is employed by letting $\gamma^+ = \gamma^+(x)$, $\gamma^- = \gamma^-(x)$ denote respectively the positive, negative orbit of (1.1) through x, $\omega(\gamma^+)$ and $\alpha(\gamma^-)$ denote the ω- and α-limit sets, respectively, of the orbit γ^+, γ^-.

LEMMA 1.1.   Suppose E is an integral of (1.1) on an open set D containing an equilibrium point x^0 of (1.1). There is no neighborhood U of x^0 for which x^0 belongs to one of the sets $\omega(\gamma^+(x))$ or $\alpha(\gamma^-(x))$ for all x in U.

PROOF.   If there are a sequence of $\{t_n\}$, $t_n \to \infty$ as $n \to \infty$, and an x^1 in D such that $x(t_n, x^1) \to x^0$ as $n \to \infty$, then continuity of E implies $E(x(t, x^1)) = E(x^0)$ for all t. The same statement is true for $t_n \to -\infty$. Therefore, if there were a neighborhood U of x^0 such that x^0 belongs to one of the sets $\omega(\gamma^+(x))$ or $\alpha(\gamma^-(x))$ for all x in U, then E would be constant on U. Since this is contrary to the definition of an integral, the lemma is proved.

Lemma 1.1 implies in particular that an equilibrium point of a conservative system can never be asymptotically stable.

LEMMA 1.2.   Suppose E is an integral of (1.1) in a bounded, open neighborhood D of an equilibrium point $x = 0$ of (1.1). If $E(0) = 0$ and $E(x) > 0$ for $x \neq 0$ in D, then $x = 0$ is a stable equilibrium point.

PROOF. If $E(0) = 0$, $E(x) > 0$ for $x \neq 0$ in D, then for any $\varepsilon > 0$, $\alpha \stackrel{\text{def}}{=} \min_{|x| = \varepsilon, x \text{ in } D} E(x) > 0$. Choose $0 < \delta < \varepsilon$ so that $\{x : |x| \leq \delta\} \subset D$, $\max_{|x| \leq \delta} E(x) < \alpha$. Since E is an integral this implies $|x(t, x^0)| < \varepsilon$, $t \geq 0$, if $|x^0| < \delta$. Thus, $x = 0$ is stable and the lemma is proved.

LEMMA 1.3. For $n = 2$, all orbits in a neighborhood of a stable isolated equilibrium point of a conservative system must be periodic orbits and the equilibrium point is a center.

PROOF. In this proof, a bar over a set denotes closure. Suppose $x = 0$ is an isolated stable equilibrium point of (1.1) for $n = 2$. Then there are neighborhoods U, V, of zero such that $\bar{V} \backslash \{0\}$ is equilibrium point free and for every x in U, the positive orbit $\gamma^+ = \gamma^+(x)$ through x belongs to $\bar{V}$. From the Poincaré-Bendixson theory, the ω-limit set $\omega(\gamma^+(x))$ of $\gamma^+(x)$ must be either $\{0\}$ or a periodic orbit. Lemma 1.1 implies $\omega(\gamma^+(x))$ cannot be $\{0\}$ for every x in U. If there is an x in U such that $\Gamma = \omega(\gamma^+(x)) = \bar{\gamma}^+(x) \backslash \gamma^+(x)$ is a periodic orbit, then Γ is asymptotically stable from either the inside or outside and, thus, there is an open set on which the integral E is constant. This contradiction shows that any trajectory which does not approach zero is a periodic orbit. Since every periodic orbit obviously has zero in its interior, this proves the lemma.

A very important class of conservative systems are Hamiltonian systems with n degrees of freedom. If $q = (q_1, \ldots, q_n)$ are the generalized position coordinates of n-particles and $p = (p_1, \ldots, p_n)$ are the generalized momentum, $H(p, q) = T(p) + V(q)$ where T is the kinetic energy and V is the potential energy, then the equations of motion are

$$(1.2) \qquad \dot{q} = \frac{\partial H}{\partial p}, \qquad \dot{p} = -\frac{\partial H}{\partial q}.$$

This is a special case of system (1.1) with $x = (q, p)$. The Hamiltonian $H(p, q)$ is an integral of the system (1.2). The kinetic energy $T(p)$ is assumed always positive if $p \neq 0$ with $T(0) = 0$, $\partial T(0)/\partial p = 0$. If $\partial T(p)/\partial p \neq 0$ for $p \neq 0$, the extreme points q^0 of $V(q)$ therefore yield the equilibrium points $(q^0, 0)$ of (1.2).

For Hamiltonian systems, we can prove

LEMMA 1.4. Suppose $q = 0$ is an extreme point of the potential energy $V(q)$ with $V(0) = 0$. If zero is locally an absolute minimum of $V(q)$, then $(0,0)$ is a stable point of equilibrium of (1.2). The equilibrium point $(0,0)$ is unstable if zero is not a minimum of $V(q)$ if $T(p)$ and $V(q)$ have the form $T(p) = T_2(p) + T_3(p)$, $V(q) = V_k(q) + V_{k+1}(q)$ where T_2 is a positive definite quadratic form, V_k is a homogeneous polynomial of degree $k \geqq 2$, $T_3(p) = o(|p|^2)$, $V_{k+1}(q) = o(|q|^k)$ as $|p|$, $|q| \to 0$ and there is a neighborhood U of $(0,0)$ such that

$$\Omega_U = \{(p,q) \in U : H(p,q) < 0\} \neq \phi$$

and $V_k(q) < 0$ if (p,q) is in Ω_U.

PROOF. The assertion concerning stability is an immediate consequence of Lemma 1.2. Suppose zero is not a minimum of $V(q)$. Let $W(p,q) = p'q$. The point $(0,0)$ is a boundary point of Ω_U. Using the fact that H is an integral of (1.2), the derivative $\dot{W}(p,q)$ along the solutions of (1.2), is easily seen to be

$$\dot{W}(p,q) = 2T_2(p) - kV_k(q) + \cdots,$$

where $\cdots$ designates terms which are $o(|p|^2)$ and $o(|q|^k)$ as $|p|$, $|q| \to 0$. In Ω_U, $T_2(p) > 0$ and $V_k(q) < 0$. Furthermore, one can choose the neighborhood U of $(0,0)$ sufficiently small that $W(p,q) > 0$ in Ω_U. Since $W(p,q) > 0$, $\dot{W}(p,q) > 0$ in Ω_U, any solution with initial value in Ω_U must leave the set Ω_U through the boundary of U since the boundary of Ω_U in the interior of U consists of points where $W(p,q) = 0$, or $H(p,q) = 0$. Since $(0,0)$ is in the boundary of Ω_U, this proves instability and the lemma.

More specific information on the nature of the integral curves for second order conservative systems can be given. Consider the second order scalar equation

$$(1.3) \qquad\qquad \ddot{u} + g(u) = 0,$$

or the equivalent system

$$(1.4) \qquad\qquad \dot{u} = v,$$
$$\dot{v} = -g(u),$$

where g is continuous and a uniqueness theorem holds for (1.4). System (1.4) is a Hamiltonian system with the Hamiltonian function or total energy given by $E(u, v) = v^2/2 + G(u)$ where $G(u) = \int_0^u g(s)ds$. The orbits of solutions of (1.4) in the (u, v)-plane must lie on the level curves of the function $E(u, v)$; that is, the curves described by $E(u, v) = h$, a constant. The equilibrium points of (1.4) are points $(u^0, 0)$ where $g(u^0) = 0$. If $G(u)$ has an absolute minimum at u^0, then $(u^0, 0)$ is stable (Lemma 1.2) and all orbits in a neighborhood of $(u^0, 0)$ must be periodic orbits (Lemma 1.3); that is, $(u^0, 0)$ is a center. Since the solutions of $E(u, v) = h$, a constant, are

$$(1.5) \qquad\qquad v = \pm\sqrt{2[h - G(u)]},$$

it follows that any isolated equilibrium point $(u^0, 0)$ such that u^0 is not a minimum of G must be unstable. In fact, the curves $v = +\sqrt{2[h - G(u)]}$,

$v = -\sqrt{2[h - G(u)]}$ are homeomorphic images of a segment of the real line in a neighborhood of $(u^0, 0)$. If a solution starting on these curves does not leave a neighborhood of $(u^0, 0)$, then the ω-limit set of the solution curve would be an equilibrium point. Since $(u^0, 0)$ is assumed isolated, this immediately gives a contradiction.

A point u^0 is a local absolute minimum of $G(u)$ if $g(u) < 0$ for $u < u^0$ and $g(u) > 0$ for $u > u^0$ and u in a neighborhood of u^0. The reverse inequalities apply for a local absolute maximum of $G(u)$. If $G(u)$ has a local absolute maximum at u^0, then the equilibrium point $(u^0, 0)$ is a saddle point in the sense that the set of all solutions which remain in a small neighborhood of $(u^0, 0)$ for $t \geq 0$ ($t \leq 0$) must lie on an arc passing through $(u^0, 0)$. This follows directly from formula (1.5). We can therefore state

LEMMA 1.5.   The stable equilibrium points of (1.4) are centers and all of the unstable equilibrium points of (1.4) are saddle points if the only extreme points of $G(u)$ are local absolute minimum and local absolute maximum.

Particular examples illustrate how easily information about a two dimensional conservative system is obtained without any computations whatsoever. The sketch of the level curves of E are easily deduced using (1.5).

Example 1.1.   Suppose the function $G(u)$ has the graph shown in Fig. 1.1a with A, B, C, D being extreme points of G. The orbits of solution curves are sketched in Fig. 1.1b, all curves of course being symmetric with respect to the u-axis. The equilibrium points corresponding to A, B, C, D are labeled as A, B, C, D on the phase plane also. The points A, C are centers, B is a saddle point and D is like the coalescence of a saddle point and a center. The curves joining B to B and D to D in Fig. 1.1b are called separatrices. A *separatrix* is

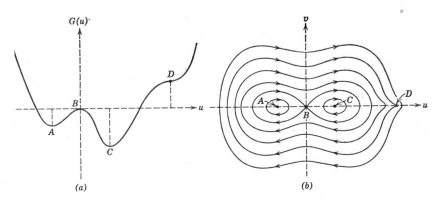

Figure V.1.1

a curve consisting of orbits of (1.3) which divides the plane into two parts and there is a neighborhood of this curve such that not all orbits in this neighborhood have the same qualitative behavior. Separatrices must therefore always pass through unstable equilibrium points.

Example 1.2. Suppose equation (1.3) is the equation for the motion of a pendulum of length l in a vacuum and u is the angle which the pendulum makes with the vertical. If g is the acceleration due to gravity, then $g(u) = k^2 \sin u$, $k^2 = g/l$, $G(u) = k^2(1 - \cos u)$ and $G(u)$ has the graph shown in Fig. 1.2a. The curves $E(u, v) = h$ clearly have the form shown in Fig. 1.2b. Explain the physical meaning of each of the orbits in Fig. 1.2b.

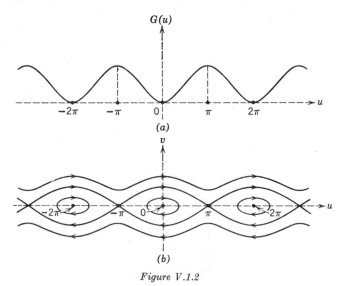

(a)

(b)

Figure V.1.2

It is not difficult to determine implicit formulas for the periods of the periodic solutions of equation (1.3) for an arbitrary $g(x)$. Any periodic solution has an orbit which must be a closed curve and conversely. From (1.5), it follows that a closed orbit must intersect the u axis at two points $(a, 0)$, $(b, 0)$, $a < b$, and must be symmetrical with respect to the u-axis. If T is the period of the periodic solution, then $T = 2\omega$ where ω is the time to traverse that part of the orbit where $v > 0$. Since v is given by (1.5) it follows from (1.4) that

(1.6)
$$T = 2\int_a^b \frac{du}{\sqrt{2(h - G(u))}}.$$

If $G(u)$ is even in u and the periodic orbit encircles the origin, then symmetry implies $a = -b$ and

$$(1.7) \qquad T = 4 \int_0^b \frac{du}{\sqrt{2(h - G(u))}}.$$

For the pendulum equation, example 1.2, $g(u) = k^2 \sin u$, $G(u) = k^2(1 - \cos u)$, and if $u(0) = b < \pi$, $v(0) = 0$, then $h = E(u(t), v(t)) = E(u(0), v(0)) = E(b, 0) = k^2(1 - \cos b)$. Consequently, the period T of the periodic orbit passing through this point $(b, 0)$ is

$$T = 4 \int_0^b \frac{du}{[2k^2(\cos u - \cos b)]^{1/2}} = \frac{2}{k} \int_0^b \frac{du}{\left[\sin^2\left(\frac{b}{2}\right) - \sin^2\left(\frac{u}{2}\right)\right]^{1/2}}.$$

If $\sin(u/2) = (\sin \psi) \sin(b/2)$, then

$$(1.8) \qquad T = \frac{4}{k} \int_0^{\pi/2} \frac{d\psi}{\left[1 - \sin^2\left(\frac{b}{2}\right)\sin^2 \psi\right]^{1/2}}.$$

This integral cannot be evaluated in terms of elementary functions, but if b is sufficiently small, then it is easy to find an approximate value of the period. Expanding in series, we obtain

$$T = \frac{2\pi}{k}\left[1 + \frac{1}{4}\sin^2\left(\frac{b}{2}\right) + \cdots\right]$$

$$= \frac{2\pi}{k}\left[1 + \frac{1}{16}b^2 + \cdots\right],$$

if b is sufficiently small. To the first approximation $T = 2\pi/k$; that is, the frequency k is approximately $\sqrt{g/l}$, which is almost the first lesson in every course in elementary mechanics. From this example and the above computation, it is seen that the *period of a periodic solution of an autonomous differential equation may vary from one solution to another* (contrast this fact to the linear equation).

EXERCISE 1.1.   For $g(u) = u + \gamma_0 u^3$, show that the period $T(b, \gamma_0)$ given by formula (1.7) is a decreasing function of b (increasing function of b) if $\gamma_0 > 0$ ($\gamma_0 < 0$). The first situation is called a *hard spring* and the second a *soft spring*. Hard implies the frequency increases with the amplitude b. Soft implies the frequency decreases with the amplitude b.

In the general case, if $(u^0, 0)$ is a stable equilibrium point of (1.3), then the restoring force $g(u)$ is said to correspond to a *hard spring (soft spring) at* u^0 if the frequency of the periodic solutions in a neighborhood of $(u^0, 0)$ is an *increasing (decreasing)* function of the distance of the periodic orbit from $(u^0, 0)$.

Example 1.3.   Consider a pendulum of mass m and length l constrained to oscillate in a plane rotating with angular velocity ω about a vertical line. If u denotes the angular deviation of the pendulum from the vertical line (see Fig. 1.3), the moment of centrifugal force is $m\omega^2 l^2 \sin u \cos u$, the

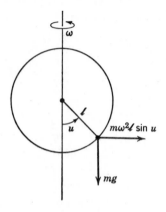

Figure V.1.3

moment of the force due to gravity is $mgl \sin u$ and the moment of inertia is $I = ml^2$. The differential equation for the motion is

(1.9) $$I\ddot{u} - m\omega^2 l^2 \sin u \cos u + mgl \sin u = 0.$$

If $\mu = m\omega^2 l^2 / I$ and $\lambda = g/\omega^2 l$, then this equation is equivalent to the system

(1.10) $$\dot{u} = v,$$
$$\dot{v} = \mu(\cos u - \lambda) \sin u,$$

which is a special case of (1.4) with $g(u) = g(u, \lambda) = -\mu(\cos u - \lambda)\sin u$. The dependence of g upon λ is emphasized since the number of equilibrium points of (1.10) depends upon λ. The equilibrium points of (1.10) are points $(u^0, 0)$ with $g(u^0, \lambda) = 0$ and are plotted in Fig. 1.4. The shaded regions correspond to $g(u, \lambda) < 0$. For any given λ, the equilibrium points are $(0, 0)$, $(\pi, 0)$ and $(\cos^{-1}\lambda, 0)$, the last one of course appearing only if $|\lambda| < 1$. For $|\lambda| \neq 1$, Lemma 1.5 implies the points on the curves labeled in Fig. 1.4 with black dots (circles) are stable (unstable), the stable points being centers and the unstable points being saddle points. From this diagram one sees that when $0 < \lambda < 1$, the stable equilibrium points are *not* $(0, 0)$ or $(\pi, 0)$ but the points $(\cos^{-1}\lambda, 0)$. Physically, one can have $\lambda < 1$ if the angular velocity ω is large enough. Analyze the behavior of the equilibrium points for $\lambda = 1$, $\lambda = -1$.

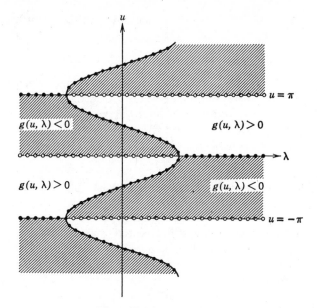

Figure V.1.4

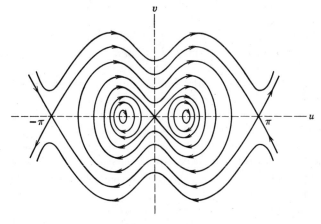

Figure V.1.5

An integral for this system is easily seen to be

$$E(u, v) = \frac{v^2}{2} - \frac{\mu}{2} \sin^2 u - \mu\lambda \cos u.$$

Suppose $0 < \lambda < 1$. The equilibrium points $(0, 0)$ and $(0, \pi)$ are saddle points and the level curves of $E(u, v)$ passing through these points are, respectively,

$$v^2 = \mu[\sin^2 u + 2\lambda(\cos u - 1)],$$

$$v^2 = \mu[\sin^2 u + 2\lambda(\cos u + 1)].$$

Both of these curves contain the points $(\cos^{-1} \lambda, 0)$ in their interiors and the latter one also passes through $(-\pi, 0)$. A sketch of the orbits is given in Fig. 1.5. The two centers correspond to the two values of u for which $\cos u = \lambda$. Interpret the physical meaning of all of the orbits in Fig. 1.5 and also sketch the orbits in the phase plane when $\lambda > 1$.

V.2. Nonconservative Second Order Equations—Limit Cycles

Up to this point, three different types of oscillations which occur in second order real autonomous differential systems have been discussed. For linear systems, there can be periodic solutions if and only if the elements of the coefficient matrix assume very special values and, in such a situation, all solutions are periodic with exactly the same period. For second order conservative systems, there generally are periodic solutions. These periodic solutions occur as a member of a family of such solutions each of which is uniquely determined by the initial conditions. The period in general varies with the initial conditions but not all solutions need be periodic. In Section I.7, artificial examples of second order systems were introduced for which there was an isolated periodic orbit to which all solutions except the equilibrium solution tend as $t \to \infty$. In Chapter II, as an application of the Poincaré-Bendixson theory, it was shown that the same situation occurs for a large class of second order systems which include as a special case the van der Pol equation. Such an isolated asymptotically stable periodic orbit was termed a limit cycle and is also sometimes referred to as a self-sustained oscillation. The phenomena exhibited by such systems is completely different from a conservative system since the periodic motion is determined by the differential equation itself in the sense that the differential equation determines a region of the plane in which the ω-limit set of any positive orbit in this region is the periodic orbit.

In many applications, these latter systems are more important since the qualitative behavior of the solutions is less sensitive to perturbations in the differential equation than conservative systems. To illustrate how the structure of the solutions change when a conservative system is subjected to small perturbations, consider the problem of the ordinary pendulum subjected to a frictional force proportional to the rate of the change of the angle u with respect to the vertical. The equation of motion is

(2.1) $\ddot{u} + \beta\dot{u} + k^2 \sin u = 0,$

where $\beta > 0$ is a constant. The equivalent system is

(2.2)
$$\dot{u} = v,$$
$$\dot{v} = -k^2 \sin u - \beta v.$$

If $E(u, v) = v^2/2 + k^2(1 - \cos u)$ is the energy of the system, then dE/dt along the solution of (2.2) is $dE/dt = -\beta v^2 \leq 0$.

We first show that $v(t) \to 0$ for all solutions of (2.2). Since $dE/dt \leq 0$, E is nonincreasing along solutions of (2.2) and $v(t)$ is bounded. Equation (2.2) and the boundedness of $\sin u(t)$ implies $\dot{v}(t)$ is bounded. If $v(t)$ does not approach zero as $t \to \infty$, then there are positive numbers ε, δ and a sequence t_n, $n = 1$, $2, \ldots, t_n \to \infty$ as $n \to \infty$ such that the intervals $I_n = [t_n - \delta, t_n + \delta]$ are non-overlapping and $v^2(t) > \varepsilon$ for t in I_n, $n = 1, 2, \ldots$. Let $p(t)$ be the integer such that $t_n < t$ for all $n \leq p(t)$. Then,

$$E(u(t), v(t)) - E(u(0), v(0)) = \int_0^t \left(\frac{dE}{dt}\right) ds$$

$$\leq -\int_0^t \beta v^2(s) ds$$

$$\leq -2\beta \delta \varepsilon p(t).$$

Since $p(t) \to \infty$ as $t \to \infty$ and $E(u(t), v(t))$ is bounded, this contradicts the fact that $v(t)$ does not approach zero as $t \to \infty$. Thus, $v(t) \to 0$ as $t \to \infty$.

With $E(u(t), v(t))$ nonincreasing and $v(t) \to 0$ as $t \to \infty$, the nature of the level curves of E implies that every solution of (2.2) is bounded. Also, since the energy is nonincreasing and bounded below, it must approach a constant as $t \to \infty$. Since E is continuous and the limit set of any solution is invariant, the limit set must lie on a level curve of E; that is, the limit set of each solution must have $\dot{E} = 0$ and, therefore, $v = 0$. Since the limit set of each solution is invariant and must have $v = 0$, it follows that u is either 0, $\pm\pi$, $\pm 2\pi$, etc. All solutions of (2.2) bounded implies they have a nonempty limit set and, therefore, each solution of (2.2) must approach one of the equilibrium points $(0, 0)$, $(\pi, 0)$, $(-\pi, 0)$, etc. To understand the qualitative behavior of the solutions, it remains only to discuss the stability properties of the equilibrium points and use the properties of the level curves of $E(u, v)$ depicted in Fig. 1.2b.

The linear variational equation relative to $(0, 0)$ is

$$\dot{u} = v,$$
$$\dot{v} = -k^2 u - \beta v,$$

and the eigenvalues of the coefficient matrix have negative real parts. Therefore, by the theorem of Liapunov (Theorem III.2.4), the origin for the nonlinear system is asymptotically stable. The linear variational equation relative to the equilibrium point $(\pi, 0)$ is

$$\dot{u} = v,$$

$$\dot{v} = k^2 u - \beta v,$$

and the eigenvalues of the coefficient matrix are real, with one positive and one negative. Since the saddle point property is preserved (Theorem III.6.1), this equilibrium point is unstable and only two orbits approach this point as $t \to \infty$. Periodicity of sin u yields the stability properties of the other equilibrium points.

This information together with the level curves of $E(u, v)$ allows one to sketch the approximate orbits in the phase plane. These orbits are shown in Fig. 2.1 for the case $\beta < 2k$.

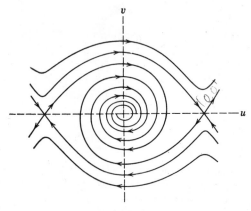

Figure V.2.1

This example illustrates very clearly that a change in the differential equation of a conservative system by the introduction of a small nonconservative term alters the qualitative behavior of the phase portrait of the solutions tremendously. It also indicates that a limit cycle will not occur by the introduction of a truly dissipative or frictional term. In such a case, energy is always taken from the system. To obtain a limit cycle in an equation, there must be a complicated transfer of energy between the system and the external forces. This is exactly the property of van der Pol's equation where the dissipative term $\phi\dot{u}$ is such that ϕ depends upon u and does not have constant sign.

Basic problems in the theory of nonlinear oscillations in autonomous systems with one degree of freedom are to determine conditions under which the differential equation has limit cycles, to determine the number of limit cycles and to determine approximately the characteristics (period, amplitude, shape) of the limit cycles. One useful tool in two dimensions is the Poincaré-Bendixson theory of Section II.1. Other important methods involve perturbation techniques, but can be proved to be applicable in general only when the equation contains either a small or a large parameter. The great advantage of such methods is the fact that the dimension of the system is not important and the equations may depend explicitly upon time in a complicated manner.

One general perturbation technique known as method of averaging will be described in the next section. As motivation for this theory, consider the van der Pol equation

$$\text{(2.3)} \qquad \ddot{u} - \varepsilon(1 - u^2)\dot{u} + u = 0,$$

where $\varepsilon > 0$ is a small parameter. From Theorem II.1.6, this equation has a unique limit cycle for every $\varepsilon > 0$. Our immediate goal is to determine the approximate amplitude and period of this solution as $\varepsilon \to 0$. Equation (2.3) is equivalent to the system

$$\text{(2.4)} \qquad \dot{u} = v,$$
$$\dot{v} = -u + \varepsilon(1 - u^2)v.$$

For $\varepsilon = 0$, the general solution of (2.4) is

$$\text{(2.5)} \qquad u = r \cos \theta, \; v = -r \sin \theta,$$

where $\theta = t + \phi$, ϕ and r are arbitrary constants. If the periodic solution of (2.4) is a continuous function of ε, then the orbit of this solution should be close to one of the circles described by (2.5) by letting $r = $ constant and θ vary from 0 to 2π. The first basic problem is to determine which constant value of r is the candidate for generating the periodic solution of (2.4) for $\varepsilon \neq 0$.

If r, θ are new coordinates, then (2.4) is transformed into the system

$$\text{(2.6)} \qquad \text{(a)} \quad \dot{\theta} = 1 + \varepsilon(1 - r^2 \cos^2 \theta) \sin \theta \cos \theta,$$

$$\text{(b)} \quad \dot{r} = \varepsilon(1 - r^2 \cos^2 \theta)r \sin^2 \theta.$$

For r in a compact set, ε can be chosen so small that $1 + \varepsilon(1 - r^2 \cos^2 \theta) \cdot \sin \theta \cos \theta > 0$ and the orbits of the solutions described by (2.6) are given by the solutions of the scalar equation

$$\text{(2.7)} \qquad \frac{dr}{d\theta} = \varepsilon g(r, \theta, \varepsilon)$$

where

$$\text{(2.8)} \qquad g(r, \theta, \varepsilon) = \frac{(1 - r^2 \cos^2 \theta)r \sin^2 \theta}{1 + \varepsilon(1 - r^2 \cos^2 \theta) \sin \theta \cos \theta}.$$

The problem of determining periodic solutions of van der Pol's equation for ε small is equivalent to finding periodic solutions $r^*(\theta, \varepsilon)$ of (2.7) of period 2π in θ. In fact, if $r^*(\theta, \varepsilon)$ is such a 2π-periodic solution of (2.7) and $\theta^*(t, \varepsilon)$, $\theta^*(0, \varepsilon) = 0$, is the solution of the equation

(2.9) $\dot\theta = 1 + \varepsilon(1 - [r^*(\theta, \varepsilon)]^2 \cos^2 \theta) \sin \theta \cos \theta,$

then $u(t) = r^*(\theta^*(t, \varepsilon), \varepsilon) \cos \theta^*(t, \varepsilon), v(t) = -r^*(\theta^*(t, \varepsilon), \varepsilon) \sin \theta^*(t, \varepsilon)$ is a solution of (2.4). Let T be the unique solution of the equation $\theta^*(T, \varepsilon) = 2\pi$. Uniqueness of solutions of equation (2.9) then implies $\theta^*(t + T, \varepsilon) = \theta^*(t, \varepsilon) + 2\pi$ for all t. Therefore, $u(t + T) = u(t)$, $v(t + T) = v(t)$ for all t and u, v is a periodic solution of (2.4) of period T. Conversely, if u, v is a periodic solution of (2.4) of period T, then $r(t + T) = r(t)$ and one can choose θ so that $\theta(t + T) = \theta(t) + 2\pi$ for all t. The functions r and θ satisfy (2.6) and, for ε small, t can be expressed as a function of θ to obtain r as a periodic function of 2π in θ. Clearly, this function satisfies (2.7).

Let us attempt to determine a solution of (2.7) of the form

(2.10) $r = \rho + \varepsilon r^{(1)}(\theta, \rho) + \varepsilon^2 r^{(2)}(\theta, \rho) + \cdots,$

where each $r^{(j)}(\theta, \rho)$ is required to be 2π-periodic in θ, and ρ is a constant. If this expression is substituted into (2.7) and like powers of ε are equated then

$$\frac{\partial r^{(1)}(\theta, \rho)}{\partial \theta} = g(\rho, \theta, 0).$$

This equation will have a 2π-periodic solution in θ if and only if $\int_0^{2\pi} g(\rho, \theta, 0)d\theta = 0$. If this expression is not zero, then it is called a *secular term*. If a secular term appears, then a solution of the form (2.10) is not possible for an arbitrary constant ρ. The constant ρ must be selected so that it at least satisfies the equation $\int_0^{2\pi} g(\rho, \theta, 0)d\theta = 0$. Having determined ρ so that this relation is satisfied (if possible), then $r^{(1)}(\theta, \rho)$ can be computed and one can proceed to the determination of $r^{(2)}(\theta, \rho)$. However, the same type of equation results for $r^{(2)}(\theta, \rho)$ and more secular terms may appear. One way to overcome these difficulties is to use a method divised by Poincaré which consists in the following: let r be expanded in a series as in (2.10) and also let ρ be expanded in a series in ε as $\rho = \rho_0 + \varepsilon\rho_1 + \varepsilon^2\rho_2 + \cdots$ and apply the same process as before successively determining $\rho_0, \rho_1, \rho_2, \cdots$ in such a way as to eliminate all secular terms. If the ρ_j can be so chosen, then one obtains a periodic solution of (2.7). This method will be discussed in a much more general setting in a later chapter.

We discuss in somewhat more detail another method due to Krylov and Bogoliubov which is generalized in the next section. Consider (2.10) as a transformation of variables taking r into ρ and try to determine $r^{(1)}(\theta, \rho)$, $r^{(2)}(\theta, \rho)$, ... and functions $R^{(1)}(\rho)$, $R^{(2)}(\rho)$, ... so that the differential equation for ρ is autonomous and given by

$$(2.11) \qquad \frac{d\rho}{d\theta} = \varepsilon R^{(1)}(\rho) + \varepsilon^2 R^{(2)}(\rho) + \cdots.$$

If such a transformation can be found, then the 2π-periodic solutions of (2.7) coincide with the equilibrium points of (2.11). Also, the transient behavior of the solutions of (2.7) will be obtainable from (2.11).

If (2.10) is substituted into (2.7), ρ is required to satisfy (2.11) and

$$g(r, \theta, \varepsilon) = \sum_{k=0}^{\infty} \varepsilon^k g^{(k)}(r, \theta),$$

then

$$(2.12) \qquad R^{(1)}(\rho) + \frac{\partial r^{(1)}(\theta, \rho)}{\partial \theta} = g^{(0)}(\rho, \theta),$$

$$R^{(2)}(\rho) + \frac{\partial r^{(2)}(\theta, \rho)}{\partial \theta} = -\frac{\partial r^{(1)}}{\partial \rho}(\theta, \rho) R^{(1)}(\rho) + g^{(1)}(\rho, \theta)$$

$$+ \frac{\partial g^{(0)}}{\partial r}(\rho, \theta) r^{(1)}(\theta, \rho).$$

Since $g^{(0)}(\rho, \theta) = g(\rho, \theta, 0)$, the first equation in (2.12) always has a solution given by

$$(2.13) \qquad R^{(1)}(\rho) = \frac{1}{2\pi} \int_0^{2\pi} g(\theta, \rho, 0)\, d\theta,$$

$$r^{(1)}(\theta, \rho) = \int [g(\rho, \theta, 0) - R^{(1)}(\rho)]\, d\theta,$$

where "$\int$" denotes the 2π-periodic primitive of the function of mean value zero. Similarly, the second equation can be solved for $R^{(2)}(\rho)$ as the mean value of the right hand side and $r^{(2)}(\theta, \rho)$ defined in the same way as $r^{(1)}(\theta, \rho)$. This process actually converges for ε sufficiently small, but this line of investigation will not be pursued here.

Let us recapitulate what happens in the first approximation. Suppose $R^{(1)}(\rho)$, $r^{(1)}(\theta, \rho)$ are defined by (2.13) and consider the exact transformation of variables

$$(2.14) \qquad r = \rho + \varepsilon r^{(1)}(\theta, \rho)$$

applied to (2.7). For ε small, the equation for ρ is

$$\left(1 + \varepsilon \frac{\partial r^{(1)}}{\partial \rho}\right) \dot{\rho} = \varepsilon[g(\rho + \varepsilon r^{(1)}, \theta, \varepsilon) - g(\rho, \theta, 0) + R^{(1)}(\rho)],$$

and thus,

(2.15) $\dot{\rho} = \varepsilon R^{(1)}(\rho) + \varepsilon^2 R^{(2)}(\rho, \theta, \varepsilon)$

where $R^{(2)}(\rho, \theta, \varepsilon)$ is continuous at $\varepsilon = 0$. Consequently, the transformation (2.14) reduces (2.7) to a higher order perturbation of the autonomous equation

(2.16) $\dot{\rho} = \varepsilon R^{(1)}(\rho),$

where $R^{(1)}(\rho)$ is the mean value of $g(\rho, \theta, 0)$.

Suppose $R^{(1)}(\rho^0) = 0$, $dR^{(1)}(\rho^0)/d\rho < 0$ for some ρ^0. Then ρ^0 is a stable equilibrium point of (2.16) and the equation (2.15) satisfies the conditions of Theorem IV.4.3 with $x = \rho$ and the equations for y and z absent. Consequently, there is an asymptotically stable 2π-periodic solution of (2.15) and therefore an asymptotically stable 2π-periodic solution of (2.7) given by (2.15).

If this is applied to van der Pol's equation where g is given by (2.8), then

$$R^{(1)}(\rho) = \frac{1}{2\pi} \int_0^{2\pi} g(\rho, \theta, 0) \, d\theta = \frac{\rho}{2}\left(1 - \frac{\rho^2}{4}\right).$$

This function $R^{(1)}(\rho)$ is such that $R^{(1)}(2) = 0$, $dR^{(1)}(2)/dp = -1 < 0$. Therefore, the van der Pol equation (2.3) has an asymptotically stable periodic solution which for $\varepsilon = 0$ is $u = 2 \cos t$. Zero is also a solution of $R^{(1)}(\rho) = 0$ and $dR^{(1)}(0)/d\rho > 0$. An application of Theorem IV.4.3 implies that zero corresponds to the unstable equilibrium point $u = 0$ of (2.3).

V.3. Averaging

One of the most important methods for the determination of periodic and almost periodic solutions of nonlinear differential equations which contain a small parameter is the so-called method of averaging. The method was motivated in the last section by trying to determine under what conditions one could perform a time varying change of variables which would have the effect of reducing a nonautonomous differential equation to an autonomous one. This idea will be developed to a further extent in this section, but our main interest will lie in the information that can be obtained from only taking the first approximation.

Suppose x in C^n, $\varepsilon \geqq 0$ is a real parameter, $f: R \times C^n \times [0, \infty) \to C^n$ is continuous, $f(t, x, \varepsilon)$ is almost periodic in t uniformly with respect to x in compact sets for each fixed ε, has continuous first partial derivatives with respect to x and $f(t, x, \varepsilon) \to f(t, x, 0)$, $\partial f(t, x, \varepsilon)/\partial x \to \partial f(t, x, 0)/\partial x$ as $\varepsilon \to 0$ uniformly for t in $(-\infty, \infty)$, x in compact sets. Along with the system of equations,

(3.1)
$$\dot{x} = \varepsilon f(t, x, \varepsilon),$$

consider the "averaged" system,

(3.2)
$$\dot{x} = \varepsilon f_0(x),$$

where

(3.3)
$$f_0(x) = \lim_{T \to \infty} \frac{1}{T} \int_0^T f(t, x, 0)\, dt.$$

The basic problem in the method of averaging is to determine in what sense the behavior of the solutions of the autonomous system (3.2) approximates the behavior of the solutions of the more complicated nonautonomous system (3.1). There are two natural types of interpretations to give to the connotation "approximates". One is to ask that the approximation be valid on a large finite interval and the other is to ask that it be valid on an infinite interval. The results given here deal only with the infinite time interval.

Every solution of the equation $\dot{x} = 0$ is a constant. Therefore, system (3.1) is a perturbation of a system which is critical with respect to the class of periodic functions of any period whatsoever and, in particular, with respect to the class of almost periodic systems. On the other hand, it will be shown below that there is an almost periodic transformation of variables which takes $x \to y$ and system (3.1) into a system of the form

$$\dot{y} = \varepsilon[f_0(y) + f_1(\varepsilon, t, y)]$$

where $f_1(0, t, y) = 0$. If there is a y_0 such that $f_0(y_0) = 0$ and $y = y_0 + z$, then this system can be written as

$$\dot{z} = \varepsilon \left[\frac{\partial f_0(y_0)}{\partial y} z + \left\{ f_0(y_0 + z) - f_0(y_0) - \frac{\partial f_0(y_0)}{\partial y} z \right\} + f_1(\varepsilon, t, y_0 + z) \right].$$

Therefore, in a neighborhood of y_0 the methods of Chapter IV can be applied if the linear system

$$\dot{z} = \varepsilon \frac{\partial f_0(y_0)}{\partial y} z$$

is noncritical with respect to the class of functions being considered. In this way, we obtain sufficient conditions for the existence and stability of periodic and almost periodic solutions of (3.1). The reasoning above is the basic idea in the method of averaging. Throughout the remainder of this chapter, we assume the norm in C^n is the Euclidean norm.

The following lemma is fundamental in the theory.

LEMMA 3.1. Suppose $g: R \times C^n \to C^n$ is continuous, $g(t, x)$ is almost periodic in t uniformly with respect to x in compact sets, has continuous first derivatives with respect to x and the mean value of $g(t, x)$ with respect to t is zero; that is,

$$(3.4) \qquad \lim_{T \to \infty} \frac{1}{T} \int_0^T g(t, x)\, dt = 0.$$

For any $\varepsilon > 0$, there is a function $u(t, x, \varepsilon)$, continuous in (t, x, ε) on $R \times C^n \times (0, \infty)$, almost periodic in t for x in compact sets and ε fixed, having a continuous derivative with respect to t and derivatives of an arbitrary specified order with respect to x such that, if

$$h(t, x, \varepsilon) = \frac{\partial u(t, x, \varepsilon)}{\partial t} - g(t, x),$$

then all of the functions h, $\partial h/\partial x$, εu, $\varepsilon \partial u/\partial x$ approach 0 as $\varepsilon \to 0$ uniformly with respect to t in R and x in compact sets.

This lemma is proved in Lemma 5 of the Appendix. It is advantageous at this point to discuss the meaning of Lemma 3.1 when the function $g(t, x)$ satisfies some additional properties. Suppose $g(i, x)$ is a finite trigonometric polynomial with coefficients which are entire functions of x. Then

$$(3.5) \qquad g(t, x) = \sum_{1 \le k \le N} a_k(x) e^{i\lambda_k t}, \ \lambda_k \ne 0,$$

where the λ_k are real and each $a_k(x)$ is an entire function of x. If

$$(3.6) \qquad u(t, x) = \sum_{1 \le k \le N} \frac{a_k(x)}{i\lambda_k} e^{i\lambda_k t},$$

then u satisfies all of the conditions stated in the lemma. In addition, u is independent of ε and

$$(3.7) \qquad \frac{\partial u(t, x)}{\partial t} - g(t, x) = 0.$$

In most applications, the manner just indicated is the method for the construction of a function u with the properties stated in the lemma. For a general function g, one may not be able to solve (3.7) in spite of the fact that the mean value of g is zero. In fact, there are almost periodic functions $g(t)$ (see the Appendix) such that $M[g] = 0$ and $\int^t g$ is an unbounded function. The lemma states that, even in this case, one can "approximately" solve (3.7) by an almost periodic function $u(t, \varepsilon)$. Of course, $u(t, \varepsilon)$ will become unbounded as $\varepsilon \to 0$, but $\varepsilon u(t, \varepsilon) \to 0$ as $\varepsilon \to 0$.

LEMMA 3.2. Suppose f satisfies the properties stated before equation (3.1), f_0 is defined in (3.3) and Ω is any compact set in C^n. Then there exist an $\varepsilon_0 > 0$ and a function $u(t, x, \varepsilon)$ continuous for (t, x, ε) in $R \times C^n \times (0, \varepsilon_0]$, almost periodic in t for x in compact sets and ε fixed and satisfying the conclusions of Lemma 3.1, with $g(t, x) = f(t, x, 0) - f_0(x)$ such that the transformation of variables

(3.8) $\qquad x = y + \varepsilon u(t, y, \varepsilon), \qquad (t, y, \varepsilon) \in R \times \Omega \times [0, \varepsilon_0]$

applied to (3.1) yields the equation

(3.9) $\qquad \dot{y} = \varepsilon f_0(y) + \varepsilon F(t, y, \varepsilon),$

where $F(t, y, \varepsilon)$ satisfies the same conditions as $f(t, y, \varepsilon)$ for $(t, y, \varepsilon) \in R \times \Omega \times [0, \varepsilon_0]$, and, in addition, $F(t, y, 0) \equiv 0$.

PROOF. If $g(t, x) = f(t, x, 0) - f_0(x)$, then the conditions of Lemma 3.1 are satisfied. Let $u(t, x, \varepsilon)$ be the function given by that lemma. Since $\varepsilon u(t, y, \varepsilon)$, $\varepsilon \partial u(t, y, \varepsilon)/\partial y$, $f(t, y, 0) - f(y) - \partial u(t, y, \varepsilon)/\partial t \to 0$ as $\varepsilon \to 0$ uniformly with respect to t in R and y in compact sets, we define $\varepsilon u(t, y, \varepsilon)$, $\varepsilon \partial u(t, y, \varepsilon)/\partial y$, $f(t, y, 0) - f_0(y) - \partial u(t, y, \varepsilon)/\partial y$ at $\varepsilon = 0$ to be zero. For any $\varepsilon_1 > 0$, let Ω_1 be a compact set in C^n containing the set $\{x: x = y + \varepsilon u(t, y. \varepsilon), (t, y, \varepsilon) \in R \times \Omega \times [0, \varepsilon_1]\}$. Choose $\varepsilon_2 > 0$ so that $I + \varepsilon \, \partial u(t, y, \varepsilon)/\partial y$ has a bounded inverse for $(t, y, \varepsilon) \in R \times \Omega \times [0, \varepsilon_2]$. This implies (3.8) can have at most one solution $y \in \Omega$ for each $(t, x, \varepsilon) \in R \times \Omega_1 \times [0, \varepsilon_2]$. For any $x_0 \in \Omega_1$ the contraction mapping principle implies there is an $\varepsilon_3(x_0) > 0$ such that (3.8) has a unique solution $y = y(t, x, \varepsilon)$ defined and continuous for $|y - x_0| \leq \varepsilon_3(x_0)$, $|x - x_0| \leq \varepsilon_3(x_0)$, $0 \leq \varepsilon \leq \varepsilon_3(x_0)$. Since Ω_1 is compact, we can choose an $\varepsilon_4 > 0$ independent of x_0 such that the same property holds with $\varepsilon_3(x_0)$ replaced by ε_4. If $\varepsilon_0 = \min(\varepsilon_1, \varepsilon_2, \varepsilon_4)$, then (3.8) does define a homeomorphism. Therefore, the transformation (3.8) is well defined for $0 \leq \varepsilon \leq \varepsilon_0$, $y \in \Omega$, $t \in R$. If $x = y + \varepsilon u(t, y, \varepsilon)$, then

$$\left(I + \varepsilon \frac{\partial u}{\partial y}\right)\dot{y} = \varepsilon f_0(y) + \varepsilon \left[f(t, y, 0) - f_0(y) - \frac{\partial u}{\partial t}\right]$$

$$+ \varepsilon [f(t, y + \varepsilon u, \varepsilon) - f(t, y, 0)].$$

From the properties of u in Lemma 3.1 and the continuity of f, the indicated property of the equation for y follows immediately.

Lemma 3.2 is most important from the point of view of the differential equations since it shows that the transformation of variables (3.8), which is almost the identity transformation for ε small, when applied to (3.1) yields a differential equation which is the same as the averaged system (3.2) up through terms of order ε. We emphasize again that for functions $f(t, x, \varepsilon)$ which are trigonometric polynomials with coefficients entire functions of x, the expression for the transformation (3.8) is obtained by taking u to be given by (3.6) where $g(t, x) = f(t, x, 0) - f_0(x)$.

Let Re $\lambda(A)$ designate the real parts of the eigenvalues of a matrix A.

THEOREM 3.1. Suppose f satisfies the conditions enumerated before (3.1). If there is an x^0 such that $f_0(x^0) = 0$, and Re $\lambda[\partial f_0(x^0)/\partial x] \neq 0$, then there are an $\varepsilon_0 > 0$ and a function $x^*(\,\cdot\,, \varepsilon)\colon R \to C^n$, satisfying (3.1), $x^*(t,\ \varepsilon)$ is continuous for (t, ε) in $R \times [0,\ \varepsilon_0]$, almost periodic in t for each fixed ε, $m[x^*(\,\cdot\,, \varepsilon)] \subset m[f(\,\cdot\,, x, \varepsilon)]$, $x^*(\,\cdot\,, 0) = x^0$. The solution $x^*(\,\cdot\,, \varepsilon)$ is also unique in a neighborhood of x^0. Furthermore, if Re $\lambda[\partial f_0(x^0)/\partial x] < 0$, then $x^*(\,\cdot\,, \varepsilon)$ is uniformly asymptotically stable for $0 < \varepsilon \leq \varepsilon_0$ and if one eigenvalue of this matrix has a positive real part, $x^*(\,\cdot\,, \varepsilon)$ is unstable.

THEOREM 3.2. Suppose f satisfies the conditions enumerated before (3.1). If $f(t + T, x, \varepsilon) = f(t, x, \varepsilon)$ for all (t, x, ε) in $R \times C^n \times [0, \infty)$ and there is an x^0 such that $f_0(x^0) = 0$, $\det[\partial f_0(x^0)/\partial x] \neq 0$, then there are an $\varepsilon_0 > 0$ and a function $x^*(t, \varepsilon)$, continuous for (t, ε) in $R \times [0, \varepsilon_0]$, $x^*(\,\cdot\,, 0) = x^0$, $x^*(t + T, \varepsilon) = x^*(t, \varepsilon)$, for all t, ε and $x^*(t, \varepsilon)$ satisfies (3.1). This solution $x^*(\,\cdot\,, \varepsilon)$ is also unique in a neighborhood of x^0.

The proofs of both of these theorems are immediate consequences of previous results. In fact, choose any compact set Ω in C^n containing x^0 in its interior. Lemma 3.2 implies that we may assume equation (3.1) is in the form (3.9) for $(t, x, \varepsilon) \in R \times \Omega \times [0, \varepsilon_0]$. If x^0 is such that $f_0(x^0) = 0$, then $y = x^0 + z$ in (3.9) implies

$$\dot{z} = \varepsilon Az + \varepsilon F(t, x^0 + z, \varepsilon) + \varepsilon[f_0(x^0 + z) - f_0(x^0) - Az]$$

$$\overset{\text{def}}{=} \varepsilon Az + \varepsilon q(t, z, \varepsilon),$$

where $A = \partial f_0(x^0)/\partial x$. One can now apply Lemma IV.4.3 and Theorem IV.4.2 directly to the equation for z to complete the proofs of Theorems 3.1 and 3.2.

In the applications, many problems cannot be phrased in such a way that the equations of motion are equivalent to a system of the form (3.1). However, the underlying ideas in the proof of Theorems 3.1 and 3.2 can be used effectively in more complicated situations. One should therefore keep in mind these basic principles and look upon the discussion here as a possible method of attack for the treatment of oscillatory phenomena in weakly nonlinear systems. Another application is now given to a class of equations which seems to occur frequently in the applications.

Consider the system

(3.10) $\dot{x} = \varepsilon f(t, x) + \varepsilon h(\varepsilon t, x),$

where $\varepsilon > 0$ is a real parameter, $f(t, x)$, $h(t, x)$ are continuous and have continuous first derivatives with respect to x, $h(t, x)$ has continuous second partial derivatives with respect to x for (t, x) in $R \times C^n$, $f(t, x)$ is almost periodic in t uniformly with respect to x in compact sets and there is a $T > 0$ such that $h(t + T, x) = h(t, x)$ for all t, x.

System (3.10) contains both a "fast" time t and a "slow" time εt. The averaging procedure is applied to only the fast time to obtain the non-autonomous "averaged" equations

(3.11) $$\dot{x} = \varepsilon f_0(x) + \varepsilon h(\varepsilon t, x) \overset{\text{def}}{=} \varepsilon G(\varepsilon t, x),$$

where

(3.12) $$f_0(x) = \lim_{T \to \infty} \frac{1}{T} \int_0^T f(t, x)\, dt.$$

If system (3.11) has a periodic solution $x^0(\varepsilon t)$ of period T/ε, then the linear variational equation for $x^0(\varepsilon t)$ is given by

(3.13) $$\frac{dy}{d\tau} = \frac{\partial G(\tau, x^0(\tau))}{\partial x} y$$

where $\tau = \varepsilon t$.

The next result states conditions under which the equation (3.10) has an almost periodic solution which approaches the periodic solution of the averaged equation (3.11) as $\varepsilon \to 0$.

THEOREM 3.3. Suppose f, h satisfy the conditions enumerated after (3.10). If $x^0(\varepsilon t)$ is a periodic solution of (3.11) of period T/ε such that no characteristic exponents of the linear variational equation (3.13) have zero real parts, then there are an $\varepsilon_0 > 0$ and a function $x^*(\,\cdot\,, \varepsilon) \colon R \to C^n$, satisfying (3.10), $x^*(t; \varepsilon)$ is continuous for (t, ε) in $R \times [0, \varepsilon_0]$, almost periodic in t for each fixed ε, $m[x^*(\,\cdot\,, \varepsilon)] \subset m[f(\,\cdot\,, x), h(\varepsilon\,\cdot\,, x)]$; and $x^*(t, \varepsilon) - x^0(\varepsilon t) \to 0$ as $\varepsilon \to 0$ uniformly on R. This solution is also unique in a neighborhood of $x^0(\,\cdot\,)$. Furthermore, if all characteristic exponents of (3.13) have negative real parts, $x^*(\,\cdot\,, \varepsilon)$ is uniformly asymptotically stable and if one exponent has a positive real part, it is unstable.

The proof of this result proceeds as follows. Choose any compact set Ω in C^n which contains $x^0(t)$, $0 \leq t \leq T$, in its interior. If $g(t, x) = f(t, x) - f_0(x)$ and $u(t, x, \varepsilon)$ is the function given by Lemma 3.1, then as in the proof of Lemma 3.2, there is an $\varepsilon_0 > 0$ such that the transformation of variables $x = y + \varepsilon u(t, y, \varepsilon)$ is well defined for $(t, y, \varepsilon) \in R \times \Omega \times [0, \varepsilon_0]$. This transformation applied to system (3.10) yields the equivalent equation

$$\dot{y} = \varepsilon G(\varepsilon t, y) + \varepsilon H(t, \varepsilon t, y, \varepsilon),$$

where $H(t, \tau, y, 0) = 0$ and $H(t, \tau, y, \varepsilon)$ satisfies the same smoothness properties in y as f, h, is almost periodic in t and periodic in τ of period T. If $y(t) = x^0(\varepsilon t) + z(t)$, then

$$\dot{z} = \varepsilon A(\varepsilon t)z + \varepsilon Z(t, \varepsilon t, z, \varepsilon),$$

where $A(\tau) = \partial G(\tau, x^0(\tau))/\partial x$ and $Z(t, \varepsilon t, z, \varepsilon)$ satisfies the conditions of Theorem IV.4.2. Suppose $P(\tau)e^{B\tau}$ is a fundamental matrix solution of (3.13) with $P(\tau + T) = P(\tau)$ and B a constant matrix. From the hypothesis of the theorem, the eigenvalues of B have nonzero real parts. If $z = P(\varepsilon t)w$, then

$$\dot{w} = \varepsilon Bw + \varepsilon W(t, \varepsilon t, w, \varepsilon)$$

where $W(t, \varepsilon t, w, \varepsilon)$ satisfies the conditions of Theorem IV.4.2. The proof of Theorem 3.3 is completed by a direct application of this result.

Another type of equation that is very common is the system

$$(3.14) \qquad \dot{x} = \varepsilon X(t, x, y, \varepsilon),$$

$$\dot{y} = Ay + \varepsilon Y(t, x, y, \varepsilon),$$

where ε is a real parameter, x, X are n-vectors, y, Y are m-vectors, A is a constant $n \times n$ matrix whose eigenvalues have nonzero real parts, X, Y are continuous on $R \times C^n \times C^m \times [0, \infty)$, have continuous first derivatives with respect to x, y and are almost periodic in t for x, y in compact sets and each fixed ε. The "averaged" equation for (3.14) is defined to be

$$(3.15) \qquad \dot{x} = \varepsilon X_0(x),$$

where

$$(3.16) \qquad X_0(x) = \lim_{T \to \infty} \frac{1}{T} \int_0^T X(t, x, 0, 0)\, dt.$$

THEOREM 3.4. Suppose X, Y satisfy the conditions enumerated after (3.14). If there is an x^0 such that $X_0(x^0) = 0$ and Re $\lambda(\partial X_0(x^0)/\partial x) \neq 0$, then there are an $\varepsilon_0 > 0$ and vector functions $x^*(\,\cdot\,, \varepsilon)$, $y^*(\,\cdot\,, \varepsilon)$ of dimensions n, m respectively, satisfying (3.14), continuous on $R \times [0, \varepsilon_0]$, almost periodic in t for each fixed ε, $x^*(\,\cdot\,, 0) = x^0$, $y^*(\,\cdot\,, 0) = 0$. This solution is also unique in a neighborhood of $(x^0, 0)$. Furthermore, if Re $\lambda(\partial X_0(x^0)/\partial x) < 0$, Re $\lambda(A) < 0$, then this solution is uniformly asymptotically stable for $0 < \varepsilon \leqq \varepsilon_0$ and if one eigenvalue of either of these matrices has a positive real part, it is unstable.

The proof follows the same lines as before. Suppose Ω is a compact set of C^n containing x^0 in its interior. If $g(t, x) = X(t, x, 0, 0) - X_0(x)$ and $u(t, x, \varepsilon)$ is the function given in Lemma 3.1, then as in the proof of Lemma 3.2, there is an $\varepsilon_0 > 0$ such that the transformation

$$x \to x + \varepsilon u(t, x, \varepsilon),\ y \to y,$$

is well defined for $(t, x, y, \varepsilon) \in R \times \Omega \times C^m \times [0, \varepsilon_0]$. This transformation applied to (3.14) yields the equivalent system

$$\dot{x} = \varepsilon X_0(x) + \varepsilon X_1(t, x, y, \varepsilon) + \varepsilon X_2(t, x, y, \varepsilon),$$

$$\dot{j} = Ay + \varepsilon Y_1(t, x, y, \varepsilon),$$

where $X_1(t, x, y, 0) = 0$ and $X_2(t, x, 0, 0) = 0$ and all functions have the same smoothness and almost periodicity properties as before. If the further transformation $x \to x$, $y \to \sqrt{\varepsilon y}$ is made, then the equations become

$$\dot{x} = \varepsilon X_0(x) + \varepsilon X^*(t, x, y, \varepsilon),$$

$$\dot{y} = Ay + Y^*(t, x, y, \varepsilon),$$

where $X^*(t, x, y, 0) = 0$, $Y^*(t, x, y, 0) = 0$. If $x \to x^0 + x$, $y \to y$, then Theorem IV.4.3 with the variable z absent applies directly to the resulting system to yield Theorem 3.4.

Using the same proof, one also obtains

THEOREM 3.5. If X, Y satisfy the conditions enumerated after (3.14), are periodic in t or period T and there is an x^0 such that $X_0(x^0) = 0$ and $\det[\partial X_0(x^0)/\partial x] \neq 0$, then there are an $\varepsilon_0 > 0$ and vector functions $x^*(\,\cdot\,, \varepsilon)$, $y^*(\,\cdot\,, \varepsilon)$ of dimensions n, m, respectively, satisfying (3.14), continuous on $R \times [0, \varepsilon_0]$, periodic in t of period T and $x^*(\,\cdot\,, 0) = x^0$, $y^*(\,\cdot\,, 0) = 0$. Furthermore this solution is unique in a neighborhood of $(x^0, 0)$.

An interesting application of Theorem 3.4 to stability of linear systems is

THEOREM 3.6. Suppose $D = \operatorname{diag}(B, A)$ where B is $n \times n$, A is $m \times m$, all eigenvalues of B have simple elementary divisors and zero real parts and all eigenvalues of A have negative real parts. If the $(n + m) \times (n + m)$ almost periodic matrix Φ is partitioned as $\Phi = (\Phi_{jk})$, $j, k = 1, 2$, where Φ_{11} is $n \times n$, Φ_{22} is $m \times m$ and if all eigenvalues of the matrix

$$(3.17) \qquad E = \lim_{T \to \infty} \frac{1}{T} \int_0^T e^{-Bt} \Phi_{11}(t) e^{Bt}\, dt,$$

have negative real parts, then there is an $\varepsilon_0 > 0$ such that the system

$$(3.18) \qquad \dot{u} = Du + \varepsilon \Phi(t) u,$$

is uniformly asymptotically stable for $0 < \varepsilon \leqq \varepsilon_0$. If one eigenvalue of E has a positive real part, then system (3.18) is unstable for $0 < \varepsilon \leqq \varepsilon_0$.

The proof of this result proceeds as follows. Let $u = (x, y)$ where x is an n-vector. The matrix e^{Bt} is almost periodic in t and therefore the bounded linear transformation $x \to e^{Bt}x$, $y \to y$ yields the equivalent system

$$\dot{x} = \varepsilon e^{-Bt} \Phi_{11}(t) e^{Bt} x + \varepsilon e^{-Bt} \Phi_{12}(t) y,$$

$$\dot{y} = Ay + \varepsilon \Phi_{21}(t) e^{Bt} x + \varepsilon \Phi_{22}(t) y.$$

This is a special case of system (3.14) and the averaged system (3.15) for this situation is $\dot{x} = Ex$. From the hypothesis on E, this averaged system satisfies the hypotheses of Theorem 3.4. It is clear from uniqueness of the solution

$x^*(\,\cdot\,,\varepsilon), y^*(\,\cdot\,,\varepsilon)$ guaranteed by Theorem 3.4 that $x^*(\,\cdot\,,\varepsilon)=0, y^*(\,\cdot\,,\varepsilon)=0,$ $0 \leqq \varepsilon \leqq \varepsilon_0$. This proves the result.

The remainder of this chapter is devoted to applications of the results in this section.

V.4. The Forced Van Der Pol Equation

Consider the equation

$$(4.1) \qquad \dot{z}_1 = z_2,$$

$$\dot{z}_2 = -z_1 + \varepsilon(1 - z_1^2)z_2 + A \sin \omega_1 t + B \sin \omega_2 t,$$

where $\varepsilon > 0$, $\omega_1 > 0$, $\omega_2 > 0$, A, B are real constants and

$$(4.2) \qquad m + m_1\omega_1 + m_2\omega_2 \neq 0,$$

for all integers m, m_1, m_2 with $|m| + |m_1| + |m_2| \leqq 4$. For $\varepsilon = 0$, the general solution of (4.1) is

$$(4.3) \qquad z_1 = x_1 \cos t + x_2 \sin t + A_1 \sin \omega_1 t + B_1 \sin \omega_2 t,$$

$$z_2 = -x_1 \sin t + x_2 \cos t + A_1\omega_1 \cos \omega_1 t + B_1\omega_2 \cos \omega_2 t,$$

where $A_1 = A(1 - \omega_1^2)^{-1}$, $B_1 = B(1 - \omega_2^2)^{-1}$ and x_1, x_2 are arbitrary constants. To discuss the existence of almost periodic solutions of system (4.1) by using the results of the previous section, consider relations (4.3) as a transformation to new coordinates x_1, x_2. After a few straightforward calculations, the new equations for x_1, x_2 are

$$(4.4) \qquad \dot{x}_1 = -\varepsilon(1 - z_1^2)z_2 \sin t,$$

$$\dot{x}_2 = \varepsilon(1 - z_1^2)z_2 \cos t,$$

where z_1, z_2 are the complicated functions given in (4.3). System (4.4) is a special case of (3.1) and the quasi-periodic coefficients in the right hand sides of (4.4) have basic frequencies 1, ω_1, ω_2.

The average of the right hand side of (4.4) with respect to t will have different types of terms depending upon whether the frequencies 1, ω_1, ω_2 satisfy (4.2) or do not satisfy (4.2). If (4.2) is satisfied, then the averaged equations of (4.4) are

$$(4.5) \qquad 8\dot{x}_1 = \varepsilon x_1[2(2 - A_1^2 - B_1^2) - (x_1^2 + x_2^2)],$$

$$8\dot{x}_2 = \varepsilon x_2[2(2 - A_1^2 - B_1^2) - (x_1^2 + x_2^2)].$$

Equations (4.5) always have the constant solution $x_1 = x_2 = 0$ and both eigenvalues of the linear variational equation are $2(2 - A_1^2 - B_1^2)$. If

$A_1^2 + B_1^2 \neq 2$, then Theorem 3.1 implies the existence of an almost periodic solution of (4.4) with frequencies contained in the module of $1, \omega_1, \omega_2$, which is zero for $\varepsilon = 0$, is uniformly asymptotically stable if $A_1^2 + B_1^2 > 2$ and unstable if $A_1^2 + B_1^2 < 2$. This implies the original equation (4.1) has an almost periodic solution which is uniformly asymptotically stable (unstable) for $A_1^2 + B_1^2 > (<)2$ and this solution for $\varepsilon = 0$ is given by

$$z_1(t) = A_1 \sin \omega_1 t + B_1 \sin \omega_2 t,$$

$$z_2(t) = \dot{z}_1(t).$$

Notice that the condition $A_1^2 + A_2^2 > 2$ can be achieved if either A or B is sufficiently large for given ω_1, ω_2, or if either ω_1 or ω_2 is sufficiently close to 1 (resonance) for a given A, B. Also, notice that $A_1^2 + A_2^2 < 2$ implies that the averaged equations have a circle of equilibrium points given by $x_1^2 + x_2^2 = 2(2 - A_1^2 - B_1^2)$. There are very interesting oscillatory phenomena associated with this set of equilibrium points, but the discussion is much more complicated and is treated in the chapter on integral manifolds.

V.5. Duffing's Equation with Small Damping and Small Harmonic Forcing

Consider the Duffing equation

(5.1) $$\ddot{u} + \varepsilon \delta \dot{u} + u + \varepsilon \gamma u^3 = \varepsilon B \cos \omega t,$$

or the equivalent system

(5.2) $$\begin{cases} \dot{u} = v \\ \dot{v} = -u - \varepsilon \gamma u^3 - \varepsilon \delta v + \varepsilon B \cos \omega t. \end{cases}$$

where $\varepsilon \geq 0$, γ, $\delta \geq 0$, B, $\omega \geq 0$ are real parameters. For $\omega^2 = 1 + \varepsilon \beta$, we wish to determine conditions on the parameters which will ensure that equation (5.1) has a periodic solution of period $2\pi/\omega$. Since for $\varepsilon = 0$, $\omega = 1$, the forcing function has a frequency very close to the free frequency of the equation. The free frequency is the frequency of the periodic solutions of the equation (5.1) when $\varepsilon = 0$. Such a situation is referred to as *harmonic forcing*. We have seen previously that the linear equation $\ddot{u} + u = \cos t$ has no periodic solutions and in fact all solutions are unbounded. This is due to the resonance effect of the forcing function. As we will see, the nonlinear equation has some fascinating properties and, in particular, more than one isolated periodic solution may exist. Contrast this statement to the results of Section IV.5. To apply the results of Section 3, we make the *van der Pol transformation*

(5.3)
$$\begin{cases} u = x_1 \sin \omega t + x_2 \cos \omega t \\ v = \omega[x_1 \cos \omega t - x_2 \sin \omega t] \end{cases}$$

in (5.2) to obtain an equivalent system

(5.4)
$$\begin{cases} \dot{x}_1 = \dfrac{\varepsilon}{\omega}[\beta u - \gamma u^3 - \delta v + B \cos \omega t] \cos \omega t \\[2mm] \dot{x}_2 = -\dfrac{\varepsilon}{\omega}[\beta u - \gamma u^3 - \delta v + B \cos \omega t] \sin \omega t \\[2mm] \beta = \dfrac{\omega^2 - 1}{\varepsilon}, \end{cases}$$

where u, v are given in (5.3). To average the right hand sides of these equations with respect to t, treating x_1, x_2 as constant, it is convenient to let

(5.5) $x_2 = r \cos \psi, \ x_1 = r \sin \psi,$

$u = x_1 \sin \omega t + x_2 \cos \omega t = r \cos (\omega t - \psi),$

$v = \omega[x_1 \cos \omega t - x_2 \sin \omega t] = -\omega r \sin(\omega t - \psi).$

This avoids cubing u and complicated trigonometric formulas. The averaged equations associated with (5.4) are now easily seen to be

(5.6)
$$\dot{x}_1 = \frac{\varepsilon}{2\omega}\left[\beta x_2 - \frac{3\gamma r^2}{4} x_2 - \delta \omega x_1 + B\right],$$

$$\dot{x}_2 = -\frac{\varepsilon}{2\omega}\left[\beta x_1 - \frac{3\gamma r^2}{4} x_1 + \delta \omega x_2\right],$$

$$r^2 = x_1^2 + x_2^2.$$

Since $x_2 = r \cos \psi$, $x_1 = r \sin \psi$, the equilibrium points of (5.6) are the solutions of the equations

$$\left(\beta - \frac{3\gamma r^2}{4}\right) r \cos \psi - \delta \omega r \sin \psi + B = 0,$$

$$\left(\beta - \frac{3\gamma r^2}{4}\right) r \sin \psi + \delta \omega r \cos \psi = 0.$$

These latter equations are equivalent to the equations

(5.7)
$$G(r, \psi, \omega) \overset{\text{def}}{=} \omega^2 - 1 - \frac{3\gamma_0 r^2}{4} + \frac{F_0}{r} \cos \psi = 0,$$

$$F(r, \psi, \omega) \overset{\text{def}}{=} F_0 \sin \psi - \delta_0 \omega r = 0,$$

where we have put $\gamma_0 = \varepsilon\gamma$, $\delta_0 = \varepsilon\delta$, $F_0 = \varepsilon B$, parameters which have a good physical interpretation in equation (5.1).

If γ_0, δ_0, F_0 are considered as fixed parameters in (5.7), then (5.7) can be considered as two equations for the three unknowns ψ, ω, r. If there exist ψ_0, ω_0, r_0 such that the matrix

(5.8)
$$\begin{bmatrix} \dfrac{\partial(rG)}{\partial r} & \dfrac{\partial(rG)}{\partial\psi} \\[2ex] \dfrac{\partial F}{\partial r} & \dfrac{\partial F}{\partial\psi} \end{bmatrix}$$

has rank 2 for $r = r_0$, $\psi = \psi_0$, $\omega = \omega_0$, then Theorem 3.2 implies there is an $\varepsilon_0 > 0$ such that equation (5.1) has a periodic solution of period $2\pi/\omega_0$ which for $\varepsilon = 0$ is given by $u = r_0 \cos(\omega_0 t - \psi_0) = r_0 \cos(t - \psi_0)$ since $\omega_0 = 1$ for $\varepsilon = 0$. In the equations (5.7), one usually considers the approximate amplitude r of the solution of (5.1) as a parameter and determines the frequency $\omega(r)$ and approximate phase $\psi(r)$ of the solution of (5.1) as functions of r. The plot in the ω, r plane of the curve $\omega(r)$ is called the *frequency response curve*.

The stability properties of the above periodic solution can sometimes be discussed by making use of Theorem 3.1. Some special cases are now treated in detail.

Case 1. $\delta = 0$ (*No Damping*). One solution of (5.7) for $\delta = \delta_0 = 0$ is given by $\psi = 0$ and

(5.9)
$$\omega^2 = 1 + \frac{3\gamma_0 r^2}{4} + \frac{F_0}{r}.$$

As mentioned earlier, relation (5.9) is called the *frequency response curve*. The rank of the matrix in (5.8) is two if $\gamma_0 = 0$, $F_0 \neq 0$ or $\gamma_0 \neq 0$, F_0/γ_0 sufficiently small. From Theorem 3.2, there is an $\varepsilon_0 > 0$ such that for $\delta = 0, 0 \leq \varepsilon \leq \varepsilon_0$, and each value of ω, r which lies on the frequency response curve, there is a $2\pi/\omega$-periodic solution of (5.1) which for $\varepsilon = 0$ is $u = r \cos \omega t = r \cos t$ since $\omega = 1$ for $\varepsilon = 0$. There is also the solution of (5.7) corresponding to $\psi = \pi$, but this corresponds to (5.9) with r replaced by $-r$. The uniqueness property guaranteed by Theorem 3.2 and the equation (5.2) for $\delta = 0$ imply this solution is the negative of the one for $\psi = 0$.

If $F_0 \leq 0$, the plots of the frequency response curves in a neighborhood of $\omega = 1$ for both the hard spring ($\gamma_0 > 0$) and the soft spring ($\gamma_0 < 0$) are given in Fig. 5.1. The pictures are indicative of what happens near $\omega = 1$.

The frequency response curves in Fig. 5.1 are usually plotted with $|r|$ rather than r and are shown in Fig. 5.2. Notice how the nonlinearity ($\gamma_0 \neq 0$) bends the response curve for the linear equation. The curve $F_0 = 0$ depicts the

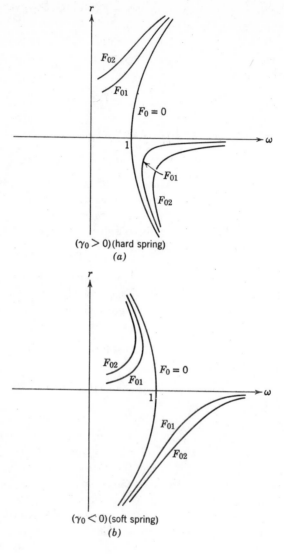

Figure V.5.1

relationship between the frequency ω and the amplitude r of the periodic solutions of the unforced conservative system $\ddot{u} + u + \gamma_0 u^3 = 0$. For each given ω near $\omega = 1$, there is exactly one periodic solution of period $2\pi/\omega$. For a given $F_0 \neq 0$, there are three such periodic solutions for some values of ω and only one for others.

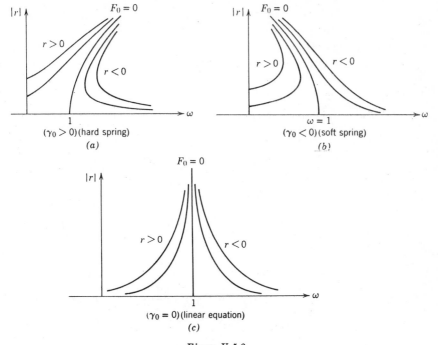

Figure V.5.2

Let us emphasize another striking property of this example which is a direct consequence of the fact that the equation is nonlinear. For $\varepsilon = 0$, system (5.1) has a free frequency which is equal to 1; that is, all solutions of the linear equation are periodic of period 2π. On the other hand, it was shown above that for $\omega^2 - 1$ of order ε there are periodic solutions of period $2\pi/\omega$. In other words, the free frequency is suppressed and has been "locked" with the forcing frequency. This phenomena is sometimes referred to as the *locking-in phenomenon* or *entrainment of frequency* and, of course, can only occur when the equations are nonlinear.

Case 2. $\delta > 0$ *(Damping)*. If $\delta_0 > 0$ and $F_0 = 0$, then using the same analysis as in the discussion of equation (2.1), one sees that equation (5.1) has no periodic solution except $u = 0$. This is reflected also in equations (5.7) which in this case, have no solution except $r = 0$. If $\delta_0 F_0 \neq 0$, there always exist values of r such that $|r| < |F_0/\delta_0 \omega|$ and the second of equations (5.7) can be solved for ψ as a function of $\omega \delta_0 r/F_0$. Using such a ψ, the first equation yields, up to terms of order ε^2,

(5.10)
$$\omega^2 = 1 + \frac{3\gamma_0 r^2}{4} \pm \sqrt{\frac{F_0^2}{r^2} - \delta_0^2}.$$

This frequency response curve for the hard spring $(\gamma_0 > 0)$ is shown in Fig. 5.3. The dotted line corresponds to the curve $\omega^2 = 1 + 3\gamma_0 r^2/4$. Thus, as for $\delta = 0$, for given F_0, δ_0 with $F_0\,\delta_0 \neq 0$, Theorem 3.2 implies there are three periodic solutions of period $2\pi/\omega$ for some values of ω and only one for others. For a given ω, which of these solutions are stable and which are unstable? To discuss this, we investigate the linear variational equation associated with these equilibrium points of the averaged equations and apply Theorem 3.1. As is to be expected the analysis will be complicated. Another type of co-ordinate system which is widely used in applying the method of averaging turns out to be useful for this discussion.

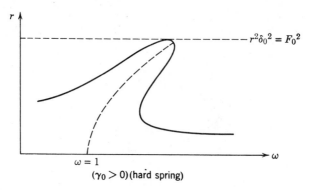

$$r^2\delta_0{}^2 = F_0{}^2$$

$$\omega = 1$$
$$(\gamma_0 > 0)(\text{hard spring})$$

Figure V.5.3

In (5.2), introduce new variables r, ψ by the relations

(5.11) $$u = r\cos(\omega t - \psi),$$

$$v = -r\omega\sin(\omega t - \psi),$$

to obtain an equivalent set of equations $(\varepsilon\beta = \omega^2 - 1)$

(5.12) $$\dot{r} = \frac{1}{\omega}\left[-\frac{\varepsilon\beta}{2}\sin 2(\omega t - \psi) - \varepsilon f\sin(\omega t - \psi)\right],$$

$$\dot{\psi} = \frac{1}{\omega}\left[\frac{\varepsilon\beta}{2} + \frac{\varepsilon\beta}{2}\cos 2(\omega t - \psi) + \frac{\varepsilon}{r}f\cos(\omega t - \psi)\right],$$

where $f = -\gamma u^3 - \delta v + B\cos\omega t$ and u, v are given in (5.11). The averaged equations associated with (5.12) are

(5.13) $$\dot{r} = -\frac{1}{2\omega}[\varepsilon\delta\omega r - \varepsilon B\sin\psi] = \frac{1}{2\omega}F(r, \psi, \omega),$$

$$\dot{\psi} = \frac{1}{2\omega}\left[\varepsilon\beta - \frac{3}{4}\varepsilon\gamma r^2 + \frac{\varepsilon B}{r}\cos\psi\right] = \frac{1}{2\omega}G(r, \psi, \omega),$$

where F, G are defined in (5.7). The equilibrium points of (5.13) are therefore the solutions r, ψ, ω of (5.7). If the eigenvalues of the coefficient matrix of the linear variational equation of any such equilibrium point have nonzero real parts, then Theorem 3.1 gives not only the existence but the stability properties of a periodic solution of period $2\pi/\omega$ of (5.1). The solution is uniformly asymptotically stable if these eigenvalues have negative real parts and unstable if one has a positive real part. A necessary and sufficient condition for the eigenvalues of this matrix to have negative real parts is that the trace of this matrix be negative and the determinant be positive. In terms of F, G in (5.13), one has an asymptotically stable solution if and only if

$$F_r + G_\psi < 0,$$

$$F_r G_\psi - F_\psi G_r > 0,$$

where the subscripts denote partial derivatives and, of course, the functions are evaluated at the equilibrium point. These partial derivatives are easily seen to be

$$F_r = -\delta_0 \omega, \ F_\psi = F_0 \cos \psi,$$

$$G_r = -\frac{3}{2}\gamma_0 r - \frac{F_0}{r^2}\cos \psi, \ G_\psi = -\frac{F_0}{r}\sin \psi.$$

Therefore, $F_r + G_\psi = -2\delta_0 \omega < 0$ at an equilibrium point and the stability or instability can be decided by investigating the sign of $F_r G_\psi - G_r F_\psi$ as long as it is different from zero.

It is possible to express this last condition for stability in terms of where the point r, ω is situated on the frequency response curve. To see this, consider the equilibrium points of (5.13); that is, the solutions of $F(r, \psi, \omega) = 0 = G(r, \psi, \omega)$, as functions of ω. Differentiating these equations with respect to ω and letting subscripts denote differentiation, one obtains

$$-F_\omega = F_r r_\omega + F_\psi \psi_\omega,$$

$$-G_\omega = G_r r_\omega + G_\psi \psi_\omega,$$

and thus,

(5.14) $$r_\omega(F_r G_\psi - G_r F_\psi) = G_\omega F_\psi - F_\omega G_\psi.$$

The stability properties of the solution can therefore be translated into the sign of $r_\omega(G_\omega F_\psi - F_\omega G_\psi)$. This expression being positive corresponds to stability and this being negative corresponds to instability. Since $F_\omega = -\delta_0 r$, $G_\omega = 2\omega$, it follows that

$$G_\omega F_\psi - F_\omega G_\psi = 2\omega F_0 \cos \psi + \delta_0 F_0 \sin \psi$$

$$= 2\omega r(\nu^2 - \omega^2) + \delta_0^2 \omega r.$$

where $\nu^2 = 1 + 3\gamma_0 r^2/4$. From (5.10), $\omega^2 - \nu^2 = O(\varepsilon)$ and since δ_0^2 is $O(\varepsilon^2)$, the sign of this expression is determined by $\omega^2 - \nu^2$ if ε is sufficiently small. Using (5.14), we finally have:

A periodic solution of Duffing's equation (5.1) with $\delta > 0$ is stable if $(dr/d\omega)$ $\cdot (\nu^2 - \omega^2) > 0$ and unstable if $(dr/d\omega)(\nu^2 - \omega^2) < 0$, where $\nu^2 = 1 + 3\gamma_0 r^2/4$ and r is given as a function of ω by the frequency response curve (5.10).

Pictorially, the situation is shown in Fig. 5.4 for a hard spring. Solid black dots represent unstable points. The arrows $\rightarrow$ represent a typical manner in which the amplitude of the stable periodic motion would change with increasing ω and the backward $\leftarrow$ depicts the situation for decreasing ω.

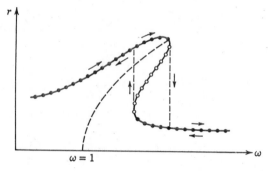

Figure V.5.4

V.6. The Subharmonic of Order 3 for Duffing's Equation

Consider the equation

(6.1)
$$\ddot{u} + \varepsilon c\dot{u} + \frac{1}{9}u + \varepsilon\gamma u^3 = B\cos\omega t$$

or the equivalent system

(6.2)
$$\dot{u} = v,$$

$$\dot{v} = -\frac{1}{9}u - \varepsilon cv - \varepsilon\gamma u^3 + B\cos\omega t,$$

where $\varepsilon \geqq 0$, $c \geqq 0$, γ, B and ω are constants, $\omega - 1 = O(\varepsilon)$ as $\varepsilon \rightarrow 0$. The problem is to determine under what conditions there exists a solution of (6.1) which is periodic of period $6\pi/\omega$ and whose zeroth order terms in ε are given by

(6.3)
$$u = r \sin\left(\frac{\omega}{3} t + \phi\right) + A \cos \omega t,$$

$$v = \frac{\omega}{3} r \cos\left(\frac{\omega}{3} t + \phi\right) - \omega A \sin \omega t.$$

Such a solution if it exists is called a subharmonic solution of order 3. If any such solution exists, then clearly A must equal $-B/(\omega^2 - 1/9)$. The free frequency of equation (6.1) is $(1/3)$ and the forcing frequency is ω which is approximately 1. Therefore, if a subharmonic solution exists, then the free frequency is again suppressed and is locked with the forcing frequency ω in the sense that a solution appears which is periodic of period three times the period of the forcing frequency. The form of (6.1) is important for the existence of a subharmonic solution. In fact, if the damping coefficient in (6.1) is a constant $\varepsilon_1 > 0$ for all $\varepsilon \geq 0$, then it was shown in Section IV.5 that no such subharmonic solution can exist.

If r, ϕ in (6.3) are chosen as new coordinates, the new equations are

(6.4)
$$\dot{r} = \frac{\varepsilon}{\omega}\left[\frac{\beta}{6} r \sin 2\left(\frac{\omega}{3} t + \phi\right) - 3(cv + \gamma u^3) \cos\left(\frac{\omega}{3} t + \phi\right)\right],$$

$$\dot{\phi} = \frac{\varepsilon}{\omega}\left[-\frac{\beta}{3} \sin^2\left(\frac{\omega}{3} t + \phi\right) + 3\frac{(cv + \gamma u^3)}{r} \sin\left(\frac{\omega}{3} t + \phi\right)\right],$$

$$\omega^2 - 1 = \varepsilon\beta,$$

where u, v are given in (6.3).

The averaged equations are (except for some terms of order ε^2)

(6.5)
$$\dot{r} = \frac{\varepsilon r}{2\omega}\left[-c + \frac{27\gamma r A}{2} \cos 3\phi\right],$$

$$\dot{\phi} = \frac{\varepsilon}{6\omega}\left[-\beta + \frac{27\gamma A}{2}(A - r \sin 3\phi) + \frac{27\gamma r^2}{4}\right]$$

Using the fact that $\omega^2 - 1 = \varepsilon\beta$ and letting $\gamma_0 = \varepsilon\gamma$, $c_0 = \varepsilon c$, the equations for the equilibrium points of (6.5) are

(6.6)   (a) $$\cos 3\phi = \frac{2c}{27\gamma A r},$$

(b) $$\omega^2 = 1 + \frac{27\gamma_0}{4}[r^2 + 2A^2] \mp \left[\left(\frac{27\gamma_0 A r}{2}\right)^2 - c_0^2\right]^{1/2},$$

where up to terms of order ε^2, the latter expressions are evaluated with $A = -9B/8$. For $c = 0$ (no damping), the formula for the frequency response curve (6.6b) simplifies to

(6.7) $$\omega^2 = 1 + \frac{27\gamma_0 A^2}{4} + \frac{27\gamma_0}{4}(r \pm A)^2.$$

A sketch of the frequency response curves is given in Fig. 6.1.

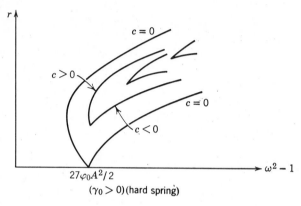

Figure V.6.1

One now uses Theorem 3.2 and Theorem 3.1 exactly as in the previous section to obtain the existence of an exact solution of (6.1) and the stability properties of such a solution for values of ω, r, φ satisfying (6.6).

The frequency response curves suggest the fact confirmed in Section IV.5 that the equation (6.1) may not have a subharmonic of order 3 if c gets too large, since the above analysis has been shown to be valid only for ε small and, thus, ω close to 1. Also, notice that once a subharmonic solution is known to exist, two other distinct ones are obtained simply by translating time by $2\pi/\omega$.

V.7. Damped Excited Pendulum with Oscillating Support

A linear damped sinusoidally excited pendulum with a rapidly vertically oscillating support of small amplitude can be represented approximately by the equation

(7.1) $$\ddot{u} + c\dot{u} + \left(1 + \varepsilon\,\frac{d^2 h(\nu t)}{dt^2}\right)\sin u - F\cos\omega t = 0,$$

where u is the angular coordinate measured from the bottom position, $h(\tau) = h(\tau + 2\pi)$, $\nu = \varepsilon^{-1}$, $0 < \varepsilon \ll 1$ and c, F, ω are real positive parameters independent of ε.

To transform (7.1) into a form (3.10) for which Theorem 3.3 is applicable, it is convenient to treat all but the term in c as derivable from the Hamiltonian

$$(7.2) \qquad H(v, u, t) = \frac{1}{2} \left[v^2 - 2\varepsilon v \frac{dh}{dt} \sin u - \varepsilon^2 \left(\frac{dh}{dt} \right)^2 \cos^2 u \right]$$
$$+ (1 - \cos u) - uF \cos \omega t,$$

where v is the momentum conjugate to u and the argument of h is νt. The equations of motion are, therefore,

$$(7.3) \qquad \dot{u} = v - \varepsilon \frac{dh}{dt} \sin u,$$

$$\dot{v} = - \left[- \varepsilon v \frac{dh}{dt} \cos u + \frac{\varepsilon^2}{2} \left(\frac{dh}{dt} \right)^2 \sin 2u + \sin u - F \cos \omega t \right]$$
$$- c \left(v - \varepsilon \frac{dh}{dt} \sin u \right),$$

where the argument of h is νt.

If $\nu t = \tau$ and differentiation with respect to τ is denoted by prime, then equations (7.3) become

$$(7.4) \qquad u' = \varepsilon [v - h' \sin u],$$

$$v' = \varepsilon \left[vh' \cos u - \frac{1}{2} (h')^2 \sin 2u - \sin u - cv + ch' \sin u \right]$$
$$+ \varepsilon F \cos \varepsilon \omega \tau.$$

where the argument of h is τ. Equations (7.4) are a special case of the system (3.10), the expressions in the square brackets being periodic in fast time τ and the other expression is periodic in the slow time $\varepsilon \tau$.

Considering the special case where $h(\tau) = A \sin \tau$ and A is a constant, the averaged equations (3.11) corresponding to equations (7.4) are

$$(7.5) \qquad u' = \varepsilon v,$$

$$v' = - \varepsilon \left[\frac{A^2}{2} \sin u \cos u + \sin u + cv \right] + \varepsilon F \cos \varepsilon \omega \tau.$$

In terms of the original time t, these equations are equivalent to the equation

$$(7.6) \qquad \dot{u} = v,$$

$$\dot{v} = - \left(1 + \frac{A^2}{2} \cos u \right) \sin u - cv + F \cos \omega t,$$

or the single second order equation

$$(7.7) \qquad \ddot{u} + c\dot{u} + \left(1 + \frac{A^2}{2} \cos u \right) \sin u = F \cos \omega t.$$

If equation (7.7) has a periodic solution of period $2\pi/\omega$ such that the characteristic exponents of its linear variational equation have negative real parts, then Theorem 3.3 implies the original equation (7.1) has an asymptotically stable, almost periodic solution which approaches this solution as $\varepsilon \to 0$. To find conditions under which (7.7) has such a periodic solution, one can use the results of Chapter IV or the method of averaging (Theorems 3.1 and 3.2) directly on (7.7). For example, if F is small, one can use the results of Chapter IV very efficiently as follows: For $F = 0$, the equilibrium solutions of (7.7) are $u = 0$, $u = \pi$ and $u = \cos^{-1}(2/A^2)$ if $A^2 > 2$. If $u = \pi + w$, then (7.7) becomes

$$(7.8) \qquad \ddot{w} + c\dot{w} + \left(\frac{A^2}{2}\cos w - 1\right)\sin w = F\cos \omega t,$$

and this may be rewritten as

$$\ddot{w} + c\dot{w} + \left(\frac{A^2}{2} - 1\right)w = F\cos \omega t + \left(\frac{A^2}{2} - 1\right)(w - \sin w)$$

$$+ \frac{A^2}{2}(1 - \cos w)\sin w.$$

If $A^2/2 > 1$ and F is small, this equation satisfies the conditions of Theorem IV.2.1 and Theorem IV.3.1. One can therefore assert the existence of an asymptotically stable periodic solution of (7.8) of period $2\pi/\omega$. This implies in turn that the pendulum described by (7.1) can execute stable motions in a neighborhood of the upright position.

V.8. Exercises

EXERCISE 8.1. An infinite conductor through which there flows a current of magnitude I attracts a conductor AB of length l and mass m along which there flows a current i. In addition, the conductor AB is attracted by a spring C whose force of attraction is proportional to its deflection with proportionality constant k. With $x = 0$ chosen as the position of AB when the spring is undeflected, the equations of motion of AB are

$$(8.1) \qquad \dot{x} = y, \ \dot{y} = -\frac{k}{m}\left(x - \frac{\lambda}{a - x}\right),$$

where $\lambda = 2Iil/k$. Discuss the behavior of the solutions of this equation along the lines of example 1.3.

EXERCISE 8.2. For $c > 0$, $h > 0$, $\omega - 1 = 0(\varepsilon)$ as $\varepsilon \to 0$, find the periodic solutions of

$$(8.2) \qquad \ddot{x} + x = \varepsilon[-c\dot{x} + hx^3 + x\cos 2\omega t].$$

Plot the frequency response curves, find the intervals on the ω-axis for which there are one, two and three periodic solutions and discuss the stability properties of these solutions.

EXERCISE 8.3.   The equation of a plane pendulum with vertically oscillating sinusoidal support can be written as

$$(8.3) \qquad \ddot{\theta} + c\dot{\theta} + \frac{g - I\omega^2 \sin \omega t}{l} \sin \theta = 0.$$

If $\tau = \omega t$, $I/l = \varepsilon$, $\varepsilon^2 k^2 = g/l\omega^2$, $c/\omega = 2\varepsilon\alpha$ we have

$$\theta'' + 2\varepsilon\alpha\theta' + (\varepsilon^2 k^2 - \varepsilon \sin \tau) \sin \theta = 0, \qquad = d/d\tau.$$

Show that the system for ε small has periodic solutions of period $2\pi/\omega$ when θ is in the neighborhood of either 0, π or 2π and $\cos^{-1}(-2k^2)$. Determine the stability properties as functions of α and k. Introduce variables ϕ and Ω by

$$\theta = \phi - \varepsilon \sin \tau \sin \phi,$$

$$\theta' = \varepsilon\Omega - \varepsilon \cos \tau \sin \phi,$$

and use averaging.

EXERCISE 8.4.   Consider the second order system

$$(8.4) \qquad \ddot{x} + x = \varepsilon[f(x) - c\dot{x} + \sin \omega t],$$

where $c > 0$, $0 < \varepsilon \ll 1$, $f(x)$ is a continuous function of x and $\omega - 1 = O(\varepsilon)$. For ε small, determine $f(x)$ so that the equation has periodic solutions of period $2\pi/\omega$. Determine the expression for the approximate frequency response. Study the stability of the solutions. Show that if $\gamma = \varepsilon^{-1}(\omega^2 - 1)$ then the periodic solutions are asymptotically stable if $c > 0$ and

$$\frac{da}{d\gamma}[\gamma + G(a) + O(\varepsilon)] < 0,$$

where $G(a) = (1/\pi a)\int_0^{2\pi} f(a \cos \phi) \cos \phi \, d\phi$. Use the transformation $x = A(t) \cdot \cos(\omega t + \Theta(t))$, $\dot{x} = -\omega A(t) \sin(\omega t + \Theta(t))$. Note that $\gamma + G(a) = 0$ is the approximate relationship between the amplitude and frequency of the periodic motion of the autonomous problem with $c = 0$.

V.9. Remarks and Suggestions for Further Study

The results of Section 1 on conservative systems are extremely simple but at the same time not very informative except for dimension 2. For high dimensional conservative systems, the theory has been and will continue to be a challenge for many years to come. Recently, Palmore [1] and Palamodov [1] have proved $H(p,q) = p'p/2 + V(q)$ real analytic and q not a minimum of V

implies the equilibrium point $(0,0)$ is unstable. Laloy [1] gave a simple example showing this result is false if one only assumes V is C^∞. For further references on this problem, see Rouche, Habets and Laloy [1]. For other fascinating aspects of conservative systems, see Poincaré [2], Birkhoff [1], Kolmogorov [1], Arnol'd [2], Moser [13], Siegel and Moser [1].

The method of averaging of Section 3 evolved from the famous paper of Krylov and Bogoliubov [1] and is contained in the book of Bogoliubov and Mitropolski [1]. The basic Lemma 3.1 is due to Bogoliubov [1]. Theorem 3.3 as well as the example in Section 7 are due to Sethna [1]. There are many variants of the method of averaging and the reader may consult Mitropolski [1] and Morrison [1] for further discussion as well as additional references. Other illustrations of equations which exhibit interesting oscillatory phenomena may be found in the books of Bogoliubov and Mitropolski [1], Malkin [2], Minorsky [1], Andronov, Vitt and Khaikin [1], Hale [7].

CHAPTER VI

Behavior Near a Periodic Orbit

In Section III.6 and Chapter IV, a rather extensive theory of the behavior of solutions of a system (and perturbations of a system) near an equilibrium point has been developed. In this and the following chapter, we consider the same questions except for a more complicated invariant set; namely, a closed curve.

Suppose $u\colon R \to R^n$ is a periodic function of period ω which is continuous together with its first derivative, $du(\theta)/d\theta \neq 0$ for all θ in $(-\infty, \infty)$ and $u\colon [0, \omega) \to R^n$ is one-to-one. If

$$(1) \qquad \Gamma = \{x \in R^n \colon x = u(\theta), \qquad 0 \leqq \theta \leqq \omega\},$$

then Γ is a closed curve and is actually a Jordan curve.

Suppose $f\colon R^n \to R^n$ is continuous and satisfies enough smoothness properties to ensure that a unique solution of

$$(2) \qquad \dot{x} = f(x)$$

passes through any point in R^n.

Throughout this chapter, it will be assumed that the curve Γ is *invariant* with respect to the solutions of (2); that is, any solution of (2) with initial value on Γ remains on Γ for all t in $(-\infty, \infty)$. If Γ is a periodic orbit of (2), then there is a periodic solution $\phi(t)$ of (2) such that

$$\Gamma = \{x\colon x = \phi(t), \qquad -\infty < t < \infty\}.$$

In this case, we may choose the parametric representation of Γ to be given by the function ϕ; that is, we may choose $u = \phi$. For periodic orbits, such a parametric representation always will be chosen in the sequel. If Γ is not a periodic orbit, then it must contain equilibrium points and orbits whose α- and ω-limit sets are equilibrium points.

From the previous remarks, if $x^*(t)$, $-\infty < t < \infty$, is any solution of (2) with $x^*(0)$ in Γ, then there are three possibilities: (i) there is a $\tau > 0$ such that $x^*(t)$ is periodic of least period τ, (ii) $x^*(t)$ is an equilibrium point; (iii) $x^*(t)$ has its α- and ω-limit sets as equilibrium points. In the first case, the curve Γ can be represented parametrically by $x = x^*(\theta)$, $0 \leqq \theta < \tau$ and any solution of (2) on Γ [such a solution must be $x^*(t + \alpha)$ for some constant α] defines a function $\theta(t)$, $-\infty < t < \infty$, such that $\dot{\theta} = 1$. In either case (ii) or (iii) any solution $x^*(t)$ of (1) on Γ defines a continuously differentiable function $\theta(t) = u^{-1}(x^*(t))$, $-\infty < t < \infty$, and $\dot{\theta} = g(\theta)$ where $g(\theta) = [\partial u^{-1}(u(\theta))/\partial x] f(u(\theta)) = [du(\theta)/d\theta]^{-1} f(u(\theta))$. In particular, in case (ii), there must be a zero of $g(\theta)$. In all three cases, we can, therefore, assert the existence of a continuous function $g(\theta)$ such that

(3) $$\frac{du(\theta)}{d\theta} g(\theta) = f(u(\theta)), \qquad g(\theta + \omega) = g(\theta), \qquad 0 \leqq \theta \leqq \omega.$$

Furthermore, if Γ is generated by a periodic solution of (2), then $g(\theta)$ can be taken $\equiv 1$ and if Γ contains an equilibrium point of (2), there must be a zero of $g(\theta)$.

In this chapter, our main concern is the behavior of the solutions of (2) near a periodic orbit (sufficient conditions for stability and the saddle point property) and the existence of a periodic orbit for the system

(4) $$\dot{x} = f(x) + F(x, t),$$

for the case in which F is a small perturbation independent of t. In the next chapter, the discussion will allow Γ to be an arbitrary invariant curve and the perturbations F to depend on t. In both cases, the first step will be the introduction of a convenient coordinate system in a neighborhood of Γ. Such a coordinate system is introduced in this chapter.

VI.1. A Local Coordinate System about an Invariant Closed Curve

By a *moving orthonormal system along* Γ is meant an orthonormal coordinate system $\{e_1(\theta), \ldots, e_n(\theta)\}$ in R^n for each θ in $[0, \omega]$ which is periodic in θ of period ω and one of the $e_j(\theta)$ is equal to $[du(\theta)/d\theta]/|du(\theta)/d\theta|$.

The first objective is to show there are moving orthonormal systems for Γ. For this purpose, the following lemma is needed.

LEMMA 1.1.   If $n \geqq 3$ and $v(\theta)$ is a unit vector in R^n which has period ω and satisfies a lipschitz condition, then there exists a unit vector ξ (independent of θ) such that $v(\theta) \neq \pm\xi$ for all θ.

PROOF.   This lemma is a consequence of well known facts from real variables. In fact, the set $x = v(\theta)$, $|v(\theta)| = 1$, $0 \leq \theta \leq \omega$, is a curve on the unit sphere S^{n-1} in R^n. Since $v(\theta)$ is lipschitzian, this curve is rectifiable and a rectifiable curve on a sphere in R^n, $n \geq 3$, covers a set of measure zero. Therefore, there always exists a vector ξ in S^{n-1} which is not on this curve or the curve defined by $-v(\theta)$.

Rather than invoke this result, a proof is given here. If S is any set on the unit sphere in R^n of diameter $<d$, then there is a constant K (independent of S) and a spherical cap S_c such that $S \subset S_c$ and the area of S_c is less than Kd^{n-1}. If M is the lipschitz constant for v; that is, $|v(\theta) - v(\theta')| \leqq M|\theta - \theta'|$, and if the interval $[0, \omega)$ is divided into N equal parts, then the curve defined by $v(\theta)$, $0 \leqq \theta < \omega$, may be covered by spherical caps whose total area is less than $NK(M\omega/N)^{n-1}$. Similarly, the curve defined by $-v(\theta)$, $0 \leqq \theta < \omega$ may be covered by spherical caps whose total area is less than the same quantity. Since $n \geqq 3$, this upper bound on the area approaches zero as $N \to \infty$, which implies that the vector ξ can be chosen from "almost" any place on the unit sphere. This proves the lemma.

For the statement of the next result, let $\mathscr{C}^p(R, R^n)$ designate the space of functions taking $R \to R^n$ which are continuous together with all derivatives up through order p.

THEOREM 1.1. If $u \in \mathscr{C}^p(R, R^n)$, $p \geqq 2$, $u(\theta + \omega) = u(\theta)$, $\omega > 0$, $du(\theta)/d\theta \neq 0$, $0 \leqq \theta < \omega$, and Γ is defined in (1), then there is a moving orthonormal system along Γ which is $\mathscr{C}^{p-1}(R, R^n)$.

PROOF. Suppose $n \geqq 3$. If $v(\theta) = [du(\theta)/d\theta]/|du(\theta)/d\theta|$, then the hypotheses on u imply that v is periodic of period ω and lipschitzian. Let e_1 be a constant unit vector (the existence of which is assured by Lemma 1.1) such that $e_1 \neq \pm v(\theta)$, $0 \leqq \theta \leqq \omega$. Adjoin to e_1 any constant vectors $e_2, \ldots, e_n$ such that $\{e_1, \ldots, e_n\}$ is an orthonormal basis for R^n. The moving orthonormal system along Γ is then obtained in the following manner: let S be the $(n-2)$-dimensional subspace of R^n orthogonal to the plane formed by e_1 and $v(\theta)$. Rotate the coordinate system about S in the positive sense until e_1 coincides with $v(\theta)$ (see Fig. 1.1). If $\xi_2(\theta), \ldots, \xi_n(\theta)$ are the rotated positions of $e_2, \ldots, e^n$, then the moving orthonormal system is given by

(1.1) $\qquad \{v(\theta), \xi_2(\theta), \ldots, \xi_n(\theta)\}, \qquad 0 \leqq \theta \leqq \omega.$

If $\gamma_j(\theta)$, $j = 1, 2, \ldots, n$, are the direction angles of $v(\theta)$, $e_j \cdot v(\theta) = \cos \gamma_j(\theta)$, $j = 1, 2, \ldots, n$, then one can show that the vectors ξ_j are given by

(1.2) $\qquad \xi_j(\theta) = e_j - \dfrac{\cos \gamma_j(\theta)}{1 + \cos \gamma_1(\theta)} (e_1 + v(\theta)), \qquad j = 2, 3, \ldots, n.$

The derivation of (1.2) proceeds as follows. Suppose

$$e_j = \bar{e}_j + \lambda_j e_1 + \mu_j v,$$

where $\bar{e}_j$ belongs to S. The final position of ξ_j is then

$$\xi_j = \bar{e}_j + \lambda_j' e_1 + \mu_j' v,$$

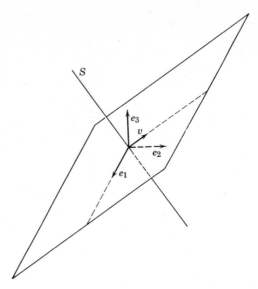

Figure VI.1.1

where λ', μ'_j are determined from λ_j, μ_j by a rotation in the e_1, v plane by an angle γ_1 with $\cos \gamma_1 = e_1 \cdot v$. Therefore

$$\lambda'_j = -\mu_j, \qquad \mu'_j = \lambda_j + 2\mu_j \cos \gamma_1,$$

and

$$\xi_j = e_j - (\lambda_j + \mu_j)e_1 + [\lambda_j + \mu_j(2 \cos \gamma_1 - 1)]v.$$

Since S is orthogonal to e_1, v, it follows that

$$e_1 \cdot [e_j - \lambda_j e_1 - \mu_j v] = 0,$$
$$v \cdot [e_j - \lambda_j e_1 - \mu_j v] = 0,$$

and this implies

$$\lambda_j = -\frac{\cos \gamma_1 \cos \gamma_j}{\sin^2 \gamma_1}, \qquad \mu_j = \frac{\cos \gamma_j}{\sin^2 \gamma_1}.$$

Substituting in the expression for ξ_j, one obtains (1.2).

For the case $n = 2$, the moving orthonormal system is easily constructed as

(1.3) $$(v(\theta), \xi_2(\theta)), \quad \xi_2(\theta) = \pm(-v_2(\theta), v_1(\theta)),$$

where the coordinates of v are v_1, v_2.

The explicit formulas (1.1)–(1.3) for the moving orthonormal system along Γ clearly imply that the system is $\mathscr{C}^{p-1}(R, R^n)$ if u is $\mathscr{C}^p(R, R^n)$. This completes the proof of the theorem.

Once a moving orthonormal system along a closed curve Γ is known, it is possible to use this system to obtain a coordinate system for a "tube" around Γ. In fact, with $u(\theta)$ as in Theorem 1.1, $v(\theta) = [du(\theta)/d\theta]/|du(\theta)/d\theta|$, let $(v, \xi_2, \ldots, \xi_n)$ be a moving orthonormal system along Γ. Consider the transformation of variables taking x into (θ, ρ), $\rho = \text{col}(\rho_2, \ldots, \rho_n)$ given by

$$(1.4) \qquad x = u(\theta) + Z(\theta)\rho, \qquad Z = [\xi_2, \ldots, \xi_n], \qquad 0 \le \theta < \omega.$$

The matrix Z is the $n \times (n-1)$ dimensional matrix whose columns are the vectors $\xi_2, \ldots, \xi_n$.

To show that this transformation is well defined in a sufficiently small neighborhood of Γ, that is, ρ sufficiently small, let

$$F(x, \theta, \rho) \overset{\text{def}}{=} u(\theta) + Z(\theta)\rho - x.$$

The partial derivatives of F with respect to θ, ρ are

$$\frac{\partial F}{\partial \theta} = \frac{du(\theta)}{d\theta} + \frac{dZ(\theta)}{d\theta}\rho,$$

$$\frac{\partial F}{\partial \rho} = Z(\theta).$$

For $\rho = 0$, $\det[\partial F/\partial \theta, \partial F/\partial \rho] = |du(\theta)/d\theta| \cdot \det[v(\theta), Z(\theta)] \ne 0$ for $0 \le \theta \le \omega$ since $du(\theta)/d\theta \ne 0$ for all θ and $v(\theta)$, $Z(\theta)$ is a moving orthonormal system. Therefore, there is a $\delta > 0$ independent of θ such that $\det[\partial F/\partial \theta, \partial F/\partial \rho] \ne 0$ for $|\rho| \le \delta$, $0 \le \theta \le \omega$. Since the closed curve Γ is compact, a finite number of applications of the implicit function theorem shows that (1.4) is a well defined transformation for $0 \le |\rho| \le \rho_1$, $\rho_1 > 0$, $0 \le \theta \le \omega$.

We now make the transformation (1.4) on the differential equation (4) using the fact that u satisfies (3) to obtain new differential equations for θ, ρ. It is assumed that f has continuous first derivatives with respect to x and u satisfies the conditions of Theorem 1.1.

If $x(t) = u(\theta(t)) + Z(\theta(t))\rho(t)$ satisfies (4), then

$$(1.5) \quad \left[\frac{du(\theta)}{d\theta} + \frac{dZ(\theta)}{d\theta}\rho\right]\dot{\theta} + Z(\theta)\dot{\rho} = f(u(\theta) + Z(\theta)\rho) + F(t, u(\theta) + Z(\theta)\rho).$$

In a ρ_1-neighborhood Γ_{ρ_1} of Γ, the coefficient matrix of $\dot{\theta}$, $\dot{\rho}$ is nonsingular. Therefore, equation (1.5) can be solved for $\dot{\theta}$, $\dot{\rho}$ as a function of θ, ρ, t. The explicit form of the equations are obtained as follows. Using (4) and projecting both sides of (1.5) onto $v(\theta)$, one obtains

$$(1.6) \qquad \dot{\theta} = g(\theta) + f_1(\theta, \rho) + h'(\theta, \rho)F(t, u(\theta) + Z(\theta)\rho),$$

where h' is the transpose of h,

(1.7) $$h(\theta, \rho) = \left[\left| \frac{du(\theta)}{d\theta} \right| + v'(\theta) \frac{dZ(\theta)}{d\theta} \rho \right]^{-1} v(\theta),$$

$$f_1(\theta, \rho) = -h'(\theta, \rho) \frac{dZ(\theta)}{d\theta} \rho g(\theta) + h'(\theta, \rho)[f(u(\theta) + Z(\theta)\rho) - f(u(\theta))],$$

By projecting both sides of (1.5) onto $Z(\theta)$, and using the fact that $Z'(\theta)f(u(\theta)) = Z'(\theta)g(\theta) \, du(\theta)/d\theta = 0$, where Z' is the transpose of Z, one obtains

(1.8) $$\dot{\rho} = A(\theta)\rho + f_2(\theta, \rho) + Z'(\theta)\left[I - \frac{dZ(\theta)}{d\theta} \rho h'(\theta, \rho) \right] F(t, u(\theta) + Z(\theta)\rho).$$

where Z' is the transpose of Z and

(1.9) $$A(\theta) = Z'(\theta)\left[-\frac{dZ(\theta)}{d\theta} g(\theta) + \frac{\partial f(u(\theta))}{\partial x} Z(\theta) \right],$$

$$f_2(\theta, \rho) = -Z'(\theta) \frac{dZ(\theta)}{d\theta} \rho f_1(\theta, \rho)$$

$$+ Z'(\theta)\left[f(u(\theta) + Z(\theta)\rho) - f(u(\theta)) - \frac{\partial f(u(\theta))}{\partial x} Z(\theta)\rho \right].$$

From the above expressions for these functions, it is easily seen that $f_1(\theta, \rho)$, $f_2(\theta, \rho)$ have continuous partial derivatives with respect to ρ up through order $k \geqq 1$ if $f(x)$ has continuous partial derivatives with respect to x up through order $k \geqq 1$. Furthermore, $f_1(\theta, \rho) = O(|\rho|)$ as $\rho \to 0$ and $f_2(\theta, 0) = 0$, $\partial f_2(\theta, 0)/\partial \rho = 0$. The number of derivatives of these functions with respect to θ is one less than the minimum of the derivatives of f and $du(\theta)/d\theta$. Also, the equations in $\dot{\theta}$, $\dot{\rho}$ contain F in a linear fashion multiplied by matrices which have an arbitrary number of derivatives with respect to ρ and $(p - 1)$ derivatives with respect to θ if u has p derivatives with respect to θ. These results are summarized in

THEOREM 1.2.   If u satisfies the conditions of Theorem 1.1 with $p \geqq 2$ and $f \in \mathscr{C}^{p-1}(R^n, R^n)$, then there exist a $\delta > 0$, n-vectors $f_2(\theta, \rho)$, $h(\theta, \rho)$ a scalar $f_1(\theta, \rho)$, an $(n - 1) \times (n - 1)$ matrix $A(\theta)$ and an $(n - 1) \times n$ matrix $B(\theta, \rho)$ with all functions being periodic in θ of period ω and having continuous derivatives of order $p - 1$ with respect to ρ and $p - 2$ with respect to θ for $0 \leqq |\rho| \leqq \delta$, $-\infty < \theta < \infty$,

(1.10) $$f_1(\theta, \rho) = O(|\rho|)   \text{ as } |\rho| \to 0,$$

$$f_2(\theta, 0) = 0, \frac{\partial f_2(\theta, 0)}{\partial \rho} = 0,$$

such that the transformation (1.4) applied to equation (4) for $|\rho| \leq \delta$ yields the equivalent system,

(1.11) $\qquad \dot{\theta} = g(\theta) + f_1(\theta, \rho) + h'(\theta, \rho)F(t, u(\theta) + Z(\theta)\rho),$

$\qquad\qquad \dot{\rho} = A(\theta)\rho + f_2(\theta, \rho) + B(\theta, \rho)F(t, u(\theta) + Z(\theta)\rho),$

where $g(\theta)$ is given in (3).

VI.2. Stability of a Periodic Orbit

In this section, the case is discussed in which the closed curve Γ is generated by a nonconstant ω-periodic solution x^0 of (2) and f has continuous first derivatives with respect to x. As mentioned before, we can assume that the parametric representation of Γ is $x = x^0(\theta)$, $0 \leq \theta < \omega$; that is, u can be chosen as x^0 and u satisfies (3) with $g(\theta) = 1$ for $0 \leq \theta \leq \omega$. Since $f(x)$ has continuous first derivatives in x, the function $u(\theta)$ has continuous second derivatives in θ. In terms of the local coordinate system (1.4), the behavior of the solutions of (2) near Γ are given by the solutions of the differential system

(2.1) $\qquad \dot{\theta} = 1 + f_1(\theta, \rho),$

$\qquad\qquad \dot{\rho} = A(\theta)\rho + f_2(\theta, \rho),$

$$A(\theta) = Z'(\theta)\left[-\frac{dZ(\theta)}{d\theta} + \frac{\partial f(u(\theta))}{\partial x} Z(\theta)) \right],$$

where $f_1(\theta, \rho), f_2(\theta, \rho)$ are continuous in θ, ρ, have continuous first derivatives, with respect to ρ, and have period ω in θ. These functions satisfy (1.10); that is,

(2.2) $\qquad |f_1(\theta, \rho)| = O(|\rho|), \qquad$ as $|\rho| \to 0,$

$\qquad\qquad f_2(\theta, 0) = 0, \qquad \dfrac{\partial f_2(\theta, 0)}{\partial \rho} = 0.$

Also associated with the periodic solution $u(\theta)$ of (2) is the linear variational equation

(2.3) $\qquad\qquad \dfrac{dy}{d\theta} = \dfrac{\partial f(u(\theta))}{\partial x} y.$

This linear system with periodic coefficients always has a nontrivial ω-periodic solution. In fact, $du(\theta)/d\theta$ is a nontrivial ω-periodic solution of this equation since (2) implies $d^2u/d\theta^2 = [\partial f(u(\theta))/\partial x] du/d\theta$. Therefore, at least one characteristic multiplier of (2.3) is equal to unity.

LEMMA 2.1. If the characteristic multipliers of the n-dimensional system (2.3) are $\mu_1, \ldots, \mu_{n-1}, 1$, then the characteristic multipliers of the $(n-1)$-dimensional system

(2.4) $$\frac{d\rho}{d\theta} = A(\theta)\rho,$$

where $A(\theta)$ is given in (2.1), are $\mu_1, \ldots, \mu_{n-1}$.

PROOF. Suppose Z is the $n \times (n-1)$ dimensional matrix defined in (1.4). The vector $du(\theta)/d\theta$ and the columns of $Z(\theta)$ are orthogonal since $v(\theta)$, $Z(\theta)$ is a moving orthonormal system. Thus, for any n-vector y, there is uniquely defined a scalar β and $(n-1)$-vector ρ such that $y = \beta \, du(\theta)/d\theta + Z(\theta)\rho$. If y is a solution of (2.3), then using the fact that $d^2u/d\theta^2 = [\partial f(u(\theta))/\partial x] \, du/d\theta$ and the columns of the matrix (v, Z) are orthogonal, one immediately deduces that ρ satisfies (2.4). This shows that the normal component of the solution of the linear variational equation coincides with the solution of the linear variational equation of the normal variation.

Now suppose $w^2, \ldots, w^n$ are $(n-1)$ solutions of (2.3), Y_1 is the $n \times n - 1$ matrix defined by $Y_1 = [w^2, \ldots, w^n]$ and $Y(\theta) = [du(\theta)/d\theta, Y_1(\theta)]$ is a fundamental matrix solution of (2.3). If K is defined by $Y(\omega) = Y(0)K$, then the eigenvalues of K are the characteristic multipliers of (2.3). From the definition of Y, it follows immediately that K must be of the form

$$K = \begin{bmatrix} 1 & K_2 \\ 0 & K_1 \end{bmatrix},$$

and, therefore, the multipliers $\mu_1, \ldots, \mu_{n-1}$ of (2.3) defined in the statement of the lemma are the eigenvalues of K_1. If $Y_1 = [du/d\theta]\alpha + ZR$, where $\alpha = \text{row}(\alpha_2, \ldots, \alpha_n)$ and R is an $(n-1) \times (n-1)$ matrix, then the remarks at the beginning of this proof imply that R is a matrix solution of (2.4) with $R(\omega) = R(0)K_1$. The matrix $R(0)$ is nonsingular. In fact, if there exists an $(n-1)$-vector c such that $R(0)c = 0$, then $Y_1(0)c - [du(0)/d\theta]\alpha c = 0$. Since $Y(0)$ is nonsingular, this implies $c = 0$ and, therefore, $R(0)$ is nonsingular. Since $R(0)$ is nonsingular, $R(\theta)$ is a fundamental matrix solution of (2.4) and K_1 is the monodromy matrix of R. Thus, the eigenvalues of K_1 are the multipliers of (2.4) and the proof of the lemma is complete.

Let us recall some of the previous concepts of stability. If M is a set in R^n, an η-neighborhood $U_\eta(M)$ is the set of x in R^n such that $\text{dist}(x, M) < \eta$. An invariant set M of (2) is said to be *stable* if for any $\varepsilon > 0$, there is a $\delta > 0$ such that, for any x^0 in $U_\delta(M)$, the solution $x(t, x^0)$ is in $U_\varepsilon(M)$ for all $t \geq 0$. An invariant set M of (2) is said to be *asymptotically stable* if it is stable and in addition there is $b > 0$ such that for any x^0 in $U_b(M)$ the solution $x(t, x_0)$ approaches M as $t \to \infty$. If $u(t)$ is a nonconstant periodic solution of (2), one says the periodic solution $u(t)$ is *orbitally stable*, *asymptotically orbitally stable* if the corresponding invariant closed curve Γ generated by u is stable, asymptotically stable, respectively. Such a periodic solution is said to be *asymptotically orbitally stable with asymptotic phase* if it is asymptotically orbitally stable and there is a $b > 0$ such that, for any x_0 with dist$(x_0, \Gamma) < b$, there is a $\tau = \tau(x_0)$ such that $|x(t, x_0) - u(t - \tau)| \to 0$ as $t \to \infty$.

THEOREM 2.1.   If f is in $\mathscr{C}^1(R^n, R^n)$, u is a nonconstant ω-periodic solution of (2), the characteristic multiplier one of (2.3) is simple and all other characteristic multipliers have modulus less than 1 (characteristic exponents have negative real parts), then the solution u is asymptotically orbitally stable with asymptotic phase.

PROOF.   The solutions in a neighborhood of the orbit Γ of u can be described by equation (2.1). The above hypotheses and Lemma 2.1 imply that the characteristic multipliers of (2.4) have modulus less than 1. For a sufficiently small neighborhood of Γ; that is, ρ sufficiently small, t may be eliminated in (2.1) so that the second equation in (2.1) has the form

$$(2.5) \qquad \frac{d\rho}{d\theta} = A(\theta)\rho + f_3(\theta, \rho),$$

where $f_3(\theta, \rho)$ has continuous first partial derivatives with respect to ρ, $f_3(\theta, 0) = 0$, $\partial f_3(\theta, 0)/\partial \rho = 0$. Theorems III.2.4 and III.7.2 imply there are $K > 0$, $\alpha > 0$, $\eta > 0$ such that for $|\rho_0| < \eta$ the solution $\rho(\theta, \theta_0, \rho_0)$, $\rho(\theta_0, \theta_0, \rho_0) = \rho_0$, of (2.5) satisfies

$$(2.6) \qquad |\rho(\theta, \theta_0, \rho_0)| \leqq K e^{-\alpha(\theta - \theta_0)} |\rho_0|, \qquad \theta \geqq \theta_0.$$

This proves the asymptotic stability of the solution $\rho = 0$ of (2.5). Since $\dot{\theta} > 1/2$, we have asymptotic orbital stability of Γ.

Equations (2.5) yield the orbits of (2) near Γ. The actual solutions of (2) near Γ are obtained from the transformation formulas (1.4) and the vector $(\theta(t), \rho(\theta(t), \theta_0, \rho_0))$ where $\rho(\theta, \theta_0, \rho_0)$ satisfies (2.5) and $\theta(t)$, $\theta(t_0) = \theta_0$, is the solution of the first equation in (2.1) with ρ replaced by $\rho(\theta, \theta_0, \rho_0)$. To prove there is an asymptotic phase shift associated with each solution of (2), it is sufficient to show that $\theta(t) - t$ approaches a constant as $t \to \infty$. If $\theta = t + \psi$, then

$$(2.7) \qquad \dot{\psi}(t) = f_1(t + \psi, \rho(t + \psi, t_0 + \psi_0, \rho_0)).$$

Since $\dot{\theta} > 1/2$, the map taking t to $\theta(t)$ has an inverse. Making use of this fact and recalling that $\theta(t) = t + \psi(t)$, we have

$$(2.8) \qquad
\begin{aligned}
\psi(t) &= \psi(t_0) + \int_{t_0}^{t} f_1(\theta(t), \rho(\theta(t), \theta_0, \rho_0))dt \\
&= \psi(t_0) + \int_{\theta(t_0)}^{\theta(t)} f_1(\theta, \rho(\theta), \theta_0, \rho_0)) \frac{dt}{d\theta} d\theta.
\end{aligned}$$

Since $|f_1(\theta,\rho)| \leq L|\rho|$, relation (2.6) and $|dt/d\theta| \leq 2$ imply

$$\left| f_1(\theta,\rho(\theta,\theta_0,\rho_0))\frac{dt}{d\theta} \right| \leq 2LKe^{-\alpha(\theta-\theta_0)}|\rho_0|.$$

Since $\theta(t) \to \infty$ if and only if $t \to \infty$ and

$$\int^\infty 2LKe^{-\alpha(\theta-\theta_0)}|\rho_0|\,d\theta < \infty,$$

we have

$$\lim_{t\to\infty} \int_{t_0}^t f_1(\theta(t),\rho(\theta(t),\theta_0,\rho_0))\,dt$$

exists. From (2.8), this implies $\psi(t)$ approaches a constant as $t \to \infty$. This completes the proof of Theorem 2.1.

Exercise 2.1. Give an example of an autonomous system which has an asymptotically orbitally stable periodic solution but there is no asymptotic phase.

Theorem 2.2. If u is a nonconstant ω-periodic solution of (2) with $(n-1)$ characteristic multipliers of (2.3) having modulii different from one, then there exist a neighborhood W_Γ of the closed orbit Γ in (1) and two sets S_Γ, U_Γ such that $S_\Gamma \cap U_\Gamma = \Gamma$ and any solution of (2) which remains in W_Γ for $t \geqq 0$ ($t \leqq 0$) must lie on S_Γ (U_Γ). If a solution x of (2) has its initial value in S_Γ (U_Γ) then $x(t) \to \Gamma$ as $t \to \infty$ ($t \to -\infty$). Furthermore, if p (q) characteristic multipliers of (2.3) have modulii <1 (>1) then S_Γ (U_Γ) is either a p-dimensional (q-dimensional) ball times a circle or a generalized Möbius band.

Proof. In a neighborhood of the periodic orbit, Γ, the orbits of (2) are given by the transformation (1.4) and the solutions $\rho(\theta)$ of the real system of differential equations (2.5). The Floquet representation theorem implies that the principal matrix solution of (2.4) can be expressed as $P(\theta)e^{B\theta}$ where $P(\theta)$ is periodic of period ω. Furthermore, Exercise III.7.2. asserts that $P(\theta)$ can always be chosen real if it is only required that P be periodic of period 2ω. If P has been chosen in this manner, the real transformation $\rho = P(\theta)r$ applied to (2.5) yields the equivalent real system

(2.9) $\dot{r} = Br + f_4(\theta, r)$,

where $f_4(\theta, 0) = 0$, $\partial f_4(\theta, 0)/\partial r = 0$, $f_4(\theta, r)$ is periodic in θ of period 2ω, and the eigenvalues of B have nonzero real parts. The conclusion of the theorem now follows immediately by the application of Theorem IV.3.1 to (2.9) and using the transformation $\rho = P(\theta)r$.

It is reasonable to call S_Γ and U_Γ the stable and unstable manifolds, respectively, of the periodic orbit Γ.

The following example illustrates that S_Γ may be a Möbius band. It looks complicated, but actually isn't if one looks at the example in the manner in which it was invented; namely, the development was from the final result desired to the differential equation. Consider the equations

(2.10) $\qquad\qquad$ (a) $\quad \dot{r} = A(\theta)r,$

$\qquad\qquad\qquad\quad$ (b) $\quad \dot{\theta} = 1,$

where $r = (r_1, r_2)$, θ are given in terms of the coordinates x, y, z in R^3 by $x = (r_1 + 1)\cos 4\pi\theta$, $y = (r_1 + 1)\sin 4\pi\theta$, $z = r_2$, and $A(\theta + 1/2) = A(\theta)$ with

$$A(\theta) = 2\begin{bmatrix} -\cos 4\pi\theta & \pi + \sin 4\pi\theta \\ -\pi + \sin 4\pi\theta & \cos 4\pi\theta \end{bmatrix}.$$

In R^3, there is a periodic solution of period $1/2$ and the periodic orbit Γ is the circle of radius one in the (x, y)-plane with center at the origin. It will be shown that this periodic orbit has stable and unstable manifolds which are Möbius bands.

The principal matrix solution of the equation (2.10a) is $P(\theta) \exp B\theta$ where

$$P(\theta) = \begin{bmatrix} \cos 2\pi\theta & \sin 2\pi\theta \\ -\sin 2\pi\theta & \cos 2\pi\theta \end{bmatrix},$$

$$B = \begin{bmatrix} -2 & 0 \\ 0 & 2 \end{bmatrix}.$$

The matrix $A(\theta)$ has period $1/2$ whereas the matrix $P(\theta)$ has period 1 in θ. Furthermore, it is easily seen that no Floquet decomposition of the principal matrix solution can have real periodic part and at the same time have period $1/2$. The characteristic multipliers of the system are the eigenvalues of $P(1/2) \exp B$ which are $-e^{-1}, -e$.

The stable and unstable manifolds of Γ are given by

$$S_\Gamma = \{(x, y, z) : r = e^{-2\theta}(\cos 2\pi\theta, -\sin 2\pi\theta)a,\ 0 \leqq \theta < \tfrac{1}{2},\ -a_0 < a < a_0\},$$

$$U_\Gamma = \{(x, y, z) : r = e^{2\theta}(\sin 2\pi\theta, \cos 2\pi\theta)b,\ 0 \leqq \theta < \tfrac{1}{2},\ -b_0 < b < b_0\},$$

where a_0, b_0 are positive constants. It is clear that these surfaces are Möbius bands.

EXERCISE 2.2. Prove that any solution of the stable (unstable) manifold $S_\Gamma(U_\Gamma)$ of Γ in Theorem 2.2 must approach the orbit Γ with asymptotic phase as $t \to \infty$ $(t \to -\infty)$.

VI.3. Sufficient Conditions for Orbital Stability in Two Dimensions

Consider the real two dimensional system

$$\text{(3.1)} \qquad \dot{x} = X(x, y),$$
$$\dot{y} = Y(x, y),$$

where X, Y are continuous together with their first partial derivatives in R^2. Suppose $x^0(t)$, $y^0(t)$ is a nonconstant periodic solution of period ω of (3.1) whose orbit in the (x, y)-plane is Γ. The linear variational equations of this solution are

$$\text{(3.2)} \qquad \dot{x} = \frac{\partial X}{\partial x}\, x + \frac{\partial X}{\partial y}\, y,$$

$$\dot{y} = \frac{\partial Y}{\partial x}\, x + \frac{\partial Y}{\partial y}\, y,$$

where the functions are evaluated at $x^0(t)$, $y^0(t)$. Lemma III.7.3 implies that the product of the characteristic multipliers of (3.2) is

$$\exp\!\left[\int_\Gamma (\partial X/\partial x + \partial Y/\partial y)\, dt \right].$$

Since 1 is a characteristic multiplier of (3.2), this exponential must be the other multiplier. Using Theorems 2.1 and 2.2, we can therefore state the following sufficient conditions for stability and instability of Γ.

LEMMA 3.1. An ω-periodic solution $x^0(t)$, $y^0(t)$ of (3.1) has a multiplier of (3.2) less than 1 and, thus, is asymptotically orbitally stable with asymptotic phase if

$$\int_0^\omega \left[\frac{\partial X(x^0(t), y^0(t))}{\partial x} + \frac{\partial Y(x^0(t), y^0(t))}{\partial y} \right] dt < 0,$$

and has a multiplier of (3.2) greater than 1 and hence unstable if this integral is positive.

To illustrate the application of this lemma, consider the scalar equation

$$\text{(3.3)} \qquad \ddot{x} + f(x)\dot{x} + g(x) = 0$$

or the equivalent system

$$\text{(3.4)} \qquad \dot{x} = y,$$
$$\dot{y} = -g(x) - f(x)y,$$

where f, g are continuous together with their first derivatives on R, and f has isolated zeros.

If $E(x, y) = y^2/2 + G(x)$, $G(x) = \int_0^x g(s)\, ds$, then the derivative $\dot{E}$ of E along the solutions of (3.4) is given by

(3.5) $$\dot{E} = -f(x)y^2 = 2f(x)[G - E].$$

If $x(t)$ is a nonconstant solution of (3.3), then there can be no $\bar{t}$ such that $\dot{x}(\bar{t}) = \ddot{x}(\bar{t}) = 0$. In fact, if this were the case, then $x^0 = x(\bar{t})$ would be a zero of g and uniqueness would imply $x(t) = x^0$ for all t. Suppose $\dot{x}(\bar{t}) = 0$ and $\ddot{x}(\bar{t}) \neq 0$. Then the solution $x(t)$ has a local extremum at $\bar{t}$ and must be either greater or less than $x(\bar{t})$ in a neighborhood of $\bar{t}$. Since the zeros of $f(x)$ are isolated, it follows that $f(x(t))$ must be of constant sign near $\bar{t}$ and consequently E cannot have an extremum at $\bar{t}$.

If Γ is a closed orbit of (3.4) determined by a periodic solution $x^0(t)$ of (3.3), then E assumes a minimum value on Γ for $t = \bar{t}$ and the above remark implies that $f(x^0(\bar{t})) = 0$, $y(\bar{t}) = \dot{x}^0(\bar{t}) \neq 0$. Therefore, if $G(x^0(\bar{t})) = h$, then $E(t) > h$ for all t. From (3.5), it follows that

$$\frac{\dot{E}}{E - h} + 2f = \frac{2(G - h)f}{E - h},$$

and integration over Γ yields

(3.6) $$\int_\Gamma f\, dt = \int_\Gamma \frac{(G - h)f}{E - h}\, dt.$$

From Lemma 3.1 and the special form of (3.4), the linear variational equation of the orbit Γ will have a multiplier <1 if (3.6) is positive and >1 if (3.6) is negative. In particular, this multiplier will be <1 if $G(x)$ takes the same value γ for all zeros of $f(x)$ and $[G(x) - \gamma]f(x)$ is positive when $f(x) \neq 0$. As a particular case which will be useful in the sequel, we state

LEMMA 3.2. If f, g have continuous first derivatives in R and
  (i) $f(x) < 0$ for $\alpha < x < \beta$, $f(x) > 0$ for $x < \alpha$, $x > \beta$ and $\alpha < 0 < \beta$,
  (ii) $xg(x) > 0$ for $x \neq 0$,
  (iii) $G(\alpha) = G(\beta)$ where $G(x) = \int_0^x g(s)\, ds$,

then any closed orbit of (3.4) has a characteristic multiplier in $(0, 1)$ and is thus asymptotically orbitally stable with asymptotic phase.

An important special case of Lemma 3.2 is the van der Pol equation $\ddot{x} - k(1 - x^2)\dot{x} + x = 0$, $k > 0$.

VI.4. Autonomous Perturbations

Consider the system of equations

(4.1) $$\dot{x} = f(x) + F(x, \varepsilon),$$

where $f\colon R^n \to R^n$, $F\colon R^{n+1} \to R^n$ are continuous, $f(x)$, $F(x, \varepsilon)$ have continuous partial derivatives with respect to x and $F(x, 0) = 0$ for all x. If system (2) has a nonconstant periodic solution $u(t)$ of period ω whose orbit is Γ, then the linear variational equation for u is

$$(4.2) \qquad \dot{y} = \frac{\partial f(u(t))}{\partial x}\, y.$$

THEOREM 4.1.   If $n - 1$ of the characteristic multipliers $\mu_1, \ldots, \mu_{n-1}$ of (4.2) are $\neq 1$, then there is an $\varepsilon_0 > 0$ and a neighborhood W of Γ such that equation (4.1) has a periodic solution $u^*(\,\cdot\,, \varepsilon)$ of period $\omega^*(\varepsilon)$, $0 \le |\varepsilon| \le \varepsilon_0$, such that $u^*(\,\cdot\,, 0) = u(\,\cdot\,)$, $\omega^*(0) = \omega$, $u^*(t, \varepsilon)$ and $\omega^*(\varepsilon)$ are continuous in t, ε for t in $R, 0 \le |\varepsilon| \le \varepsilon_0$. If none of $\mu_1, \ldots, \mu_{n-1}$ is a root of one, then for any constant $k > 0$, the neighborhood W can be chosen so that $u^*(\,\cdot\,, \varepsilon)$ is the only periodic solution of (4.1) of period $\le k$ in W. If $\mu_1, \ldots \mu_{n-1}$ have modulii < 1, then $u^*(\,\cdot\,, \varepsilon)$ is asymptotically orbitally stable with asymptotic phase and is unstable if any μ_j has modulus > 1.

PROOF.   Theorem 1.2 and the transformation (1.4) imply that system (4.1) is equivalent to the system (1.11) with $F(t, x)$ replaced by $F(x, \varepsilon)$ above. Elimination of t in these equations yields a system

$$(4.3) \qquad \frac{d\rho}{d\theta} = A(\theta)\rho + R(\theta, \rho, \varepsilon),$$

where $A(\theta)$, $R(\theta, \rho, \varepsilon)$ are ω-periodic in θ and R is in $\mathscr{L}i\not{p}(\eta, M)$ where $\mathscr{L}i\not{p}(\eta, M)$ is defined in Section IV.2. Lemma 2.1 implies that the characteristic multipliers of $d\rho/d\theta = A(\theta)\rho$ are the numbers $\mu_1, \ldots, \mu_{n-1}$. Theorem IV.2.1 implies the existence of an $\varepsilon_0 > 0$, $\rho_0 > 0$ and an ω-periodic solution $\rho^*(\theta, \varepsilon)$ of (4.3) continuous in θ, ε, $0 \le |\varepsilon| \le \varepsilon_0$, $\rho^*(\theta, 0) = 0$, and $\rho^*(\theta, \varepsilon)$ is the only ω-periodic solution in the region $|\rho| < \rho_0$. Theorem IV.3.1 yields the stability properties of $\rho^*(\theta, \varepsilon)$ when $\mu_1, \ldots, \mu_{n-1}$ have modulii different from one. Using this $\rho^*(\theta, \varepsilon)$ in the first equation in (1.11) and letting $\theta(t)$ be the solution of this equation with $\theta(0) = 0$, one finds there is a unique continuous $\omega^*(\varepsilon)$, $0 \le |\varepsilon| \le \varepsilon_0$, $\omega^*(0) = \omega$, such that $\theta(\omega^*(\varepsilon)) = \omega$, $0 \le |\varepsilon| \le \varepsilon_0$. The function $\theta(t)$, $\rho^*(\theta(t), \varepsilon)$, and the transformation (1.4) yield the desired periodic solution of (4.1) and the stability properties are as stated in the theorem.

Suppose now that no μ_j, $j = 1, 2, \ldots, n-1$, is a root of 1. For any integer k, the periodic solution u of (2) has period $k\omega$ and, therefore, all of the functions in the equations of the above proof can be considered as periodic in θ of period $k\omega$. The hypothesis on the μ_j imply by Theorem IV.2.1 that there is a periodic solution of (4.3) of period $k\omega$ for $0 \le |\varepsilon| \le \varepsilon_0$ and this solution is unique in the region $|\rho| < \rho_0$. Therefore, this solution must be the

$\rho^*(\theta, \varepsilon)$ given above. But, any periodic solution x of (4.1) which lies in a sufficiently small neighborhood of Γ must define a closed curve in R^n with a parametric representation of the form $x = u(\theta) + Z(\theta)\rho(\theta, \varepsilon)$ where $\rho(\theta, \varepsilon)$ is a periodic solution of (4.3) of period $k\omega$ for some integer k. This completes the proof of the theorem.

EXERCISE 4.1. Suppose $g(x, y)$ has continuous first derivatives with respect to x, y. Show there is an $\varepsilon_0 > 0$ such that for $|\varepsilon| < \varepsilon_0$ the equation

$$\ddot{x} - k(1 - x^2)\dot{x} + x = \varepsilon g(x, \dot{x}), \ k > 0,$$

has a unique periodic solution in a neighborhood of the unique periodic orbit of the van der Pol equation

$$\ddot{x} - k(1 - x^2)\dot{x} + x = 0.$$

Can you prove that this orbit is asymptotically orbitally stable with asymptotic phase?

EXERCISE 4.2. Give an example of an autonomous system $\dot{x} = f(x)$ which has an ω-periodic orbit Γ whose linear variational equation has 1 as a simple multiplier and yet, in any neighborhood of Γ, there are other periodic orbits.

EXERCISE 4.3. Is it possible for an equation $\dot{x} = f(x)$ to have an isolated ω-periodic orbit Γ whose linear variational equation has 1 as a simple multiplier and yet, there is a perturbation $\varepsilon g(x)$ such that in any neighborhood of Γ there is more than one periodic orbit?

VI.5. Remarks and Suggestions for Further Study

The particular construction of the moving orthonormal system given in Section 1 is based on a presentation given by Urabe [1] for the case in which the closed curve Γ is a periodic orbit. The manner in which the coordinate system is used to obtain a set of differential equations equivalent to (4) in a neighborhood of Γ is different. Since Urabe discusses only the case of a periodic orbit, it is possible to obtain the equations for the normal variation explicitly in terms of t and not θ as in (1.11). Lemma 1.1 was given by Diliberto and Hufford [1]. Lemma 3.2 is due to Coppel [1, p. 86]. The stability Theorem 2.1 was known to Poincaré for analytic systems.

In general, it is very difficult to discuss the behavior of solutions of (2) in a neighborhood of an orbit when more than one characteristic multiplier of (2.3) has modulus equal to one. Hale and Stokes [1] have given the following result. If system (1) has a k-parameter family of periodic solutions, then the linear variational equation has k characteristic multipliers equal to one. If the remaining multipliers have modulus less than one, then every orbit in this family is stable and, in fact, the solutions of (2) near this family approach a member of this family asymptotically with asymptotic phase.

Autonomous perturbations of nonlinear systems which possess a k-parameter family of periodic solutions is extremely difficult and is discussed at great length in the book of Urabe [1]. Under the hypothesis that only one multiplier of the linear variational equation has modulus one, the behavior of solutions near a periodic orbit of a nonautonomous perturbation of (2) will be discussed in the next chapter. If the nonautonomous perturbation is "small" for large t; that is, the system is "asymptotically" autonomous, then one can discuss the qualitative relationship between all solutions of the nonautonomous equation and the autonomous one without any hypotheses regarding the solutions of the autonomous one (see Markus [1], Opial [1], Yoshizawa [1]). For nonautonomous perturbations of k-parameter families of periodic solutions, see Hale and Stokes [1], Yoshizawa and Kato [1].

CHAPTER VII

Integral Manifolds of Equations with a

Small Parameter

Assuming that system (VI.2) has an invariant set which is a smooth Jordan curve Γ, it was shown in Chapter VI that a transformation of variables could be introduced in such a way that the behavior near Γ of the solutions of a perturbation of (VI.2), namely (VI.4), is reduced to a study of the equations

$$\dot{\theta} = g(\theta) + \Theta(t, \theta, \rho),$$
$$\dot{\rho} = C(\theta)\rho + R(t, \theta, \rho),$$

where ρ represents the normal deviation from Γ and $g(\theta)$ represents the behavior of the solutions of the unperturbed equation on Γ. If $g(\theta) \equiv 1$, the curve Γ corresponds to a periodic orbit and if g vanishes at some point, then an equilibrium point lies on Γ.

For $g(\theta) \equiv 1$ and the perturbations independent of t, some specific results were given in Chapter VI for the above equation. The analysis for this case is extremely simple since the problem is easily reduced to the study of the behavior near an equilibrium point of a nonautonomous system. On the other hand, if Θ, R contain t explicitly, then one cannot reduce the question to such a local problem since t cannot be eliminated from the equations even if $g(\theta) \equiv 1$. One must study the behavior near the invariant set itself. If $g(\theta)$ has a zero for some value of θ, then the problem remains nonlocal even if Θ, R are independent of t. Other difficulties also arise in this latter case and will be discussed later.

The purpose of this chapter is to study such nonlocal problems for even more general systems than the above. More specifically, we will be concerned with equations of the form

(1) $\dot{\theta} = \Theta^*(t, \theta, x, y, \varepsilon) = w(t, \theta, \varepsilon) + \Theta(t, \theta, x, y, \varepsilon),$

   $\dot{x} = A(\theta, \varepsilon)x + F(t, \theta, x, y, \varepsilon),$

   $\dot{y} = B(\theta, \varepsilon)y + G(t, \theta, x, y, \varepsilon),$

229

where $(\varepsilon, \theta, x, y)$ is in $R \times R^k \times R^n \times R^m$, the matrix $A(\theta, \varepsilon)$ is of such a nature that the solutions of $\dot{x} = A(\theta, \varepsilon)x$ approach zero exponentially as $t \to \infty$, the solutions of $\dot{y} = B(\theta, \varepsilon)y$ approach zero exponentially as $t \to -\infty$, for some class of functions $\theta(t)$, and F, G, Θ are sufficiently small in some sense for x, y, ε small. These statements will be made more precise later.

It will not be assumed that the functions in (1) are periodic in the vector θ although it will be assumed they are bounded for x, y in compact sets. In specific applications, a frequently occurring special case of (1) does have all functions periodic in θ. Such problems arise naturally from the study of the local perturbation theory of differential equations near an invariant torus. In many important situations, the flow on the invariant torus is parallel in the sense that all solutions are either periodic or quasiperiodic and then $w(t, \theta, \varepsilon)$ in (1) is a constant vector. If equations (1) arise from the local perturbation theory of a periodic orbit, then θ is a scalar, w is a constant and the Floquet theory for periodic systems permits one to take the matrices $A(\theta, \varepsilon)$, $B(\theta, \varepsilon)$ independent of θ. For the general perturbation theory of invariant torii with parallel flow, the matrices $A(\theta, \varepsilon)$, $B(\theta, \varepsilon)$ cannot be taken independent of θ because there is no Floquet theory for general almost periodic systems. The exercises in Section 8 illustrate the many ways in which equations of type (1) arise.

Definition 1. A surface S in (z, t)-space is an *integral manifold* of a system of ordinary differential equations $\dot{z} = Z(z, t)$ if for any point P in S, the solution $z(t)$ of the equation through P is such that $(z(t), t)$ is in S for all t in the domain of definition of the solution $z(t)$.

Our interest in this chapter lies in determining in what sense the qualitative behavior of the solutions of (1) and the system

$$(2) \qquad \begin{aligned} \dot{\theta} &= w(t, \theta, 0), \\ \dot{x} &= A(\theta, 0)x, \\ \dot{y} &= B(\theta, 0)y, \end{aligned}$$

are the same. This system has an integral manifold S in $R \times R^k \times R^n \times R^m$ whose parametric representation is given by

$$S = \{(t, \theta, x, y) \colon x = 0, y = 0\}.$$

Furthermore, under the stability properties alluded to earlier, the manifold S has a type of saddle point structure associated with it. In fact, there is an $R \times R^k \times R^n$ dimensional manifold of solutions of (2) which approach S as $t \to \infty$ and an $R \times R^k \times R^m$ dimensional manifold of solutions of (2) which approach S as $t \to -\infty$. Intuitively, one expects that Θ, F, and G small enough for ε, x, y small will imply the existence of some other integral manifold S_ε of (1) which is close to S for ε small and has the same type of stability properties as the integral manifold S of (2).

Therefore, in this chapter, we determine conditions on Θ, F, G which ensure that (1) has an integral manifold of the form $x = f(t, \theta, \varepsilon)$, $y = g(t, \theta, \varepsilon)$, (t, θ) in $R \times R^k$, which for $\varepsilon = 0$ reduce to $x = 0$, $y = 0$. For this specific case, such a surface will be an integral manifold for (1) if the triple of functions

$$[\theta(t) = \theta(t, \theta_0, t_0), \qquad x(t) = f(t, \theta(t), \varepsilon), \qquad y(t) = g(t, \theta(t), \varepsilon)],$$

is a solution of (1) for every t_0 in R, θ_0 in R^k.

Section 1 consists of a historical and intuitive discussion of the problems involved in determining integral manifolds for (1) as well as possible approaches to the solutions of these problems. The main theorems are stated in Section 2 with the proof being delegated to Sections 3, 4, 5, 6. In Section 7, applications of the results of Section 2 are given for equations which are perturbations of equations possessing an elementary periodic orbit. Also, in Section 7, the method of averaging of Chapter V is extended to averaging with respect to t as well as the variables θ in (1). The exercises in Section 8 illustrate more fully the implications of the results of this chapter.

VII.1. Methods of Determining Integral Manifolds

This section is devoted to an intuitive discussion of integral manifolds as well as some methods that have proved successful in the determination of integral manifolds. We choose for our discussion the problem of perturbing an autonomous system

$$(1.1) \qquad \dot{x} = X(x),$$

which has a nonconstant ω_0-periodic solution $u(t)$ such that $n - 1$ of the characteristic exponents of the linear variational equation

$$(1.2) \qquad \dot{y} = \frac{\partial X(u(t))}{\partial x} y$$

have negative real parts. As we have seen in Chapter VI, this implies the orbit described by u is asymptotically stable and this in turn implies that the cylinder

$$(1.3) \qquad S = \{(t, x): x = u(\theta), 0 \leq \theta \leq \omega_0, -\infty < t < \infty\},$$

in (t, x)-space is asymptotically stable. The cylinder S is an integral manifold of (1.1) in $R \times R^n$.

If we introduce the coordinate system

$$(1.4) \qquad x = u(\theta) + Z(\theta)\rho$$

of Chapter VI, then in a neighborhood of S the solutions of the equation are described by

(1.5)
$$\dot{\theta} = 1 + \Theta(\theta, \rho),$$
$$\dot{\rho} = A(\theta)\rho + R(\theta, \rho),$$

where $\Theta(\theta, \rho) = O(|\rho|)$, $R(\theta, \rho) = O(|\rho|^2)$ as $|\rho| \to 0$, all functions are ω_0-periodic in θ, and the characteristic exponents of $d\rho/d\theta = A(\theta)\rho$ have negative real parts. From the Floquet theory, a fundamental system of solutions of this linear equation is of the form $P(\theta)e^{B\theta}$, $P(\theta + \omega_0) = P(\theta)$. If we let $\rho(t) = P(\theta)z(t)$ and use the fact that $\dot{\theta} = 1 + O(|\rho|)$, we obtain an equivalent system

(1.6)
$$\dot{\theta} = 1 + \Theta_1(\theta, z),$$
$$\dot{z} = Bz + Z(\theta, z),$$

where $\Theta_1(\theta, z) = O(|z|)$, $Z(\theta, z) = O(|z|^2)$ as $|z| \to 0$ and the eigenvalues of B have negative real parts.

For the sake of our intuitive exposition, we need the following elementary result proved in Chapter X, Lemma 1.6. There is a positive definite matrix C such that any solution of the linear system $\dot{z} = Bz$ with initial value on the ellipsoid $z'Cz = c > 0$, a constant, must enter the interior of this ellipsoid for increasing time. In Lemma X.1.6, it is shown that the matrix $C = \int_0^\infty e^{B't}e^{Bt}dt$ satisfies the desired properties. Since $Z(\theta, z) = O(|z|^2)$ as $|z| \to 0$, there is a $c_0 > 0$ sufficiently small such that any solution of (1.6) with initial value on the set

$$U_s(c) = \{(t, x): x = u(\theta) + Z(\theta)P(\theta)z, 0 \leq \theta \leq \omega_0, z'Cz = c,$$
$$-\infty < t < \infty\}, 0 < c \leq c_0,$$

must enter this set for increasing t (and therefore remain inside this set). The set $U_s(c)$ projected into the x-space is a tube surrounding the closed curve $\mathscr{C} = \{x : x = u(\theta), 0 \leq \theta \leq \omega_0\}$. In two dimensions, it is an annulus around $\mathscr{C}$. In this case, the geometry is very simple since $U_s(c)$ represents a cylinder inside S and a cylinder outside S.

Now if the original differential equation (1.1) is perturbed to the form

(1.7)
$$\dot{x} = X(x) + \varepsilon X^*(t, x)$$

where $X^*(t, x)$ is bounded in a neighborhood of S, then for a given $c > 0$ and ε sufficiently small, the solutions will still be entering $U_s(c)$ as time increases. Therefore, for a fixed $c > 0$, we would expect some kind of integral surface inside $U_s(c)$ for ε small and it should look similar to a cylinder. However, even if such a surface exists we would not expect the solutions on this surface to behave in a manner similar to the behavior of the solutions on the original

S since these solutions have no strong stability properties associated with them. This is precisely why we discuss the preservation of the structure and stability properties of the set S rather than the preservation of any such properties for a particular solution on S.

Some methods for asserting the existence of integral manifolds for the perturbed equation are now given. If the transformation (1.4) is applied to (1.7), the equivalent equations are

(1.8)
$$\dot{\theta} = 1 + \Theta(\theta,\rho) + \varepsilon\Theta^*(t,\theta,\rho),$$
$$\dot{\rho} = A(\theta)\rho + R(\theta, \rho) + \varepsilon R^*(t, \theta, \rho).$$

Method 1 (*Levinson-Diliberto*). If we assume that the perturbation term is ω_1-periodic in t, then the above discussion of the geometric implications of stability allows one to obtain an annulus map via the differential equation (1.7) in the following way. Let $U_{s\tau}(c)$ be $U_s(c) \cap \{t = \tau\}$; that is, the cross section of $U_s(c)$ at $t = \tau$. Actually, all of these cross sections are the same and equal to $U_{s0}(c)$. If $x(t, x_0)$ is the solution of the perturbed equation (1.7) with initial value x_0 at $t = 0$ and ε is sufficiently small, then for any $\tau > 0$, the function $x(\tau, \cdot)$ is a mapping of $U_{s0}(c)$ into the interior of $U_{s\tau}(c) = U_{s0}(c)$; that is, a mapping of $U_{s0}(c)$ into itself. Because of the strong stability properties of the curve $\mathscr{C}$ and the fact that solutions "rotate" around this curve, one would suspect that there is a curve $\mathscr{C}_{\varepsilon\tau}$ such that $x(\tau, \mathscr{C}_{\varepsilon\tau}) = \mathscr{C}_{\varepsilon\tau}$ and $\mathscr{C}_{0\tau} = \mathscr{C}$. If this is true and if τ is chosen to be ω_1, then the periodicity in the equation implies $x(k\omega_1, \mathscr{C}_{\varepsilon\omega_1}) = \mathscr{C}_{\varepsilon\omega_1}$ for every integer k. By considering the "cylinder" generated by the solutions which start on $\mathscr{C}_{\varepsilon\omega_1}$, we obtain an invariant manifold of the equation. In this case, the solutions on the cylinder can be described by the solutions of an equation on a torus, the torus being obtained by identifying the cross sections of the cylinder at $t = 0$ and $t = \omega_1$. The basis of this idea was proposed and exploited in a beautiful paper of Levinson (1950). Using the idea of the proof of Levinson, Diliberto and his colleagues greatly simplified and improved the work of Levinson as well as discussing integral manifolds of a much more complicated type. In fact, if the perturbation term X^* in (1.7) is an arbitrary quasi-periodic function, then it can be written in the form $F(t, t, \ldots, t, x)$ where $F(t_1, \ldots, t_p, x)$ is periodic in t_j of period $\omega_j, j = 1, \ldots, p$. The functions Θ^*, R^* in (1.8) have this same periodicity structure in t. By artificially introducing variables ζ_j such that $\dot{\zeta}_j = 1$, we obtain from (1.8) a system

$$\dot{\zeta}_j = 1, \qquad j = 1, 2, \ldots, p,$$
$$\dot{\theta} = 1 + \Theta(\theta, \rho) + \varepsilon P^*(\theta, \zeta, \rho),$$
$$\dot{\rho} = A(\theta)\rho + R(\theta, \rho) + \varepsilon Q^*(\theta, \zeta, \rho),$$

where $P^*(\theta, \zeta, \rho)$, $Q^*(\theta. \zeta, \rho)$, $\zeta = (\zeta_1, \ldots, \zeta_p)$, are periodic in θ and ζ. This is a special case of (1) with θ in (1) a $(p + 1)$-vector consisting of the scalar θ above and the p-vector ζ. Notice that the equations no longer contain t explicitly. Diliberto has introduced an ingenious device to discuss integral manifolds for such equations by generalizing the idea of the period map. The interested reader may consult the references for the details of this method as well as the applications.

Method 2 (Partial Differential Equations).   Suppose we have a system

(1.9)   (a) $\dot{\theta} = w(\theta) + \Theta(\theta, x)$,

                          (b) $\dot{x} = A(\theta)x + F(\theta, x)$,

where θ in R^k, x in R^n are vectors, and all functions are periodic in θ of vector period ω. If $S = \{(\theta, x) : x = f(\theta), f(\theta + \omega) = f(\theta), \theta$ in $R^k\}$ is to be an integral manifold of this equation, then $x(t) = f(\theta(t))$, where $\theta(t)$ is a solution of

(1.10) $\dot{\theta} = w(\theta) + \Theta(\theta, f(\theta))$,

must satisfy (1.9b). Performing the differentiation, we find that f must satisfy the partial differential equation

(1.11) $\dfrac{\partial f}{\partial \theta} [w(\theta) + \Theta(\theta, f)] - A(\theta)f = F(\theta, f), \; f(\theta + \omega) = f(\theta)$.

This method has not been completely developed at this time but a beginning has been made by Sacker [1]. The case where w, A are independent of θ is not too difficult with this approach since one can integrate along characteristics to obtain essentially the same method presented below. If there are values of θ for which $w(\theta) = 0$, then the problem is much more difficult since the solution will not in general be as smooth as the coefficients in the equation. The smoothness properties depend in a delicate manner upon w, A. Another difficulty that arises in this problem is that the usual iteration procedures involve a loss of derivatives. To circumvent this difficulty, Sacker solves in each iteration the elliptic equation

$$\mu\Delta_\theta f + \dfrac{\partial f}{\partial \theta} [w(\theta) + \Theta(\theta, f)] - A(\theta)f = F(\theta, f), \qquad f(\theta + \omega) = f(\theta),$$

for μ small and Δ_θ the Laplacian operator. The solutions then have as many derivatives as desired, but there is a delicate analysis involved in choosing $\mu \to 0$ in such a way as to obtain a solution of the original equation. Further research needs to be done with this method to see if it is possible to obtain results when the perturbation terms depend upon t in a general way and also to discuss the stability properties of the manifold.

Method 3. (Krylov-Bogoliubov-Mitropolski).   Circa 1934, Krylov and Bogoliubov made a significant contribution to this problem in the following way. They defined a mapping of "cylinders" into "cylinders" via the differential equation (1.8) such that the fixed points of this map are integral manifolds of the equation which for $\varepsilon = 0$ reduce to S. They also discussed the stability properties of this manifold. These methods do not use any particular properties of the dependence of the perturbation terms upon t.

To illustrate the ideas, consider the equation

(1.12)
$$\dot{\theta} = 1 = \Theta(t,\theta,\rho,\epsilon)$$
$$\dot{\rho} = A\rho + R(t,\theta,\rho,\epsilon)$$

where A is an $n \times n$ constant matrix, $Re\lambda A < 0$, $\Theta(t,\theta,0,0) = 0$, $R(t,\theta,0,0) = 0$, $\partial R(t,\theta,0,0)/\partial\rho = 0$. We look for integral manifolds of (1.12) of the form $\rho = f(t,\theta,\epsilon)$ with $|f(t,\theta,\epsilon)|$ bounded for $(t,\theta) \in R^2$, $f(t,\theta,\epsilon)$ lying in a small neighborhood of $\rho = 0$ for ϵ small.

Let $S = \{f{:}R \times R \to R^n$ continuous and bounded) with the topology of uniform convergence. Suppose (1.12) has an integral manifold $f(t,\theta,\epsilon)$, with $f(\cdot,\cdot,\epsilon) \in S$. If $\theta(t,t_0,\theta_0,f)$ is the solution of the equation

(1.13) $$\dot{\theta} = 1 + \Theta(t,\theta,f(t,\theta,\epsilon),\epsilon), \qquad \theta(t_0,t_0,\theta_0,f) = \theta_0,$$

then $(\theta(t,t_0,\theta_0,f), f(t,\theta(t,t_0,\theta_0,f),\epsilon)$ must satisfy (1.12) for all $(t_0,\theta_0) \in R^2$. Therefore, the variation of constants formula implies

$$f(t,\theta(t,t_0,\theta_0,f),\epsilon) = e^{A(t-t_0)}f(t_0,\theta_0,\epsilon)$$
$$+ \int_{t_0}^{t} e^{A(t-s)}R(s,\theta(s,t_0,\theta_0,f),f(s,\theta(s,t_0,\theta_0,f),\epsilon),\epsilon)ds$$

and

$$e^{-A(t-t_0)}f(t,\theta(t,t_0,\theta_0,f),\epsilon) = f(t_0,\theta_0,\epsilon)$$
$$+ \int_{t_0}^{t} e^{A(t_0-s)}R(s,\theta(s,t_0,\theta_0,f),$$
$$f(s,\theta(s,t_0,\theta_0,f),\epsilon),\epsilon)ds.$$

Since $f(t,\theta,\epsilon)$ is bounded and $\exp(-A(t-t_0)) \to 0$ as $t \to -\infty$, we have a formula for the function $f(t_0,\theta_0,\epsilon)$; namely,

(1.14) $$f(t_0,\theta_0,\epsilon) = \int_{-\infty}^{t_0} e^{A(t_0-s)}R(s,\theta(s,t_0,\theta_0,f),f(s,\theta(s,t_0,\theta_0,f),\epsilon),\epsilon)ds.$$

For a given $f \in S$, the right hand side of Eq. (1.14) can be considered as a transformation $\mathscr{F}{:}S \to S$ and the sought for integral manifolds are fixed points of $\mathscr{F}$. By careful estimation, one can apply the contraction mapping principle to obtain fixed points of $\mathscr{F}$ in an appropriate bounded subset of S consisting of functions with a sufficiently small lipschitz constant in θ.

VII.2. Statement of Results

In this section, we give detailed results on the existence and stability of integral manifolds of (1). Let $\Omega(\sigma, \varepsilon_0) = \{(x, y, \varepsilon): |x| < \sigma, \ |y| < \sigma, \ 0 < \varepsilon \leqq \varepsilon_0\}$. The following hypotheses on the functions in (1) will be used:

(H_1) All functions A, B, w, Θ, F, G are continuous and bounded in $R \times R^k \times \Omega(\sigma, \varepsilon_0)$.

(H_2) The functions A, B, w, Θ, F, G are lipschitzian in θ with lipschitz constants $r(\varepsilon_1)$, $r(\varepsilon_1)$, $L(\varepsilon_1)$, $\eta(\rho, \varepsilon_1)$, $\gamma(\rho, \varepsilon_1)$, $\gamma(\rho, \varepsilon_1)$ respectively, in $R \times R^k \times \Omega(\rho, \varepsilon_1)$ where $r(\varepsilon_1)$, $L(\varepsilon_1)$, $\eta(\rho, \varepsilon_1)$, $\gamma(\rho, \varepsilon_1)$ are continuous and non-decreasing for $0 \leqq \rho \leqq \sigma$, $0 < \varepsilon_1 \leqq \varepsilon_0$.

(H_3) The functions Θ, F, G are lipschitizian in x, y with lipschitz constants $\mu(\rho, \varepsilon_1)$, $\delta(\rho, \varepsilon_1)$, $\delta(\rho, \varepsilon_1)$, respectively, in $R \times R^k \times \Omega(\rho, \varepsilon_1)$, where $\mu(\rho, \varepsilon_1)$, $\delta(\rho, \varepsilon_1)$ are continuous and nondecreasing for $0 \leqq \rho \leqq \sigma$, $0 < \varepsilon_1 \leqq \varepsilon_0$.

(H_4) The functions $|F(t, \theta, 0, 0, \varepsilon)|$, $|G(t, \theta, 0, 0, \varepsilon)|$ are bounded by $N(\varepsilon)$ for (t, θ) in $R \times R^k$, $0 < \varepsilon \leqq \varepsilon_0$ where $N(\varepsilon)$ is continuous and nondecreasing for $0 < \varepsilon \leqq \varepsilon_0$.

(H_5) There exist a positive constant K and a continuous positive function $\alpha(\varepsilon)$, $0 < \varepsilon \leqq \varepsilon_0$, such that, for any continuous function $\theta(t)$ defined on $(-\infty, \infty)$, and any real number τ, the principal matrix solution $\Phi(t, \tau)$, $\Psi(t, \tau)$ of $\dot{x} = A(\theta(t), \varepsilon)x$, $\dot{y} = B(\theta(t), \varepsilon)y$, respectively, satisfy

$$(2.1) \qquad |\Phi(t, \tau)| \leqq Ke^{-\alpha(\varepsilon)(t-\tau)}, \qquad t \geqq \tau,$$

$$|\Psi(t, \tau)| \leqq Ke^{\alpha(\varepsilon)(t-\tau)}, \qquad t \leqq \tau.$$

(H_6) Let $p(\Delta, D, \varepsilon)$, $q(\Delta. D, \varepsilon)$ be defined by

$$(2.2) \quad p(\Delta, D, \varepsilon) = \gamma(D, \varepsilon) + \delta(D, \varepsilon)\Delta + \frac{Kr(\varepsilon)}{\alpha}[\delta(D, \varepsilon)D + N(\varepsilon)],$$

$$q(\Delta, D, \varepsilon) = L(\varepsilon) + \eta(D, \varepsilon) + \mu(D, \varepsilon)\Delta,$$

and suppose that

$(2.3) \qquad$ (a) $\quad \alpha(\varepsilon) - q(\Delta, D, \varepsilon) > 0,$

$\qquad\qquad$ (b) $\quad \delta(D, \varepsilon)D + N(\varepsilon) < \alpha(\varepsilon)D/K,$

$\qquad\qquad$ (c) $\quad Kp(\Delta, D, \varepsilon) < [\alpha(\varepsilon) - q(\Delta, D, \varepsilon)]\Delta,$

$\qquad\qquad$ (d) $\quad \dfrac{p(\Delta, D, \varepsilon)\mu(D, \varepsilon)}{\alpha(\varepsilon) - q(\Delta, D, \varepsilon)} + \delta(D, \varepsilon) < \dfrac{\alpha(\varepsilon)}{2K},$

for $0 < \varepsilon \leqq \varepsilon_0$.

We are now in a position to state

THEOREM 2.1.   If system (1) satisfies hypotheses $(H_1) - (H_6)$, then there exist functions $f(t, \theta, \varepsilon)$ in R^n, $g(t, \theta, \varepsilon)$ in R^m which are continuous in $R \times R^k \times (0, \varepsilon_1]$, bounded by D, lipschitzian in θ with lipschitz constant Δ, such that the set

$$S_\varepsilon = \{(t, \theta, x, y) : x = f(t, \theta, \varepsilon), y = g(t, \theta, \varepsilon),   (t,\theta) \text{ in } R \times R^k\}$$

is an integral manifold of (1). If the functions in (1) are periodic in θ with vector period ω, then $f(t, \theta, \varepsilon)$, $g(t, \theta, \varepsilon)$ are also periodic in θ with vector period ω. If the functions in (1) are T-periodic in t then $f(t, \theta, \varepsilon)$, $g(t, \theta, \varepsilon)$ are T-periodic in t. If the functions in (1) are almost periodic in t, then so are f, g.

Before proving this theorem, we state some immediate Corollaries which we again state as theorems because of their importance in the applications.

THEOREM 2.2.   Suppose system (1) satisfies (H_1)-(H_5) even for $\varepsilon = 0$ and $\alpha(\varepsilon) = \alpha$ a constant. If $\eta(0, 0) = \gamma(0, 0) = \delta(0, 0) = N(0) = 0$ and $\alpha - L(0) > 0$, then there are $\varepsilon_1 > 0$ and continuous functions $D(\varepsilon)$, $\Delta(\varepsilon)$, $0 < \varepsilon \leqq \varepsilon_1$, approaching zero as $\varepsilon \to 0$ such that the conclusions of Theorem 2.1 are valid for this $D(\varepsilon)$, $\Delta(\varepsilon)$.

PROOF.   From the hypothesis of Theorem 2.2, $\alpha - L(0) > 0$. Therefore, we may take ε_0 such that $\alpha - L(\varepsilon) > 0$, $0 \leqq \varepsilon \leqq \varepsilon_0$. Since $\eta(0, 0) = 0$, there are positive $\varepsilon_1, \Delta_1, D_1$ so that (2.3a) is satisfied for $0 < \varepsilon \leqq \varepsilon_1$, $\Delta \leqq \Delta_1$, $D \leqq D_1$. Since $\delta(0, 0) = 0$ and $N(0) = 0$, it follows that one can further restrict ε_1 and choose $D(\varepsilon)$ so that $D(\varepsilon) \to 0$ as $\varepsilon \to 0$ and (2.3b) is satisfied for $0 < \varepsilon \leqq \varepsilon_1$. Since $\gamma(0, 0) = 0$, $\delta(0, 0) = 0$ it follows that ε_1 and a function $\Delta(\varepsilon) \to 0$ as $\varepsilon \to 0$ can be chosen so that (2.3c) is satisfied for $0 < \varepsilon \leqq \varepsilon_1$. Since $p(\Delta(\varepsilon), D(\varepsilon), \varepsilon)$, $\delta(D(\varepsilon), \varepsilon) \to 0$ as $\varepsilon \to 0$ if $\Delta(\varepsilon)$, $D(\varepsilon)$ are chosen as above, it follows that one can further restrict $\varepsilon_1 > 0$ such that (2.3d) is satisfied for $0 < \varepsilon \leqq \varepsilon_1$. Theorem 2.1 is therefore applicable to complete the proof of Theorem 2.2.

If $w(t, \theta, \varepsilon)$ is a constant in (1) then $L = 0$ and the condition $\alpha - L > 0$ is automatically satisfied from (H_5). The hypotheses (H_1)-(H_4) in Theorem 2.2 merely express smoothness and smallness conditions on the perturbation functions Θ, F, G.

COROLLARY 2.1.   Suppose Θ, F, G satisfy hypotheses $(H_1) - (H_4)$ even for $\epsilon = 0$. Suppose w is a constant k-vector, A is an $n \times n$ constant matrix whose eigenvalues have negative real parts, B is an $m \times m$ constant matrix whose eigenvalues have positive real parts. Then the conclusions of Theorem 2.2 are valid for the system

$$\dot{\theta} = w + \Theta(t, \theta, x, y, \epsilon)$$
$$\dot{x} = Ax + F(t, \theta, x, y, \epsilon)$$
$$\dot{y} = By + G(t, \theta, x, y, \epsilon)$$

Consider the system

(2.4)
$$\dot{\theta} = w + \epsilon\Theta(t, \theta, x, y, \epsilon),$$
$$\dot{x} = \epsilon A(\theta)x + \epsilon F(t, \theta, x, y, \epsilon),$$
$$\dot{y} = \epsilon B(\theta)y + \epsilon G(t, \theta, x, y, \epsilon),$$

where w is a constant.

THEOREM 2.3.   Suppose A, B, w, Θ, F, G in system (2.4) satisfy hypotheses (H_1)-(H_4) even for $\epsilon = 0$ and $\gamma(0, 0) = \delta(0, 0) = N(0) = 0$. If $\alpha = \epsilon\alpha_1$, $\alpha_1 > 0$, a constant, in (H_5) and $\alpha_1 - \overline{\lim}_{\epsilon \to 0}\eta(0, \epsilon) > 0$, then the conclusions of Theorem 2.2 are valid for system (2.4).

PROOF.   It is sufficient to satisfy relations (2.3) with $L = 0$ and $\alpha(\epsilon)$ replaced by α_1 since the form of equation (2.4) implies that ϵ is a common factor of all functions in (2.3). The proof proceeds now exactly as in the proof of Theorem 2.2.

Consider the system

(2.5)
$$\epsilon\dot{\theta} = w + \Theta(t, \theta, x, y, \epsilon),$$
$$\epsilon\dot{x} = A(\theta)x + F(t, \theta, x, y, \epsilon),$$
$$\epsilon\dot{y} = B(\theta)x + G(t, \theta, x, y, \epsilon),$$

where w is a constant.

THEOREM 2.4.   Suppose A, B, w, Θ, F, G in system (2.5) satisfy (H_1)-(H_4) even for $\epsilon = 0$, $\gamma(0, 0) = \delta(0, 0) = N(0) = 0$. If $\alpha = \alpha_1/\epsilon$, $\alpha_1 > 0$, a constant in (H_5) and $\alpha_1 - \overline{\lim}_{\epsilon \to 0}\eta(0. \epsilon) > 0$, then the conclusions of Theorem 2.2 are valid for system (2.5).

PROOF.   The proof is essentially the same as the proof of Theorem 2.3.
The proof of Theorem 2.1 will be broken down into simple steps in order to clarify the basic ideas. Some of these steps are of interest in themselves and are stated as lemmas. One could use the same method of proof as given below to state results on integral manifolds involving systems which are combinations of systems (1), (2.4) and (2.5). These results are easily obtained when the need arises and it does not seem worthwhile to state them in detail.

VII.3. A "Nonhomogeneous Linear" System

In the proof of Theorem 2.1, we will use successive approximations in a manner very similar to that used in Chapter IV for the results on the behavior near an equilibrium point. Let $S^n = \{f\colon (-\infty, \infty) \times R^k \to R^n$ which are continuous and bounded$\}$. For any f in S^n, define $\|f\| = \sup\{|f(t, \theta)|, (t, \theta)$ in $R \times R^k\}$. The idea for successive approximations in (1) for an integral manifold is to let $x = f(t, \theta)$, f in S^n, $y = g(t, \theta)$, g in S^m, in the first equation in (1) to obtain an equation in θ alone say $\dot{\theta} = h(t, \theta, f, g)$. This equation can be solved for $\theta(t, \tau, \zeta, f, g)$ where $\theta(\tau) = \zeta$ for any τ, ζ in $R \times R^k$. This function of θ is then substituted in the second and third equations in (1) to obtain equations of the form

$$\dot{x} = A(\theta(t))x + \hat{f}(t, \theta(t)),$$

$$\dot{y} = B(\theta(t))y + \hat{g}(t, \theta(t)),$$

where $\hat{f}, \hat{g}$ depend upon f, g, but, more importantly, the initial data for $\theta(t)$; namely τ, ζ. The problem is then to determine a bounded solution of this system as a function of (τ, ζ).

Thus, we are led to a type of "nonhomogeneous linear" system. We refer to the equations in this manner because at each stage in the iteration process the equations are linear in x, y; that is, linear in the coordinates through which the integral manifold is defined. For simplicity in the notation, we will assume that the y-equation in (1) is absent. The general case follows along the same lines. For any P in S^k, Q in S^n, we first consider the "nonhomogeneous linear" system

(3.1)   (a) $\dot{\theta} = P(t, \theta),$

          (b) $\dot{x} = A(\theta)x + Q(t, \theta).$

For any (τ, ζ) in $R \times R^k$, let $\theta^*(t) = \theta^*(t, \tau, \zeta, P)$ be the solution of (3.1a) with $\theta^*(\tau) = \zeta$. Since $P(t, \theta)$ is bounded, such a solution always exists on $(-\infty, \infty)$. We wish to find a function $X(\cdot, \cdot, P, Q)$ in S^n such that

$$(\theta^*(t), X(t, \theta^*(t), P, Q)), \qquad t \text{ in } (-\infty, \infty),$$

is a solution (3.1) for t in $(-\infty, \infty)$ and all (τ, ζ) in $R \times R^k$. If we can accomplish this, then the set $\mathscr{S} = \{(t, \theta, x)\colon x = X(t, \theta, P, Q), (t, \theta) \text{ in } R \times R^k\}$ is an integral manifold of (3.1). To derive the equation for such a function X, let $\Phi(t, \tau, \zeta, P)$ be the principal matrix solution of the linear system

(3.2) $$\dot{x} = A(\theta^*(t, \tau, \zeta, P))x$$

The variation of constants formula applied to (3.1b) yields

$$X(t, \theta^*(t, \tau, \zeta, P), P, Q) = \Phi(t, \tau, \zeta, P)X(\tau, \zeta, P, Q)$$

$$+ \int_\tau^t \Phi(t, s, \zeta, P)Q(s, \theta^*(s, \tau, \zeta, P))\, ds, \qquad t \in (-\infty, \infty).$$

If A is independent of θ, then this relation is much simpler since $\Phi(t, \tau, \zeta, P)$ $= \exp[A(t - \tau)]$. Multiplying the above relation by $\Phi^{-1}(t, \tau, \zeta, P)$ and using the properties of a principal matrix solution, one sees that the variation of constants formula can be written as

$$X(\tau, \zeta, P, Q) = \Phi(\tau, t, \zeta, P)X(t, \theta^*(t, \tau, \zeta, P), P, Q)$$

$$- \int_\tau^t \Phi(\tau, s, \zeta, P)Q(s, \theta^*(s, \tau, \zeta, P))\, ds, \qquad t \in (-\infty, \infty).$$

Since X is assumed to belong to S^n and, in particular, is bounded, we can let t approach $-\infty$ in this relation and use hypothesis (H_5) to obtain

$$(3.3) \quad X(t, \theta, P, Q) = \int_{-\infty}^0 \Phi(t, u + t, \theta, P)Q(u + t, \theta^*(u + t, t, \theta, P))\, du,$$

where we have replaced τ, ζ in the end result by t, θ.

In the manner in which relation (3.3) was obtained, it follows that (3.3) defines an integral manifold of (3.1) and, furthermore, it is the only such integral manifold which remains in a region for which the x-coordinate is bounded. These facts are summarized in

LEMMA 3.1.   If (H_5) is satisfied, then for any P in S^k, Q in S^n, equation (3.1) has an integral manifold defined parametrically by $X(t, \theta, P, Q)$ in (3.3) and it is the only integral manifold of (3.1) for which the x coordinate is bounded.

Our initial goal is to derive some properties of the function $X(t, \theta, P, Q)$ defined by (3.3). In particular, we wish to discuss bounds and smoothness properties of X as a function of the bounds and smoothness properties of P, Q. The basic result for the "nonhomogeneous linear" system (3.1) is the following:

LEMMA 3.2.   Suppose $A(\theta), P(t, \theta), Q(t, \theta)$ are lipschitzian in θ with lipschitz constants $r, L(P), M(Q)$, respectively, and hypotheses (H_5) is satisfied with $\alpha > L(P)$. If $X(\cdot, \cdot, P, Q)$ is defined by (3.3), then $X(\cdot, \cdot, P, Q)$ belongs to S^n, defines an integral manifold of (3.1) and satisfies

(3.4)

$$(a) \quad |X(t, \theta, P, Q) - X(t, \bar{\theta}, P, Q)| \leq \frac{K}{\alpha - L(P)}\left[M(Q) + \frac{Kr}{\alpha}\|Q\|\right]|\theta - \bar{\theta}|,$$

(b) $\|X(\cdot,\ \cdot,\ P, Q) - X(\cdot,\ \cdot,\ P, \bar{Q})\| \leq \dfrac{K}{\alpha}\ \|Q - \bar{Q}\|, X(\cdot,\ \cdot,\ P, 0) = 0,$

(c) $\|X(\cdot,\ \cdot,\ P, Q) - X(\cdot,\ \cdot,\ \bar{P}, Q)\| \leq \dfrac{K}{\alpha(\alpha - L(P))}$

$$\times \left[M(Q) + \frac{Kr}{\alpha}\ \|Q\| \right] \| P - \bar{P} \|,$$

for all t in R, θ, $\bar{\theta}$ in R^k, P, $\bar{P}$ in S^k, Q, $\bar{Q}$ in S^n. If $P(t, \theta)$, $Q(t, \theta)$ are periodic in θ with vector period ω, then $X(t, \theta, P, Q)$ is periodic in θ with vector period ω. If $P(t, \theta)$, $Q(t, \theta)$ are T-periodic (or almost periodic) in t, then $X(t, \theta, P, Q)$ is T-periodic (or almost periodic) in t. If $P(t, \theta)$, $Q(t, \theta)$ are independent of t, then $X(t, \theta, P, Q)$ is independent of t.

PROOF. Relation (3.4b) is almost immediate since $X(\cdot,\ \cdot,\ \cdot, Q)$ is linear in Q and hypothesis (H_5) implies that

$$\|X(\cdot,\ \cdot,\ \cdot, Q)\| \leq \frac{K\|Q\|}{\alpha}.$$

The relations (3.4a), (3.4c) are more difficult to obtain since $\theta^*(\cdot,\ \cdot,\ \theta, P)$ depends in a nonlinear fashion on θ and P. This nonlinear dependence is also reflected in the function $\Phi(\cdot,\ \cdot,\ \theta, P)$ if A depends on θ. Our first objective is to obtain estimates of this dependence on θ, P.

Since $P(t, \theta)$ is lipschitzian in θ with lipschitz constant $L(P)$, a simple application of differential inequalities to (3.1a) yields the estimates

(3.5) (a) $|\theta^*(t, \tau, \zeta, P) - \theta^*(t, \tau, \bar{\zeta}, P)| \leq e^{L(P)|t-\tau|}|\zeta - \bar{\zeta}|,$

(b) $|\theta^*(t, \tau, \zeta, P) - \theta^*(t, \tau, \zeta, \bar{P})| \leq \dfrac{e^{L(P)|t-\tau|} - 1}{L(P)}\ \|P - \bar{P}\|,$

for all t, τ in R and P, $\bar{P}$ in S^k. In fact, both of these relations are easily obtained from the relation

$$D^+\gamma(u) \leq L(P)\gamma(u) + \|P - \bar{P}\|$$

where D^+ is the right hand derivative, $\gamma(u) = |\theta^*(u, \tau, \zeta, P) - \theta^*(u, \tau, \zeta, \bar{P})|$ to obtain (3.5b) and $\gamma(u) = |\theta^*(u, \tau, \zeta, P) - \theta^*(u, \tau, \bar{\zeta}, P)|$, $P = \bar{P}$ to obtain (3.5a)

We now obtain estimates for the dependence of $\Phi(t, \tau, \theta, P)$ on θ, P. If we use the fact that this is a principal matrix solution of (3.2) and the difference $\Phi(u, \tau, \theta, P) - \Phi(u, \tau, \bar{\theta}, \bar{P})$ is a solution of the matrix equation

$$\frac{dx}{du} = A(\theta^*(u, \tau, \theta, P))\,x + [A(\theta^*(u, \tau, \theta, P))$$

$$- A(\theta^*(u, \tau, \bar{\theta}, \bar{P}))]\Phi(u,, \tau, \bar{\theta}, \bar{P}),$$

then the variation of constants formula yields

$$\Phi(t, u+t, \theta, P) - \Phi(t, u+t, \bar{\theta}, \bar{P}) = \int_u^0 \Phi(t, v+t, \theta, P)$$
$$\times [A(\theta^*(v+t, u+t, \theta, P)) - A(\theta^*(v+t, u+t, \bar{\theta}, \bar{P}))]$$
$$\times \Phi(v+t, u+t, \bar{\theta}, \bar{P})\, dv$$

for all $u \in R$. Therefore, from (H_5) and the lipschitzian hypothesis on A, we obtain

$$(3.6)\qquad |\Phi(t, u+t, \theta, P) - \Phi(t, u+t, \bar{\theta}, \bar{P})|$$
$$\le K^2 r e^{\alpha u} \int_u^0 |\theta^*(v+t, u+t, \theta, P) - \theta^*(v+t, u+t, \bar{\theta}, \bar{P})|\, du$$

for $u \le 0$.

If we let $P = \bar{P}$ and use (3.5a) and (3.6), then

$$(3.7)\quad |\Phi(t, u+t, \theta, P) - \Phi(t, u+t, \bar{\theta}, \bar{P})| \le \frac{K^2 r}{L(P)} e^{\alpha u}[e^{-L(P)u} - 1]|\theta - \bar{\theta}|$$

for all $u \le 0$. If we let $\theta = \bar{\theta}$ and use (3.5b) and (3.6), then

$$(3.8)\qquad |\Phi(t, u+t, \theta, P) - \Phi(t, u+t, \theta, \bar{P})| \le \frac{K^2 r}{L^2(P)} e^{\alpha u}[e^{-L(P)u} - 1$$
$$+ L(P)u]\|P - \bar{P}\|$$

for all $u \le 0$.

From (3.3) and (H_5) we have

$$|X(t, \theta, P, Q) - X(t, \bar{\theta}, P, Q)|$$
$$\le \int_{-\infty}^0 K e^{\alpha u} M(Q)|\theta^*(u+t, t, \theta, P) - \theta^*(u+t, t, \bar{\theta}, P|\, du$$
$$+ \int_{-\infty}^0 |\Phi(t, u+t, \theta, P) - \Phi(t, u+t, \bar{\theta}, P)|\, \|Q\|\, du.$$

Using (3.5a) and (3.7), we obtain relation (3.4a). Similar estimates using (3.5b) and (3.8) give (3.4c). This completes the proof of the first part of the lemma.

Now suppose that P, Q in (3.1) have vector period ω in θ. The uniqueness theorem implies $\theta^*(t, \tau, \zeta + \omega, P) = \theta^*(t, \tau, \zeta, P) + \omega$ and, thus $\Phi(t, \tau, \theta + \omega, P) = \Phi(t, \tau, \theta, P)$. From formula (3.3), one obtains $X(t, \theta + \omega, P, Q) = X(t, \theta, P, Q)$. If P, Q in (3.1) are T-periodic in t, then uniqueness of solutions yields $\theta^*(t + \tau + T, \tau + T, \zeta, P) = \theta^*(t + \tau, \tau, \zeta, P)$. This fact in turn implies $\Phi(t + \tau, \tau, \theta, P)$ is T-periodic in τ. Since

$\Phi(\tau, t + \tau, \theta, P)\Phi(t + \tau, \tau, \theta, P) = I$, the function $\Phi(\tau, t + \tau, \theta, P)$ is T-periodic in τ. Using (1.3), one has $X(t, \theta, P, Q)$ is T-periodic in t. If $P(t, \theta)$, $Q(t, \theta)$ are independent of t, then the same argument yields $X(t, \theta, P, Q)$ is periodic in t with an arbitrary period and, therefore, must be independent of t.

The case where P, Q are almost periodic in t is handled as follows. If δ is any real number, we define P_δ in S^k, Q_δ in S^n by $P_\delta(t, \theta) = P(t + \delta, \theta)$, $Q_\delta(t, \theta) = Q(t + \delta, \theta)$. Using the same process as in the estimation of the lipschitz function of $\theta^*(t, \tau, \zeta, P)$ in P, one easily obtains

$$|\theta^*(u + t + \delta, t + \delta, \theta, P) - \theta^*(u + t, t, \theta, P)| \leqq \frac{e^{L(P)|u|} - 1}{L(P)} \, \|P_\delta - P\|$$

for all u, t, θ. Now using the same type of argument that was used to obtain (3.8), one obtains

$$|\Phi(t + \delta, u + t + \delta, \theta, P) - \Phi(t, u + t, \theta, P)|$$

$$\leqq \frac{K^2 r}{L^2(P)} \, e^{\alpha u}[e^{-L(P)u} - 1 + L(P)u] \, \|P_\delta - P\|$$

for all $u \leqq 0$. Using these two relations in (3.3), we arrive at the following inequality

$$|X(t + \delta, \theta, P, Q) - X(t, \theta, P, Q)|$$

$$\leqq \frac{K}{\alpha} \, \|Q_\delta - Q\| + \frac{K}{\alpha(\alpha - L(P))} \, [M(Q) + \frac{Kr}{\alpha} \, \|Q\|] \, \|P_\delta - P\|$$

for all t, δ, θ. This inequality and the same type of argument as used in the proof of Theorems 1.1 and 2.1 of Chapter IV complete the proof of Lemma 3.2.

The constant α in the statement of Lemma 3.2 is a measure of the rate of approach of the solutions of (3.1) to the integral manifold and the constant $L(P)$ is a measure in some sense of the maximum rate at which solutions on the manifold can converge or diverge from one another. A natural question to ask is whether the condition $\alpha - L(P)$ is necessary to obtain lipschitz smoothness of the parametric representation of the manifold.

To understand some of the difficulties that might be encountered if $\alpha - L(P) < 0$, let us discuss the following example in the two-dimensional (u, v)-space. If $u = r \cos \theta$, $v = r \sin \theta$, the system is

$$\dot\theta = k \sin \theta - \frac{\cos \theta}{r \sin \theta} \, (r - 1)^2,$$

$$\dot r = r(1 - r).$$

The circle $r = 1$ is an integral manifold of this system. If $r = 1 + \rho$, then the linear variational equation for the manifold $r = 1$ is $\dot\rho = -\rho$ and the constant

α in hypothesis H_5) is $+1$. On the other hand, on the manifold $r = 1$, the Liptschitz constant $L(P)$ can be taken to be k. Also, on $r = 1$, there are the stable node $(-1, 0)$ and the saddle point $(1, 0)$. The trajectories of this system near $r = 1$ for $k < 1$ and > 1 are shown in Fig. 3.1. For $k < 1$ all

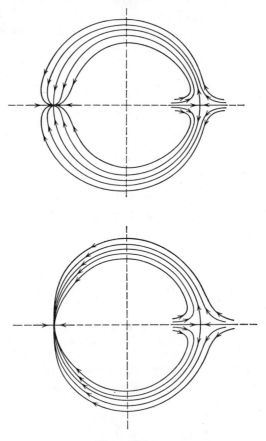

Figure VII.3.1

orbits enter the node $(-1, 0)$ tangent to the circle $r = 1$ and for $k > 1$ all orbits enter the node $(-1, 0)$ perpendicular to the circle $r = 1$. A small perturbation in the equation for $k > 1$ could possibly lead to an invariant curve which has a cusp at $(-1, 0)$.

The following example shows that a cusp can arise. Consider the system

$$\dot{\theta} = k \sin \theta,$$
$$\dot{x} = -x + f(\theta),$$

where f has a continuous first derivative and $f(\theta + 2\pi) = f(\theta)$. The first equation has the solution

$$\theta^*(t, \tau, \zeta) = 2 \arctan\left(e^{k(t-\tau)} \tan \frac{\zeta}{2}\right),$$

and therefore the integral manifold is given from (3.3) by

$$X(t, \theta, k) \equiv X(\theta, k) = \int_{-\infty}^{0} e^{u} f\left[2 \arctan\left(e^{ku} \tan \frac{\theta}{2}\right)\right] du.$$

For any closed interval in the interior of $(-\pi, \pi)$,

$$\frac{\partial X(\theta, k)}{\partial \theta} = \int_{-\infty}^{0} e^{(1-k)u} \frac{df}{d\theta}\left[\sin^2 \frac{\theta}{2} + e^{-2ku} \cos^2 \frac{\theta}{2}\right]^{-1} du$$

exists and is continuous for any k. If $k > 1$, this integral becomes unbounded as $\theta \to \pi$ or $-\pi$ if $df/d\theta$ is different from zero in a neighborhood of these points and $X(\theta, k)$ is not lipschitzian.

VII.4. The Mapping Principle

To apply Lemma 3.2 to prove Theorem 2.1, we define a mapping whose fixed points coincide with the integral manifold of (1) and prove this mapping is a contraction. It is convenient to formulate this mapping in more general terms and specialize it to system (1) later. For given constants Δ, D, let

$$(4.1) \qquad S^n(\Delta, D) = \{f \text{ in } S^n : \|f\| \leq D, |f(t, \theta) - f(t, \bar{\theta})| \leq \Delta|\theta - \bar{\theta}|$$

$$\text{for all } (t, \theta, \bar{\theta}) \text{ in } R \times R^k \times R^k\}.$$

Let $P : R \times R^k \times S^n(\Delta, D) \to R^k$, $Q : R \times R^k \times S^n(\Delta, D) \to R^n$ and for any f in $S^n(\Delta, D)$ suppose that $P(\cdot, \cdot, f)$ is in S^k and $Q(\cdot, \cdot, f)$ is in S^n. For any f in $S^n(\Delta, D)$, we know from the discussion of system (3.1) that the system

$$(4.2) \qquad \dot{\theta} = P(t, \theta, f),$$
$$\dot{x} = A(\theta)x + Q(t, \theta, f),$$

has an integral manifold defined parametrically by (3.3). We rewrite (3.3) again to emphasize the dependence upon f as

$$(4.3) \qquad X^*(t, \theta, f) = \int_{-\infty}^{0} \Phi_1(t, u + t, \theta, f)Q(u + t, \theta_1(u + t, t, \theta, f)) \, du,$$

where we are using the simplified but hopefully not confusing notation

$$\theta_1(t, \tau, \zeta, f) = \theta^*(t, \tau, \zeta, P(\cdot, \cdot, f)),$$
$$\Phi_1(t, \tau, \zeta, f) = \Phi(t, \tau, \zeta, P(\cdot, \cdot, f)),$$
$$X^*(t, \theta, f) = X(t, \theta, P(\cdot, \cdot, f), Q(\cdot, \cdot, f)).$$

If there exists an f in $S^n(\Delta, D)$ such that $f(t, \theta) = X^*(t, \theta, f)$, then f will define an integral manifold of (4.2). It is clear that such a procedure will be applicable to system (1) without the y equation simply by defining

(4.4) $$P(t, \theta, f) = w(t, \theta, \varepsilon) + \Theta(t, \theta, f(t, \theta), \varepsilon),$$
$$Q(t, \theta, f) = F(t, \theta, f(t, \theta), \varepsilon).$$

We now derive some conditions on P, Q which will ensure that the map $X^*(\cdot, \cdot, f), f$ in $S^n(\Delta, D)$ is a contraction.

Assume the following: there are constants $\beta(D), M(\Delta, D), L(\Delta, D)$ and $a(D), b(D)$ such that

(4.5) $$\|Q(\cdot, \cdot, f)\| \leq \beta(D),$$
$$\|Q(\cdot, \cdot, f) - Q(\cdot, \cdot, \bar{f})\| \leq a(D)\|f - \bar{f}\|,$$
$$\|P(\cdot, \cdot, f) - P(\cdot, \cdot, \bar{f})\| \leq b(D)\|f - \bar{f}\|,$$
$$|Q(t, \theta, f) - Q(t, \bar{\theta}, f)| \leq M(\Delta, D)|\theta - \bar{\theta}|,$$
$$|P(t, \theta, f) - P(t, \bar{\theta}, f)| \leq L(\Delta, D)|\theta - \bar{\theta}|,$$

for all $(t, \theta, \bar{\theta})$ in $R \times R^k \times R^k$ and $f, \bar{f}$ in $S^n(\Delta, D)$.

If we now use (3.4) with $M(Q), L(P)$ replaced by $M(\Delta, D), L(\Delta, D)$, respectively,

$$(\|X^*\cdot, \cdot, f)\| \leq \frac{K}{\alpha} \beta(D),$$

$$|X^*(t, \theta, f) - X^*(t, \bar{\theta}, f)| \leq \frac{KM_1(\Delta, D)}{\alpha - L(\Delta, D)} |\theta - \bar{\theta}|,$$

$$\|X^*(\cdot, \cdot, f) - X^*(\cdot, \cdot, \bar{f})\| \leq \frac{K}{\alpha}\left[\frac{M_1(\Delta, D)b(D)}{\alpha - L(\Delta, D)} + a(D)\right]\|f - \bar{f}\|,$$

$$M_1(\Delta, D) = M(\Delta, D) + \frac{Kr}{\alpha} \beta(D).$$

for all $(t, \theta, \bar{\theta})$ in $R \times R^k \times R^k$ and $f, \bar{f}$ in $S^n(\Delta, D)$ provided that $\alpha - L(\Delta, D) > 0$. From these relations we can state the following

LEMMA 4.1.   Suppose Q, P satisfy (4.5) and for any f in $S^n(\Delta, D)$, the unique integral manifold $X^*(\cdot, \cdot, f)$ of (4.2) is defined by (4.3). The mapping

$T : S^n(\Delta, D) \to S^n(\Delta, D)$ defined by $Tf = X^*(\cdot, \cdot, f)$ is a contraction provided that Δ, D satisfy the relations

(4.7)

(a) $\alpha - L(\Delta, D) > 0$,

(b) $K\beta(D) < \alpha D$,

(c) $KM_1(\Delta, D) < [\alpha - L(\Delta, D)]\Delta$,

(d) $\dfrac{K}{\alpha}\left[\dfrac{M_1(\Delta, D)b(D)}{\alpha - L(\Delta, D)} + a(D)\right] < \dfrac{1}{2}$.

If conditions (4.7) are satisfied, then $X^*(\cdot, \cdot, f)$ has a unique fixed point in $S^n(\Delta, D)$.

VII.5. Proof of Theorem 2.1

We are now in a position to prove Theorem 2.1 for the case when y is absent. The only thing to do is to find the constants in (4.5) with P, Q defined by (4.4) and substitute these in (4.7). From hypotheses (H_2)-(H_4), one obtains, for $0 < \varepsilon \leqq \varepsilon_0$,

(5.1)

(a) $\beta(D) = \delta(D, \varepsilon)D + N(\varepsilon)$,

(b) $a(D) = \delta(D, \varepsilon)$,

(c) $b(D) = \mu(D, \varepsilon)$,

(d) $M(\Delta, D) = \gamma(D, \varepsilon) + \delta(D, \varepsilon)\Delta$,

(e) $L(\Delta, D) = L(\varepsilon) + \eta(D, \varepsilon) + \mu(D, \varepsilon)\Delta$.

If (5.1) is used in (4.7), one obtains relation (2.3) in hypotheses (H_6).

This completes the proof of Theorem 2.1 when the vector y is absent. The case with y present follows along the same lines if use is made of the fact that the equation

$$\dot\theta = P(t, \theta),$$
$$\dot x = A(\theta)x + Q(t, \theta),$$
$$\dot y = B(\theta)y + R(t, \theta),$$

has a unique integral manifold given by $X(t, \theta, P, Q)$, $Y(t, \theta, P, R)$ with X given by (3.3) and

$$Y(t, \theta, P, R) = \int_0^\infty \Psi(t, u+t, \theta, P)R(u+t, \theta^*(u + t, t, \theta, P))\, du$$

where $\Psi(t, \tau, \zeta, P)$ is the principal matrix solution of the linear system

$$\dot y = B(\theta^*(t, \tau, \zeta, P))y.$$

VII.6. Stability of the Perturbed Manifold

For the unperturbed part of system (1), namely,

(6.1)
$$\dot{\theta} = w(t, \theta, \varepsilon),$$
$$\dot{x} = A(\theta, \varepsilon)x,$$
$$\dot{y} = B(\theta, \varepsilon)y,$$

there is a unique integral manifold with the x and y coordinates bounded and this is given by $x = 0$, $y = 0$. This manifold has a saddle point structure in the sense that any solution of (6.1) is such that $x \to 0$ exponentially as $t \to \infty$ and $y \to 0$ exponentially as $t \to -\infty$. One can prove that the saddle point structure is also preserved for the perturbed manifold whose existence is assured by Theorem 2.1.

We do not prove this fact in detail, but only indicate the proof for the special case

(6.2)
$$\dot{\theta} = 1 + \Theta(t,\theta,\rho,\epsilon)$$
$$\dot{\rho} = A\rho + R(t,\theta,\rho,\epsilon)$$

where $\Theta(t,\theta,\rho,\epsilon)$, $R(t,\theta,\rho,\epsilon)$ are continuous, bounded and continuously differentiable in θ, ρ in $R \times R \times \Omega(\rho,\epsilon_0)$, periodic in θ of period ω, $\Theta(t,\theta,0,0) = 0$, $R(t,\theta,0,0) = 0$, $\partial R(t,\theta,0,0)/\partial \rho = 0$, and the eigenvalues of A have negative real parts.

Corollary 2.1 implies there exists an integral manifold $S_\epsilon = \{(t,\theta,x): x = f(t,\theta,\epsilon), (t,\theta) \in R^2\}$, $0 \le |\epsilon| \le \epsilon_0$, $f(t,\theta,\epsilon)$ is periodic in θ of period ω and $f(t,\theta,0) = 0$. Our first objective is to show there is a neighborhood $V \subset R^2 \times R^n$ of $\{(t,\theta,\rho):\rho = 0\}$ such that if $(t_0,\theta_0,\rho_0) \in V$, then the solution $(\theta(t),\rho(t))$ of (6.2) through (t_0,θ_0,ρ_0) satisfies

$$|\rho(t) - f(t,\theta(t),\epsilon)| \to 0 \qquad \text{as } t \to \infty.$$

This shows that solutions go back to the integral manifold as $t \to \infty$. We will also show that this approach is at an exponential rate.

To do this, we consider all solutions of the equations (6.2). The variation of constants formula gives

(6.3)
$$\rho(t) = e^{A(t-t_0)}\rho_0 + \int_{t_0}^{t} e^{A(t-s)}R(s,\theta(s),\rho(s),\epsilon)ds$$
$$\dot{\theta}(t) = 1 + \Theta(t,\theta(t),\rho(t),\epsilon), \qquad \theta(t_0) = \theta_0.$$

We first show there is a function $\Psi(t,\theta,t_0,\rho_0,\epsilon)$, ω-periodic in θ, such that the solution of (6.3) can be represented as

(6.4) $\rho(t) = \Psi(t, t_0, \theta(t), \rho_0, \epsilon).$

As in the proof of the existence of the integral manifold S_ϵ, we transform this to a problem of finding a fixed point of a certain mapping. Let

$$\tilde{S}^1 = \{\Psi : \{(t, t_0) : t \geq t_0\} \times R \times R^n \to R^n, \text{ continuous, bounded,}$$

$$\text{together with first derivatives in } \theta, \rho\}.$$

For Ψ in $\tilde{S}^1$, let

$$|\Psi| = \sup\{|\Psi(t, t_0, \theta, \rho)| + |\partial\Psi(t, t_0, \theta, \rho)/\partial\theta|$$

$$+ |\partial\Psi(t, t_0, \theta, \rho)/\partial\rho| : t \geq t_0, (\theta, \rho) \in R \times R^n\}.$$

The smoothness assumptions here are only to make the notation simpler. For Ψ in $\tilde{S}^1$ and $t \leq \tau$, let

$$(T\Psi)(\tau, t, \theta, \rho) = e^{A(\tau-t)}\rho + \int_t^\tau e^{A(\tau-s)} R(s, \theta^*(s), \Psi(s, t, \theta^*(s), \rho), \epsilon)$$

$$\dot{\theta}^*(s) = 1 + \Theta(s, \theta(s), \Psi(s, t, \theta^*(s), \rho), \epsilon), \qquad \theta^*(t) = \theta.$$

One can now proceed as before to show that T is a contraction on an appropriate subset of $\tilde{S}^1$ to obtain a fixed point of T (or one can use the implicit function theorem). If $\Psi = T\Psi$, we have shown that (6.4) is satisfied. There is a continuous function $M(\sigma, \epsilon_0)$, $\sigma \geq 0$, $|\epsilon| \leq \epsilon_0$, $M(0, 0) = 0$, constants $k > 0$, $\alpha > 0$ such that, for $|\rho_0| \leq \sigma$, $|\epsilon| \leq \epsilon_0$,

$$|\partial\Psi(t, t_0, \theta, \rho_0, \epsilon)/\partial\theta| \leq M(\sigma, \epsilon_0)$$

$$|\partial\Psi(t, t_0, \theta, \rho_0, \epsilon)/\partial\rho| \leq Ke^{-\alpha(t-t_0)/2}, \qquad t \geq t_0$$

$$|e^{At}| \leq Ke^{-\alpha t}, \qquad t \geq 0.$$

For ρ_0 sufficiently small, we know that the solution of (6.2) through (t_0, θ_0, ρ_0) satisfies (6.4) for $t \geq t_0$. Thus, our integral manifold S_ϵ must also have this property; that is, for any $t \geq t_0$, $\theta \in R$, there is a $\theta_0' = \theta_0'(t, \theta)$ such that the solution $(\theta'(s), \rho'(s))$ of (6.2) through $(t_0, \theta_0', f(t_0, \theta_0', \epsilon))$ satisfies $\theta'(t) = \theta$, $\rho'(t) = f(t, \theta, \epsilon)$. To obtain this assertion, we have simply integrated backwards from t to t_0 starting from the point $(t, \theta, f(t, \theta, \epsilon))$. Thus, if $\rho_0' = f(t_0, \theta_0', \epsilon)$, then

$$f(t, \theta, \epsilon) = \Psi(t, t_0, \theta, \rho', \epsilon)$$

for all t, θ and

$$|f(t, \theta, \epsilon) - \Psi(t, t_0, \theta, \rho_0, \epsilon)| \leq Ke^{-\alpha(t-t_0)/2}|\rho_0' - \rho_0|.$$

If $(\theta(t), \rho(t))$ is any solution of (6.2) with $\theta(t_0) = \theta_0$, $\rho(t_0) = \rho_0$, ρ_0 sufficiently small, we have

$$|f(t,\theta(t),\epsilon) - \Psi(t,t_0,\theta(t),\rho_0,\epsilon)|$$

$$\leq Ke^{-\alpha(t-t_0)/2}|f(t_0,\theta'_0(t,\theta(t),\epsilon) - \rho_0|$$

$$\leq 2Ke^{-\alpha(t-t_0)/2}, \qquad t \geq t_0.$$

This proves the asserted stability of the integral manifold S_ϵ.

VII.7. Applications

In this section, we give some applications of the theory of this chapter. There is such a large variety of applications that it is impossible to do them justice without devoting a treatise to the subject. However, it is hoped that the few mentioned below together with the ones delegated to the exercises in Section 8 may indicate the possible scope of the theory and stimulate further reading in the literature.

Suppose the equation

$$(7.1) \qquad\qquad \dot{x} = f(x)$$

where f has continuous first derivatives in R^n, has a nonconstant ω-periodic solution u whose linear variational equation

$$(7.2) \qquad\qquad \dot{y} = \frac{\partial f(u(t))}{\partial x} y$$

has $n - 1$ characteristic multipliers not on the unit circle. Suppose also that $g(t, x)$ is continuous for (t, x) in $R \times R^n$, has continuous first derivatives with respect to x and is bounded for t in R and x in any compact set.

THEOREM 7.1   Under the above hypotheses, there are a neighborhood U of the periodic orbit $\mathscr{C} = \{x : x = u(\theta), 0 \leq \theta \leq \omega\}$ and an $\varepsilon_1 > 0$ such that the system

$$(7.3) \qquad\qquad \dot{x} = f(x) + \varepsilon g(t, x)$$

has an integral manifold S_ε in $R \times U$, $0 \leq |\varepsilon| \leq \varepsilon_1$, $S_0 = R \times \mathscr{C}$ (a cylinder), S_ε is asymptotically stable if $n - 1$ multipliers of (7.2) are inside the unit circle and unstable if one is outside the unit circle. The set S_ε has a parametric representation given by

$$S_\varepsilon = \{(t, x): x = u(\theta) + v(t, \theta, \varepsilon), (t, \theta) \text{ in } R \times R\},$$

where $v(t, \theta, 0) = 0$, $v(t, \theta, \varepsilon) = v(t, \theta + \omega, \varepsilon)$ and is almost periodic (T-periodic) in t if $g(t, \theta)$ is almost periodic (T-periodic) in t.

PROOF.   This is a simple consequence of the previous results. In fact, using the coordinate transformation in Chapter VI, one obtains a special case of system (1). The above hypotheses permit a direct application of Theorem 2.2 to obtain existence of the manifold for the interval $0 < \varepsilon \leqq \varepsilon_1$. Replacing ε by $-\varepsilon$ one obtains the result on $-\varepsilon_1 \leqq -\varepsilon < 0$. Defining the manifold of (1) to be zero at $\varepsilon = 0$ completes the proof of existence. Stability is a consequence of the remarks in Section 6.

In case g in (7.3) is periodic in t of period T, then the integral manifold S_ε given in Theorem 7.1 has a parametric representation which is periodic in t of period T. The cross section of S_ε at $t = 0$ and $t = T$ are therefore the same and the fact that it is an integral manifold implies this cross section is mapped into itself through the solutions of the differential equation. Since the differential equation is T-periodic, following a solution $x(t, x_0)$, $x(0, x_0) = x_0$, from $t = 0$ to $t = 2T$ is the same as following the solution from $t = 0$ to $t = T$ and then following the solution $x(t, x(T, x_0))$ from $t = 0$ to $t = T$. The solutions of (7.3) on S_ε in this case can therefore be considered as a differential equation without critical points on the torus obtained by taking the section of the surface S_ε from $t = 0$ to $t = T$ and identifying the ends of this section. The theory of Chapter II then gives the possible behavior of the solutions on S_ε.

THEOREM 7.2.   Suppose $f(x)$ satisfies the conditions of Theorem 7.1 and let $g(t, x)$ be continuous and uniformly bounded together with its first partial derivatives with respect to x in a neighborhood of $\mathscr{C}$. If $g(t, x)$ is almost periodic in t uniformly with respect to x and

$$M[g(\cdot, x)] = \lim_{T \to \infty} \frac{1}{T} \int_0^T g(t, x)\, dt = 0,$$

then there are a neighborhood U of $\mathscr{C}$ and an $\omega_0 > 0$ such that the system

(7.4) $$\dot{x} = f(x) + g(\omega t, x)$$

has an integral manifold S_ω in $R \times U$, $\omega \geqq \omega_0$, $S_\omega \to R \times \mathscr{C}$ as $\omega \to \infty$, S_ω is asymptotically stable if $n - 1$ multipliers of (7.2) are inside the unit circle and unstable if one is outside the unit circle. The set S_ω has a parametric representation given by

$$S_\omega = \{(t, x): x = u(\theta) + v(\omega t, \theta, \omega^{-1}),\ (t, \theta)\ \text{in}\ R \times R\},$$

where $v(t, \theta, 0) = 0$, $v(t, \theta, \alpha) = v(t, \theta + \omega, \alpha)$ and is almost periodic in t.

PROOF.   From Lemma 5 of the Appendix, for any $\eta > 0$, there are a function $w(t, x, \eta)$ with as many derivatives in x as desired and a function $\sigma(\eta) \to 0$ as $\eta \to 0$ such that

$$|\partial w(t, x, \eta)/\partial t - g(t, x)| < \sigma(\eta)$$

for all $t \in R$ and x in a neighborhood of $\mathscr{C}$. Also ηw, $\eta \partial w/\partial x \to 0$ as $\eta \to 0$. Therefore, as in the proof of Lemma V.3.2, there is an $\omega_0 > 0$ such that the transformation

$$x = y + \frac{1}{\omega}\, w\left(\omega t,\, y,\, \frac{1}{\omega}\right)$$

is a homeomorphism in a neighborhood of $\mathscr{C}$. If this transformation is applied to (7.4) one obtains a system

$$\dot{y} = f(y) + G(\omega t,\, y,\, \omega^{-1}),$$

where $G(\tau,\, y,\, \omega^{-1})$ is continuous in τ, y, ω together with its first derivative with respect to y, is uniformly bounded for τ in R, y in a neighborhood of $\mathscr{C}$ and $\omega \geqq \omega_0$, and $G(\tau,\, y,\, 0) = 0$. If one uses the coordinate transformation in Chapter VI and lets $\omega t = \tau$, $\omega^{-1} = \varepsilon$, then the new system is a special case of system (1) for which Theorem 2.3 applies directly. This will complete the proof of existence of an integral manifold. The stability follows from Section 6.

Another application of the previous results concerns a generalization of the method of averaging. Consider the system

(7.5)
$$\dot{\psi} = e + \varepsilon \Psi(\psi,\, \rho),$$
$$\dot{\rho} = \varepsilon R(\psi,\, \rho),$$

where ψ is in R^k, ρ is in R^p, $e = (1,\, \ldots,\, 1)$, and $\Psi(\psi,\, \rho)$, $R(\psi,\, \rho)$ are multiply periodic in ψ of period ω and have continuous first derivatives with respect to ρ. Let

(7.6)
$$\Psi_0(\psi,\, \rho) = \lim_{T \to \infty} \frac{1}{T} \int_0^T \Psi(\psi + et,\, \rho)\, dt,$$

$$R_0(\psi,\, \rho) = \lim_{T \to \infty} \frac{1}{T} \int_0^T R(\psi + et,\, \rho)\, dt.$$

From the Appendix, for any $\eta > 0$, there are functions $u(\psi,\, \rho,\, \eta)$, $v(\psi,\, \rho,\, \eta)$ with as many derivatives in ψ, ρ as desired and a function $\sigma(\eta) \to 0$ as $\eta \to 0$ such that

$$\left| \frac{\partial u}{\partial \psi}\, e - \Psi(\psi,\, \rho) + \Psi_0(\psi,\, \rho) \right| < \sigma(\eta),$$

$$\left| \frac{\partial v}{\partial \psi}\, e - R(\psi,\, \rho) + R_0(\psi,\, \rho) \right| < \sigma(\eta),$$

and the functions ηu, ηv, $\eta \partial u/\partial \psi$, $\eta \partial v/\partial \psi$, $\eta \partial u/\partial \rho$, $\eta \partial v/\partial \rho \to 0$ as $\eta \to 0$ uniformly for ψ in R^k, ρ in a bounded set. Therefore, as in the proof of Theorem V.3.2, there is an $\varepsilon_0 > 0$ such that the transformation

(7.7)
$$\psi = \phi + \varepsilon u(\phi,\, r,\, \varepsilon),$$
$$\rho = \phi + \varepsilon v(\phi,\, r,\, \varepsilon),$$

is a homeomorphism for $|\varepsilon| < \varepsilon_0$, ψ, ϕ in R^k and ρ, r is a bounded set.

If the transformation (7.7) is applied to (7.5), then a few simple computations yield

(7.8)
$$\dot{\phi} = e + \varepsilon\Psi_0(\phi, r) + \varepsilon\Psi_1(\phi, r, \varepsilon),$$
$$\dot{r} = \varepsilon R_0(\phi, r) + \varepsilon R_1(\phi, r, \varepsilon),$$

where Ψ_1, R_1 have the same smoothness properties as Ψ, R, but now satisfy $\Psi_1(\phi, r, 0) = 0$, $R_1(\phi, r, 0) = 0$. In other words, by a transformation (7.7) which is essentially the identity transformation near $\varepsilon = 0$, the system (7.5) is transformed into an equation which is a higher order perturbation of the *averaged equations*

(7.9)
$$\dot{\theta} = e + \varepsilon\Psi_0(\theta, \rho),$$
$$\dot{\rho} = \varepsilon R_0(\theta, \rho).$$

One is now in a position to state results concerning the existence of integral manifolds by asserting the averaged equations have certain properties. For example, suppose there is a ρ_0 such that $R_0(\theta, \rho_0) = 0$ and furthermore,

(7.10) $R_0(\theta, \rho_0 + z) = C(\theta)z + H(\theta, z),$

$$C(\theta) = \begin{bmatrix} A(\theta) & 0 \\ 0 & B(\theta) \end{bmatrix}, \qquad H(\theta, z) = \begin{bmatrix} F(\theta, z) \\ G(\theta, z) \end{bmatrix}, \qquad z = \begin{bmatrix} x \\ y \end{bmatrix},$$

$$\Psi_0(\theta, \rho_0 + z) = \Theta(\theta, z),$$

where all matrices are partitioned so that any matrix operations will be compatible. As an immediate consequence of Theorem 2.3 and the form of transformed equations (7.8), one can state

THEOREM 7.3.   Suppose A, B in (7.10) satisfy hypothesis (H_5) in Section 1 with $\alpha = \varepsilon\alpha_1$, $\alpha_1 > 0$, a constant. If the lipschitz constant L of $\Theta(\theta, 0)$ satisfies $\alpha_1 - L > 0$, then there are $\varepsilon_1 > 0$, continuous functions $D(\varepsilon)$, $\Delta(\varepsilon)$, $0 < \varepsilon \leq \varepsilon_1$, approaching zero as $\varepsilon \to 0$ and a function $f(\psi, \varepsilon)$ in R^p which is continuous in $R^k \times [0, \varepsilon_1]$,

$$|f(\psi, \varepsilon) - \rho_0| < D(\varepsilon),$$
$$|f(\psi, \varepsilon) - f(\bar{\psi}, \varepsilon)| < \Delta(\varepsilon)|\psi - \bar{\psi}|,$$

such that $f(\psi, \varepsilon)$ is multiply periodic in ψ of vector period ω and the set

$$S_\varepsilon = \{(\psi, \rho) : \rho = \rho_0 + f(\psi, \varepsilon), \psi \text{ in } R^k\}$$

is an integral manifold of (7.5). If the y component of z is present in (7.10), then the manifold is unstable and if it is absent, the manifold is stable.

A simple corollary of Theorem (7.3) which is useful for the applications is

COROLLARY 7.1.   Suppose the averaged equations (7.9) are independent of ψ; that is, the averaged equations are

(7.11) $\dot\theta = e + \varepsilon \Psi_0(\rho),$

$\dot\rho = \varepsilon R_0(\rho).$

Also, suppose there is a ρ_0 such that $R_0(\rho_0) = 0$ and the real parts of the eigenvalues of the matrix $C = \partial R_0(\rho_0)/\partial \rho$ are nonzero. Then the conclusions of Theorem 7.3 remain valid with the manifold being stable if all eigenvalues of C have negative real parts and unstable if one eigenvalue has a positive real part.

VII.8. Exercises

Some of the exercises listed below are rather difficult and a complete discussion of some could lead to interesting new results in the theory of oscillations.

EXERCISE 8.1.   Use Section VI.3 to show that the hypotheses of Theorem 7.1 imposed on (7.1) are satisfied for the van der Pol equation

(8.1) $\dot x_1 = x_2,$

$\dot x_2 = -x_1 + k(1 - x_1^2)x_2, \qquad k > 0.$

Use Theorems 7.1 and 7.2 to discuss the existence and stability of integral manifolds of the system

(8.2) $\dot x_1 = x_2,$

$\dot x_2 = -x_1 + k(1 - x_1^2)x_2 + \varepsilon \sin \omega t,$

when either ε is small or ω is large.

Use the theory of Chapter IV to discuss the existence and stability properties of a periodic solution of (8.2) of period $2\pi/\omega$ for either $|\varepsilon|$ small or ω large.

In Chapter II it was shown that every solution of (8.1) except the zero solution approaches the periodic orbit $\mathscr{C}$ as $t \to \infty$. Can you use the previous discussion to give the qualitative behavior of all solutions of (8.2) in some large domain of R^2 for $|\varepsilon|$ small? Hint: Show there is an open set U in R^2 such that the tangent vector to the solution curves of (8.1) on the boundary of U point toward the interior of U. This implies that the tangent vector to the solutions of (8.1) in (t, x)-space is pointing toward the interior of the "cylinder" $R \times \partial U$. Therefore for $|\varepsilon|$ small or ω large, the same will be true of the solutions of (8.2). Now use continuity with respect to the vector field.

EXERCISE 8.2. Suppose both of the vector systems $\dot{x} = f(x)$, $\dot{y} = F(y)$ satisfy the condition of Theorem 7.1. What types of integral manifolds does the system

$$\dot{x} = f(x) + \varepsilon g(t, x, y),$$
$$\dot{y} = F(y) + \varepsilon G(t, x, y)$$

possess for ε small if g, G are bounded for t in R and x, y in compact sets? What are the stability properties of each of the manifolds? Generalize this result. Hint: Use the coordinate system of Chapter VI for each periodic orbit.

EXERCISE 8.3. For ω large, discuss the existence and nature of integral manifolds of the system

$$\dot{x}_1 = x_2,$$
$$\dot{x}_2 = -x_1 + k[1 - (x_1 + B \sin \omega t)^2]x_2,$$

as a function of B.

EXERCISE 8.4. For ω large, discuss the question of the existence and the dependence of integral manifolds of the system

$$\dddot{x} + 2\ddot{x} + \dot{x} + Kf(x + B \sin \omega t) = 0$$

on the constants K and B for odd functions $f(x)$ which are monotone nondecreasing and approach a limit as $x \to \infty$. This problem is difficult and one could never solve it in this general context. Take special functions f.

EXERCISE 8.5. Discuss the existence of integral manifolds of the equation

$$\dot{x}_1 = x_2,$$
$$\dot{x}_2 = -x_1 + \varepsilon(1 - x_1^2)x_2 + A \sin \omega t,$$

for ε small and various values of A and ω. Let

$$x_1 = \rho \sin \theta_1 + A(1 - \omega^2)^{-1} \sin \omega t$$
$$x_2 = \rho \cos \theta_1 + A\omega(1 - \omega^2)^{-1} \cos \omega t$$

and $\theta_2 = t$ to obtain a system of differential equations for θ_1, θ_2, ρ and then apply Corollary 7.1. What are the stability properties of the manifolds? What happens geometrically as A, ω vary?

EXERCISE 8.6. For ε small, discuss the existence and stability properties of integral manifolds of the system

$$\ddot{x} - \varepsilon(1 - x^2 - ay^2)\dot{x} + x = 0,$$
$$\ddot{y} - \varepsilon(1 - y^2 - \alpha x^2)\dot{y} + \sigma^2 y = 0,$$

as a function of a, α, σ. Let $x = \rho_1 \cos \theta_1$, $\dot{x} = -\rho_1 \sin \theta_1$, $y = \rho_2 \cos \sigma\theta_2$, $\dot{y} = -\sigma\rho_2 \sin \sigma\theta_2$ to obtain a system of the same form as system (7.8). Now apply the method of averaging described above and, in particular, Corollary 7.1 for the case when $k + l\sigma \neq 0$ for all integers k, l for which $|k| + |l| \leq 3$. What are the periodic solutions? Can you describe geometrically what happens when the stability properties of the periodic orbits change under variation of the constants a and α? What happens when $k + l\sigma = 0$ for some integers k and l with $|k| + |l| \leq 3$? In the equation, change $1 - x^2 - ay^2$ to $1 - x^2 - ay^2 + bx^2y^2$ and discuss what happens as a function of a, b, α.

EXERCISE 8.7. Carry out the same analysis as in Exercise 8.6 for the equations

$$\ddot{x} - \varepsilon\left(\dot{x} - \frac{\dot{x}^3}{3}\right) + \sigma x + \mu y = 0,$$

$$\ddot{y} - \varepsilon\left(\dot{y} - \frac{\dot{y}^3}{3}\right) - \mu x + \nu y = 0,$$

for those values of the parameters σ, μ, ν for which the characteristic roots of the equation for $\varepsilon = 0$ are simple and purely imaginary, say $\pm i\omega_1$, $\pm i\omega_2$. Under this hypothesis, this system can be transformed by a linear transformation to a system which for $\varepsilon = 0$ is given by $\ddot{u} + \omega_1^2 u = 0$, $\ddot{v} + \omega_2^2 v = 0$. Now apply the same type of argument as in Exercise 8.6.

VII.9. Remarks and Suggestions for Further Study

Detailed references to the method of Krylov-Bogoliubov may be found in the books of Bogoliubov and Mitropolski [1] or Hale [3]. The original results on integral manifolds using this method considered only the case in which the flow on the unperturbed manifold was a parallel flow; that is, $w(t, \theta, \varepsilon)$ in (1) is a constant. However, the method of proof given in the text is basically the same as the original proof for parallel flow except for the technical details. Other results along this line were obtained by Diliberto [3] and Kyner [1]. Kurzweil [1, 2] has given another method for obtaining the existence of integral manifolds and has the problem formulated in such a general framework as to have applications to partial differential equations, difference equations and some types of functional differential equations. Further results may be found in Pliss [1].

It is not assumed in system (1) that the functions are periodic in the vector θ. The results have implications to the theory of center manifolds and stability theory (Pliss [2], Kelley [1]) and the theory of bifurcation (Chafee [1], Lykova [1]).

Hypothesis (H_5) in Section 2 is too strong. For example, if $w(t,\theta,\epsilon) = 1$, one need only assume the exponential estimates are valid for functions $\theta(t) = t + \theta_0$ (see Montandon [1]). For a more general case, see Coppel and Palmer [1], Henry [1].

The basic result in the proof of the theorems on integral manifolds was Lemma 3.2. The proof of this lemma shows that the Lipschitz constant $L(P)$ was used only to obtain estimates on the dependence of the solution $\theta^*(t, \tau, \theta, P)$ of $\dot{\theta} = P(t, \theta)$ upon θ. There are many ways to obtain estimates of this dependence without using the lipschitz constant of P. For example, if $P(t, \theta)$ has continuous partial derivatives with respect to θ, then $\partial\theta^*(t, \tau, \theta, P)/\partial\theta$ is a solution of the equation

$$\zeta = \left[\frac{\partial P(t, \theta^*)}{\partial\theta} \right] \zeta.$$

Consequently, one can use the eigenvalues of the symmetric part of $\partial P/\partial\theta$ to estimate the rate of growth of ζ. The hypothesis (H_5) can also be verified by using the eigenvalues of the symmetric part of the matrices $A(\theta)$, $B(\theta)$. Using the method of partial differential equations mentioned in Section 1, Sacker [1] has exploited these concepts to great advantage to discuss the existence and smoothness properties of integral manifolds for equations (1) which are independent of t and periodic in θ. Diliberto [3] has also used these same concepts and the method of the text to find integral manifolds.

The first example in Section 3 after Lemma 3.2 is due to McCarthy [1] and the second to Kyner [1]. The generalized average defined in relation (5.6) was first introduced by Diliberto [1]. Some of the exercises in Section 8 can be found in Hale [7].

CHAPTER VIII

Periodic Systems with a Small Parameter

In Chapter IV, we discussed the existence of periodic solutions of equations containing a small parameter in noncritical cases; that is systems

(1) $$\dot{x} = Ax + \varepsilon\, f(t, x),$$

where $f(t + T, x) = f(t, x)$ and no solution except $x = 0$ of the unperturbed equation

(2) $$\dot{x} = Ax,$$

is T-periodic. In Chapter V, the method of averaging was applied to some systems for which (2) has nontrivial T-periodic solutions. The basis of this method is to make a change of variables which transforms the system into one which can be considered as a perturbation of the averaged equations. If the averaged equations are noncritical with respect to the class of T-periodic functions, then the results of Chapter IV can be applied. If the averaged equations are critical with respect to T-periodic functions, then the process can be repeated. In addition to being a very cumbersome procedure, it is very difficult to use averaging and reflect any qualitative information contained in the differential equation itself into the iterative scheme. For example, if system (1) has a first integral, what is implied for the iterations?

For periodic systems, other more efficient procedures are available. The present chapter is devoted to giving a general method for determining periodic solutions of equations including (1) which may be critical with respect to T-periodic functions. This method gives necessary and sufficient conditions for the existence of a T-periodic solution of (1) for ε small. These conditions consist of transcendental equations (the bifurcation or determining equations) for the determination of a T-periodic function which is a solution of (2). The bifurcation equations are given in such a way as to permit a qualitative discussion of their dependence upon properties of the right hand side of (1). This is illustrated very well when f in (1) enjoys some even and oddness properties or system (1) possesses a first integral. In general, such systems have families of T-periodic solutions.

In addition to the advantage mentioned in the previous paragraph, the method of this chapter can be generalized to arbitrary nonlinear systems. This topic as well as a more general formulation of the method in Banach spaces will be treated in the next chapter.

For $A = 0$, the, basic ideas are very elementary and easy to understand geometrically. For this reason, this case is treated in detail in Section 1. Also, in Section 1, it is shown how to reduce the study of periodic solutions of many equations as well as the determination of characteristic exponents of linear periodic systems to this simple form. Section 2 is devoted to a discussion of the general system (1) as well as results on systems possessing either symmetry properties or first integrals. In Section 3, we reprove a result of Chapter VI using the method of this chapter rather than a coordinate system around a periodic orbit.

VIII.1. A Special System of Equations

Suppose $f\colon R \times C^n \to C^n$ is a continuous function with $\partial f(t, x)/\partial x$ also continuous, $f(t + T, x) = f(t, x)$, ε is a parameter and consider the system of equations

(1.1) $$\dot{x} = \varepsilon f(t, x).$$

Our problem is to determine whether or not system (1.1) has any T-periodic solutions for ε small. For $\varepsilon = 0$ all solutions of (1.1) are T-periodic; namely, they are constant functions. The basic question is the following: if there are T-periodic solutions of (1.1) which are continuous in ε, which solutions of the degenerate equation do they approach as $\varepsilon \to 0$? One precedure was indicated for attacking this question in Chapter V. Another method is due to Poincaré in which a periodic power series expansion is assumed for the solution as well as the initial data. The initial data is then used to eliminate the secular terms that naturally arise in the determination of the coefficients in the power series of the solutions.

In this section, we indicate another method for solving this problem which seems to have some qualitative advantages over the method of Poincaré. Let $\mathscr{P}_T = \{g\colon R \to C^n,\ g \text{ continuous},\ g(t + T) = g(t)\}$, $\|g\| = \sup_{0 \le t \le T} |g(t)|$. For any g in $\mathscr{P}_T$, define Pg to be the mean value of g; that is,

(1.2) $$Pg = \frac{1}{T} \int_0^T g(t)\, dt, \qquad \|Pg\| \le \|g\|.$$

If g is T-periodic, then the system

(1.3) $$\dot{x} = g(t)$$

has a T-periodic solution if and only if $Pg = 0$. Furthermore, if $Pg = 0$, let $\mathcal{K}g$ be the unique T-periodic solution of (1.3) which has mean value zero; that is,

$$(1.4) \qquad \mathcal{K}g = (I - P)\int_0^{\cdot} g(s)\,ds, \qquad \|\mathcal{K}g\| \leq K\|g\|, \qquad K = 2T.$$

Every T-periodic solution of (1.3) can then be written as

$$x = a + \mathcal{K}g$$

where a is a constant n-vector and $a = Px$.

These simple remarks imply the following:

LEMMA 1.1. Suppose P and $\mathcal{K}$ are defined in (1.2), (1.4). Then

(i) $x(t)$ is a T-periodic solution of (1.1) only if $Pf(\cdot, x(\cdot)) = 0$; that is, only if $f(\cdot, x(\cdot)) = (I - P)f(\cdot, x(\cdot))$.

(ii) System (1.1) has a T-periodic solution x if and only if x satisfies the system of equations

$$(1.5) \qquad \begin{array}{ll} \text{(a)} & x = a + \varepsilon\mathcal{K}(I - P)f(\cdot, x); \\ \text{(b)} & \varepsilon Pf(\cdot, x) = 0, \end{array}$$

where a is a constant n-vector given by $a = Px$.

PROOF. If x is a T-periodic solution of (1.1), let $g(t) = f(t, x(t))$ and assertion (i) follows immediately. The fact that x satisfies (1.5) is just as obvious. If x is a solution of (1.5), then $f(\cdot, x) = (I - P)f(\cdot, x)$ and, therefore, x is a T-periodic solution of (1.1). This proves the lemma.

LEMMA 1.2. For any $\alpha > 0$, there is an $\varepsilon_0 > 0$ such that for any a in C^n with $|a| \leq \alpha$, $|\varepsilon| \leq \varepsilon_0$, there is a unique function $x^* = x^*(a, \varepsilon)$ which satisfies (1.5a). Furthermore, $x^*(a, \varepsilon)$ has a continuous first derivative with respect to a, ε and $x^*(a, 0) = a$. If there is an $a = a(\varepsilon)$ with $|a(\varepsilon)| \leq \alpha$ for $0 \leq |\varepsilon| \leq \varepsilon_0$ and

$$(1.6) \qquad G(a, \varepsilon) \overset{\text{def}}{=} Pf(\cdot, x^*(a, \varepsilon)) = 0,$$

then $x^*(a, \varepsilon)$ is a T-periodic solution of (1.1). Conversely, if (1.1) has a T-periodic solution $\bar{x}(\varepsilon)$ which is continuous in ε and has $P\bar{x}(\varepsilon) = a(\varepsilon)$, $|a(\varepsilon)| \leq \alpha$, $0 \leq |\varepsilon| \leq \varepsilon_0$, then $\bar{x}(\varepsilon) = x^*(a(\varepsilon), \varepsilon)$ where $x^*(a, \varepsilon)$ is the function given above and $a(\varepsilon)$ satisfies (1.6) for $0 < |\varepsilon| \leq \varepsilon_0$.

PROOF. Suppose $\alpha > 0$ is given, β is any number > 0 and a is an n-vector with $|a| \leq \alpha$. Define $\mathcal{S}(\beta) = \{y \text{ in } \mathcal{P}_T: \|y\| \leq \beta, \ Py = 0\}$ and the operator $\mathcal{F}: \mathcal{S}(\beta) \to \mathcal{P}_T$ by

$$(1.7) \qquad \mathcal{F}y = \varepsilon\mathcal{K}(I - P)f(\cdot, y + a)$$

for y in $\mathscr{S}(\beta)$. If y is a fixed point of $\mathscr{F}$, then $y + a$ satisfies (1.5a). Let M, N be bounds on $|f(t, x)|$, $|\partial f(t, x)/\partial x|$, respectively, for t in R, $|x| \leq \beta + \alpha$ and choose ε_1 so that $4\varepsilon_1 TM < \beta$, $4\varepsilon_1 TN < 1/2$. One easily shows that $\mathscr{F}: \mathscr{S}(\beta) \to \mathscr{S}(\beta)$ and is a uniform contraction with respect to a, ε for $|a| \leq \alpha$, $|\varepsilon| \leq \varepsilon_1$. Let $y^*(a, \varepsilon)$ be the unique fixed point of $\mathscr{F}$ in $\mathscr{S}(\beta)$. The existence and properties of $x^*(a, \varepsilon) = y^*(a, \varepsilon) + a$ stated in the lemma are now a consequence of the uniform contraction principle. If $a(\varepsilon)$ satisfies (1.6), then $x^*(a(\varepsilon), \varepsilon)$ is a solution of (1.1) from Lemma 1.1. Conversely, if $\bar{x}(\varepsilon)$ satisfies the properties in the statement of the lemma, then there is a $\beta > \alpha$ such that $|\bar{x}(\varepsilon) - a(\varepsilon)| \leq \beta$ for $0 \leq |\varepsilon| \leq \varepsilon_1$. For this β choose $\varepsilon_0 \leq \varepsilon_1$ so that $\mathscr{F}$ is a contraction on $\mathscr{S}(\beta)$ for $0 \leq |\varepsilon| \leq \varepsilon_0$. Then $\bar{x}(\varepsilon) - a(\varepsilon) = y^*(a(\varepsilon), \varepsilon)$ and $\bar{x}(\varepsilon) = x^*(a(\varepsilon), \varepsilon)$. The conclusion of the lemma follows from Lemma 1.1.

Lemma 1.2 asserts the following: one can specify an arbitrary n-vector a and then determine uniquely a T-periodic function $x^*(a, \varepsilon)$ with mean value a in such a way that all of the Fourier coefficients of the function $\dot{x}^*(t) - f(t, x^*(t))$ are zero except for the constant term (this is the same as saying (1.5a) is satisfied). The n-vector a is then used to try to make the constant term equal to zero (that is, satisfy (1.6)). Equations (1.6) are sometimes referred to as the *determining equations* or *bifurcation equations* of (1.1).

As a consequence of the above proof, the function $x^*(a, \varepsilon)$ can be obtained as the limit of the sequence $\{x^{(k)}\}$, $x^{(k)} = y^{(k)} + a$ where the $y^{(k)}$ are defined successively by $y^{(k+1)} = \mathscr{F}y^{(k)}$, $k = 0, 1, 2, \ldots$, $y^{(0)} = 0$, and $\mathscr{F}$ is defined in (1.7). Using only the first approximation, one arrives at

THEOREM 1.1. Suppose $x^*(a, \varepsilon)$ is defined as in Lemma 1.2 and let $G(a, \varepsilon)$ be defined by (1.6). If there is an n-vector a_0 such that

$$(1.8) \qquad G(a_0, 0) = 0, \qquad \det\left[\frac{\partial G(a_0, 0)}{\partial a}\right] \neq 0,$$

then there are an $\varepsilon_1 > 0$ and a T-periodic solution $x^*(\varepsilon)$, $|\varepsilon| \leq \varepsilon_1$, of (1.1), $x^*(0) = a_0$, and $x^*(\varepsilon)$ is continuously differentiable in ε.

PROOF. The hypothesis (1.8) and the implicit function theorem imply there is an ε_1, $0 < \varepsilon_1 \leq \varepsilon_0$, such that equation (1.6) has a continuously differentiable solution $a(\varepsilon)$, $|a(\varepsilon)| < \alpha$, $0 \leq |\varepsilon| \leq \varepsilon_1$. Lemma 1.2 implies the remaining assertions of the theorem.

Notice that

$$(1.9) \qquad G(a, 0) = \frac{1}{T}\int_0^T f(t, a)\, dt,$$

and therefore can be calculated without knowing anything about the solutions of (1.1). Compare Theorem 1.1 with Theorem V.3.2. Can you prove Theorem V.3.2 by the above method?

In the applications, Theorem 1.1 is not sufficiently general even for equation (1.1). More specifically, it is sometimes necessary to take the terms in the Taylor expansion of $G(a, \varepsilon)$ in (1.6) of the first, second, or even higher order in ε. Except for the numerical computations involved, there is conceptually no difficulty in obtaining these terms. To obtain the first order terms of $G(a, \varepsilon)$ in ε, one must compute the first order terms in ε in $x^*(a, \varepsilon)$, etc. This is accomplished by the iteration procedure $x^{(k)} = y^{(k)} + a$, $y^{(k+1)} = \mathscr{F} y^{(k)}$, $k = 0, 1, 2, \ldots$, $y^{(0)} = a$, where the mapping $\mathscr{F}$ is defined in (1.7).

It is easy to generalize the above results to a system of the form

$$(1.10) \qquad \dot{z} = Bz + \varepsilon h(t, z),$$

where $h(t + T, z) = h(t, z)$, h and $\partial h/\partial z$ are continuous in $R \times C^{m+n}$, $B = \mathrm{diag}(0_n, B_1)$, 0_n is the $n \times n$ zero matrix, B_1 is an $m \times m$ constant matrix such that $e^{B_1 T} - I$ is nonsingular. This latter condition states that the system $\dot{y} = B_1 y$ is noncritical with respect to $\mathscr{P}_T$. If $z = (x, y)$, $h = (f, g)$ where x, f are n-vectors, then the above system is equivalent to

$$(1.10)' \qquad \dot{x} = \varepsilon f(t, x, y),$$
$$\dot{y} = B_1 y + \varepsilon g(t, x, y).$$

For any $h = (f, g)$ in $\mathscr{P}_T$ define Ph to be the $(n + m)$-dimensional constant vector given by

$$(1.11) \qquad Ph = \begin{bmatrix} \dfrac{1}{T} \displaystyle\int_0^T f(t)\, dt \\ 0 \end{bmatrix}.$$

The nonhomogeneous linear system

$$(1.12) \qquad \dot{z} = Bz + h(t)$$

has a T-periodic solution if and only if $Ph = 0$. Furthermore, if $Ph = 0$, let $\mathscr{K} h$ be the unique solution z of (1.12) with $Pz = 0$. For any h in $\mathscr{P}_T$, one can therefore define $\mathscr{K}(I - P)h$ and there is a constant $K > 0$ such that

$$(1.13) \qquad \|\mathscr{K}(I - P)h\| \leq K \|h\|;$$

that is $\mathscr{K}(I - P)$ is a continuous linear mapping of $\mathscr{P}_T$ into $\mathscr{P}_T$. Every T-periodic solution of (1.12) can be written as

$$z = a^* + \varepsilon \mathscr{K} h, \quad a^* = \begin{bmatrix} a \\ 0 \end{bmatrix},$$

where a is a constant n-vector with $a^* = Pz$.

LEMMA 1.3. Suppose P is defined by (1.11), H is in $\mathscr{P}_T$ and $\mathscr{K}(I - P)H$ is the unique T-periodic solution z of

$$\dot{z} = Bz + (I - P)H$$

with $Pz = 0$. Then $z(t)$ is a T-periodic solution of (1.10) only if $Ph(\cdot, z(\cdot)) = 0$; that is, only if $h(\cdot, z(\cdot)) = (I - P)h(\cdot, z(\cdot))$. Also, system (1.10) has a T-periodic solution z if and only if the system of equations

(1.14) (a) $z = a^* + \varepsilon \mathcal{K}(I - P)h(\cdot, z(\cdot)), \qquad a^* = (a, 0),$

      (b) $\varepsilon Ph(\cdot, z(\cdot)) = 0,$

is satisfied, where a is a constant n-vector given by $a^* = Pz$.

LEMMA 1.4. For any $\alpha > 0$, there is an $\varepsilon_0 > 0$ such that for any a in R^n with $|a| \leq \alpha$, $|\varepsilon| \leq \varepsilon_0$, there is a unique function $z^* = z^*(a, \varepsilon)$ which satisfies (1.14a). Furthermore, $z^*(a, \varepsilon)$ has a continuous first derivative with respect to a, ε and $z^*(a, 0) = a^*$. If there is an $a(\varepsilon)$ with $|a(\varepsilon)| \leq \alpha$ for $0 \leq |\varepsilon| \leq \varepsilon_0$ and

(1.15) $$G(a, \varepsilon) = \frac{1}{T} \int_0^T f(t, z^*(a, \varepsilon)(t))\, dt = 0,$$

then $z^*(a, \varepsilon)$ is a T-periodic solution of (1.10). Conversely, if (1.10) has a T-periodic solution $\tilde{z}(\varepsilon)$ which is continuous in ε and has $P\tilde{z}(\varepsilon) = a^*(\varepsilon)$, $a^* = (a, 0)$, $|a(\varepsilon)| \leq \alpha$, $0 \leq |\varepsilon| \leq \varepsilon_0$, then $\tilde{z}(\varepsilon) = z^*(a(\varepsilon), \varepsilon)$ where $z^*(a, \varepsilon)$ is the function given above and $a(\varepsilon)$ satisfies (1.15) for $0 < |\varepsilon| \leq \varepsilon_0$.

THEOREM 1.2. Suppose $z^*(a, \varepsilon)$ is defined as in Lemma 1.4 and $G(a, \varepsilon)$ is defined as in (1.15). If there is an n-vector a_0 such that

$$G(a_0, 0) = 0, \qquad \det\left[\frac{\partial G(a_0, 0)}{\partial a}\right] \neq 0,$$

then there is an $\varepsilon_1 > 0$ and a T-periodic solution $z^*(\varepsilon)$, $|\varepsilon| \leq \varepsilon_1$, of (1.10), $z^*(0) = a_0^*$, $a_0^* = (a_0, 0)$ and $z^*(\varepsilon)$ is continuously differentiable in ε.

EXERCISE 1.1 Verify all statements made above concerning system (1.10) and prove in detail Lemmas 1.3, 1.4, and Theorem 1.2.

EXERCISE 1.2. Consider the equations

$$\dot{x}_1 = x_2,$$
$$\dot{x}_2 = -x_1 + \varepsilon(1 - x_1^2)x_2 + \varepsilon p \cos(\omega t + \alpha),$$

where $p \neq 0$, $\varepsilon > 0$, ω, α are real numbers with $\omega^2 = 1 + \varepsilon\beta$, $\beta \neq 0$. Find conditions on p, ω, α which will ensure that this equation has a periodic solution of period $2\pi/\omega$ for ε small. Draw the frequency response curve. Let

$$x_1 = z_1 \sin \omega t + z_2 \cos \omega t,$$
$$x_2 = \omega(z_1 \cos \omega t - z_2 \sin \omega t),$$

and apply Theorem 1.1.

EXERCISE 1.3. Discuss the possibility of the existence of a subharmonic solution of order 2 (that is, a solution whose period is twice the period of the vector field) for the equation

(1.16) $\dot{x}_1 = x_2$,

$$\dot{x}_2 = -\sigma^2 x_1 + \varepsilon(3\nu \cos 2t - x_1^2) - \varepsilon^2(\lambda x_2 + \mu x_1) - \varepsilon^3 \mu x_1^2,$$

where $\sigma = 1$. Let

$$x_1 = z_1 \sin t + z_2 \cos t,$$
$$x_2 = z_1 \cos t - z_2 \sin t,$$

and apply Theorem 1.1.

EXERCISE 1.4. For $\sigma = 2/3$, $\lambda = \varepsilon\lambda_1$ in (1.16) discuss the existence of a subharmonic solution of order 3 of (1.16). Let

$$x_1 = z_1 \sin \sigma t + z_2 \cos \sigma t,$$
$$x_2 = \sigma(z_1 \cos \sigma t - z_2 \sin \sigma t),$$

and determine the Taylor expansion of $G(a, \varepsilon)$ in (1.6) up through terms of order ε and apply an appropriate implicit function theorem.

EXERCISE 1.5. For $\sigma = 2/n$, n a positive integer, show that (1.16) possessing a subharmonic of order n implies that λ must be $O(\varepsilon^{n-2})$ as $\varepsilon \to 0$. You cannot possibly find n terms in the Taylor expansion of $G(a, \varepsilon)$ so you must discuss the qualitative properties of the expansion.

EXERCISE 1.6. Using the method of Exercise 1.5, show that the Duffing equation

$$\dot{x}_1 = x_2,$$
$$\dot{x}_2 = -(2n+1)^{-2}x_1 - \varepsilon^s c x_2 + \varepsilon a x_1 + \varepsilon b x_1^3 + B \cos t,$$

can have a subharmonic solution of order $2n + 1$ only if $s \geq n$.

EXERCISE 1.7. Suppose $f(x, y)$ is a continuously differentiable scalar function of the scalars x, y, $f(0, 0) = 0$ and either $f(x, -y) = f(x, y)$ or $f(-x, y) = -f(x, y)$ for all x, y. Given any $\alpha > 0$, show there is an $\varepsilon_0 > 0$ such that for any (x_0, y_0) with $x_0^2 + y_0^2 < \alpha^2$, there is a periodic solution of the system

(1.17) $\dot{x} = y$,

$$\dot{y} = -x + \varepsilon f(x, y),$$

for $|\varepsilon| \leq \varepsilon_0$. This implies the equilibrium point $(0, 0)$ is a center. Let

$x = \rho \sin \theta$, $y = \rho \cos \theta$ to obtain the equation

$$\frac{d\rho}{d\theta} = \varepsilon F(\rho, \theta, \varepsilon)$$

for the orbits of (1.17). If $f(-x, y) = -f(x, y)$, observe that $F(\rho, -\theta, \varepsilon) = -F(\rho, \theta, \varepsilon)$. Show that the transformation $\mathscr{F}$ in (1.7) in this special case maps even 2π-periodic functions of θ into even functions of θ and, therefore, the fixed point must be even. This implies $G(a, \varepsilon) \equiv 0$. If $f(x, -y) = f(x, y)$, then let $\theta = \phi + \pi/2$ and apply the same argument.

EXERCISE 1.8. Consider the system

$$\dot{x}_1 = x_2,$$
$$\dot{x}_2 = -\mu x_1^3,$$

where $\mu > 0$ is a small parameter. A first integral of this equation is $E(x_1, x_2) = x_2^2/2 + \mu x_1^4/4$ and, thus, all of the orbits are periodic orbits. For $\mu = 0$, the only periodic orbits are the equilibrium points which lie on the x_1-axis. Can you deduce this result by using the above perturbation theory? If $\varepsilon = \sqrt{\mu}$, $x_1 = z_1$, and $x_2 = \varepsilon z_2$, then $\dot{z}_1 = \varepsilon z_2$, $\dot{z}_2 = -\varepsilon z_1^3$ which is a special case of system (1.1).

The above theory is also useful for determining the characteristic exponents of certain types of linear systems with periodic coefficients. More specifically, consider the system of equations

(1.18) $\dot{w} = Cw + \varepsilon \Phi(t)w$

where ε is a parameter, $0 \leq |\varepsilon| \leq \varepsilon_0$, w is an $(n + m)$-vector, $\Phi(t) = \Phi(t + T)$, $T = 2\pi/\omega$, is a continuous[1] $(n + m) \times (n + m)$ matrix and $C = \text{diag}(C_1, C_2)$ is an $(n + m) \times (n + m)$ constant matrix such that all eigenvalues of the $n \times n$ matrix $e^{C_1 T}$ are ρ_0 and no eigenvalue of $e^{C_2 T}$ is ρ_0. The problem is to determine the characteristic multipliers of (1.18) which are close to ρ_0 for $\varepsilon \neq 0$.

Suppose λ_1 is an eigenvalue of C_1 and then $\rho_0 = e^{\lambda_1 T}$. From the continuity of the principal matrix solution of (1.18) in ε and the Floquet theory, it follows that there is a multiplier $\rho(\varepsilon) = e^{\mu(\varepsilon)T}$ which is continuous in ε and $\mu(0) = \lambda_1$. Furthermore, to this multiplier there must exist a T-periodic $(n + m)$-vector $p(t, \varepsilon)$ such that $e^{\mu(\varepsilon)t}p(t, \varepsilon)$ is a solution of (1.18). Conversely, any such solution of (1.18) yields a characteristic exponent of (1.18). Therefore, if one makes the transformation $w = e^{\mu t}p$ in (1.18), then

(1.19) $\dot{p} = (C - \mu I)p + \varepsilon \Phi(t)p,$

---

[1] Φ could be an integrable function.

and the problem is to determine μ in such a way that (1.19) has a T-periodic solution.

In order to make the previous theory directly applicable, suppose that the eigenvalues of C_1 have simple elementary divisors. Then the matrix $e^{(C_1-\lambda_1 I)t}$ is T-periodic. If $w = (u, v)$, where u is an n-vector, v is an m-vector, and β is any complex number, the transformation

$$(1.20) \qquad u = e^{(\lambda_1+\varepsilon\beta)t}e^{(C_1-\lambda_1 I)t}x,$$
$$v = e^{(\lambda_1+\varepsilon\beta)t}y,$$

in (1.18) yields

$$(1.21) \qquad \dot{x} = -\varepsilon\beta x + \varepsilon e^{-(C_1-\lambda_1 I)t}\Phi_{11}(t)e^{(C_1-\lambda_1 I)t}x + \varepsilon e^{-(C_1-\lambda_1 I)t}\Phi_{12}y,$$
$$\dot{y} = [C_2 - (\lambda_1 + \varepsilon\beta)I]y + \varepsilon\Phi_{21}(t)e^{(C_1-\lambda_1 I)t}x + \varepsilon\Phi_{22}(t)y,$$

where we have partitioned Φ as $\Phi = (\Phi_{ij})$.

Since $e^{(C_1-\lambda_1 I)t}$ is T-periodic, system (1.21) can be written as

$$(1.22) \qquad \dot{z} = Bz + \varepsilon\Psi(t, \beta)z,$$

where $B = \text{diag}(0_n, C_2 - \lambda_1 I)$, $z = (x, y)$ and $\Psi(t + T, \beta) = \Psi(t, \beta)$ for all t, β. The explicit expression for Ψ is easily obtained from (1.21). If β can be determined in such a way that system (1.22) has a T-periodic solution $p(t)$, then this solution yields from (1.20) a solution of (1.18) of the form

$$w(t) = e^{(\lambda_1+\varepsilon\beta)t}p(t), \qquad p(t + T) = p(t).$$

Therefore, $\lambda_1 + \varepsilon\beta$ is a characteristic exponent of (1.18). Conversely, we have seen above that every such characteristic exponent can be obtained in this way.

Since system (1.22) is a special case of system (1.10), we may apply the preceding theory. The function z^* given in Lemma 1.4 will now be a function of a, β and ε and the linearity of (1.22) implies that $z^*(a, \beta, \varepsilon) = Z^*(\beta, \varepsilon)a$, where $Z^*(\beta, \varepsilon)$ is a T-periodic $(n + m) \times n$ matrix. If $Z^* = \text{col}(X^*, Y^*)$ where X^* is $n \times n$ then the function $G(a, \beta, \varepsilon)$ in (1.15) is also linear in a; that is, $G(a, \beta, \varepsilon) = D(\beta, \varepsilon)a$, where $D(\beta, \varepsilon)$ can be computed directly from (1.22) as

$$(1.23) \qquad D(\beta, \varepsilon) = \frac{1}{T}\int_0^T [\Psi_{11}(t, \beta)X^*(\beta, \varepsilon)(t) + \Psi_{12}(t, \beta)Y^*(\beta, \varepsilon)(t)]\, dt,$$

where Ψ is partitioned as $\Psi = (\Psi_{ij})$. Lemma 1.4 then implies that any β for which $\det D(\beta, \varepsilon) = 0$ will yield a characteristic exponent $\lambda_1 + \varepsilon\beta$ of (1.18).

EXERCISE 1.9. Suppose $D(\beta, \varepsilon)$ is given in (1.23). Prove that the n characteristic multipliers of (1.18) which for $\varepsilon = 0$ are equal to ρ_0 are the n roots of the equation $\det D(\beta, \varepsilon) = 0$.

EXERCISE 1.10.   Consider the system

$$(1.24) \qquad \dot{w} = Cw + \varepsilon\Phi(t)w,$$

where $\Phi(t + T) = \Phi(t)$ is continuous, $C = \mathrm{diag}(\lambda_1, \ldots, \lambda_n)$ and for some fixed j,

$$e^{\lambda_j T} \neq e^{\lambda_k T}, \qquad k = 1, 2, \ldots, n, \qquad k \neq j.$$

Show that the characteristic multiplier $\rho(\varepsilon)$ of (1.24) such that $\rho(0) = e^{\lambda_j T}$ is given by

$$\rho(\varepsilon) = e^{\mu(\varepsilon)T}, \qquad \mu(\varepsilon) = \lambda_j + \frac{\varepsilon}{T}\int_0^T \phi_{jj}(t)\, dt + o(\varepsilon) \qquad \text{as } \varepsilon \to 0,$$

where $\Phi = (\phi_{ik})$, $i, k = 1, 2, \ldots, n$.

EXERCISE 1.11.   Suppose that the system

$$\ddot{x} + x = \varepsilon f(x, \dot{x}, y, \dot{y}),$$
$$\ddot{y} + \sigma^2 y = \varepsilon g(x, \dot{x}, y, \dot{y}),$$

has a periodic solution of period $2\pi/\omega(\varepsilon)$, $\omega(0) = 1$, which for $\varepsilon = 0$ is given by $x = a\sin t$, $y = 0$. Prove that this solution is asymptotically orbitally stable with asymptotic phase if

$$\varepsilon \int_0^{2\pi} \frac{\partial f}{\partial \dot{x}}(a\sin t, a\cos t, 0, 0)\, dt < 0,$$

$$\varepsilon \int_0^{2\pi} \frac{\partial g}{\partial \dot{y}}(a\sin t, a\cos t, 0, 0)\, dt < 0,$$

and $\sigma \neq$ an integer. The characteristic multipliers of the linear variational equation relative to this periodic solution are given by $1, 1, e^{2\pi i\sigma}, e^{-2\pi i\sigma}$ for $\varepsilon = 0$. The hypothesis on σ implies that the roots $e^{2\pi i\sigma}$ and $e^{-2\pi i\sigma}$ are simple. Therefore, Exercise 1.10 can be applied to obtain the first order change in these multipliers with ε provided the constant part of the variational equation is transformed to a diagonal form. Since one multiplier remains identically one for $\varepsilon \neq 0$ and the product of the multipliers is given by a well known formula, one can evaluate the first order change in the other multiplier.

EXERCISE 1.12.   Generalize the result of Exercise 1.11 to a system of n-second order equations.

EXERCISE 1.13.   Show that the system

$$\ddot{x} + x = \varepsilon(1 - x^2 - y^2)\dot{x},$$
$$\ddot{y} + 2y = \varepsilon(1 - x^2 - y^2)\dot{y}, \qquad \varepsilon > 0,$$

has two nonconstant periodic solutions both of which are asymptotically orbitally stable with asymptotic phase.

EXERCISE 1.14.   Consider the Mathieu equation

(1.25) $\dot{x}_1 = x_2$,

                                $\dot{x}_2 = -\sigma^2 x_1 - \varepsilon(\cos 2t)x_1$,

where $\sigma = \sigma(\varepsilon)$ and $\sigma(0) = m$, a nonnegative integer. From the general theory of this equation in Chapter III, we know that these values of σ are precisely the ones which may give rise to instability. In fact, the instability zones are determined from those values of σ for which equation (1.25) has a periodic solution of period π or 2π; that is, the multipliers are $+1$ or -1. Determine approximately these values of $\sigma(\varepsilon)$ as a function of ε for the case when $\sigma(0) = 1$, $\sigma(0) = 2$. Are the solutions unbounded in a neighborhood of the points $(1, 0)$, $(2, 0)$ in the (σ, ε)-plane? Let

$$x_1 = z_1 \sin mt + z_2 \cos mt,$$

$$x_2 = m[z_1 \cos mt - z_2 \sin mt],$$

and $m^2 - \sigma^2 = \varepsilon\beta$ and apply the above theory for determining β in such a way that the resulting equations have a periodic solution.

EXERCISE 1.15.   For $\sigma(0) = m$ in Exercise 1.14, show that $\sigma^2(\varepsilon) = m^2 + 0(\varepsilon^m)$ as $\varepsilon \to 0$.

EXERCISE 1.16.   Discuss as in Exercise 1.14 the equation (1.25) with $\sigma(0) = 0$. Suppose $\varepsilon \geqq 0$, let $\sigma^2 = \varepsilon\beta$, $x_1 = z_1$, $x_2 = \sqrt{\varepsilon}z_2$ and analyze the resulting equations for periodic solutions.

EXERCISE 1.17.   For what values of ω are all of the solutions of the equation

$$\ddot{x} + \sigma^2 x = \varepsilon(\sin \omega t)y,$$

$$\ddot{y} + \mu^2 y = \varepsilon(\cos \omega t)x,$$

bounded for ε small and $\neq 0$?

Let us use the same ideas as above to discuss the existence of periodic solutions for equations which may contain several independent parameters rather than just a single parameter ϵ as in (1.1), (1.10). Consider the equation

(1.26) $\dot{x} = f(t, x, \lambda)$

where λ in R^k is a parameter, x is in R^n, $f(t, x, \lambda)$ is continuous in t, x, λ together with at least first derivatives in x, λ, $f(t + T, x, \lambda) = f(t, x, \lambda)$,

(1.27) $f(t, 0, 0) = 0$, $\partial f(t, 0, 0)/\partial x = 0$

Equation (1.26) is a generalization of (1.1) since, for example, for $\lambda = (\lambda_1, \lambda_2)$, it could be of the form

$$\dot{x} = \lambda_1 f^{(1)}(t,x) + \lambda_2 f^{(2)}(t,x)$$

which makes (1.1) correspond to $f^{(2)}(t,x) = 0$. Equation (1.26) could also be of the form

$$\dot{x} = f^{(0)}(x) + \sum_{j=1}^{k} \lambda_j f^{(j)}(t,x)$$

where $f^{(0)}(0) = 0$, $\partial f^{(0)}(0)/\partial x = 0$. If $f^{(0)}(x) \not\equiv 0$, this latter equation corresponds to an $f(t,x,\lambda)$ which can be made small together with its first derivative by choosing λ small and x small. In problems of this type, we will be interested in T-periodic solutions with small norm.

Lemmas 1.1, 1.2 have an analogue for system (1.26), (1.27) which we state without proof since the proof will be the same.

LEMMA 1.3. Suppose $P, \mathscr{H}$ are defined in (1.2), (1.4). Then $x(t)$ is a T-periodic solution of (1.26) if and only if

(1.28)
$$\begin{aligned} &\text{(a)} \quad x = a + \mathscr{H}(I - P)f(\cdot, x(\cdot), \lambda) \\ &\text{(b)} \quad Pf(\cdot, x(\cdot), \lambda) = 0 \end{aligned}$$

where a is a constant n-vector given by $a = Px$.

LEMMA 1.4. If (1.27) is satisfied, then there are $\lambda_0 > 0$, $\alpha_0 > 0$ such that, for any a in R^n, $|a| \leq \alpha_0$, any λ in R^k, $|\lambda| \leq \lambda_0$, there is a unique T-periodic function $x^* = x^*(a,\lambda)$ continuous together with its first derivatives, satisfying (1.28a), $x^*(0,0) = 0$. If there exist (a,λ) such that $|a| \leq \alpha_0$, $|\lambda| \leq \lambda_0$ and

(1.29)
$$G(a,\lambda) \overset{\text{def}}{=} \frac{6}{T} \int_0^T f(t, x^*(a,\lambda)(t), \lambda) dt = 0$$

then $x^*(a,\lambda)$ is a T-periodic solution of (1.26). Conversely, if (1.26) has a T-periodic solution $\bar{x}(\lambda)$ which is continuous in λ and has $P\bar{x}(\lambda) = a(\lambda)$, $|a(\lambda)| \leq \alpha_0$, $|\lambda| \leq \lambda_0$, then $\bar{x}(\lambda) = x^*(a(\lambda),\lambda)$ where $x^*(a,\lambda)$ is the function given above and $a(\lambda)$ satisfies (1.29) for $0 \leq |\lambda| \leq \lambda_0$.

One can also give an analogous extension of Lemmas 1.1, 1.2 to a system of the form

(1.30)
$$\dot{x} = f(t,x,y,\lambda)$$
$$\dot{y} = B_1 y + g(t,x,y,\lambda)$$

where $e^{B_1 T} - I$ is nonsingular if one supposes f,g and their derivatives with respect to x,y vanish at $x = 0$, $y = 0$, $\lambda = 0$. We do state the results explicitly.

Having the method formulated so as to apply to (1.26) and (1.30) gives the opportunity to discuss more general problems as well as to discuss old problems more completely. We illustrate this with some examples.

The simplest example is the so-called *Hopf bifurcation*. Suppose λ is a scalar, $A(\lambda)$ is a x X 2 matrix continuously differentiable in λ with the eigenvalues of $A(0)$ being $\pm i$. Then there is a neighborhood of $\lambda = 0$ for which the matrix $A(\lambda)$ has two complex conjugate eigenvalues $\mu(\lambda) \pm i\nu(\lambda)$, continuous and continuously differentiable in λ such that $\mu(0) = 0$, $\nu(0) = i$. Let us suppose $d\mu(0)/d\lambda \neq 0$. This hypothesis implies that we may redefine the eter λ in a neighborhood of zero by replacing λ by $\mu(\lambda)$, $\beta(\lambda) = -[\nu(\mu^{-1}(\lambda))]^2$ and assume $A(\lambda)$ has the form

(1.31)
$$A(\lambda) = \begin{pmatrix} \lambda & \beta(\lambda) \\ -\beta(\lambda) & \lambda \end{pmatrix}, \qquad \beta(0) = 1$$

Consider the second order equation

(1.32)
$$\dot{x} = A(\lambda)x + f(x,\lambda)$$

where

(1.33)
$$f(0,\lambda) = 0, \qquad \partial f(0,\lambda)/\partial x = 0$$

and $A(\lambda)$ is given in (1.31).

The *Hopf Bifurcation Theorem* is stated explicitly in the following way.

THEOREM 1.3.   Consider system (1.31), (1.32), (1.33) and suppose f has continuous second derivatives in x. Then there are $\lambda_0 > 0$, $a_0 > 0$, $\delta_0 > 0$, scalar functions $\lambda^*(a)$, $\omega^*(a)$ and an $\omega^*(a)$-periodic function $x^*(a)$ with all functions being continuously differentiable in a for $|a| < a_0$ such that $x^*(a)$ is a solution of the equation

$$\dot{x} = A(\lambda^*(a))x + f(x,\lambda^*(a))$$

and

$$\lambda^*(0) = 0, \qquad \omega^*(0) = 2\lambda,$$

$$x^*(a) = \begin{pmatrix} a\cos t \\ -a\sin t \end{pmatrix} + o(|a|).$$

Furthermore, for $|\lambda| < \lambda_0$, $|\omega - 2\pi| < \delta_0$, every ω-periodic solution of (1.32) with norm $< \delta_0$ must be of the above type except for a translation in phase.

PROOF.   Since we are in the plane, every periodic orbit of (1.32) must encircle an equilibrium point. The hypothesis on $A(\lambda)$ implies $\det A(\lambda) \neq 0$ for λ close to zero. Thus, the solution $x = 0$ of (1.32) is isolated for $|\lambda|$ sufficiently small and any solution of (1.32) with sufficiently small norm must encircle $x = 0$. This justifies the introduction of polar coordinates

(1.34) $$x_1 = \rho \cos \theta, \qquad x_2 = -\rho \sin \theta$$

to obtain the equivalent system

(1.35)
$$\dot{\theta} = \beta(\lambda) - \frac{1}{\rho}[f_1 \sin \theta + f_2 \cos \theta]$$
$$\dot{\rho} = \lambda \rho + f_1 \cos \theta - f_2 \sin \theta$$

For $|\lambda|, |\rho|$ small, $\dot{\theta} > 0$ and so we can eliminate t to obtain

(1.36) $$\frac{d\rho}{d\theta} = \alpha(\lambda)\rho + R(\theta, \rho, \lambda)$$

where $R(\theta + 2\pi, \rho, \lambda) = R(\theta, \rho, \lambda)$, $R(\theta, 0, \lambda) = 0$, $\partial R(\theta, 0, \lambda)/\partial \rho = 0$, $\alpha(0) = 0$, $\alpha'(0) = 1$.

Finding periodic solutions of period 2π of (1.36) is equivalent to finding periodic orbits of the original equation (1.32). If $\rho(\theta)$ is a 2π-periodic solution of (1.36), one obtains the period of the periodic orbit $x_1 = \rho(\theta)\cos\theta$, $x_2 = -\rho(\theta)\sin \theta$ by finding that value of ω for which $\theta(\omega) = 2\pi$ where $\theta(t), \theta(0) = 0$, satisfies the first equation in (1.35) with $\rho = \rho(\theta)$.

Lemma 1.4 is applicable directly to (1.36). Let $\rho^*(\theta, a, \lambda)$ be the unique solution for a, λ small of the equation

$$\rho = a + \mathscr{K}(I - P)[\alpha(\lambda)\rho + R(\cdot, \rho, \lambda)]$$

Then $\rho^*(\theta, 0, \lambda) = 0$ since $x = 0$ is a solution of (1.32). The bifurcation function $G_H(a, \lambda)$ in (1.29) is

$$G_H(a, \lambda) = \frac{1}{2\pi}\int_0^{2\pi}[\alpha(\lambda)\rho^*(\theta, a, \lambda) + R(\theta, \rho^*(\theta, a, \lambda), \lambda)]\, d\theta.$$

Since $\rho^*(\theta, 0, \lambda) = 0$, we have $G_H(0, \lambda) = 0$. If we define the function $\bar{G}_H(a, \lambda) = G_H(a, \lambda)/a$, then $\bar{G}_H(0, 0) = 0$, $\partial \bar{G}_H(0, 0)/\partial \lambda = 1$. The Implicit Function Theorem implies there is a unique $\lambda^*(a)$, $|a| < a_0$ such that $\lambda^*(0) = 0$ and $\bar{G}_H(\lambda^*(a), a) = 0$. The fact that this function $\lambda^*(a)$ leads to functions $\omega^*(a), x^*(a)$ as stated in the theorem is left for the reader to verify.

EXERCISE 1.18.   Verify that $G_H(a, \lambda)$ in (1.37) is an odd function of a.

From Exercise 1.18, $G_H(a, \lambda) = a g_H(a^2, \lambda)$. Suppose $A(\lambda), f(x, \lambda)$ in (1.32) have derivatives continuous up through order four and

(1.38) $$g_H(r, \lambda) = \alpha(\lambda) - \gamma(\lambda)r + o(|r|) \qquad \text{as } |r| \to 0$$

where $\gamma(0) = \gamma_0 \neq 0$; that is, $G_H(a, \lambda) = \alpha(\lambda)a - \gamma(\lambda)a^3 + o(|a|^3)$ as $|a| \to 0$. Then, $g_H(r, \lambda)$ satisfies $g_H(0, 0) = 0$, $\partial g_H(0, 0)/\partial r = -\gamma_0 \neq 0$. The Implicit

Function Theorem implies there is a unique $r^*(\lambda), |\lambda| < \lambda_0$, continuously differentiable in λ, $r^*(0) = 0$, $r^*(\lambda) \neq 0$ for $0 < |\lambda| < \lambda_0$ and $g_H(r^*(\lambda), \lambda) = 0$. If $r^*(\lambda) > 0$ for $0 < |\lambda| < \lambda_0$, then we can define $a^*(\lambda)$ by the relation $a^2 = r^*(\lambda)$ to obtain $G_H(a^*(\lambda), \lambda) = 0$ for $|\lambda| < \lambda_0$. Furthermore, $\pm\sqrt{r^*(\lambda)}$ and $a = 0$ are the only solutions of $G(a, \lambda) = 0$ for $|\lambda| < \lambda_0$, $|a| < a_0$. The condition $r^*(\lambda) > 0$ for $0 < |\lambda| < \lambda_0$ is equivalent to $(\operatorname{sgn}\gamma_0)\lambda > 0$, $0 < |\lambda| < \lambda_0$. Furthermore, $+\sqrt{r}(\lambda)$ and $-\sqrt{r^*(\lambda)}$ correspond to the same periodic orbit of (1.32). Consequently, there is a unique nonconstant periodic solution of (1.32) in a neighborhood of $x = 0$ if the following conditions are satisfied

(1.39)
$$\begin{aligned} \gamma_0 > 0, \qquad & 0 < \lambda < \lambda_0 \\ \gamma_0 < 0, \qquad & -\lambda_0 < \lambda < 0 \end{aligned}$$

Let us now investigate the stability of this periodic orbit. It is sufficient to investigate the stability of the corresponding 2π-periodic solution $\rho^*(a^*(\lambda), \lambda)$ of Equation (1.36). We will use averaging to determine the stability properties.

Make a 2π-periodic transformation of variables in (1.36) of the form

(1.40)
$$\rho = s + u_1(\theta)s^2 + u^2(\theta)s^3$$

and choose $u_1(\theta)$, $u_2(\theta)$ to be 2π-periodic and choose a constant γ_0 so that the differential equation for s has the form

(1.41)
$$\frac{ds}{d\theta} = \lambda s - \gamma_0 s^3 + S(\theta, s, \lambda)$$

where $S(\theta, s, \lambda) = o(|s|(|\lambda| + s^2))$ as $|\lambda| \to 0$, $|s| \to 0$. This is always possible.

To be specific in the discussion of stability suppose $\gamma_0 > 0$. Equation (1.41) can be considered as a perturbation of the equation

(1.42)
$$\dot{y} = -\gamma_0 y^3$$

which has $y = 0$ uniformly asymptotically stable. Thus, there is a neighborhood of $y = 0$ say $|y| < \delta$ such that for $|\lambda| < \lambda_0(\delta)$, every solution $y(\theta)$ of (1.41) with initial value satisfying $|y(0)| \leq \delta$ must satisfy $|y(\theta)| \leq \delta$ for $t \geq 0$. Suppose $|y(0)| \leq \delta$. Since (1.41) is 2π-periodic, $y(\theta + 2\pi)$ is also a solution and $y(0) < y(2\pi)$ implies $y(2\pi k) < y(2\pi(k + 1))$ for $k = 0,1,2,\ldots$. Thus, $y(2\pi k) \to \bar{y}_0$ as $k \to \infty$ and it is easy to show that the solution of (1.41) with initial value $\bar{y}_0$ is 2π-periodic. A similar argument holds if $y(0) > y(2\pi)$. Since the equation is 2π-periodic, this implies that, for every solution $y(t)$ of (1.41) with $|y(0)| \leq \delta$, $|\lambda| \leq \lambda_0$, there is a 2π-periodic solution $\phi_y(\theta)$ such that $y(\theta) - \phi_y(\theta) \to 0$, as $\theta \to \infty$. This means the corresponding periodic orbit of (1.32) in R^2 corresponding to $\phi_y(\theta)$ is attracting the orbit corresponding to $y(\theta)$. Since $\gamma_0 \neq 0$, there is only one nontrivial periodic orbit. Since $\gamma_0 > 0$, the trivial solution is always unstable for $0 < \lambda < \lambda_0$ and we have the nontrivial periodic orbit is asymptotically orbitally stable.

If $\gamma_0 < 0$, one argues in a similar way to obtain the nontrivial periodic orbit is unstable. We summarize these remarks as follows.

COROLLARY 1.1.   Suppose $A(\lambda), f(z,\lambda)$ have derivatives continuous up through order four and suppose $G_H(a,\lambda)$ in (1.37) satisfies $G_H(a,\lambda) = ag_H(a^2,\lambda)$ where $g_H(r,\lambda)$ satisfies

$$g_H(r,\lambda) = \lambda - \gamma(\lambda)r + o(|r|) \qquad \text{as } r \to 0$$

$$\gamma(0) = \gamma_0 \neq 0.$$

Then there is a neighborhood of $\lambda = 0$, $x = 0$ in which equation (1.32) has the following properties:

(i) $\gamma_0 > 0, -\lambda_0 < \lambda \leq 0,$ $x = 0$ is asymptotically stable and there is no periodic orbit.

$\qquad\quad$ $0 < \lambda < \lambda_0,$ $x = 0$ is unstable and there is an asymptotically orbitally stable periodic orbit.

(ii) $\gamma_0 < 0, -\lambda_0 < \lambda < 0,$ $x = 0$ is asymptotically stable and there is an unstable periodic orbit.

$\qquad\quad$ $0 \leq \lambda < \lambda_0,$ $x = 0$ is unstable and there is no periodic orbit.

EXERCISE 1.19.   Prove there exists asymptotic phase in Corollary 1.1. *Hint:* Show the characteristic exponent for the solution s of (1.41) is not zero.

EXERCISE 1.20.   Suppose $G_H(a,\lambda)$ in (1.37) satisfies $G_H(a,\lambda) = ag_H(a^2,\lambda)$ with $g(r,0) = cr^m + o(|r|^m)$ as $r \to 0$ for some $c \neq 0$ and integer $m \geq 1$. State and prove the analogue of Corollary 1.1.

EXERCISE 1.21.   Generalize the Hopf bifurcation theorem to a system

$$\dot{x} = A(\lambda)x + f(x,y,\lambda)$$

$$\dot{y} = B(\lambda)y + g(x,y,\lambda)$$

where f, g, and their partial derivatives vanish for $x = 0$, $y = 0$, x in R^2, y in R^n. Suppose $A(\lambda)$ as before and $\det[I - \exp B(0)2\pi] \neq 0$. Give conditions which will ensure there is a unique nonconstant period orbit for a given λ sufficiently small. Discuss the stability properties of this orbit.

EXERCISE 1.22.   Consider the scalar equation $\dot{x} = x^2 + f(t,x,\lambda)$ where $f(t + 1,x,\lambda) = f(t,x,\lambda)$ has continuous derivatives up through order two in

$x, \lambda, f(t, x, 0) = o(|x|^2)$ as $|x| \to 0$, $f(t, 0, \lambda) = 0(|\lambda|)$ as $\lambda \to 0$. Prove there is a $\lambda_0 > 0$, $\delta_0 > 0$, and a continuously differentiable function $\gamma(\lambda)$, $|\lambda| < \lambda_0$, $\gamma(0) = 0$, such that the following properties hold with respect to 1-periodic solutions of the above equation with norm $< \delta_0$:

(i) $\gamma(\lambda) > 0$ implies no 1-periodic solution
(ii) $\gamma(\lambda) = 0$ implies one 1-periodic solution
(iii) $\gamma(\lambda) < 0$ implies two 1-periodic solutions.

Note that your results are true for λ a parameter in a Banach space. Discuss the stability properties in case (ii). In case (iii), if $x_1(t, \lambda) < x_2(t, \lambda)$ are the 1-periodic solutions, show that $x_1(t, \lambda)$ is uniformly asymptotically stable and $x_2(t, \lambda)$ is unstable. *Hint*: Obtain the bifurcation function $G(a, \lambda)$ from (1.29) and observe that $G(a, 0) = a^2 + o(|a|^2)$ as $|a| \to 0$. Thus, $G(a, \lambda)$ has a minimum at some $a^*(\lambda)$ for $|\lambda|$ small, $a^*(0) = 0$. Let $\gamma(\lambda) = G(a^*(\lambda), \lambda)$. For case (ii), note that every bounded solution of the differential equation must approach a 1-periodic solution. At a zero a of $G(a, \lambda)$, note that the sign of $\partial G(a, \lambda)/\partial a$ is the same as the sign of the characteristic exponent of the corresponding periodic solution if this characteristic exponent is not zero. If two distinct periodic solutions exist and one characteristic exponent is zero, show that an appropriate small perturbation of the vector field will yield at least three periodic solutions near zero, which is a contradiction.

EXERCISE 1.23.   Give an appropriate generalization of Exercise 1.22 to the equation in R^2,

$$\dot{x} = x^2 + f(t, x, y, \lambda)$$
$$\dot{y} = -y + g(t, x, y, \lambda)$$

EXERCISE 1.24.   Consider the scalar equation

$$\dot{x} = -x^3 + \lambda_1 f_1(t, x) + \lambda_2 f_2(t, x)$$

where $\lambda = (\lambda_1, \lambda_2)$ is in R^2, $f_j(t + 1, x) = f_j(t, x)$ for all $t, x, j = 1, 2$,

$$\gamma_1 = \int_0^1 f_2(t, 0) dt \neq 0$$

$$f_1(t, 0) = 0, \qquad \gamma_2 = \int_0^1 [\partial f_1(t, 0)/\partial x] \, dt \neq 0$$

Discuss the existence of 1-periodic solutions in a neighborhood of $x = 0$, $\lambda = 0$. Show there is a cusp Γ in λ-space such that on one side of Γ there is one 1-periodic solution and the other side there are three. Show the cusp is approximately given by $\lambda_2^2 = (4\gamma_1^3/27\gamma_2^2)\lambda_1^3$. *Hint*: Show that the bifurcation function $G(a, \lambda)$ is approximately given by $-a^3 + \lambda_2\gamma_2 a + \lambda_1\gamma_1$ and determine those λ_1, λ_2 as functions of a so that $G(a, \lambda) = 0$, $\partial G(a, \lambda) = 0$.

EXERCISE 1.25.   Discuss the stability of the periodic solution in Exercise 1.24.

EXERCISE 1.26.   Generalize Exercise (1.24) to the equation

$$\dot{x} = -x^3 + f(t,x,\lambda)$$

where λ is a parameter varying in a Banach space.

EXERCISE 1.27.   Generalize Exercise (1.24) to the planar system

$$\dot{x} = -x^3 + f(t,x,\lambda)$$
$$\dot{y} = -y + g(t,x,\lambda)$$

where $\lambda = (\lambda_1, \lambda_2)$ is in R^2.

VIII.2. Almost Linear Systems

In this section, the general perturbation scheme given in Section 1 will be generalized to the system

$$(2.1) \qquad\qquad \dot{x} = B(t)x + \varepsilon f(t, x),$$

where $B(t + T) = B(t), f(t + T, x) = f(t, x)$ and $\partial f(t, x)/\partial x$ are continuous for all t in R, x in C^n.

Let the columns of $\Phi(t)$, a matrix of dimension $n \times p$, be a basis for the T-periodic solutions of

$$(2.2) \qquad\qquad \dot{x} = B(t)x,$$

and let the rows of $\Psi(t)$, a matrix of dimension $p \times n$, be a basis for the T-periodic solutions of the adjoint equation

$$(2.3) \qquad\qquad \dot{y} = -yB(t).$$

The $p \times p$ matrices ($'$ denotes transpose)

$$(2.4) \qquad C = \int_0^T \Phi'(t)\Phi(t)\, dt, \qquad D = \int_0^T \Psi(t)\Psi''(t)\, dt$$

are nonsingular. In fact, if $q(t) = \Phi(t)a$ and $Ca = 0$, then $\int_0^T q'(t)q(t)\, dt = 0$ which implies $q(t) = 0$ for all t in $[0, T]$. But this implies $a = 0$. The same argument holds for the matrix D.

As before, let $\mathscr{P}_T$ be the space of continuous T-periodic n-vector functions with the uniform topology. The fact that C, D in (2.4) are nonsingular allows one to define two projection operators P, Q on $\mathscr{P}_T$ in the following way:

$$(2.5) \qquad Pf = \Phi(\cdot)a, \qquad a = C^{-1} \int_0^T \Phi'(t) f(t) \, dt,$$

$$Qf = \Psi'(\cdot)b, \qquad b = D^{-1} \int_0^T \Psi(t) f(t) \, dt.$$

It is easy to check that these are projections. Notice that P takes $\mathscr{P}_T$ onto the subspace of $\mathscr{P}_T$ spanned by the T-periodic solutions of (2.2) and Q takes $\mathscr{P}_T$ onto the subspace of $\mathscr{P}_T$ spanned by the transpose of the T-periodic solutions of (2.3). The justification for these definitions lies in the following lemma.

LEMMA 2.1. If f is a given element of $\mathscr{P}_T$, then a necessary and sufficient condition that the equation

$$(2.6) \qquad \dot{x} = B(t)x + f(t)$$

has a T-periodic solution is that $Qf = 0$. If $Qf = 0$, then there is a unique T-periodic solution $\mathscr{K}f$ such that $P\mathscr{K}f = 0$. Furthermore, $\mathscr{K}(I - Q)$ is a continuous linear operator taking $\mathscr{P}_T$ into $\mathscr{P}_T$.

PROOF. The first part of the lemma is a restatement of Lemma IV.1.1. If $Qf = 0$, then there is a solution of (2.6) in $\mathscr{P}_T$. If x_0 is the initial value of any solution in $\mathscr{P}_T$, then x_0 must satisfy the equation $Ex_0 = b$, where $E = X(T) - I$, $b = \int_0^T X(T)X^{-1}(s)f(s)ds$ and $X(t)$ is the principal matrix solution of $\dot{x} = B(t)x$. Let E^* be a right inverse of E; that is, a matrix taking the range of E into C^n such that $EE^* = I$. Since $Qf = 0$ implies b is in the range of E, the vector $x_0 = E^*b$ corresponds to the initial value of a solution $x^*(f)$ of (2.6) in $\mathscr{P}_T$. If $\mathscr{K}f = (I - P)x^*(f)$ then the operator $\mathscr{K}$ takes $\mathscr{P}_T$ into $\mathscr{P}_T$ and is clearly linear and continuous. Also, $P\mathscr{K}f = 0$ and $\mathscr{K}f$ is the only solution of (2.6) in $\mathscr{P}_T$ whose P-projection is zero. This completes the proof of Lemma 2.1.

COROLLARY 2.1. If f is in $\mathscr{P}_T$ and a is a given p-dimensional vector, then the unique solution of $\dot{x} = B(t)x + (I - Q)f$ with $Px = \Phi(\cdot)a$ is given by $\Phi a + \mathscr{K}(I - Q)f$.

Example 2.1. Suppose $B = B(t)$ is a constant $n \times n$ matrix, $B = \text{diag}(0_p, B_1)$ where 0_p is the p-dimensional zero matrix and $e^{B_1 T} - I$ is nonsingular. Then Φ, Ψ may be chosen as

$$\Phi = \begin{bmatrix} I_p \\ 0 \end{bmatrix}, \qquad \Psi = \Phi' = [I_p, 0],$$

where I_p is the $p \times p$ identity matrix. It is easy to see that the matrices C, D in (2.4) and operators P, Q in (2.5) are given by $C = D = TI_p$

$$Pf = Qf = \begin{bmatrix} \dfrac{1}{T} \displaystyle\int_0^T f_p(t)\,dt \\[2mm] 0 \end{bmatrix},$$

where f_p denotes the first p components of f. This is the same operator defined in Section 1 for equation (1.10). If $Qf = 0$, then the explicit expression for $\mathscr{K}f$ is

$$(\mathscr{K}f)(t) = \begin{bmatrix} \displaystyle\int^t f_p(s)\,ds \\[3mm] (e^{-B_1 T} - I)^{-1} \displaystyle\int_0^T e^{-B_1 s} f_{n-p}(t+s)\,ds \end{bmatrix},$$

where $\int^t f_p$ is the unique primitive of f_p of mean value zero, and f_{n-p} denotes the last $n - p$ components of f.

Example 2.2. Suppose $B = B(t)$ is constant and

$$B = \begin{bmatrix} 0 & 1 \\ 0 & 0 \end{bmatrix}.$$

We may choose $\Phi = \mathrm{col}(1, 0)$ and $\Psi = \mathrm{row}(0, 1)$. Then $C = D = T$ and

$$Pf = \Phi(\cdot)a = \begin{bmatrix} 1 \\ 0 \end{bmatrix}\left(\frac{1}{T}\int_0^T f_1(s)\,ds\right),$$

$$Qf = \Psi'(\cdot)b = \begin{bmatrix} 0 \\ 1 \end{bmatrix}\left(\frac{1}{T}\int_0^T f_2(s)\,ds\right),$$

where $f = (f_1, f_2)$. If $Qf = 0$, one easily computes $\mathscr{K}f$ to be

$$(\mathscr{K}f)(t) = \begin{bmatrix} \displaystyle\int^t \left[f_1(s) - \frac{1}{T}\int_0^T f_1 + \int^s f_2 \right] ds \\[4mm] -\dfrac{1}{T}\displaystyle\int_0^T f_1 + \int^t f_2 \end{bmatrix},$$

where "$\int^t$" again denotes the primitive of the integrand of mean value zero.

LEMMA 2.2. *If the operators P, Q and $\mathscr{K}$ are defined as in (2.5) and Lemma 2.1, then system (2.1) has a T-periodic solution x if and only if x satisfies the system of equations*

(2.7)   (a) $x = Px + \varepsilon\mathscr{K}(I - Q)f(\cdot, x),$

          (b) $\varepsilon Qf(\cdot, x) = 0.$

PROOF. Suppose x is any element of $\mathscr{P}_T$. We first show that $Q(\dot{x} - Bx) = 0$. In fact, from (2.5),

$$Q(\dot{x} - Bx) = \Psi' D^{-1} \int_0^T \Psi(t)[\dot{x}(t) - B(t)x(t)]\, dt$$

$$= \Psi' D^{-1} \int_0^T [\Psi(t)\dot{x}(t) + \dot{\Psi}(t)x(t)]\, dt$$

$$= \Psi' D^{-1} [\Psi(t)x(t)]_0^T = 0.$$

An element x of $\mathscr{P}_T$ is a solution of (2.1) if and only if

$$Q(\dot{x} - Bx) = \varepsilon Q f(\cdot, x),$$
$$(I - Q)(\dot{x} - Bx) = \varepsilon(I - Q)f(\cdot, x).$$

Since $Q(\dot{x} - Bx) = 0$, these equations are equivalent to (2.7b) and $\dot{x} - Bx = (I - Q)f(\cdot, x)$. Corollary 2.1 implies this latter equation is equivalent to (2.7a) and this proves the lemma.

LEMMA 2.3. For any $\alpha > 0$, there is an $\varepsilon_0 > 0$ such that for any constant p-vector a, $|a| \leqq \alpha$, $|\varepsilon| \leqq \varepsilon_0$, there is a unique function $x^* = x^*(a, \varepsilon)$ which satisfies

$$(2.8) \qquad\qquad x^* = \Phi a + \varepsilon \mathscr{K}(I - Q)f(\cdot, x^*).$$

Furthermore, $x^*(a, \varepsilon)$ has a continuous first derivative with respect to a, ε and $x^*(a, 0) = a$. If there is an $a = a(\varepsilon)$ with $|a(\varepsilon)| \leqq \alpha$, $0 \leqq |\varepsilon| \leqq \varepsilon_0$ and

$$(2.9) \qquad\qquad G(a, \varepsilon) \overset{\text{def}}{=} Q f(\cdot, x^*(a, \varepsilon)) = 0,$$

then $x^*(a, \varepsilon)$ is a T-periodic solution of (2.1). Conversely, if (2.1) has a T-periodic solution $\bar{x}(\varepsilon)$ which is continuous in ε and has $P\bar{x}(\varepsilon) = \Phi a(\varepsilon)$, $|a(\varepsilon)| \leqq \alpha$, $0 \leqq |\varepsilon| \leqq \varepsilon_0$, then $\bar{x}(\varepsilon) = x^*(a(\varepsilon), \varepsilon)$ where $x^*(a, \varepsilon)$ is the function defined above and $a(\varepsilon)$ satisfies (2.9) for $0 < |\varepsilon| \leqq \varepsilon_0$.

PROOF. Suppose $\alpha > 0$ is given and a is any p-vector with $|a| \leqq \alpha$. Let β be a positive number such that $\|\Phi a\| \leqq \beta$ for $|a| \leqq \alpha$. For any $\gamma > 0$, define $\mathscr{S}(\gamma) = \{y \text{ in } \mathscr{P}_T : Py = 0, \|y\| \leqq \gamma\}$ and the operator $\mathscr{F}: \mathscr{S}(\gamma) \to \mathscr{P}_T$ by

$$(2.10) \qquad\qquad \mathscr{F}y = \varepsilon\, \mathscr{K}(I - Q)f(\cdot, y + \Phi a)$$

for y in $\mathscr{S}(\gamma)$. If $y^*(a, \varepsilon)$ is a fixed point of $\mathscr{F}$ in $\mathscr{S}(\gamma)$, then $x^*(a, \varepsilon) = \Phi a + y^*(a, \varepsilon)$ is a solution of (2.8). The proof now proceeds exactly as in the proof of Lemma 1.2.

From the point of view of applications, it is convenient to observe that relation (2.5) implies that $G(a, \varepsilon) = 0$ in (2.9) is equivalent to

$$(2.11) \qquad\qquad F(a, \varepsilon) \overset{\text{def}}{=} \int_0^T \Psi(t)f(t, x^*(a, \varepsilon)(t))\, dt = 0.$$

Lemma 2.3 asserts the following: one can arbitrarily preassign an element $x_P = \Phi a$ of the subspace of $\mathscr{P}_T$ defined by $\{f \text{ in } \mathscr{P}_T \colon Pf = f\}$ and then uniquely determine a solution of (2.7a) with Px replaced by x_P. The element x_P is then used to attempt to solve the remaining equation (2.7b).

Equations (2.9) or (2.11) are called the *determining equations* or *bifurcation equations of* (2.1) and can be determined approximately by successive approximations since $\mathscr{F}$ in (2.10) is a contraction operator. Taking only the first approximation Φa for $x^*(a, \varepsilon)$ and using the implicit function theorem, one arrives at

THEOREM 2.1. Suppose $x^*(a, \varepsilon)$ is defined as in Lemma 2.3 and let $F(a, \varepsilon)$ be defined by (2.11). If there is a p-vector a_0 such that

$$F(a_0, 0) = 0, \qquad \det\left[\frac{\partial F(a_0, 0)}{\partial a}\right] \neq 0,$$

then there are an $\varepsilon_1 > 0$ and a T-periodic solution $x^*(\varepsilon)$, $0 \leqq |\varepsilon| \leqq \varepsilon_1$, of (2.1), $x^*(0) = \Phi a_0$ and $x^*(\varepsilon)$ is continuously differentiable in ε.

EXERCISE 2.1. If f is in $\mathscr{P}_T$ and $Qf = 0$, give a constructive procedure for determining the function $\mathscr{K}f$ in $\mathscr{P}_T$ given in Lemma 2.1.

EXERCISE 2.2. State and prove the appropriate generalizations of the results of this section for the equation

$$\dot{x} = B(t)x + f(t, x, \varepsilon),$$

where $B(t + T) = B(t), f(t + T, x, \varepsilon) = f(t, x, \varepsilon)$ are continuous for (t, x, ε) in $R \times C^n \times C$, and

$$f(t, 0, 0) = 0,$$
$$|f(t, x, \varepsilon) - f(t, y, \varepsilon)| \leqq \eta(|\varepsilon|, \sigma)|x - y|,$$

for t in R, ε in C, x, y in C^n, $|x| < \sigma$, $|y| < \sigma$ and $\eta(\alpha, \sigma)$ is a continuous nondecreasing function for $\alpha \geqq 0$, $\sigma \geqq 0$, $\eta(0, 0) = 0$.

EXERCISE 2.3. Consider the system

$$(2.12) \qquad \dot{x}_1 = x_2,$$
$$\dot{x}_2 = \varepsilon[(\sigma + \alpha \cos 2t)x_1 + bx_1^3 + cx_2],$$

where σ, α, b, c are constants. Find conditions on these constants which will ensure that π-periodic solutions exist. What are their stability properties?

EXERCISE 2.4. Find conditions on σ so that (2.12) for $b = c = 0$ has a π-periodic solution. Interpret your result in the light of the theory of the Mathieu equation.

If the system (2.1) possesses some additional properties, it is sometimes possible to make qualitative statements about the bifurcation equations (2.9) or (2.11) without any successive approximations. The next few pages are devoted to this question.

Definition 2.1.   A system of differential equations $\dot{x} = g(t,\ x)$ where x, g are n-vectors is said to have *property* (E) *with respect to* S if there exists a symmetric, constant $n \times n$ matrix S such that $S^2 = I$ and

$$Sg(-t, Sx) = -g(t, x)$$

for all t, x.

If $x(t)$ is a solution of a system which has property (E) with respect to S, notice that $Sx(-t)$ is also a solution. In fact, if x is a solution of such a system and $w(t) = Sx(t)$, then

$$\dot{w}(t) = S\dot{x}(t) = Sg(t, x(t)) = Sg(t, S^{-1}w(t)) = Sg(t, Sw(t)) = -g(-t, w(t)).$$

This implies $w(-t) = Sx(-t)$ is a solution of the original equation.

LEMMA 2.4.   Suppose S is an $n \times n$ symmetric matrix, $S^2 = I$, $B(t) = B(t + T)$ is an $n \times n$ continuous matrix and $f(t) = f(t + T)$ is a continuous n-vector such that

$$(2.13) \qquad\qquad \text{(a)} \quad SB(-t) = -B(t)S,$$
$$\text{(b)} \quad Sf(-t) = -f(t).$$

Let the columns of the $n \times p$ matrix Φ be a basis for the T-periodic solutions of (2.2), $\mathscr{K}$ be defined as in Lemma 2.1 and Q as in (2.5). Then

$$(2.14) \qquad \text{(a)} \quad S\Phi(-t)a = \Phi(t)a \text{ for all } t \text{ if } S\Phi(0)a = \Phi(0)a,$$
$$\text{(b)} \quad S(Qf)(-t) = -(Qf)(t),$$
$$\text{(c)} \quad S[\mathscr{K}(I - Q)f](-t) = [\mathscr{K}(I - Q)f](t).$$

PROOF.   The fact that B satisfies (2.13) implies that system (2.2) has property (E) with respect to S and thus, if $S\Phi(0)a = \Phi(0)a$, then uniqueness of the solutions implies $S\Phi(-t)a = \Phi(t)a$ for all t. This proves (2.14a). The matrix $S\Phi(-t)$ is a basis for the T-periodic solutions of (2.2) and, consequently, there is a $p \times p$ nonsingular matrix Γ such that $S\Phi(-t) = \Phi(t)\Gamma$ for all t. In the same way, there is a $p \times p$ nonsingular matrix M such that $\Psi(-t)S = M\Psi(t)$ for all t.

For any f satisfying (2.13b), we have from (2.5), and $S' = S$ that

$$S(Qf)(-t) = S\Psi'(-t)D^{-1}\int_0^T \Psi(\alpha)f(\alpha)\, d\alpha$$

$$= \Psi'(t)M'D^{-1}\int_0^T \Psi(-\alpha)f(-\alpha)\, d\alpha$$

$$= -\Psi''(t)M'D^{-1}\int_0^T \Psi'(-\alpha)Sf(\alpha)\,d\alpha$$

$$= -\Psi''(t)M'D^{-1}M\int_0^T \Psi'(\alpha)f(\alpha)\,d\alpha$$

On the other hand, from (2.4) and $S' = S$, $S^2 = I$, one shows that $MDM' = D$. Therefore, $M'D^{-1}M = D^{-1}$ and

$$S(Qf)(-t) = -\Psi''(t)D^{-1}\int_0^T \Psi'(\alpha)f(\alpha)\,d\alpha$$

$$= -(Qf)(t).$$

This proves (2.14b).

If f satisfies (2.13b), then (2.14b) implies that $(I-Q)f$ satisfies

$$S(I-Q)f(-t) = -(I-Q)f(t).$$

Therefore, to prove (2.14c), it is sufficient to show that (2.14c) is satisfied with $(I-Q)f$ replaced by f and $Qf = 0$. We first show that $PS(\mathscr{K}f)(-\cdot) = 0$. From (2.5), this is true if and only if

$$0 = \int_0^T \Phi'(t)S(\mathscr{K}f)(-t)\,dt$$

$$= \Gamma'\int_0^T \Phi'(-t)(\mathscr{K}f)(-t)\,dt$$

$$= \Gamma'\int_0^T \Phi'(t)(\mathscr{K}f)(t)\,dt.$$

Since $P\mathscr{K}f = 0$ it follows that $\int_0^T \Phi'(t)(\mathscr{K}f)(t)\,dt = 0$. Therefore, $PS(\mathscr{K}f)(-\cdot)$ $= 0$. Since $\mathscr{K}f$ is the unique solution of (2.6) with $P\mathscr{K}f = 0$ and $S\mathscr{K}f(-\cdot)$ is also a solution of this same equation with P-projection zero, it follows that (2.14c) is true. This proves the lemma.

THEOREM 2.2. Let the columns of the $n \times p$ matrix Φ be a basis for the T-periodic solutions of (2.2) and let the rows of the $p \times n$ matrix Ψ be a basis for the T-periodic solutions of (2.3). Let $x^*(a, \varepsilon)$ be the function determined in Lemma 2.3 and suppose system (2.1) has property (E) with respect to S. Then

(2.15) $$Sx^*(a, \varepsilon)(-t) = x^*(a, \varepsilon)(t)$$

for all t provided that $(I-S)\Phi(0)a = 0$. In addition, if $F(a, \varepsilon)$ is defined as in (2.11), then

(2.16) $$(I+M)F(a, \varepsilon) = 0.$$

where M is the matrix such that $\Psi(-t)S = M\Psi(t)$.

PROOF.   Suppose $S\Phi(0)a = \Phi(0)a$; let $\mathscr{S}(\gamma)$ be defined as in the proof of Lemma 2.3 and $\mathscr{S}^*(\gamma)$ be the subset of $\mathscr{S}(\gamma)$ consisting of those y in $\mathscr{S}(\gamma)$ which satisfy $Sy(-t) = y(t)$. From Lemma 2.4, $\mathscr{F}y = \varepsilon\mathscr{K}(I - Q)f(\cdot, y + \Phi a)$ belongs to $\mathscr{S}^*(\gamma)$ for $|\varepsilon| \leq \varepsilon_0$ where ε_0 is given in Lemma 2.3. Since this operator has a unique fixed point $y^*(a, \varepsilon)$ in $S(\gamma)$, $Sy^*(a, \varepsilon)(-t) = y^*(a,\varepsilon)(t)$. Therefore (2.14a) implies $x^*(a, \varepsilon) = y^*(a, \varepsilon) + \Phi a$ satisfies relation (2.15). Using (2.15), property (E) and the fact that $\Psi(-t)S = M\Psi(t)$, one obtains

$$F(a, \varepsilon) = \int_0^T \Psi(t)f(t, x^*(a, \varepsilon)(t))\,dt$$

$$= \int_0^T \Psi(t)f(t, Sx^*(a, \varepsilon)(-t))\,dt$$

$$= -\int_0^T \Psi(t)Sf(-t, x^*(a, \varepsilon)(-t))\,dt$$

$$= -M\int_0^T \Psi(-t)f(-t, x^*(a, \varepsilon)(-t))\,dt$$

$$= -MF(a, \varepsilon).$$

This proves the theorem.

In the applications, Theorem 2.2 can be very useful and especially the conclusion expressed in (2.16). This relation says that the bifurcation functions given by the components of the vector F in (2.11) are dependent provided that the p-vector a satisfies $(I - S)\Phi(0)a = 0$. This can lead to results for (2.1) involving the existence of families of periodic solutions. The following exercises illustrate this point and involve situations in which property (E) expresses some even and oddness properties of the functions in (2.1).

EXERCISE 2.5.   Suppose $f(-t, x) = -f(t, x)$ for all t in R, x in R^n. Show that the system $\dot{x} = \varepsilon f(t, x)$ has an n-parameter family of periodic solutions for ε small. If $f(t, x) = A(t)x$ and $A(-t) = -A(t)$, this implies in particular all characteristic multipliers are 1.

EXERCISE 2.6.   Reconsider Exercise 1.7 in the light of Theorem 2.2.

EXERCISE 2.7.   Show that all solutions of the equation

$$\ddot{y} + \sigma^2 \dot{y} = \varepsilon g(y, \dot{y}, \ddot{y}), \qquad \sigma \neq 0,$$

in a neighborhood of $y = \dot{y} = \ddot{y} = 0$ are periodic for ε small if $g(y, -\dot{y}, \ddot{y}) = -g(y, \dot{y}, \ddot{y})$. Write the equation as a third order system $\dot{x} = A(\sigma^2)x + \varepsilon f(t, x)$, let $x = [\exp A(\tau^2)t]z$, $\tau^2 = \sigma^2 + \varepsilon\beta$ and apply the previous results to the equation for z using β as one of the undetermined parameters.

EXERCISE 2.8.   Consider the system of second order equations

$$\ddot{u} + \sigma^2 u = \varepsilon g_1(u, \dot{u}, v, \dot{v}),$$
$$\ddot{v} + \mu^2 v = \varepsilon g_2(u, \dot{u}, v, \dot{v}),$$

where $\sigma \neq m\mu$, $\mu \neq m\sigma$ for $m = 0, \pm 1, \ldots$ and

$$g_1(-u, \dot{u}, v, -\dot{v}) = -g_1(u, \dot{u}, v, \dot{v}),$$
$$g_2(-u, \dot{u}, v, -\dot{v}) = g_2(u, \dot{u}, v, \dot{v}).$$

Show that this system has two 2-parameter families of periodic solutions. Let $x_1 = u$, $x_2 = \sigma \dot{u}$, $x_3 = v$, $v_4 = \mu \dot{v}$ to obtain a fourth order system $\dot{x} = A(\sigma, \mu)x + \varepsilon f(x)$ where $A(\sigma, \mu) = \text{diag}(B(\sigma), B(\mu))$, $x = (y, z)$, y, z two-vectors and let $y = [\exp B(\tau)t]Y$, $z = Z$, $\tau = \sigma + \varepsilon\beta$ and apply Theorem 2.2 to the equation for (Y, Z) using β as one of the undetermined parameters. Generalize this result to other types of symmetry as well as a system of n-second order equations.

Definition 2.2.   A *first integral* of a system

(2.17) $$\dot{x} = g(t, x)$$

is a function $u: R \times C^n \to C$ which has continuous first partial derivatives such that

$$\frac{\partial u(t, x)}{\partial x} g(t, x) + \frac{\partial u(t, x)}{\partial t} = 0$$

for all t, x.

If $x(t, t_0, x_0)$, $x(t_0, t_0, x_0) = x_0$ is a solution of (2.17) and u is a first integral of (2.17), then $u(t, x(t, t_0, x_0)) = u(t_0, x_0)$ for all t for which $x(t, t_0, x_0)$ is defined. If $t_0 = 0$, we write $x(t, x_0)$, $x(0, x_0) = x_0$, for a solution of (2.17).

Let $V(t, x_0) = \partial x(t, x_0)/\partial x_0$. In Chapter I, we have seen that V is a principal matrix solution of the equation

(2.18) $$\dot{z} = Hz, \quad H = H(t, x_0) = \partial g(t, x(t, x_0))/\partial x.$$

LEMMA 2.5.   If u is a first integral of (2.17) which has continuous second partial derivatives, then $u_x(t, x_0) \overset{\text{def}}{=} \partial u(t, x(t, x_0))/\partial x$ is a solution of the adjoint equation

(2.19) $$\dot{w} = -wH,$$

where H is defined in (2.18).

PROOF.   Since $u(t, x(t, x_0)) = u(0, x_0)$ for all t and all x_0, it follows that $u_x(t, x(t, x_0))V(t, x_0) = u_x(0, x_0)$ for all t. Differentiating this relation with respect to t, one obtains

$$0 = \dot{u}_x V + u_x \dot{V} = [\dot{u}_x + u_x H] V$$

for all t. Since V is nonsingular, u_x must satisfy (2.19) and the lemma is proved.

LEMMA 2.6.   If $u(t + T, x, \varepsilon) = u(t, x, \varepsilon)$ is a first integral of (2.1) which has continuous second partial derivatives with respect to t and x and $x^*(a, \varepsilon)$ is the T-periodic function given in Lemma 2.3, then

(2.20)
$$\int_0^T [u_x Q f]_{x=x*(a, \varepsilon)}(t) \, dt = 0$$

for $\varepsilon \neq 0$, where Q is defined in (2.5).

PROOF.   Since u is a first integral of (2.1), we have in particular that

$$\int_0^T [u_x(Bx + \varepsilon f) + u_t]_{x=x*(a, \varepsilon)}(t) \, dt = 0.$$

Since $x^*(a, \varepsilon)(t)$ and $u(t, x, \varepsilon)$ are T-periodic in t, we also have

$$0 = \int_0^T \frac{d}{dt} [u(t, x^*(a, \varepsilon)(t), \varepsilon)] \, dt$$

$$= \int_0^T [u_x \dot{x}^* + u_t]_{x=x*(a, \varepsilon)}(t) \, dt$$

$$= \int_0^T [u_x(Bx + \varepsilon f - \varepsilon Q f) + u_t]_{x=x*(a, \varepsilon)}(t) \, dt.$$

Subtracting these two expressions gives (2.20) and proves the lemma.

Suppose $x^*(a, \varepsilon)$ is the T-periodic function given in Lemma 2.3 and suppose $u(t, x, \varepsilon) = u(t + T, x, \varepsilon)$ is a first integral of (2.1) with $u_x(t, 0, 0) \neq 0$. From Lemma 2.5, it follows that $u_x(t, 0, 0)$ is a nontrivial T-periodic solution of $\dot{w} = -wB(t)$. Since $x^*(0, 0) = 0$, $u_x(t, x^*(0, 0)(t), 0) = u_x(t, 0, 0) \neq 0$ is a T-periodic solution of the adjoint equation $\dot{w} = -wB(t)$. With the rows of the $p \times n$ matrix Ψ defining a basis for the T-periodic solutions of this equation, this implies there is a p-dimensional row vector $h \neq 0$ such that $u_x(t, 0, 0) = h\Psi(t)$. Obviously, there is a continuous function $\mu(t, a, \varepsilon) = \mu(t + T, a, \varepsilon)$ such that $\mu(t, 0, 0) = 0$ and

$$u_x(t, x^*(a, \varepsilon)(t), \varepsilon) = u_x(t, 0, 0) + \mu(t, a, \varepsilon)$$
$$= h\Psi(t) + \mu(t, a, \varepsilon).$$

If $F(a, \varepsilon)$ is defined as in (2.11), then (2.5) implies that equation (2.20) can be written as

$$0 = \int_0^T [h\Psi'(t) + \mu(t, a, \varepsilon)]\Psi''(t)D^{-1}F(a, \varepsilon)\,dt$$

$$= \left[h + \int_0^T \mu(t, a, \varepsilon)\Psi''(t)D^{-1}\,dt \right] F(a, \varepsilon)$$

for $0 < |\varepsilon| \leq \varepsilon_0$. Since h is nonzero and $\mu(t, 0, 0) = 0$, it follows there are $\alpha_1 \neq 0$, $\varepsilon_1 \neq 0$ such that $h + \int_0^T \mu(t, a, \varepsilon)\Psi''(t)D^{-1}\,dt = h_1$ is $\neq 0$ for $|a| \leq \alpha_1$, $|\varepsilon| \leq \varepsilon_1$. Therefore, there is a linear relation among the components of F; that is, a linear relation among the bifurcation functions of (2.1). *One of the bifurcation functions is therefore redundant if there is a first integral $u(t, x, \varepsilon) = u(t + T, x, \varepsilon)$ of (2.1) with $u_x(t, 0, 0) \neq 0$.*

If $u_1, \ldots, u_k$ are first integrals of (2.1), we say they are *linearly independent* if the matrix $u_x(t, 0, 0) = \partial u(t, 0, 0)/\partial x$, $u = (u_1, \ldots, u_k)$ has rank k. Using the same argument as above, one obtains

THEOREM 2.3.   Let $u = (u_1, \ldots, u_k)$ be $k \leq p$ linearly independent first integrals of (2.1), $u(t + T, \varepsilon) = u(t, x, \varepsilon)$ be continuous and have continuous second partial derivatives with respect to t, x. Then there are an $\alpha_1 > 0$, $\varepsilon_1 > 0$ and a $k \times p$ matrix H of rank k such that $HF(a, \varepsilon) = 0$ for $|a| \leq \alpha_1$, $|\varepsilon| \leq \varepsilon_1$ and $F(a, \varepsilon)$ defined in (2.11). If $k = p$, then there exists a p-parameter family of T-periodic solutions $x^*(a, \varepsilon)$, $|a| \leq \alpha_1$, $|\varepsilon| \leq \varepsilon_1$, where $x^*(a, \varepsilon)$ is given in Lemma 2.3.

The proof of the first part of this theorem is essentially the same as the above argument for $k = 1$. If $k = p$, then H is a nonsingular matrix and $HF(a, \varepsilon) = 0$ implies $F(a, \varepsilon) \equiv 0$ for $|a| \leq \alpha_1$, $|\varepsilon| \leq \varepsilon_1$; that is, the bifurcation equations are automatically satisfied. An application of Lemma 2.3 completes the proof of the theorem.

EXERCISE 2.9.   (Liapunov's theorem)   Suppose the system

(2.21) $\dot{x}_1 = \sigma x_2 + f_1(x_1, x_2, y)$,

$\dot{x}_2 = -\sigma x_1 + f_2(x_1, x_2, y)$,

$\dot{y} = Cy + g(x_1, x_2, y)$,

where $\sigma > 0$, x_1, x_2 are scalars, y is an m-vector, f_1, f_2, g are analytic in a neighborhood of $x_1 = 0 = x_2$, $y = 0$ with the power series beginning with terms of degree at least two. Suppose this system has a first integral $W(x_1, x_2, y)$ which has continuous second derivatives with respect to x_1, x_2, y and the eigenvalues $\lambda_1, \ldots, \lambda_m$ of the constant matrix C satisfy $\lambda_k \neq i\sigma n$ for $k = 1, 2, \ldots, m$, and every integer n. Prove that this system has a two parameter family of periodic solutions. Introduce polar type coordinates $x_1 = \rho \cos \sigma\theta$, $x_2 = -\rho \sin \sigma\theta$ in (2.21) and eliminate t in the resulting equations to obtain differential equations for ρ, y as functions of θ. Let $\rho \to \varepsilon\rho$, $y \to \varepsilon y$ and apply Theorem 2.3. If all eigenvalues of C are purely imaginary, do there exist other families of periodic solutions? What additional conditions are sufficient for the existence of other families of periodic solutions? Is it necessary to assume that f_1, f_2, g are analytic?

EXERCISE 2.10. Interpret the general theory of this section for the particular n^{th} order scalar equation

$$\frac{d^n y}{dt^n} = \varepsilon f(t, y).$$

EXERCISE 2.11. Consider the system

(2.22) $$\dot{w} = A_0 w + \varepsilon \Phi(t)w,$$

where $\Phi(t + T) = \Phi(t)$ is a continuous $n \times n$ matrix and A_0 is a constant matrix. Give a constructive procedure which is in the spirit of this section for obtaining a principal matrix solution of (2.22) of the form $U(t, \varepsilon) \exp[B(\varepsilon)t]$ with $U(t + T, \varepsilon) = U(t, \varepsilon)$, $U(t, 0) = I$, $B(0) = A_0$. Consider the $n \times n$ matrix equation

(2.23) $$\dot{W} = AW + \varepsilon \Phi(t)W$$

and, for any given $n \times n$ constant matrix A_1, let $W = U \exp[A_0 - A_1]t$. The differential equation for U is

(2.24) $$\dot{U} = A_0 U - U A_0 + U A_1 + \varepsilon \Phi(t)U.$$

Determine necessary and sufficient conditions that the nonhomogeneous matrix equation

$$\dot{U} = A_0 U - U A_0 + F(t).$$
$$F(t + T) = F(t),$$

have a T-periodic solution. Use this fact to obtain a set of matrix equations $\Gamma(A_1, \varepsilon) = 0$ whose solution is necessary and sufficient for obtaining a T-periodic matrix solution of (2.24). Show there is always a matrix function $A_1(\varepsilon)$ such that $\Gamma(A_1(\varepsilon), \varepsilon) = 0$ for ε small.

EXERCISE 2.12. Generalize exercise 2.11 to a class of almost periodic matrix perturbations $\Phi(t)$.

EXERCISE 2.13. If A_0 in (2.22) is a T-periodic matrix function, how could the procedure of exercise (2.11) be modified to obtain the principal matrix solution?

In Exercise 2.2, the reader was supposed to prove that one could use the methods of this chapter to obtain T-periodic solutions for equations which contained nonlinear terms when $\epsilon = 0$. Of course, one must then discuss those solutions which remain in a sufficiently small neighborhood of $\epsilon = 0$, $x = 0$. As in Section 1, a further generalization is needed to discuss equations which contain several independent parameters. In the next pages, we give this generalization and a few illustrations.

Consider the T-periodic system

$$(2.25) \qquad \dot{x} = B(t)x + f(t,x,\lambda)$$

where $B(t + T) = B(t)$ is a continuous $n \times n$ matrix, λ is in R^k, $f(t + T,x,\lambda) = f(t,x,\lambda)$ has continuous derivatives up through order one in x,λ,

$$(2.26) \qquad f(t,0,0) = 0, \ \partial f(t,0,0)/\partial x = 0.$$

The analogue of Lemma 2.3 is the following which is stated without proof.

LEMMA 2.7. *If P,Q are defined in* (2.5), $\mathscr{K}:\mathscr{P}_T \to \mathscr{P}_T$ *in Lemma 2.1, and* (2.26) *is satisfied, then there are $\lambda_0 > 0$, $\alpha_0 > 0$, such that, for any a in R^n, $|a| \leq \alpha_0$, any λ in R^k, $|\lambda| \leq \lambda_0$, there is a unique T-periodic function $x^* = x^*(a,\lambda)$ continuous together with its first derivative satisfying $x^*(0,0) = 0$,*

$$x^* = \Phi_a + \mathscr{K}(I - Q)f(\cdot,x^*,\lambda).$$

If there exist a,λ such that

$$G(a,\lambda) \overset{\text{def.}}{=} \frac{1}{T} \int_0^T \Psi(t)f(t,x^*(a,\lambda)(t),\lambda)dt = 0$$

then $x^*(a,\lambda)$ is a T-periodic solution of (2.25). *Conversely, if* (2.25) *has a T-periodic solution $\bar{x}(\lambda)$ which is continuous in λ and has $P\bar{x}(\lambda) = \Phi a(\lambda)$, $|a(\lambda)| \leq \alpha_0$, $|\lambda| \leq \lambda_0$, then $\bar{x}(\lambda) = x^*(a(\lambda),\lambda)$ where $x^*(a,\lambda)$ is the function given above and $a(\lambda)$ satisfies* (2.27) *for $0 \leq |\lambda| \leq \lambda_0$.*

Let us illustrate the use of this lemma for Duffing's equation. In Section V.5, we have discussed Duffing's equation with a small harmonic forcing by the method of averaging. In order to do this, we assumed all of the parameters were multiples of a single small parameter ϵ. Also, because of the nature of averaging, we could only discuss a periodic solution if its characteristic multipliers were not on the unit circle. This implies that the information obtained is incomplete in the neighborhood of the points in parameter space where the number of periodic solutions changes from one to three. A more complete discussion can be given based on Lemma 1.4. We give an indication of the procedure below for the case of no damping.

Consider the equation

$$(2.28) \qquad \frac{d^2u}{d\tau^2} + u + \gamma u^3 = F\cos \omega\tau$$

where $\gamma \neq 0$ is a fixed real number, $F, \omega - 1$ are small real parameters. The objective is to obtain the $2\pi/\omega$-periodic solutions of (2.28) which have small norm. If

$$\omega^2 = 1 + \beta, \quad \omega\tau = t, \quad \lambda_1 = \frac{\beta}{1 + \beta}, \quad \mu_0 = \frac{\gamma}{1 + \beta}, \quad \mu_2 = \frac{F}{1 + \beta}$$

then

(2.29) $$\frac{d^2u}{dt^2} + u = -\mu_0 u^3 + \lambda_1 u + \mu_2 \cos t$$

and we determine 2π-periodic solutions of (2.29) of small norm. Since $\gamma \neq 0$, the transformation $u = |\mu_0|^{-1/2}x$ for β small can be made to obtain

(2.30) $$\ddot{x} + x = -(\text{sgn } \gamma)x^3 + \lambda_1 x + \lambda_2 \cos t,$$

where $\lambda_2 = |\mu_0|^{1/2}\mu_2$. Therefore, we must analyze the existence of small 2π-periodic solutions of (2.30) for the parameter $\lambda = (\lambda_1, \lambda_2)$ varying in a neighborhood of $\lambda = 0$.

We could transform (2.30) to a system of two first order equations, but it is actually more convenient to work directly with (2.30). Let $\mathscr{P}_{2\pi} = \{x : R \to R,$ continuous, 2π-periodic$\}$ with the usual norm. If

(2.31)
$$P : \mathscr{P}_{2\pi} \to \mathscr{P}_{2\pi}$$
$$(Ph)(t) = \frac{1}{\pi}\cos t \int_0^{2\pi} h(s)\cos s\,ds + \frac{1}{\pi}\sin t \int_0^{2\pi} h(s)\sin s\,ds$$

then P is a continuous projection and the equation

(2.32) $$\ddot{x} + x = h(t)$$

for h in $\mathscr{P}_{2\pi}$ has a 2π-periodic solution if and only if $Ph = 0$. If $Ph = 0$, then we can obtain a unique 2π-periodic solution $\mathscr{K}h$ of (2.32) which is orthogonal to $\sin t$, $\cos t$. The general 2π-periodic solution of (2.32) is

(2.32) $$x(t) = a\cos(t - \phi) + (\mathscr{K}h)(t)$$

where a, ϕ are arbitrary real numbers.

Therefore, a 2π-periodic solution of (2.30) is given by

$$x(t) = a\cos(t - \phi) + \mathscr{K}(I - P)[-(\text{sgn } \gamma)x^3 + \lambda_1 x + \lambda_2 \cos \cdot](t)$$
$$P[-(\text{sgn } \gamma)x^3 + \lambda_1 x + \lambda_2 \cos \cdot] = 0$$

for some a, ϕ. The parameters a, ϕ must be determined through the bifurcation equations $P[\ ] = 0$. For the method, it is convenient to put the parameter ϕ in the perturbation term by replacing t by $t + \phi$; that is, consider the equation

(2.33) $$\ddot{x} + x = -(\text{sgn } \gamma)x^3 + \lambda_1 x + \lambda_2 \cos(t + \phi)$$

and a 2π-periodic solution x so that $Px = a \cos t$ for some a. Such a solution x must satisfy

(2.34)
(a) $x(t) = a\cos t + \mathcal{H}(I - P)[-(\mathrm{sgn}\,\gamma)x^3 + \lambda_1 x + \lambda_2 \cos(\cdot + \phi)](t)$

(b) $\qquad\qquad P[-(\mathrm{sgn}\,\gamma)x^3 + \lambda_1 x + \lambda_2 \cos(\cdot + \phi)] = 0$

In the usual way, solve equation (2.34a) for a 2π-periodic function, $x^*(\phi, a, \lambda)$ for a, λ small, $x^*(\phi, a, \lambda) = a\cos t + o(|a|)$ as $a \to 0$.

Then the bifurcation equations are

(a) $\lambda_2 \cos\phi + \dfrac{1}{\pi}\displaystyle\int_0^{2\pi} [\lambda_1 x^*(t) - (\mathrm{sgn}\,\gamma)x^{*3}(t)] \cos t\,dt = 0$

(b) $\lambda_2 \sin\phi + \dfrac{1}{\pi}\displaystyle\int_0^{2\pi} [\lambda_1 x^*(t) - (\mathrm{sgn}\,\gamma)x^{*3}(t)] \sin t\,dt = 0$

EXERCISE 2.14. Prove $x^*(\phi, a, \lambda)$ is an even function of t.

From Exercise (2.14), Equation (2.35b) is equivalent to $\lambda_2 \sin\phi = 0$ or $\phi = 0$ if $\lambda_2 \neq 0$. If $\lambda_2 = 0$, then the only 2π-periodic solution of (2.33) near $x = 0$, $\lambda_1 = 0$ is $x = 0$. Thus, we may assume $\phi = 0$ and have the result that there is a 2π-periodic solution of Equation (2.30) near $x = 0$, $\lambda = 0$ if and only if this solution is even, $x(t) = x^*(0, a, \lambda)(t)$ and (a, λ) satisfy the equation

(2.36) $G(a, \lambda) \overset{\text{def}}{=} \lambda_2 + \dfrac{1}{\pi}\displaystyle\int_0^{2\pi} [\lambda_1 x^*(0, a, \lambda)(t) - (\mathrm{sgn}\,\gamma)x^{*3}(0, a, \lambda)(t)] \cos t\,dt = 0$

It remains to analyze Equation (2.36). It is not difficult to show that

$$G(a, \lambda) = \alpha_0(\lambda) + \alpha_1(\lambda)a + \alpha_2(\lambda)a^2 + \alpha_3(\lambda)a^3 + o(|a|^3)$$

where
$$\alpha_0(\lambda) = \lambda_2 + o(|\lambda_1| + |\lambda_2|)$$

$$\alpha_1(\lambda) = \lambda_1 + o(|\lambda_1| + |\lambda_2|)$$

$$\alpha_2(\lambda) = 0(|\lambda|)$$

$$\alpha_3(0) = -\frac{3}{4}\,\mathrm{sgn}\,\gamma$$

The first objective is to obtain the values of λ which give rise to multiple solutions of $G(a, \lambda) = 0$; that is, those values of λ for which

(2.37) $\qquad\qquad G(a, \lambda) = 0, \qquad \partial G(a, \lambda)/\partial a = 0$

The Jacobian of these two functions with respect to λ is given by

$$\det \begin{bmatrix} 0 & 1 \\ 1 & 0 \end{bmatrix} = -1.$$

Consequently, there are unique solutions $\lambda_1^*(a), \lambda_2^*(a)$ of (2.37) near $a = 0$ and they are given approximately by

$$(2.38) \qquad \lambda_1(a) = \frac{9}{4}(\operatorname{sgn}\gamma)a^2 + o(|a|^2), \qquad \lambda_2(a) = -\frac{3}{2}(\operatorname{sgn}\gamma)a^3 + o(|a|^3)$$

Formulas (2.38) are the parametric representation of a cusp given approximately by $\lambda_2^2 = (16/81)(\operatorname{sgn}\gamma)\lambda_1^3$ in the (λ_1,λ_2)-plane. Since $\partial G(0,0)/\partial\lambda_2 = 1$, it follows that there is one solution on one side of this cusp and three on the other. This analysis gives a complete picture of the small 2π-periodic solutions in a neighborhood of $x = 0$ for $\lambda = (\lambda_1,\lambda_2)$ varying independently over a full neighborhood of zero.

EXERCISE 2.15. (General) Analyze all the exercises in this chapter in the spirit of the above discussion for Duffing's equation.

Let us consider another example which will illustrate in a more dramatic way the importance of considering independent variations in the parameters. Suppose $g(x)$ has continuous derivatives up through order 2 in the scalar variable x and the equation

$$(2.39) \qquad \ddot{x} + g(x) = 0$$

has a family of periodic solutions $\phi(\omega(a)t + \alpha, a)$ for α in R, $|a - a_0| < \delta$, for some fixed $a_0, \delta, \omega(a_0) = 1$, $\phi(\theta, a) = \phi(\theta + 2\pi, a)$; that is, the solutions have period $2\pi/\omega(a)$ where a can be considered as the amplitude of the solution and α the phase. We assume the period $2\pi/\omega(a)$ is the least period. Suppose $f(t)$ is a continuous 2π-periodic function (2π is not necessarily the least period), λ_1, λ_2 are small real parameters and consider the equation

$$(2.40) \qquad \ddot{x} + g(x) = -\lambda_1\dot{x} + \lambda_2 f(t).$$

Let $\Gamma = \{(\phi(t,a_0), \dot{\phi}(t,a_0)), t \text{ in } R\}$ be the 2π-periodic orbit in the phase space $(x, \dot{x})$ corresponding to the amplitude a_0. The problem is to determine all the 2π-periodic solutions of (2.40) which lie in a neighborhood of Γ for $\lambda = (\lambda_1,\lambda_2)$ in a neighborhood of zero. If the least period of Γ is 2π and the least period of f is 2π, then we are looking for harmonic solutions near Γ. If the least period of Γ is 2π and the least period of f is $2\pi/k$ for some integer $k \geq 1$, we are looking for subharmonics of order k near Γ.

We need the following pypotheses:

$$(H_1) \qquad\qquad \omega(a_0) = 1, \qquad \omega'(a_0) \neq 0$$

(H_2) If $p(t) = \phi(t,a_0)$ and $h(\alpha) = \int_0^{2\pi} \dot{p}(t)f(t-\alpha)dt \Big/ \int_0^{2\pi} \dot{p}^2(t)dt$

then $h(\alpha)$ has a unique maximum at α_M in $[0,2\pi)$, a unique minimum at α_m in $[0,2\pi)$, $h''(\alpha_M) \neq 0, h''(\alpha_m) \neq 0$.

We can then state

THEOREM 2.4. If (H_1), (H_2) are satisfied, then there exist enighborhoods
U of Γ, V of $\lambda = 0$ and curves $C_m \subseteq V$, $C_M \subseteq V$, defined respectively by
$\lambda_1 = c_m(\lambda_2)$, $\lambda_1 = c_M(\lambda_2)$ with c_m, c_M continuously differentiable in λ_2,
C_m, C_M respectively are tangent to the lines $\lambda_1 = h(\alpha_m)\lambda_2$, $\lambda_1 = h(\alpha_M)\lambda_2$
at $\lambda_1 = \lambda_2 = 0$, these curves intersect only at zero, the sectors between these
curves containing $\lambda_2 = 0$ has no 2π-periodic solutions and the other sectors
contain at least two solutions. If the least period of $f(t)$ is k^{-1}, $k \geq 1$ an
integer, then $x(t)$ a 2π-periodic solution implies $x(t + m/k)$, $m = 0,1,\ldots,$
$k-1$ is also a 2π-periodic solution.

PROOF. Our first objective is to obtain a convenient coordinate system
near Γ. The vector $\tau(\theta) = (\dot{p}(\theta), p(\theta))$ is tangent to Γ and $\gamma(\theta) = (\dot{p}(\theta), -\dot{p}(\theta))$
is orthogonal to $\tau(\theta)$. For a fixed α_0, an application of the Implicit Function
Theorem shows that the mapping

(2.41)
$$x = p(\alpha) + a\ddot{p}(\alpha)$$
$$y = \dot{p}(\alpha) - a\dot{p}(\alpha)$$

is a homeomorphism of a neighborhood of $(\alpha_0, 0)$ into a neighborhood of
$(p(\alpha_0), \dot{p}(\alpha_0))$. By using the compactness of Γ and further restricting the size
of a, one easily concludes that there is an $a_0 > 0$ such that the set

$$U = \{(x, y) \text{ given by (2.41) for } 0 \leq \alpha < 2\pi, |a| < a_0\}$$

is an open neighborhood of Γ and, for any $(x, y) \in U$, there is a unique (α, a)
such that (x, y) satisfies (2.41); that is, (α, a), $0 \leq \alpha < 2\pi, |a| < a_0$, serves as a
coordinate system for a neighborhood of Γ.

If $(x(t), y(t)), y(t) = \dot{x}(t)$, is a 2π-periodic solution of (2.40) which lies in a
sufficiently small neighborhood of Γ, then the initial value $(x(0), y(0))$ uniquely
determines an (α, a) such that $(x(0), y(0))$ is given by (2.41). Therefore, it is
sufficient to consider only those 2π-periodic solutions of (2.41) of the form

$$x(t) = p(t + \alpha) + z(t + \alpha)$$
$$y(t) = \dot{p}(t + \alpha) + \dot{z}(t + \alpha)$$

where $(z(\alpha), \dot{z}(\alpha)) = a\gamma(\alpha)$ for some constant a; that is, $(z(0), \dot{z}(0))$ is orthogonal
to $\tau(\alpha) = (\dot{p}(\alpha), \ddot{p}(\alpha))$. If $x(t + \alpha) = p(t + \alpha) + z(t + \alpha)$, $t + \alpha$ is replaced by t
and $f_\alpha(t) = f(t - \alpha)$, then

(2.42) $$\ddot{z} + g'(p)z = \lambda_1 \dot{z} + \lambda_1 \dot{p} + \lambda_2 f_\alpha(t) + G(t, z)$$

where $G(t + 2\pi, z) = G(t, z)$ and $G(t, z) = 0(|z|^2)$ as $|z| \to 0$.

We now apply our basic method to Equation (2.42).

Hypothesis (H_1) implies that all 2π-periodic solutions of the linear equation

$$(2.43) \qquad\qquad \ddot{v} + g'(p)v = 0$$

are constant multiples of $\dot{p}$. Let us define $\int_0^{2\pi} \dot{p}^2 = \eta$ and define the continuous projection $P : \mathscr{P}_{2\pi} \to \mathscr{P}_{2\pi}$ by the relation

$$(2.44) \qquad\qquad P\phi = \dot{p} \int_0^{2\pi} \phi\dot{p}/\eta.$$

For a given $\phi \in \mathscr{P}_{2\pi}$, the Fredholm alternative implies that the nonhomogeneous equation

$$(2.45) \qquad\qquad \ddot{v} + g'(p)v = \phi(t)$$

has a 2π-periodic solution if and only if $P\phi = 0$. If $P\phi = 0$ and $v_0(t) = v_0(t;\phi)$ is any particular solution of (2.45), then each 2π-periodic solution of (2.45) is given by

$$v(t) = \beta\dot{p}(t) + v_0(t,\phi)$$

for some constant β. The solution v will have $(v(0), \dot{v}(0))$ orthogonal to $(\dot{p}(0), \ddot{p}(0))$ if and only if

$$(2.46) \qquad \beta[\dot{p}(0)^2 + \ddot{p}(0)^2] = -\dot{p}(0)v_0(0;\phi) - \ddot{p}(0)\dot{v}_0(0;\phi).$$

This relation uniquely defines $\beta = \beta(\phi)$ as a function of ϕ. If we define

$$(2.47) \qquad (\mathscr{K}\phi)(t) = \beta(\phi)\dot{p}(t) + v_0(t,\phi), \quad \phi \in (I-P)\mathscr{P}_{2\pi},$$

then one can easily verify that $\mathscr{K} : (I - P)\mathscr{P}_{2\pi} \to \mathscr{P}_{2\pi}$ is a continuous linear operator.

With this definition of $\mathscr{K}$, (2.40) will have a solution of the form, $x = p + z$, if and only if the following equations are satisfied:

$$(2.48a) \qquad z = \mathscr{K}(I-P)[-\lambda_1 \dot{z} - \lambda_1 \dot{p} + \lambda_2 f_\alpha + G(\cdot, z)]$$

$$(2.48b) \qquad P[-\lambda_1 \dot{z} - \lambda_1 \dot{p} + \lambda_2 f_\alpha + G(\cdot, z)] = 0.$$

An application of the Implicit Function Theorem to (2.48a) shows there is a neighborhood $U \subseteq \mathscr{P}_{2\pi}$ of zero and a neighborhood $V \subseteq R^2$ of $\lambda = 0$ such that (2.48a) has a solution $z^*(\alpha, \lambda)$ for λ in V, $0 \leq \alpha \leq 2\pi$, this solution is unique in U, has continuous second derivatives with respect to α, λ and $z^*(\alpha, 0) = 0$ for all α. Therefore, (2.40) will have a 2π-periodic solution in a sufficiently small neighborhood of Γ for λ in a sufficiently small neighborhood of zero if and only if (α, λ) satisfy Equation (2.48b) with z replaced by $z^*(\alpha, \lambda)$; that is, if and only if (α, λ) satisfy the bifurcation equation

$$(2.49) \qquad F(\alpha, \lambda) \overset{\text{def}}{=} \int_0^{2\pi} \dot{p}[-\lambda_1 \dot{z}^*(\alpha, \lambda) - \lambda_1 \dot{p} + \lambda_2 f_\alpha + G(\cdot, z^*(\alpha, \lambda))]/\eta = 0.$$

The function $F(\alpha, \lambda)$ satisfies $F(\alpha, 0) = 0$ and

(2.50) $F(\alpha,\lambda) = -\lambda_1 + h(\alpha)\lambda_2 + \text{h.o.t.}$

where $h(\alpha)$ is given in (H$_2$) and h.o.t. designates terms which are $O(|\lambda|^2)$ as $|\lambda| \to 0$.

For any $\lambda_2 \neq 0$, the Bifurcation Equation (2.50) is equivalent to

(2.51) $h(\alpha) - \dfrac{\lambda_1}{\lambda_2} + Q\left(\alpha, \dfrac{\lambda_1}{\lambda_2}, \lambda\right) = 0$

where $Q(\alpha,\beta,\lambda)$ has continuous derivatives up through order two and $Q(\alpha,\beta,0) = 0$.

It remains to analyse the bifurcation equation (2.51). Finding all possible solutions of (2.51) in a neighborhood of $\lambda = 0$ is equivalent to finding all possible solutions of the equation

(2.52) $H(\alpha,\beta,\lambda_2) \overset{\text{def}}{=} h(\alpha) - \beta + Q(\alpha,\beta,\lambda_2) = 0$

for all $\alpha \in R$, $\beta \in R$, and λ_2 in a small neighborhood of zero. For $\lambda_2 = 0$, the only possible solutions (α_0,β_0) are those for which $h(\alpha_0) = \beta_0$. If α_0 is such that $h'(\alpha_0) \neq 0$, then the Implicit Function Theorem implies there is a $\delta(\alpha_0,\beta_0) > 0$ and unique solution $\alpha^*(\beta,\lambda_2)$ of (2.52) for $|\beta - \beta_0|, |\lambda_2| < \delta(\alpha_0,\beta_0)$, $\alpha^*(\beta_0,0) = \alpha_0$.

It $h'(\alpha_0) = 0$, then Hypothesis (H$_2$) implies $h''(\alpha_0) \neq 0$. Therefore, there is a $\delta(\alpha_0,\beta_0) > 0$ and a function $\alpha^*(\beta,\lambda_2)$, $\alpha^*(\beta_0,0) = \alpha_0$, such that

$$\partial H(\alpha^*(\beta,\lambda_2),\beta,\lambda_2)/\partial\alpha = 0$$

for $|\beta - \beta_0|, |\lambda_2| < \delta(\alpha_0,\beta_0)$ and $\alpha^*(\beta,\lambda_2)$ is unique in the region $|\alpha - \alpha_0| < \delta(\alpha_0,p_0)$. Thus, the function $M(\beta,\lambda_2) \overset{\text{def}}{=} H(\alpha^*(\beta,\lambda_2),\beta,\lambda_2)$ is a maximum or a minimum of $H(\alpha,\beta,\lambda_2)$ with respect to α for β,λ_2 fixed. A few elementary calculations shows that $M(\beta_0,0) = 0$, $\partial M(\beta_0,0)/\partial\beta = -1$. Therefore, the Implicit Function Theorem implies there is a $\delta(\beta_0)$ and a unique function $\beta^*(\lambda_2), \beta^*(0) = \beta_0$, such that $M(\beta^*(\lambda_2),\lambda_2) = 0$ for $|\lambda_2| < \delta(\beta_0)$. There are two simple solutions of (2.52) near α_0 on one side of the curve $\beta = \beta^*(\lambda_2)$ and no solutions on the other. In terms of the original coordinates λ, this implies there are two solutions of (2.51) near α_0 on one side of the curve $\lambda_1 = \beta^*(\lambda_2)\lambda_2$ and none on the other. The curve $\lambda_1 = \beta^*(\lambda_2)\lambda_2$ is a bifurcation curve and is tangent to the line $\lambda_1 = \beta_0\lambda_2$ at $\lambda = 0$.

The above analysis can be applied to each of the points α_j in Hypothesis (H$_2$). One can choose a $\delta > 0$ such that all solutions of (2.51) for $|\alpha - \alpha_j| < \delta$, $|\lambda| < \delta$ are obtained with the argument above.

The complement of these small regions $|\alpha - \alpha_j| < \delta$ in $[0,2\pi]$ is compact and $h'(\alpha) \neq 0$ in this set. A repeated application of the Implicit Function Theorem now shows that no further bifurcation takes place and all solutions are obtained for $|\lambda|, |\lambda_2| < \delta$. This shows there exist two 2π-periodic solutions in the sectors mentioned in the theorem. If the least period of $f(t)$ is $1/k$, $k \geq 1$ an integer, then we can obtain other 2π-periodic solutions by replacing t by

$t + m/k$, $k = 0,1,2, \ldots, k - 1$. This completes the proof of the theorem.

EXERCISE 2.16. Let S be a sector in Theorem 2.4 which contains 2π-periodic solutions. Suppose γ is a continuous curve in S defined parametrically by $\lambda_j = \lambda_j(\beta)$, $0 \leq \beta \leq 1$, $\lambda_1^2(\beta) + \lambda_2^2(\beta) = 0$ implies $\beta = 0$ and let $x(t,\beta)$ be a 2π-periodic solution of (2.40) for $\lambda_j = \lambda_j(\beta)$ which is continuous in β for $0 < \beta \leq 1$. Prove that $x(\cdot,\beta)$ is continuous at $\beta = 0$ if and only if

$$\lim_{\beta \to 0} \lambda_1(\beta)/\lambda_2(\beta)$$

exists. If this limit is h_0 then $x(\cdot,\beta) \to p(\cdot + \alpha_0)$ where α_0 is a solution of $h(\alpha_0) = h_0$.

Exercise 2.16 shows the difference between considering parameters independently and considering them varying only along some straight line in λ-space. In fact, if one considers $\lambda^0 = (\lambda_1^0, \lambda_2^0)$ fixed in R^2, puts $\lambda = \epsilon\lambda^0$ and considers Equation (2.40) for ϵ small, then the most interesting part of the qualitative behavior of the solutions is lost.

EXERCISE 2.17.   Generalize Theorem 2.4 to the case where $h(\alpha)$ has a finite number of extreme values α_j on $[0,2\pi]$ with $h''(\alpha_j) \neq 0$.

EXERCISE 2.18.   Generalize Theorem 2.4 to the equation

$$\ddot{x} + g(x) = \lambda_1 f_1(t)x + \lambda_2 f_2(t)$$

VIII.3. Periodic Solutions of Perturbed Autonomous Equations

In this section, part of Theorem VI.4.1 is reproved by using the methods of this chapter rather than a coordinate system near a periodic orbit. Consider the equation

$$(3.1) \qquad \dot{x} = f(x) + F(x, \varepsilon),$$

where $f: R^n \to R^n$, $F: R^{n+1} \to R^n$ are continuous, $f(x)$, $F(x, \varepsilon)$ have continuous first partial derivatives with respect to x and $F(x, 0) = 0$ for all x. If the system

$$(3.2) \qquad \dot{x} = f(x)$$

has a nonconstant periodic solution $u(t)$ of period ω whose orbit is $\mathscr{C}$, then the linear variational equation for u is

$$(3.3) \qquad \dot{y} = A(t)\, y, \qquad A(t) = \frac{\partial f(u(t))}{\partial x},$$

and this linear periodic system always has at least one characteristic multiplier equal to 1.

THEOREM 3.1. If 1 is a simple multiplier of (3.3), then there are an $\varepsilon_0 \neq 0$ and a neighborhood W of $\mathscr{C}$ such that equation (3.1) has a periodic solution $u^*(\cdot, \varepsilon)$ of period $\omega^*(\varepsilon)$, $0 \leq |\varepsilon| \leq \varepsilon_0$, such that $u^*(\cdot, 0) = u(\cdot)$, $\omega^*(0) = \omega$, $u^*(t, \varepsilon)$ and $\omega^*(\varepsilon)$ are continuous in t, ε for t in R, $0 \leq |\varepsilon| \leq \varepsilon_0$.

PROOF. Let $\mathscr{P}_\omega$ be the space of continuous ω-periodic n-vector functions with the supremum norm. For any real number β, consider the transformation $t = (1 + \beta)\tau$ in (3.1). If $x(t) = y(\tau)$, then y satisfies the equation

$$(3.4) \qquad \frac{dy}{d\tau} = (1 + \beta)[f(y) + F(y, \varepsilon)].$$

If (3.4) has a solution in $\mathscr{P}_\omega$, then (3.1) has a solution in $\mathscr{P}_{(1+\beta)\omega}$. If $y(\tau) = u(\tau) + z(\tau)$, then z satisfies the equation

$$(3.5) \qquad \frac{dz}{d\tau} = A(\tau)z + G(\tau, z, \varepsilon, \beta),$$

$$G(\tau, z, \varepsilon, \beta) = (1 + \beta)[f(u + z) + F(u + z, \varepsilon)] - A(\tau)z - f(u).$$

Since 1 is a simple characteristic multiplier of (3.3), it follows that $\dot{u}$ is a basis for the ω-periodic solutions of (3.3) and there is an ω-periodic row vector ψ which is a basis for the ω-periodic solutions of the system adjoint to (3.3). Also, it can be assumed that $\int_0^\omega |\psi(t)|^2 \, dt = 1$. We now apply the preceding theory to determine ω-periodic solutions of (3.5). For any h in $\mathscr{P}_\omega$, let

$$\gamma(h) = \int_0^\omega \psi(t)h(t) \, dt.$$

From Section 2, for any h in $\mathscr{P}_\omega$, we know that the equation

$$\frac{dz}{d\tau} = A(\tau)z + h(\tau) - \gamma(h)\psi'(\tau)$$

has a unique solution $\mathscr{M}h$ in $\mathscr{P}_\omega$ such that $\int_0^\omega \dot{u}'(t)\mathscr{M}h(t) \, dt = 0$ and $\mathscr{M}: \mathscr{P}_\omega \to \mathscr{P}_\omega$ is a continuous linear operator. Let $\mathscr{F}: \mathscr{P}_\omega \to \mathscr{P}_\omega$ be defined by

$$\mathscr{F}z = \mathscr{M}G(\cdot, z, \varepsilon, \beta).$$

In the usual way, one shows there are $\varepsilon_1 > 0$, $\beta_1 > 0$, $\delta_1 > 0$ such that $\mathscr{F}$ is uniformly contracting on $\mathscr{P}_\omega(0) = \{z \text{ in } \mathscr{P}_\omega: \|z\| \leq \delta_1\}$. Therefore, $\mathscr{F}$ has a unique fixed point $z^*(\varepsilon, \beta)$ which is continuous in ε, β and continuously differentiable in β. Also $z^*(0, 0) = 0$. This fixed point $z^*(\varepsilon, \beta)$ is a solution of the relation

(3.6) $$\frac{dz}{d\tau} = A(\tau)z + G(\tau, z, \varepsilon, \beta) - B(\varepsilon, \beta)\psi'(\tau),$$

$$B(\varepsilon, \beta) = \gamma(G(\tau, z^*(\varepsilon, \beta), \varepsilon, \beta)).$$

The function $B(\varepsilon, \beta)$ is continuous in ε, β and satisfies $B(0, 0) = 0$. Also, since $z^*(\varepsilon, \beta)$ is continuously differentiable with respect to β, it follows that $B(\varepsilon, \beta)$ is continuously differentiable with respect to β. Furthermore, one can conclude that $\partial B(\varepsilon, \beta)/\partial\beta = \int_0^\omega \psi(\tau)\dot{u}(\tau)\,d\tau$ for $\varepsilon = 0$, $\beta = 0$. This last integral must be different from zero for otherwise the equation

$$\frac{dz}{d\tau} = A(\tau)z + \dot{u}(\tau)$$

would have a solution in $\mathscr{P}_\omega$. This is impossible since this equation always has the solution $z(\tau) = \tau\dot{u}(\tau)$. Since $B(0, 0) = 0$ and $\partial B(0, 0)/\partial\beta \neq 0$, the implicit function theorem implies the existence of a $\beta = \beta(\varepsilon)$ such that $B(\varepsilon, \beta(\varepsilon)) = 0$ for $|\varepsilon| \leqq \varepsilon_0$. Thus, the function $z^*(\varepsilon, \beta(\varepsilon))$ is an ω-periodic solution of (3.1) and the theorem is proved.

VIII.4. Remarks and Suggestions for Further Study

Poincaré [2], in his famous treatise on celestial mechanics, was the first to describe a systematic method for the determination of periodic solutions of differential equations containing a small parameter. In 1940, Cesari [3] gave a method for determining characteristic exponents for linear periodic systems which is in the spirit of the presentation given in this chapter although different in detail. The method of Cesari was extended by Hale [1] and Gambill and Hale [1] to apply to periodic solutions of nonlinear differential equations with small parameters. Further modifications of this basic method by Cesari [3] and [5] led to the presentation given in this chapter. Concurrent with this development, Friedrichs [1], Lewis [1] and Bass [1, 2] gave methods which are very closely related to the above.

Exercise 1.3 is due to Reuter [1]; Exercises 1.5 and 1.6 to Hale [6]; Exercise 1.15 to Hale [7]; Exercise 1.17 to Bailey and Cesari [1]. The Hopf bifurcation theorem was first stated by Hopf [1] although the phenomena was known to many people (see, for example, Poincaré: [2], Minorsky [1], [2]). For a recent book on Hopf bifurcation, see Marsden and MacCracken [1]. The result is also known for equations with delays (see Hale [10]). Exercises 1.22 through 1.27 are in the spirit of bifurcation problems discussed from the general point of view. An extensive literature is available (see, for example, Andronov,

Leontovich, Gordon and Maier [1], Sotomayor [1], Newhouse [1]). Except for notational changes, Lemma 2.3 is due to Lewis [1]. A less general form of property (E) of Section 2 is given by Hale [7]. Theorem 2.3 was stated by Lewis in [1]. Exercises 2.7 and 2.8 are special cases of results by Hale [2]. Exercise 2.9 is due to Lyapunov [1] for analytic systems. Exercise 2.10 is due to Bogoliubov and Sadovnikov [1] while Exercises 2.11 and 2.12 to Golomb [1]. The treatment of Duffing's equation in Section 2 follows Hale and Rodrigues [1]. For the case where damping is also considered, see Hale and Rodrigues [2]. The discussion of periodic solutions near a periodic orbit in Section 2 follows Hale and Táboas [1] (see also Loud [2]). For the case where hypothesis (H_1) is violated, see Hale and Táboas [2]. For an abstract version of this problem, see Hale [11].

An interesting extension of the idea of periodicity as well as property (E) is contained in the theory of autosynartetic solutions of differential equations of Lewis [2]. For small perturbations of nonlinear second order equations see Loud [1].

CHAPTER IX

Alternative Problems for the Solution of

Functional Equations

To motivate the discussion in this chapter, let us interpret the procedure of the previous chapter in a more general setting. Let $\mathscr{P}_T$ be the Banach space of T-periodic n-vector functions with the supremum norm; let A be an $n \times n$ matrix whose columns are in $\mathscr{P}_T$; let L be the linear operator defined on continuously differentiable functions in $\mathscr{P}_T$ by $(Lx)(t) = \dot{x}(t) - A(t)x(t)$; let $N \colon \mathscr{P}_T \to \mathscr{P}_T$ be defined by $(Nx)(t) = \varepsilon f(t, x(t))$ where f is a continuously differentiable function of x and T-periodic in t. The problem of finding a solution of the differential equation

$$\dot{x} - A(t)x = \varepsilon f(t, x)$$

in $\mathscr{P}_T$ is then equivalent to finding an x in $\mathscr{P}_T$ which is in the domain of L such that $Lx = Nx$. With $P, Q, \mathscr{K}$ defined as in relation (VIII.2.5) and Lemma VIII.2.1, the Lemma VIII.2.2 asserts that the equation $Lx = Nx$ is equivalent to the equations

(1) (a) $x = Px + \mathscr{K}(I - Q)Nx$

(b) $QNx = 0$.

Equations (1) have a distinct advantage over the original differential equation. In fact, for N "small" and a fixed x_0 with $Px_0 = x_0$, one can determine a solution $x^*(x_0)$ of (1a) such that $Px^*(x_0) = x_0$. The existence of a solution of $Lx = Nx$ is then equivalent to the determination of an x_0 such that $QNx^*(x_0) = 0$. We have referred to the equations for x_0 as the determining or bifurcation equations, but, more appropriately, we are going to say that these equations represent an *alternative problem* for $Lx = Nx$.

It is the purpose of this chapter to determine larger classes of equations which are equivalent to $Lx = Nx$ for x in a Banach space and then specify alternative problems when N is not necessarily small. This general approach is taken because the ideas are applicable to problems in fields other than

298

ordinary differential equations; for example, integral equations, functional differential equations and partial differential equations. However, our applications are confined to ordinary differential equations.

IX.1. Equivalent Equations

If X, Z are Banach spaces and B is an operator which takes a subset of X into Z, we let $\mathscr{D}(B)$, $\mathscr{R}(B)$, $\mathscr{N}(B)$ denote the domain, range and null space respectively of B. If E is a projection operator defined on a Banach space Z we denote $\mathscr{R}(E)$ by Z_E and Z_E will always denote a subspace which is obtained through a projection operator E in this way. The symbol I will denote the identity. If $L: \mathscr{D}(L) \subset X \to Z$ is a linear operator, then K is said to be a bounded right inverse of L if K is a bounded linear operator taking $\mathscr{R}(L)$ onto $\mathscr{D}(L)$ and $LKz = z$ for z in $\mathscr{R}(L)$.

Let X, Z be Banach spaces; let $N: X \to Z$ be an operator which may be linear or nonlinear; let $L: \mathscr{D}(L) \subset X \to Z$ be a linear operator which may have a nontrivial null space and may have range deficient in Z.

LEMMA 1.1.   Suppose $\mathscr{N}(L)$ and $\mathscr{R}(L)$ admit projections, $\mathscr{N}(L) = X_P$, $\mathscr{R}(L) = Z_{I-Q}$ and suppose L has a bounded right inverse K with $PK = 0$. The equation

(1.1) $Lx = Nx$

is equivalent to the equations

(1.2)   (a) $x = Px + K(I - Q)Nx$,

                         (b) $QNx = 0$.

PROOF.   Equation (1.1) is clearly equivalent to the equations

(1.3)   (a) $(I - Q)(Lx - Nx) = 0$,

                         (b) $Q(Lx - Nx) = 0$.

Since $LX = (I - Q)Z$ and Q is a projection, $QL = 0$. Therefore, (1.3b) is equivalent to (1.2b) and (1.3a) is equivalent to $Lx = (I - Q)Nx$. Since $(I - Q)Nx$ belongs to the range of L and K is a bounded right inverse of L, this latter equation is equivalent to $x = x_0 + K(I - Q)Nx$ where x_0 is in $\mathscr{N}(L)$. But, $PK = 0$ implies $x_0 = Px$ and the lemma is proved.

The condition $PK = 0$ in Lemma 1.1 is no restriction. In fact, if M is any bounded right inverse of L, then $K \overset{\text{def}}{=} (I - P)M$ is also a bounded right inverse and $PK = 0$.

Even with these few elementary remarks, one is in a position to state some alternative problems for (1.1). More specifically, if N were small enough in some sphere so that the contraction principle is applicable to (1.2a) with $Px = x_0$ fixed, then (1.2a) can be solved for an $x^*(x_0)$ and an alternative problem for (1.1) is $QNx^*(x_0) = 0$.

In other words, one can fix an arbitrary element x_0 of $\mathscr{N}(L)$, solve (1.2a) for $x^*(x_0)$ and try to determine x_0 so that (1.2b) is satisfied. The alternative problem has the same "dimension" as the null space of L. In many cases, this dimension is finite whereas the original equation $Lx = Nx$ is infinite dimensional. This result will be stated precisely in the next section, but now we want to obtain other sets of equations which are equivalent to (1.1) and at the same time permit the discussion of cases when N is not small. The idea is quite simple. If one wishes to apply the contraction principle to (1.2a), then $K(I - Q)N$ must be small in some sense. However, if N is large on the whole space, then one should be able to make the product small by choosing the projection operator Q so that fewer values of x are being considered. This is the idea which now will be made precise.

The accompanying Fig. 1 is useful in visualizing the next lemma.

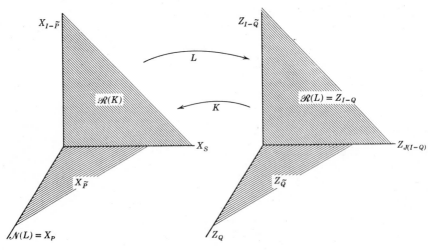

Figure IX.1.1

LEMMA 1.2.   Suppose P, Q, K are as in Lemma 1.1 and let S be any projection operator on X such that $X_S \subset \mathscr{R}(K)$, $SP = 0$. The following conclusions are then valid:

(i) $\tilde{P} = P + S$ is a projection operator.

(ii)   The preimage in Z_{I-Q} under K of X_S, $X_{I-\tilde{P}} \cap \mathscr{D}(L)$ induces a projection $J: Z_{I-Q} \to Z_{I-Q}$. If $X_S = KZ_{J(I-Q)}$, let $I - \tilde{Q} = (I - J)(I - Q)$. Then $\tilde{Q}: Z \to Z$ is a projection and $X_{I-\tilde{P}} \cap \mathscr{D}(L) = K(Z_{I-\tilde{Q}})$.

(iii)   For any x in $\mathscr{D}(L)$,

(1.4)
$$x = K(I - \tilde{Q})Lx + \tilde{P}x.$$

(iv) $\tilde{Q}L = L\tilde{P}$.

PROOF.   (i)   Since $\mathscr{R}(S) \subset \mathscr{R}(K)$ and $PK = 0$ it follows that $PS = 0$. A direct computation now shows that $\tilde{P} = P + S$ is a projection.

(ii) K is a one-to-one map of Z_{I-Q} onto $X_{I-P} \cap \mathscr{D}(L)$ since $Kz_1 = Kz_2$ implies $K(z_1 - z_2) = 0$ and $0 = LK(z_1 - z_2) = z_1 - z_2$. If Z_1, Z_2 are the primages under K of X_S, $X_{I-\tilde{P}} \cap \mathscr{D}(L)$, respectively, then $Z_1 \cap Z_2 = \{0\}$ and $Z_{I-Q} = Z_1 \oplus Z_2$. If $z_n \in Z_2$, $z_n \to z$ as $n \to \infty$, then $Kz_n \to Kz$ as $n \to \infty$ since K is continuous. Furthermore, there are $x_n \in X_{I-\tilde{P}} \cap \mathscr{D}(L)$ and $x \in \mathscr{D}(L)$ such that $Kz_n = x_n$, $Kz = x$. Since $x_n \to x$ and $X_{I-\tilde{P}}$ is closed, we have $x \in X_{I-\tilde{P}}$. Thus $x \in X_{I-\tilde{P}} \cap \mathscr{D}(L)$ and Z_2 is closed. In the same way or using the fact that X_S is closed and K is continuous, one sees that Z_1 is closed. Therefore, a projection J is induced on Z_{I-Q}. If we let $X_S = KZ_{J(I-Q)}$ and $I - \tilde{Q} = (I - J)(I - Q)$ and use the fact that $(I - Q)(I - J) = (I - J)$, it is it is clear that $I - Q$ is a projection. Since $X_S = KZ_{J(I-Q)}$, it is also obvious that $X_{I-\tilde{P}} \cap \mathscr{D}(L) = KZ_{I-\tilde{Q}}$.

(iii)   For any x in $\mathscr{D}(L)$, $x = KLx + Px$. Since $\tilde{Q}$ is a projection operator,

(1.5)
$$x = K(I - \tilde{Q})Lx + K\tilde{Q}Lx + Px.$$

From property (ii), $K(I - \tilde{Q})Lx$ belongs to X_{I-S} since $X_{I-\tilde{P}} \subset X_{I-S}$. Also, a direct computation shows that $\tilde{Q}(I - Q) = J(I - Q)$ and, therefore, $K\tilde{Q}Lx = K\tilde{Q}(I - Q)Lx$ belongs to X_S. Operating on (1.5) with S and using the fact that $SP = 0$, we obtain $Sx = SK\tilde{Q}Lx = K\tilde{Q}Lx$. Since $\tilde{P} = P + S$, this proves relation (iii).

(iv)   Applying L to (1.4) and using the fact that K is a right inverse for L, we have $Lx = (I - \tilde{Q})Lx + L\tilde{P}x$. Therefore, property (iv) is satisfied. This completes the proof of the lemma.

LEMMA 1.3.   Suppose X, Z are Banach spaces, $L: \mathscr{D}(L) \subset X \to Z$ is a linear operator with $\mathscr{R}(L)$, $\mathscr{N}(L)$ admitting projections by $I - Q$, P, respectively, L has a bounded right inverse K, $PK = 0$, and $\tilde{P}$, $\tilde{Q}$ are the operators defined in Lemma 1.2. For any operator $N: X \to Z$, the equation $Lx - Nx = 0$ has a solution if and only if

(1.6)
(a) $x = \tilde{P}x + K(I - \tilde{Q})Nx$,

(b) $\tilde{Q}(Lx - Nx) = 0$.

PROOF.   With $\tilde{P}$, $\tilde{Q}$ as in Lemma 1.2, the equation $Lx - Nx = 0$ is equivalent to the equations $\tilde{Q}(Lx - Nx) = 0$, $(I - \tilde{Q})(Lx - Nx) = 0$. If $(I - \tilde{Q})(Lx - Nx) = 0$, then relation (1.4) implies that $K(I - \tilde{Q})Nx =$

$K(I - \tilde{Q})Lx = (I - \tilde{P})x$ and, thus, (1.6a) is satisfied. If $Lx - Nx = 0$, then (1.6b) is automatically satisfied. Conversely, if (1.6a) is satisfied, then $L(I - \tilde{P})x = (I - \tilde{Q})Nx$. Since relation (iv) of Lemma 1.2 is satisfied, this implies $(I - \tilde{Q})(Lx - Nx) = 0$. If (1.6b) is also satisfied, then $Lx - Nx = 0$. This proves the lemma.

IX.2. A Generalization

As we have seen in Lemma 1.3, the geometric Lemma 1.3 permits the establishment of many sets of equations which are equivalent to equation (1.1). Some assumptions on the linear operator L were imposed in order to obtain these results. However, once the basic relations between the operators $\tilde{P}, \tilde{Q}$ are obtained, one can give an abstract formulation of the basic processes involved. To do this, let X, Z be Banach spaces; let $N: \mathscr{D}(N) \subset X \to Z$ be an operator which may be linear or nonlinear; let $L: \mathscr{D}(L) \subset X \to Z$ be a linear operator and let $F = L - N$. A solution of $Fx = 0$ will be required to belong to $\mathscr{D}(L) \cap \mathscr{D}(N)$. The following hypotheses are made:

H_1:   There are projection operators $\tilde{P}: X \to X$, $\tilde{Q}: Z \to Z$ such that $\tilde{Q}L = L\tilde{P}$.
H_2:   There is a linear map $K: Z_{I-\tilde{Q}} \to X_{I-\tilde{P}}$ such that
    (i) $K(I - \tilde{Q})Lx = (I - \tilde{P})x$, x in $\mathscr{D}(L)$,
    (ii) $LK(I - \tilde{Q})Nx = (I - \tilde{Q})Nx$, x in $\mathscr{D}(N)$.
H_3:   All fixed points of the operator $\Delta = \tilde{P} + K(I - \tilde{Q})N$ belong to $\mathscr{D}(L) \cap \mathscr{D}(N)$.

For the operators L, N satisfying the hypotheses of Lemma 1.3, it was demonstrated that the hypotheses H_1-H_3 can be satisfied by a large class of operators $\tilde{P}, \tilde{Q}$. In fact, hypothesis H_1 is just relation (iv) of Lemma 1.2 and the operators $\tilde{P}, \tilde{Q}$ depend upon a rather arbitrary subspace of X. Hypothesis H_2 (ii) corresponds to the existence of a right inverse of L. Hypothesis H_2 (i) is relation (1.4) of Lemma 1.2. Hypotheses H_3 was automatically satisfied for the particular $\tilde{P}, \tilde{Q}$ constructed and the bounded right inverse considered. If L, N are as specified above and $K, \tilde{P}, \tilde{Q}$ exist so that H_1-H_3 are satisfied, then it is not true that $\tilde{P}, \tilde{Q}$ can be obtained by the construction of Lemma 1.2. In fact, in that construction it was assumed that L had a bounded right inverse and $\mathscr{R}(L), \mathscr{N}(L)$ admitted projections. There is no way to deduce these properties from H_1-H_3. Of course, in the applications, Lemma 1.2 is a rather natural way to obtain $\tilde{P}, \tilde{Q}$.

LEMMA 2.1. If hypotheses H_1-H_3 are satisfied, then the equation $Fx = 0$ has a solution x in $\mathscr{D}(L) \cap \mathscr{D}(N)$ if and only if

(2.1)   (a) $x = \Delta x \overset{\text{def}}{=} \tilde{P}x + K(I - \tilde{Q})Nx$,

               (b) $\tilde{Q}Fx = 0$.

PROOF. The relation $Fx = 0$ implies $\tilde{Q}Fx = 0$, $(I - \tilde{Q})Fx = 0$. There-fore, $(I - \tilde{Q})Lx = (I - \tilde{Q})Nx$ and hypothesis H_2 (i) implies

$$K(I - \tilde{Q})Nx = K(I - \tilde{Q})Lx = (I - \tilde{P})x.$$

Thus, $x = \Delta x$. Conversely, suppose (2.1) is satisfied. If $x = \Delta x$, then $(I - \tilde{P})x = K(I - \tilde{Q})Nx$. Hypothesis H_3 implies x in $\mathscr{D}(L) \cap \mathscr{D}(N)$ and H_2 (ii) implies that

$$L(I - \tilde{P})x = LK(I - \tilde{Q})Nx = (I - \tilde{Q})Nx.$$

But this fact together with H_1 implies that $(I - \tilde{Q})Fx = 0$. By hypothesis, $\tilde{Q}Fx = 0$ and the lemma is proved.

IX.3. Alternative Problems

In addition to the hypotheses H_1-H_3 imposed on the operators $\tilde{P}$, $\tilde{Q}$, K, L, N of the previous section, we also suppose $\mathscr{D}(N) = X$ and

H_4: There exist a constant μ and a continuous nondecreasing function $\alpha(\rho)$, $0 \leqq \rho < \infty$ such that

$$|K(I - \tilde{Q})Nx_1 - K(I - \tilde{Q})Nx_2| \leqq \alpha(\rho)|x_1 - x_2|,$$

$$|K(I - \tilde{Q})Nx_1| \leqq \alpha(\rho)|x_1| + \mu \quad \text{for} \quad |x_1|, |x_2| < \rho.$$

For any positive constants c, d with $c < d$, let

(3.1)   (a) $V(c) = \{x \text{ in } X_{\tilde{P}} : |x| \leqq c\}$.

        (b) $\mathscr{S}(\tilde{x}, c, d) = \{x \text{ in } X : \tilde{P}x = \tilde{x}, \tilde{x} \text{ in } V(c), |x| \leqq d\}$.

where $\tilde{x}$ is a fixed element of $V(c)$.

THEOREM 3.1. Suppose $\mathscr{D}(N) = X$, $V(c)$, $\mathscr{S}(\tilde{x}, c, d)$, $c < d$, are defined by (3.1), the hypotheses H_1-H_3 of the previous section and hypothesis H_4 above are satisfied with $\alpha(d) < 1$, $\alpha(d) d < d - c - \mu$. Then there exists a unique continuous function $G \colon V(c) \to X$, $G\tilde{x}$ in $\mathscr{S}(\tilde{x}, c, d)$ such that $G\tilde{x}$ satisfies the equation

(3.2) $$x = \Delta x \overset{\text{def}}{=} \tilde{P}x + K(I - \tilde{Q})Nx.$$

If there is an $\tilde{x}$ in $V(c)$ such that

(3.3) $$\tilde{Q}FG\tilde{x} = 0,$$

then $FG\tilde{x} = 0$. Conversely, if there is an x such that $Fx = 0$, $|x| \leqq d$, $|\tilde{P}x| \leqq c$, then $x = G\tilde{P}x$ and $\tilde{x} = \tilde{P}x$ is a solution of (3.3).

PROOF.   For any $\tilde{x}$ in $V(c)$ and x in X, let $H(x, \tilde{x}) = \tilde{x} + K(I - \tilde{Q})Nx$.
For any x in $\mathscr{S}(\tilde{x}, c, d)$, $\tilde{P}H(x, \tilde{x}) = \tilde{x}$,

$$|H(x, \tilde{x})| \leqq c + \alpha(d)|x| + \mu \leqq c + \alpha(d)\ d + \mu < d,$$

and

$$|H(x_1, \tilde{x}) - H(x_2, \tilde{x})| \leqq \alpha(d)|x_1 - x_2|.$$

Therefore, $H(\,\cdot\,, \tilde{x})\colon \mathscr{S}(\tilde{x}, c, d) \to \mathscr{S}(\tilde{x}, c, d)$ is a contraction and there is a unique fixed point $G\tilde{x}$ of $H(\,\cdot\,, \tilde{x})$ in $\mathscr{S}(\tilde{x}, c, d)$. The function G is obviously continuous and satisfies (3.2) since $\tilde{x} = \tilde{P}\tilde{x}$. If $\tilde{x}$ satisfies (3.3), then Lemma 2.1 implies that $FG\tilde{x} = 0$.

Conversely, if x is a solution of $Fx = 0$, $\tilde{x} = \tilde{P}x$, $|x| \leqq d$, $|\tilde{x}| \leqq c$, then Lemma 2.1 implies that $x = \Delta x$ and $\tilde{Q}Fx = 0$. But the proof of the first part of the lemma showed that Δ had a unique fixed point with $\tilde{P}x = \tilde{x}$. Therefore, the solution x of $Fx = 0$ must satisfy $x = G\tilde{x}$, $\tilde{x} = \tilde{P}x$. Since x must also satisfy $\tilde{Q}Fx = 0$, it follows that $\tilde{x}$ satisfies (3.3). This completes the proof of the theorem.

In the sense of Theorem 3.1, finding a solution $\tilde{x}$ in $V(c)$ of (3.3) is equivalent to finding a solution of equation (1.1) with $|x| \leqq d$, $|\tilde{P}x| \leqq c$. Therefore, we refer to equation (3.3) as an alternative problem for equation (1.1).

IX.4. Alternative Problems for Periodic Solutions

In this section, we go into some detail on the construction and more detailed meaning of alternative problems in connection with the existence of periodic solutions of ordinary differential equations.

Suppose $g\colon (-\infty, \infty) \times C^n \to C^n$ is continuous, $g(t, u)$ is locally lipschitzian in u, $g(t, u) = g(t + 2\pi, u)$ for all $t \in (-\infty, \infty)$, $u \in C^n$. Our problem is the determination of 2π-periodic solutions of the equation

$$(4.1) \qquad\qquad \dot{u} = g(t, u).$$

Let us reformulate this problem in terms of our previous notation.

Let Y be the Banach space of continuous functions taking $[0, 2\pi]$ into C^n with the uniform norm and let $B\colon \mathscr{D}(B) \subset Y \to Y$ be the linear operator whose domain $\mathscr{D}(B) = \{y \in Y$ such that $\dot{y} \in Y\}$ and $By(t) = \dot{y}(t)$, $0 \leqq t \leqq 2\pi$, for $y \in \mathscr{D}(B)$. Let $M\colon Y \to Y$ be the operator defined by $(My)(t) = g(t, y(t))$, $0 \leqq t \leqq 2\pi$, $y \in Y$; $\Gamma\colon Y \to C^n$ be the bounded linear operator defined by $\Gamma y = y(0) - y(2\pi)$, $y \in Y$. Finding a 2π-periodic solution of (4.1)i s now equivalent to solving the boundary value problem

$$(4.2) \qquad\qquad By = My,$$
$$\Gamma y = 0, \qquad y \in Y.$$

If $X = \mathcal{N}(\Gamma)$ and L, N are the restrictions of B, M to X, then the boundary value problem (4.2) is equivalent to finding a solution of the equation

$$(4.3) \qquad\qquad Lx = Nx, \qquad x \in X.$$

Let $P\colon X \to X$ be the projection operator defined by

$$(4.4) \qquad\qquad Px = \frac{1}{2\pi} \int_0^{2\pi} x(t)\,dt;$$

that is, Px is the mean value of the 2π-periodic function x in X. With this notation, $\mathcal{N}(L) = X_P = \{$all constant functions in $X\}$, $\mathcal{R}(L) = X_{I-P} = \{$all 2π-periodic functions with mean value zero$\}$. Therefore, the operator Q in Section 1 is P. If $z \in X_{I-P}$, then the equation $Lx(t) = \dot{x}(t) = z(t)$ has a solution in X and a unique solution with mean value zero which depends continuously upon z. If we designate this solution by Kz, then $K\colon X_{I-P} \to X_{I-P}$, $PK = 0$, and K is a bounded right inverse of L.

Any x in X has a Fourier series

$$(4.5) \qquad\qquad x \sim \sum_{k=-\infty}^{\infty} a_k e^{ikt}, \qquad a_k = \frac{1}{2\pi} \int_0^{2\pi} e^{-ikt} x(t)\,dt.$$

For any given integer $m > 0$, let $S_m\colon X \to X$ be defined by

$$S_m x = \sum_{0 < |k| \le m} a_k e^{ikt}.$$

One easily checks that S_m is a projection operator and Lemma 1.2 implies $\tilde{P}_m, \tilde{Q}_m$ of that lemma are given by $\tilde{P}_m = P + S_m$, $\tilde{Q}_m = \tilde{P}_m$. Equation (4.3) is then equivalent to the equations

$$(4.6) \qquad\qquad \text{(a)} \quad x = \tilde{P}_m x + K(I - \tilde{P}_m)Nx,$$

$$\qquad\qquad\qquad \text{(b)} \quad \tilde{P}_m(Lx - Nx) = 0,$$

for every integer $m \ge 0$.

From the definitions of $\tilde{P}_m$, K and the form of x in (4.5),

$$K(I - \tilde{P}_m)x(t) = \sum_{|k| > m} \frac{a_k}{ik} e^{ikt}.$$

Recall that Parseval's relation implies that $\int_0^{2\pi} |x(t)|^2\,dt = \sum_{k=-\infty}^{\infty} |a_k|^2$. Since

$$\sum_{|k| > m} \left| \frac{a_k}{k} \right| \le \left(\sum_{|k| > m} |a_k|^2 \right)^{1/2} \left(\sum_{|k| > m} \frac{1}{k^2} \right)^{1/2}$$

$$\le \left(\int_0^{2\pi} |x(t)|^2\,dt \right)^{1/2} \left(\sum_{|k| > m} \frac{1}{k^2} \right)^{1/2}$$

$$\le (2\pi)^{1/2} \left(\sum_{|k| > m} \frac{1}{k^2} \right)^{1/2} |x|$$

$$\overset{\text{def}}{=} \gamma(m)|x|,$$

it follows that the Fourier series for $K(I - \tilde{P}_m)x(t)$ is absolutely convergent and

$$|K(I - \tilde{P}_m)x| \leq \gamma(m)|x|$$

or all $x \in X$. Note that $\gamma(m) \to 0$ as $m \to \infty$ since the series $\sum_k k^{-2} < \infty$.

Since $g(t, u)$ is assumed to be locally lipschitzian in u, there is a constant $\nu > 0$ and a continuous nondecreasing function $\beta(\rho)$, $0 \leq \rho < \infty$, such that $Nx(t) = g(t, x(t))$ satisfies

$$|Nx_1 - Nx_2| \leq \beta(\rho)|x_1 - x_2|$$

$$|Nx_1| \leq \beta(\rho)|x_1| + \nu$$

for all $|x_1| \leq \rho$, $|x_2| \leq \rho$.

For any constant $d > 0$ and any m, it follows that

$$|K(I - P_m)(Nx_1 - Nx_2)| \leq \gamma(m)\beta(d)|x_1 - x_2|$$

$$|K(I - P_m)Nx_1| \leq \gamma(m)\beta(d)|x_1| + \gamma(m)\nu$$

for all $|x_1| \leq d$, $|x_2| \leq d$. Suppose $0 < c < d$ are arbitrary constants and m is chosen so large that

$$\gamma(m)\beta(d) < 1, \qquad \gamma(m)\beta(d)\, d + \gamma(m)\nu < d - c$$

The operator Δ defined by (3.2) is then a contraction mapping of the set $\mathscr{S}(\tilde{x}, c, d)$ in (3.1) into itself. This implies that the equation

(4.7) $$x = \tilde{x} + K(I - \tilde{P}_m)Nx$$

has a solution $G\tilde{x}$ for every $\tilde{x} \in \{x \text{ in } X_{\tilde{P}_m} : |x| \leq c\}$. An alternative problem for (4.1) is then the equation

(4.8) $$\tilde{P}_m(L - N)G\tilde{x} = 0.$$

In words, the existence of this alternative problem implies the following. If (4.1) is to have a 2π-periodic solution u, then all of the Fourier coefficients of the function $v(t) = \dot{u}(t) - g(t, u(t))$ must be zero. Let

$$u(t) = u_m(t) + u_m^*(t),$$

$$u_m(t) = \sum_{|k| \leq m} a_k e^{ikt},$$

$$u_m^*(t) = \sum_{|k| > m} a_k e^{ikt}.$$

The above remarks imply there is always an integer m such that one can fix $\tilde{x} = u_m$ and determine $G\tilde{x} = u_m^*(u_m)$ in such a way that the Fourier series for $v(t)$ contains only the harmonics e^{ikt}, $|k| \leq m$ (this is the meaning of the solution $G\tilde{x}$ of (4.6a)). The alternative problem then involves the determination of $u_m(t)$ in such a way that the remaining Fourier coefficients of $v(t)$ vanish.

There are two serious questions remaining concerning the determination of periodic solutions of (4.1) by this method. First, the size of the finite dimensional problem cannot be determined a priori but involves very careful estimates of the norm of the operator $K(I - \tilde{P}_m)$. Also, the finite dimensional problem (4.8) involves a function G which is only implicitly known and trying to assert the actual existence of an $\tilde{x}$ is extremely difficult.

In fact, one cannot hope to solve (4.7) directly, but must use some approximation scheme. Let us write $G\tilde{x} = \tilde{x} + \tilde{y}$ where $\tilde{y}$ is in $(I - \tilde{P}_m)X$ and is given by $\tilde{y} = K(I - \tilde{P}_m)NG\tilde{x}$. It is impossible to determine $\tilde{y}$, but if the existence of $G\tilde{x}$ is proved by the contraction principle, then one has obtained a priori estimates on $\tilde{y}$ as

$$(4.9) \qquad |\tilde{y}| \leq \delta_m ,$$

for some constant δ_m. The alternative problem (4.8) will certainly have a solution if one can show that the equation

$$(4.10) \qquad \tilde{P}_m(L - N)(\tilde{x} + \tilde{y}) = 0$$

has a solution $\tilde{x}$ for every given $\tilde{y}$ which satisfies the bounds (4.9). To show this latter property, one would naturally look first at the approximate equation obtained by setting $\tilde{y} = 0$; namely,

$$(4.11) \qquad \tilde{P}_m(L - N)\tilde{x} = 0.$$

Equation (4.11) is the mth *Galerkin approximation* for the solution of $Lx = Nx$. Methods for the determination of explicit solutions of (4.11) and for showing that (4.10) has a solution for every $\tilde{y}$ satisfying (4.9) would take us too far afield. The interested reader may consult the references for the practical applicability of this method.

IX.5. The Perron-Lettenmeyer Theorem

In this section, we illustrate how the techniques of Sections 1 and 3 can be used to give a proof of a theorem of Perron and Lettenmeyer concerning the number of analytic solutions of a system of linear differential equations of the form

$$(5.1) \qquad t^{\sigma_j}\dot{x}_j = N_j(t)x, \qquad j = 1, 2, \ldots, n,$$

where $x = \mathrm{col}(x_1, \ldots, x_n)$, $N_j = \mathrm{row}(N_{j1}, \ldots, N_{jn})$, the $N_{jk}(t)$ are analytic for $|t| \leq \delta$, $\delta > 0$, and each σ_j is a nonnegative integer.

THEOREM 5.1. If $\mu = n - \sum_{j=1}^{n}\sigma_j \geq 0$, then there are at least μ linearly independent solutions of (5.1) which are analytic for $|t| \leq \delta$.

PROOF. Let $\mathscr{B}$ be the set of all scalar functions which are analytic for $|t| \leqq \delta$. If b is in $\mathscr{B}$, then $b(t) = \sum_{m=0}^{\infty} b_m t^m$ and we define $|b| = \sum_{m=0}^{\infty} |b_m| \delta^m$. The function $|\cdot|$ is a norm on $\mathscr{B}$ and $\mathscr{B}$ is a Banach space. For any nonnegative integer σ, let $l_\sigma : \mathscr{B} \to \mathscr{B}$ be the linear operator defined by

(5.2)
$$l_\sigma b(t) = t^\sigma db(t)/dt.$$

It is clear that

$$\mathscr{N}(l_\sigma) = \{\text{constant functions in } \mathscr{B}\},$$

$$\mathscr{R}(l_\sigma) = \{c \text{ in } \mathscr{B} : c = t^\sigma b, \; b \text{ in } \mathscr{B}\}.$$

If the projection operators p, q_σ on $\mathscr{B}$ are defined by

(5.3)
$$pb = b_0, \qquad q_\sigma b = \sum_{m=0}^{\sigma-1} b_m t^m,$$

then p, $1 - q_\sigma$ are projections onto $\mathscr{N}(l_\sigma)$, $\mathscr{R}(l_\sigma)$, respectively. The right inverse k_σ of l_σ on $\mathscr{R}(l_\sigma)$ such that $pk_\sigma = 0$ is given by

(5.4)
$$k_\sigma(1 - q_\sigma)b(t) = \sum_{m=0}^{\infty} \frac{b_{m+\sigma}}{m+1} t^{m+1},$$

and $|k_\sigma(1 - q_\sigma)b| \leqq \delta^{1-\sigma}|(1 - q_\sigma)b|$. This proves k_σ is continuous.

Let X be the Banach space which is the cross product of n copies of $\mathscr{B}$ with $|x|_X = \max_j |x_j|_\mathscr{B}$, $x = \mathrm{col}(x_1, \ldots, x_n)$, x_j in $\mathscr{B}$. We write $|x|$ for $|x|_X$ since no confusion will arise. Let $\sigma_1, \ldots, \sigma_n$ be the nonnegative integers in (5.1), $\sigma = (\sigma_1, \ldots, \sigma_n)$ and define operators L_σ, P, Q_σ, K_σ, N by

(5.5)
$$L_\sigma x = \mathrm{col}(l_{\sigma_1} x_1, \ldots, l_{\sigma_n} x_n),$$

$$Px = \mathrm{col}(px_1, \ldots, px_n),$$

$$Q_\sigma x = \mathrm{col}(q_{\sigma_1} x_1, \ldots, q_{\sigma_n} x_n),$$

$$K_\sigma x = \mathrm{col}(k_{\sigma_1} x_1, \ldots, k_{\sigma_n} x_n),$$

$$Nx = \mathrm{col}(N_1 x, \ldots, N_n x),$$

where the l_{σ_j} are defined by (5.2), the p, q_{σ_j} by (5.3), k_{σ_j} by (5.4), and $N_j x$ in (5.1). Equation (5.1) may now be written as

(5.6)
$$L_\sigma x = Nx, \qquad x \text{ in } X,$$

and from Lemma 1.1 is equivalent to the linear equations

(5.7)
$$x = Px + K_\sigma(I - Q_\sigma)Nx,$$

$$Q_\sigma Nx = 0, \qquad x \text{ in } X.$$

If the linear operator $K_\sigma(I - Q_\sigma)N$ were a contraction operator, then one could fix a constant n-vector x_0 and determine a unique solution $x^* = x^*(x_0)$,

$Px^* = x_0$, of the equation $x = x_0 + K_\sigma(I - Q_\sigma)Nx$. Equations (5.7) would then have a solution if and only if x_0 satisfied the equations $Q_\sigma Nx^*(x_0) = 0$. It is clear that such an $x^*(x_0)$ is linear in x_0. Thus there would be $\sum_{j=1}^{n} \sigma_j$ homogeneous linear equations for n unknown scalars (the components of x_0) and, therefore $n - \sum_{j=1}^{n} \sigma_j$ solutions of (5.7). This explains the possible appearance of μ in the statement of the theorem except for the fact that the operator $K_\sigma(I - Q_\sigma)N$ may not be contracting. On the other hand, there is a natural way to use Lemma 1.2 to circumvent this difficulty.

For any given nonnegative integer r, let $\tilde{P} = P_{r+1}$ be the projection operator on X which takes each component of x in X into the polynomial of degree r consisting of the first $(r + 1)$ terms of its power series expansion. The operator $\tilde{Q}$ in Lemma 1.2 is easily seen to be $Q_{r+\sigma}, r + \sigma = \text{col}(r + \sigma_1, \ldots, r + \sigma_n)$. Thus, Lemma 1.3 implies that (5.6) is equivalent to the equations

(5.8)
$$x = P_{r+1}x + K_\sigma(I - Q_{r+\sigma})Nx,$$

$$Q_{r+\sigma}(L_\sigma - N)x = 0.$$

Furthermore from the definition of K_σ, it follows immediately that there is a $\beta > 0$ independent of r such that $|K_\sigma(I - Q_{r+\sigma})Nx| \leq \beta|x|/(r + 1)$ for all x in X. Consequently, the operator $K(I - Q_{r+\sigma})N$ is contracting for r sufficiently large. Finally, fixing an n-vector polynomial $f(t)$ of degree r, one can determine a unique solution $x^* = x^*(f)$ of $x = f + K_\sigma(I - Q_{r+\sigma})Nx$ which is linear and continuous in f. Therefore, the equations (5.8) have a solution if and only if f satisfies $Q_{r+\sigma}(L_\sigma - N)x^*(f) = 0$. These represent $nr + \sum_{j=1}^{n} \sigma_j$ homogeneous linear equations for the $n(r + 1)$ coefficients of the polynomial f. Therefore there are always $\mu = n - \sum_{j=1}^{n} \sigma_j$ solutions. This proves the theorem.

If all $\sigma_j - 1$ in (5.1), the equation is said to have a regular singular point at $t = 0$. In this case, the above theorem says nothing.

XI.6. Remarks and Suggestions for Further Study

For the case in which $\tilde{P} = P$, $\tilde{Q} = Q$ and N in (1.1) is small and has a small lipschitz constant, Theorem 3.1 has appeared either explicitly or implicitly in many papers; in particular, Cesari [5, 6], Cronin [1], Bartle [1], Graves [1], Nirenberg [1], Vainberg and Tregonin [1], Antosiewicz [1]. However, Cesari [5, 6] injected a significant new idea when he observed that finite dimensional alternative problems could always be associated with certain types of equations (1.1) even when the nonlinearities N are not small. The construction of $\tilde{P}$, $\tilde{Q}$ given in Section 1 follows Bancroft, Hale and

Sweet [1] and was motivated by Cesari [6]. The abstract formulation in Section 2 was independently discovered by Locker [1] proceeding along the lines of Section 4.

Cesari has applied the methods of Section 4 to prove in [5] the existence and bound of a 2π-periodic solution of

$$\ddot{x} + x^3 = \sin t$$

by using the second Galerkin approximation $x = a \sin t + b \sin 3t$ and to show in [6] that the boundary value problem

$$\ddot{x} + x + \alpha x^3 = \beta t, \qquad 0 \leq t \leq 1,$$

$$x(0) = 0, \qquad x(1) + \dot{x}(1) = 0,$$

has a solution by using the first Galerkin approximation. Borges [1] has applied the same process to obtain the existence and bounds for periodic solutions of nonlinear (periodic or autonomous) second order differential equations by using only the first Galerkin procedure. Knobloch [1, 2] has also used the method for existence of periodic solutions of second order equations. The papers of Urabe [2] discuss similar procedures for multipoint boundary value problems. Williams [1] has discovered interesting connections between Section 4 and the Leray-Schauder degree when the alternative problem is finite dimensional.

When N is small enough to make $K(I - Q)N$ a contraction operator on some subset of the basic space X, Theorem 3.1 is applicable to hyperbolic partial differential equations; see Cesari [7], Hale [9], Rabinowitz [1], Hall [1]. Cesari [8] has also obtained results for elliptic partial differential equations. For applications to functional differential equations, see Perelló [1, 2] and to integral equations, see Vainberg and Tregonin [1].

The idea for the proof of the Perron-Lettenmeyer Theorem of Section 5 was communicated to the author by Sibuya. Amplifications of this idea appear in Harris, Sibuya and Weinberg [1]. A paper which is not unrelated to the approach of this chapter is the paper of McGarvey [1] on asymptotic solutions of linear equations with periodic coefficients.

CHAPTER X

The Direct Method of Liapunov

In the previous chapters, we have repeatedly asserted the stability of certain solutions or sets of solutions of a differential equation. The proofs of these results in most cases were based upon an application of the variation of constants formula, and, as a consequence, the analysis was confined to a small neighborhood of the solution or set under discussion. In his famous memoir, Liapunov gave some very simple geometric theorems (generally referred to as the direct method of Liapunov) for deciding the stability or instability of an equilibrium point of a differential equation. The idea is a generalization of the concept of energy and its power and usefulness lie in the fact that the decision is made by investigating the differential equation itself and not by finding solutions of the differential equation. These basic ideas of Liapunov have been exploited extensively with many books in the references being devoted entirely to this subject. The purpose of the present chapter is to give an introduction to some of the fundamental ideas and problems in this field. There is also one section devoted to a generalization known as the principle of Wazewski.

X.1. Sufficient Conditions for Stability and Instability in Autonomous Systems

Let $\Omega \subset R^n$ be an open set in R^n with 0 in Ω. A scalar function $V(x)$, x in Ω, is *positive semidefinite* on Ω if it is continuous on Ω and $V(x) \geqq 0$, x in Ω. A scalar function $V(x)$ is *positive definite* on Ω if it is positive semidefinite, $V(0) = 0$, $V(x) > 0$, $x \neq 0$. A scalar function $V(x)$ is *negative semidefinite* (*negative definite*) on Ω if $-V(x)$ is positive semidefinite (positive definite) on Ω.

The function $V(x_1, x_2) = x_1^2 + x_2^2$ is positive definite on R^2, $V(x_1, x_2) = x_1^2 + x_2^2 - x_2^3$ is positive definite on a sufficiently small strip about the x_1-axis, $V(x_1, x_2) = x_2^2 + x_1^2/(1 + x_1^4)$ is positive definite on R^2. In each of these cases, there is a $c_0 > 0$ such that $\{(x_1, x_2): V(x_1, x_2) = c\}$ is a closed curve for every nonnegative constant $c \leqq c_0$. Near $(x_1, x_2) = (0, 0)$, each of these functions have the qualitative properties depicted in Fig. 1.1. Any positive definite function in R^n has some of these same qualitative properties, but a precise characterization of the level curves of V has not yet been solved. However, the following lemma holds.

LEMMA 1.1.   If V is positive definite on Ω, then there is a neighborhood U of $x = 0$ and a constant $c_0 > 0$ such that any continuous curve from the origin to ∂U must intersect the set $\{x: V(x) = c\}$ provided that $0 \leq c \leq c_0$.

PROOF.   Let U be a bounded open neighborhood of 0 in U with $\bar{U} \subset \Omega$. If

$$l = \min_{x \text{ in } \partial U} V(x),$$

then $l > 0$. Since $V(0) = 0$, $V(x) \geq l$ for x in ∂U, and V is continuous, it follows that the function $V(x)$ must take on all values between 0 and l along any continuous curve going from 0 to ∂U.

LEMMA 1.2.   If $V(x) = V_p(x) + W(x)$ where $W(x) = o(|x|^p)$ as $|x| \to 0$ is continuous and V_p is a positive definite homogeneous polynomial of degree p, then $V(x)$ is positive definite in a neighborhood of $x = 0$.

PROOF.   If $\min_{|x|=1} V_p(x) = k$, then $k > 0$. Therefore

$$V_p(x) = |x|^p V(x/|x|) \geq k|x|^p$$

for $x \neq 0$. For any $\varepsilon > 0$, there is a $\delta(\varepsilon) > 0$ such that $|W(x)| < \varepsilon|x|^p$ for $|x| < \delta(\varepsilon)$. If $\varepsilon = k/2$, then

$$V(x) \geq V_p(x) - |W(x)| \geq k|x|^p - \left(\frac{k}{2}\right)|x|^p = \left(\frac{k}{2}\right)|x|^p$$

if $0 < |x| < \delta(k/2)$. This proves the lemma.

LEMMA 1.3.   Suppose $V_p(x)$ is a homogeneous polynomial of degree p. If p is odd, then V_p cannot be sign definite.

PROOF.   If $x = (x_1, \ldots, x_n)$, $x_1 \neq 0$, let $x = x_1 u$. Then $V_p(x) = x_1^p V_p(u)$. For a given value of u for which $V_p(u) \neq 0$, say positive, the function $V_p(x)$ has the sign of x_1^p. If p is odd, this implies $V_p(x)$ must change sign in every neighborhood of $x = 0$.

General criteria for determining the positive definiteness of an arbitrary homogeneous polynomial of degree p are not known, but for $p = 2$, we have

LEMMA 1.4.   (Sylvester).   The quadratic form $x'Ax = \sum_{i,j=1}^{n} a_{ij} x_i x_j$, $A' = A$, is positive definite if and only if

$$\det(a_{ij}, i, j = 1, 2, \ldots, s) > 0, \qquad s = 1, 2, \ldots, n.$$

A proof of this lemma may be found in Bellman [2].

Consider the differential equation

(1.1) $$\dot{x} = f(x),$$

where $f: R^n \to R^n$ is continuous and satisfies enough smoothness conditions to ensure that a solution of (1.1) exists through any point, is unique and depends continuously upon the initial data.

For any scalar function V defined and continuous together with $\partial V/\partial x$ on an open set Ω of R^n, we define $\dot{V} = \dot{V}_{(1.1)}$ as

(1.2) $$\dot{V}(x) = \frac{\partial V(x)}{\partial x} f(x).$$

If $x(t)$ is a solution of (1.1), then $dV(x(t))/dt = \dot{V}(x(t))$; that is, $\dot{V}$ is the derivative of V along the solutions of (1.1). Notice that $\dot{V}$ can be computed directly from $f(x)$ and, therefore, involves no integrations.

The theory below also remains valid for scalar functions V defined and only continuous on an open set Ω of R^n provided that $\dot{V}$ is given as

$$\dot{V}(\xi) = \overline{\lim_{h \to 0^+}} \frac{1}{h} [V(x(h, \xi)) - V(\xi)],$$

where $x(h, \xi)$ is the solution of (1.1) with initial value ξ at $h = 0$. If V is locally lipschitz continuous, then this latter definition can be shown to be equivalent to

$$\dot{V}(\xi) = \overline{\lim_{h \to 0^+}} \frac{1}{h} [V(\xi + hf(\xi)) - V(\xi)].$$

In the applications, it is sometimes necessary to consider functions $V(x)$ which do not have continuous first partial derivatives at all points x. On the other hand, the functions $V(x)$ are usually piecewise smooth with the sets of discontinuities in the derivatives of V occurring on surfaces of lower dimension than the basic space R^n.

Since the proofs of the theorems below do not use the differentiability of V, they are stated without this hypothesis. However, in the majority of the applications, $\dot{V}$ will be given by (1.2).

THEOREM 1.1.   If there is a continuous positive definite function $V(x)$ on Ω with $\dot{V} \leq 0$, then $x = 0$ is a solution of (1.1) and it is stable. If, in addition, $\dot{V}$ is negative definite on Ω, then the solution $x = 0$ is asymptotically stable.

PROOF.   Let $B(r)$ be the ball in R^n of radius r with center at the origin. There is an $r > 0$ such that $B(r) \subset \Omega$. For $0 < \varepsilon < r$, $0 < k = \min_{|x| = \varepsilon} V(x)$. Suppose that $0 < \delta \leq \varepsilon$ is such that $V(x) < k$ for $|x| \leq \delta$. Such a δ always exist since $V(0) = 0$ and V is continuous. If x_0 is in $B(\delta)$, then the solution $x(t)$ of (1.1) with $x(0) = x_0$ is in $B(\varepsilon)$ for $t \geq 0$ since $\dot{V}(x(t)) \leq 0$ implies $V(x(t)) \leq V(x_0)$, $t \geq 0$. This proves $x = 0$ is a solution and it is stable.

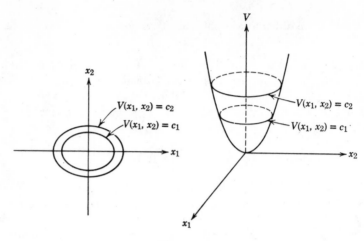

Figure X.1.1

Since $x = 0$ is stable, there is a $b_0 > 0$, $H > 0$, such that the solution $x(t)$ exists and satisfies $|x(t)| < H$ for $t \geqq 0$ provided that $|x_0| < b_0$. Also, for any $\varepsilon > 0$, there is a $\delta(\varepsilon) > 0$ such that $|x(t)| < \varepsilon$ for $t \geqq 0$, $|x_0| < \delta(\varepsilon)$. To prove asymptotic stability, it is sufficient to show there is a $T > 0$ such that $|x(t)| < \delta(\varepsilon)$ for $t \geqq T$, $|x_0| < b_0$. Suppose there is an x_0 with $|x_0| < b_0$ and $|x(t)| \geqq \delta(\varepsilon)$ for $t \geqq 0$. If $\gamma > 0$ is such that $\dot{V}(x(t)) \leqq - \gamma$ for $\delta(\varepsilon) \leqq x < H, t \geqq 0$, then

$$V(x(t)) \leqq V(x_0) - \gamma t.$$

If $\beta > 0$, k satisfy $0 < \beta \leqq V(x) \leqq k$ for $\delta(\varepsilon) \leqq |x| < H$, choose $T = (k - \beta)/\gamma$. For $t \geqq T$, $V(x(t)) < \beta$. Thus, there must exist a t_0, $0 \leqq t_0 \leqq T$ such that $|x(t_0)| < \delta(\varepsilon)$. Stability implies $|x(t)| < \varepsilon$ for $t \geqq t_0$ and, in particular, for $t \geqq T$. This completes the proof of the theorem.

THEOREM 1.2. Suppose $x = 0$ is an equilibrium point of (1.1) contained in the closure of an open set U and let Ω be a neighborhood of $x = 0$. Suppose V is a scalar function on Ω which satisfies:

(i) V, $\dot{V}$ positive on $U \cap \Omega \backslash \{0\}$, $V(0) = 0$,

(ii) $V = 0$ on that part of the boundary of U inside Ω.

Then the solution $x = 0$ of (1.1) is unstable. More specifically, if Ω_0 is any bounded neighborhood of $x = 0$ with $\bar{\Omega}_0$ in Ω, then any solution with initial value in $(U \cap \Omega_0) \backslash \{0\}$ leaves Ω_0 in finite time.

PROOF. Let $r > 0$ be such that $B(r) \subset \Omega$. For any $0 < s \leqq r$, there is an $x_0 \neq 0$ in $U \cap B(s)$ and thus $V(x_0) > 0$. Since $\dot{V} \geqq 0$ in $U \cap \Omega$, it follows that the solution $x(t)$ with $x(0) = x_0$ satisfies $V(x(t)) \geqq V(x_0) > 0$ for $x(t)$ in $U \cap \Omega$.

If $\alpha = \min\{\dot{V}(x) \colon x$ in $U \cap \Omega$ with $V(x) \geqq V(x_0)\}$, then $\alpha > 0$ and

$$V(x(t)) = V(x_0) + \int_0^t \dot{V}(x(t))\, dt \geqq V(x_0) + \alpha t$$

for all t for which $x(t)$ remains in $U \cap \Omega$. If Ω_0 is any bounded neighborhood of $x = 0$ which is contained in Ω, this implies $x(t)$ reaches $\partial\Omega_0$ since $V(x)$ is bounded on $U \cap \Omega_0$. This completes the proof of the theorem.

Theorems 1.1 and 1.2 give an indication of a procedure for determining stability or instability of the equilibrium point of an autonomous equation without explicitly solving the equation. On the other hand, there is no general way for constructing the functions V and the ingenuity of the investigator must be used to its fullest extent. For linear systems (1.1), the following lemma is useful. In this lemma, $Re\ \lambda(A)$ designates the real parts of the eigenvalues of a matrix A.

LEMMA 1.5. Suppose A is a real $n \times n$ matrix. The matrix equation

(1.3) $$A'B + BA = -C$$

has a positive definite solution B for every positive definite matrix C if and only if $Re\ \lambda(A) < 0$.

PROOF. Consider the linear differential equation

(1.4) $$\dot{x} = Ax,$$

and the scalar function $V(x) = x'Bx$ where B is a symmetric matrix.

(1.5) $$\dot{V}_{(1.4)}(x) = x'[A'B + BA]x.$$

If there exists a positive definite B such that B satisfies (1.3) with C positive definite, then Theorem 1.1 implies that all solutions of (1.4) approach zero as $t \to \infty$. Of course, this implies $Re\ \lambda(A) < 0$.

Conversely, if $Re\ \lambda(A) < 0$ and C is a positive definite matrix, define

(1.6) $$B = \int_0^\infty e^{A't} C e^{At}\, dt.$$

The matrix B is well defined since there are positive constants K, α such that $|e^{At}| \leqq Ke^{-\alpha t}$, $t \geqq 0$. Furthermore, it is clear that B is positive definite. Also,

$$A'B + BA = \int_0^\infty \frac{d}{dt}(e^{A't} C e^{At})dt$$
$$= -C.$$

This proves the lemma.

From the proof of Lemma 1.5, it is clear that we have proved the following converse theorem of asymptotic stability for the linear system (1.4).

LEMMA 1.6. If the system (1.4) is asymptotically stable, then there is a positive definite quadratic form whose derivative along the solutions of (1.4) is negative definite.

Let us now apply Lemma 1.5 to the equation

$$(1.7) \qquad \dot{x} = Ax + f(x)$$

where f has continuous first derivatives in R^n with $f(0) = 0$, $\partial f(0)/\partial x = 0$.

If $Re\ \lambda(A) < 0$ then Lemma 1.5 implies there is a positive definite matrix B such that $A'B + BA = -I$. Let $V(x) = x'Bx$. Then $\dot{V} = \dot{V}_{(1.7)}$ is given by

$$\dot{V} = -x'x + 2x'Bf(x).$$

Lemma 1.2 implies that $-\dot{V}$ is positive definite in a neighborhood of $x = 0$ and Theorem 1.1 implies the solution $x = 0$ of (1.7) is asymptotically stable. This is the same result as obtained in Chapter III using the variation of constants formula.

If $Re\ \lambda(A) \neq 0$ and an eigenvalue of A has a positive real part, then we can assume without loss of generality that $A = \operatorname{diag}(A_-, A_+)$ where $Re\ \lambda(A_-) < 0$, $Re\ \lambda(A_+) > 0$. Let B_1 be the positive definite solution of $A'_- B_1 + B_1 A_- = -I$ and B_2 be the positive definite solution of $(-A'_+)B_2 + B_2(-A'_+) = -I$ which are guaranteed by Lemma 1.5. If $x = (u, v)$ where u, v have the same dimensions as B_1, B_2, respectively, let $V(x) = -u'B_1u + v'B_2v$. Then $\dot{V}(x) = \dot{V}_{(1.7)}(x) = x'x + o(|x|^2)$ as $|x| \to 0$. Lemma 1.2 implies $\dot{V}(x)$ is positive definite in a neighborhood of $x = 0$. On the other hand, the region U where V is positive obviously satisfies the conditions of Theorem 1.2. Thus, the solution $x = 0$ of (1.7) is unstable if there is an eigenvalue of A with a positive real part. This result was also obtained in Chapter III using the variation of constants formula.

The above results are easily generalized. To simplify the presentation, we say a scalar function V is a *Liapunov function on an open set* G in R^n if V is continuous on $\bar{G}$, the closure of G, and $\dot{V}(x) = [\partial V(x)/\partial x]f(x) \leq 0$ for x in G. Let

$$S = \{x \text{ in } \bar{G} \colon \dot{V}(x) = 0\},$$

and let M be the largest invariant set of (1.1) in S.

THEOREM 1.3. If V is a Liapunov function on G and $\gamma^+(x_0)$ is a bounded orbit of (1.1) which lies in G, then the ω-limit set of γ^+ belongs to M; that is, $x(t, x_0) \to M$ as $t \to \infty$.

PROOF. Since $\gamma^+(x_0)$ is bounded, $V(x(t, x_0))$ is bounded below for $t \geq 0$ and $\dot{V}(x(t, x_0)) \leq 0$ implies $V(x(t, x_0))$ is nonincreasing. Therefore, $V(x(t, x_0)) \to$ a constant c as $t \to \infty$ and continuity of V implies $V(y) = c$ for any y in $\omega(\gamma^+)$. Since $\omega(\gamma^+)$ is invariant, $V(x(t, y)) = c$ for all t and y in $\omega(\gamma^+)$. Therefore, $\omega(\gamma^+)$ belongs to S. This proves the theorem.

COROLLARY 1.1. If V is a Liapunov function on $G = \{x \text{ in } R^n : V(x) < \rho\}$ and G is bounded, then every solution of (1.1) with initial value in G approaches M as $t \to \infty$.

COROLLARY 1.2. If $V(x) \to \infty$ as $|x| \to \infty$ and $\dot{V} \leq 0$ on R^n, then every solution of (1.1) is bounded and approaches the largest invariant set M of (1.1) in the set where $\dot{V} = 0$. In particular, if $M = \{0\}$, then the solution $x = 0$ is *globally asymptotically stable*.

PROOF. For any constant ρ, V is a Liapunov function on the bounded set $G = \{x : V(x) < \rho\}$. Furthermore, for any x_0 in R^n there is a ρ such that x_0 belongs to G. Corollary 1.1 therefore implies the result.

Notice that $\dot{V} < 0$ in $G\backslash\{0\}$ implies $M = \{0\}$ and one obtains from Theorem 1.3 the asymptotic stability theorem in Theorem 1.1.

THEOREM 1.4. Suppose $x = 0$ is an equilibrium point of (1.1) contained in the closure of an open set U and let Ω be a neighborhood of $x = 0$. Assume that

    (i) V is a Liapunov function on $G = U \cap \Omega$,

    (ii) $M \cap G$ is either the empty set or zero,

    (iii) $V(x) < \eta$ on G when $x \neq 0$,

    (iv) $V(0) = \eta$ and $V(x) = \eta$ when x is on that part of the boundary of G in Ω.

Then $x = 0$ is unstable. More precisely, if Ω_0 is a bounded neighborhood of zero with $\overline{\Omega}_0$ contained in Ω, then any solution with initial value in $(U \cap \Omega_0)\backslash\{0\}$ leaves Ω_0 in finite time.

PROOF. If x_0 is in $(U \cap \Omega_0)\backslash\{0\}$, then $V(x_0) < \eta$. Furthermore, $\dot{V} \leq 0$ implies $V(x(t, x_0)) \leq V(x_0) < \eta$ for all $t \geq 0$. If $x(t, x_0)$ does not leave Ω_0, then $\omega(\gamma^+(x_0))$ is not empty and in $\overline{\Omega}_0$. As in the previous proof, $\omega(\gamma^+(x_0)) \subset M$. Since $\omega(\gamma^+(x_0))$ is nonempty, this implies $\omega(\gamma^+(x_0)) = \{0\}$ and belongs to the set where $V(x) = \eta$. This contradicts the fact that $V(x(t, x_0)) \leq V(x_0) < \eta$ and proves the theorem.

Example 1.1. Consider the van der Pol equation

(1.8) $$\ddot{x} + \varepsilon(x^2 - 1)\dot{x} + x = 0,$$

and its equivalent Lienard form

(1.9) $$\dot{x} = y - \varepsilon\left(\frac{x^3}{3} - x\right),$$

$$\dot{y} = -x.$$

In Chapter II, it was shown that this equation has a unique asymptotically stable limit cycle for every $\varepsilon > 0$. The exact location of this cycle in the (x, y)-plane is extremely difficult to obtain but the above theory allows one to determine a region near $(0, 0)$ in which the limit cycle cannot lie. Such a region can be found by determining the stability region of the solution $x = 0$ of (1.8) with t replaced by $-t$. This has the same effect as taking $\varepsilon < 0$.

Therefore, suppose $\varepsilon < 0$ and let $V(x, y)$ be the total energy of (1.9); that is, $V(x, y) = (x^2 + y^2)/2$. Then

$$\dot{V}(x, y) = -\varepsilon x^2 \left(\frac{x^2}{3} - 1 \right)$$

and $\dot{V}(x, y) \leq 0$ if $x^2 < 3$. Consider the region $G = \{(x, y): V(x, y) < 3/2\}$. It is clear that G is bounded and V is a Liapunov function on G. Furthermore, $S = \{(x, y): \dot{V} = 0\} = \{(0, y), y^2 < 3\}$. Also, from (1.9), $M = \{(0, 0)\}$ and Corollary 1.1 implies every solution starting in the circle $x^2 + y^2 < 3$ approaches zero as $t \to \infty$. Finally, the limit cycle of (1.8) for $\varepsilon > 0$ must be outside this circle.

Example 1.2. Consider the equation

(1.10) $$\ddot{x} + f(x)\dot{x} + h(x) = 0,$$

where $xh(x) > 0$, $x \neq 0$, $f(x) > 0$, $x \neq 0$ and $H(x) = \int_0^x h(s)\, ds \to \infty$ as $|x| \to \infty$. Equation (1.10) is equivalent to the system

(1.11) $$\dot{x} = y,$$
$$\dot{y} = -h(x) - f(x)y.$$

Let $V(x, y)$ be the total energy of the system; that is, $V(x, y) = y^2/2 + H(x)$. Then $\dot{V}(x, y) = -f(x)y^2 \leq 0$. For any ρ, the function V is a Liapunov function on the bounded set $G = \{(x, y): V(x, y) < \rho\}$. Also, the set S where $\dot{V} = 0$ belongs to the union of the x-axis and y-axis. From (1.11), this implies $M = \{(0, 0)\}$ and Corollary 1.2 implies that $x = 0$ is globally asymptotically stable.

EXERCISE 1.1. Use Theorems 1.1 and 1.2 to prove Lemmas V.1.2 and V.1.4.

EXERCISE 1.2. Consider the second order system

$$\dot{x} = y - xf(x, y),$$
$$\dot{y} = -x - yf(x, y).$$

Discuss the stability properties of this system when f has a fixed sign.

EXERCISE 1.3. Consider the equation

$$\ddot{x} + a\dot{x} + 2bx + 3x^2 = 0, \qquad a > 0, b > 0.$$

Determine the maximal region of asymptotic stability of the zero solution which can be obtained by using the total energy of the system as a Liapunov function.

EXERCISE 1.4. Consider the system $\dot{x} = y$, $\dot{y} = z - ay$, $\dot{z} = -cx - F(y)$, $F(0) = 0$, $a > 0$, $c > 0$, $aF(y)/y > c$ for $y \neq 0$ and $\int_0^y [F(\xi) - c\xi/a]\, d\xi \to \infty$ as $|y| \to \infty$. If $F(y) = ky$ where $k > c/a$, verify that the characteristic roots of the linear system have negative real parts. Show that the origin is asymptotically stable even when F is nonlinear. Hint: Choose V as a quadratic form plus the term $\int_0^y F(s)\, ds$.

EXERCISE 1.5. Suppose there is a positive definite matrix Q such that $J'(x)Q + QJ(x)$ is negative definite for all $x \neq 0$, where $J(x)$ is the Jacobian matrix of $f(x)$. Prove that the solution $x = 0$ of $\dot{x} = f(x)$, $f(0) = 0$, is globally asymptotically stable. Hint: Prove and make use of the fact that $f(x) = \int_0^1 J(sx)x\, ds$.

EXERCISE 1.6. Suppose you have the function $V = y^2 e^{-x}$ defined in the whole (x, y)-plane and that relative to some differential equation $\dot{V} = -y^2 V$. Can one conclude anything about the solutions of the original differential equation? If not, what is the trouble?

EXERCISE 1.7. Suppose $h(x, y)$ is a positive definite function such that $h(x, y) \to \infty$ as $x^2 + y^2 \to \infty$. Discuss the behavior in the phase plane of the solutions of the equations

$$\dot{x} = \varepsilon x + y - xh(x, y),$$
$$\dot{y} = \varepsilon y - x - yh(x, y),$$

for all values of ε in $(-\infty, \infty)$.

EXERCISE 1.8. Consider the n-dimensional system $\dot{x} = f(x) + g(t)$ where $x'f(x) \leq -k|x|^2$, $k > 0$, for all x and $|g(t)| \leq M$ for all t. Find a sphere of sufficiently large radius so that all trajectories enter this sphere. Show this equation has a T-periodic solution if g is T-periodic. If, in addition, $(x - y)'[f(x) - f(y)] < 0$ for all $x \neq y$ show there is a unique T-periodic solution. Hint: Use Brouwer's fixed point theorem.

EXERCISE 1.9. Suppose f, g are as in Exercise 1.8 except $g(t)$ is almost periodic. Does the equation $\dot{x} = f(x) + g(t)$ have an almost periodic solution?

EXERCISE 1.10. Prove the zero solution of (1.7) is unstable if there is an eigenvalue of A with a positive real part even though some eigenvalues may have zero real parts.

X.2. Circuits Containing Esaki Diodes

In this section, we give an example which illustrates many of the previous ideas. Consider the circuit shown in Fig. 2.1. The square box in this diagram represents an Esaki diode with the characteristic function $f(v)$ representing the current flow as a function of the voltage drop v. Kirchoff's

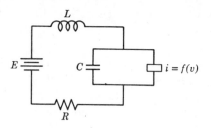

Figure X.2.1

laws imply that the relation between the current i and voltage v are given by

$$(2.1) \qquad L\frac{di}{dt} = E - Ri - v \overset{\text{def}}{=} I(i, v),$$

$$-C\frac{dv}{dt} = f(v) - i \overset{\text{def}}{=} V(i, v),$$

where E, R, C, L are positive constants and $vf(v) \geqq 0$ for all v.

LEMMA 2.1. *If there is an $A > 0$ such that $xf(x) > E^2/R$ for $|x| > A$, then every solution of (2.1) is bounded. In fact, every solution is ultimately in a region bounded by a circle.*

PROOF. If $W(i, v) = (Li^2 + Cv^2)/2$, then the derivative of W along solutions of (2.1) is

$$\dot{W} = -\left[Ri\left(i - \frac{E}{R}\right) + vf(v) \right].$$

Let $W_0 = [L(E/R)^2 + CA^2]/2$. If $W(i, v) > W_0$, then either $|i| > E/R$ or $|v| > A$. If $|i| > E/R$, then $\dot{W} < 0$ and if $|i| \leqq E/R$, $|v| > A$, then

$$\dot{W} < -\left[Ri^2 - Ei + \frac{E^2}{R} \right] = -\left[Ri^2 - E\left(i - \frac{E}{R}\right) \right] \leqq -Ri^2 \leqq 0.$$

For $i = 0$, $|v| > A$, we have also $\dot{W} < 0$. Therefore, $\dot{W} < 0$ in the region

$W(i, v) > W_0$. Since the region $W < \rho$ is bounded for any ρ and $W(i, v) \to \infty$ as $|i|, |v| \to \infty$, it follows that every solution of (2.1) is bounded. This proves the lemma.

The problem at hand is to find conditions on f and the parameters in (2.1) which will ensure that every solution of (2.1) approaches an equilibrium point as $t \to \infty$. Let $f'(v) = df(v)/dv$.

LEMMA 2.2. If the conditions of Lemma 2.1 are satisfied and $f'(v) > 0$ for all v, then every solution of (2.1) approaches the unique equilibrium point of (2.1).

PROOF. First of all, it is clear there is only one equilibrium point of (2.1) if $f'(v) > 0$ for all v. If

$$(2.2) \qquad Q(i, v) = \frac{1}{2L} I^2 + \frac{1}{2C} V^2$$

then the derivative of Q along the solutions of (2.1) is

$$(2.3) \qquad \dot{Q} = -(RL^{-2}I^2 + f'C^{-2}V^2) \leqq 0.$$

Since Lemma 2.1 implies all solutions of (2.1) are bounded and $\dot{Q} = 0$ only at the equilibrium point, it follows from Theorem 1.3 that the assertion of Lemma 2.2 is true.

The most interesting cases in the applications are when f' changes sign and, in fact, can take on values $< -R^{-1}$ so that equation (2.1) has three equilibrium points. However, if $f' > -R^{-1}$ (only one equilibrium point), then a limit cycle may appear unless there are other restrictions on f. In fact, one can prove

THEOREM 2.1. If $-f' < R^{-1}$, $\max_v(-f'/C) > R/L$, then there is a value of E such that equation (2.1) has at least one periodic orbit.

PROOF. Choose E so that the equilibrium point (i_0, v_0) is such that $-f'(v_0)/C > R/L$. From Lemma 2.1, there is a circle Ω with center at $(0, 0)$ such that the trajectories of (2.1) cross Ω from the outside to the inside. If $i = i_0 + u$, $v = v_0 + w$, $x = (u, w)$, then

$$(2.4) \qquad \dot{x} = Ax + \ldots \qquad A = \begin{bmatrix} -RL^{-1} & -L^{-1} \\ C^{-1} & -f_0'C^{-1} \end{bmatrix},$$

where ... represents higher order terms in x and $f_0' = f'(v_0)$. The hypotheses of the theorem imply that the eigenvalues of A have positive real parts. Replacing t by $-t$ in (2.4) has the same effect as replacing A by $-A$, a matrix whose eigenvalues have negative real parts. Lemma 1.6 implies there exist a positive definite matrix B such that the derivative of $W(x) = x'Bx$ along the solutions of (2.4) satisfies $\dot{W}(x) = -x'x + o(|x|^2)$ as $|x| \to 0$. Returning to the original time scale and using Lemma 1.2, one sees that the trajectories of (2.4) are crossing the ellipses $x'Bx = c$ for $c > 0$ sufficiently small from the inside to the outside. The annulus bounded by one of these ellipses and the circle Ω contains a positive semiorbit of (2.1). The Poincaré-Bendixson theorem implies the result.

To obtain more information for the case when f' changes sign, observe first that system (2.1) can be written as

$$(2.5) \qquad\qquad L\frac{di}{dt} = \frac{\partial P}{\partial i},$$

$$-C\frac{dv}{dt} = \frac{\partial P}{\partial v},$$

where

$$(2.6) \qquad P(i, v) = Ei - \frac{Ri^2}{2} - iv + \int_0^v f(s)\,ds$$

$$= -\frac{I^2}{2R} + U(v)$$

and

$$(2.7) \qquad U(v) = \frac{(E - v)^2}{2R} + \int_0^v f(s)\,ds.$$

THEOREM 2.2. If there is an $A \geqq 0$ such that $vf(v) \geqq 0$, $vf(v) > E^2/R$ for $|v| > A$ and

$$(2.8) \qquad\qquad \frac{f'(v)}{C} + \frac{R}{L} > 0$$

for all v, then each solution of (2.1) approaches an equilibrium point of (2.1) as $t \to \infty$.

PROOF. Consider the function $S = Q + \lambda P$ where Q is defined in (2.2), P is defined in (2.6) and

$$(2.9) \qquad\qquad \frac{-f'}{C} < \lambda < \frac{R}{L}.$$

Some straightforward but tedious calculations show that $\dot{P}$ along the solutions of (2.1) is given by

$$(2.10) \qquad\qquad \dot{P} = L^{-1}I^2 - C^{-1}V^2.$$

Thus, relation (2.3) and (2.10) imply that $\dot{S} = \dot{Q} + \lambda\dot{P}$ satisfies

$$\dot{S} = -[(R - \lambda L)L^{-2}I^2 + (f' + \lambda C)C^{-2}V^2] \leqq 0$$

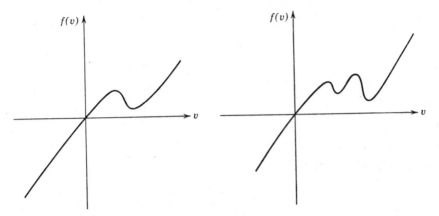

Figure X.2.2

by our choice of λ. Furthermore $\dot{S} = 0$ if and only if $I = 0$, $V = 0$; that is, only at the equilibrium points of (2.1). Since all solutions of (2.1) are bounded from Lemma 2.1, the conclusion of the theorem follows from Theorem 3.1.

It is of interest to determine which equilibrium points of (2.1) are stable. Let S be defined as in the proof of Theorem 2.2. If λ satisfies (2.9), then one can show that *the extreme points of S are the equilibrium points of (2.1)*. Since $\dot{S} \leqq 0$, it follows from Theorems 1.1 and 1.2 that the stable equilibrium points of (2.1) are the minima of S and the unstable equilibrium points are the other extreme points of S.

LEMMA 2.3. *If λ satisfies (2.9), $\lambda \neq 0$, then the extreme points of $S(i, v)$ coincide with the extreme points of $U(v)$ in the sense that the extreme points of S coincide with the solutions of $I = 0$, $\partial U/\partial v = 0$. Every strict local minimum of $U(v)$ represents a local minimum of S and hence a stable equilibrium point of (2.1).*

PROOF. A few elementary calculations yields

$$\frac{\partial S}{\partial i} = -(RL^{-1} - \lambda)I + C^{-1}V,$$

$$\frac{\partial S}{\partial v} = -R^{-1}(RL^{-1} - \lambda)I - f'C^{-1}V + \frac{\lambda \partial U}{\partial v}.$$

If $\partial U/\partial v = 0$ then $R^{-1}(E - v) = f(v)$. If $I = 0$, then $i = R^{-1}(E - v)$. Thus, $I = 0$, $\partial U/\partial v = 0$ implies $V = 0$ and this in turn implies an extreme point of S. If $\partial S/\partial i = 0$, $\partial S/\partial v = 0$, then $I = V = 0$ and this in turn implies $\partial U/\partial v = 0$. This proves the first part of the lemma.

The proof of the second part of the lemma is left as an exercise and involves showing that $\partial^2 U/\partial v^2 > 0$ implies the quadratic form

$$\frac{\partial^2 S}{\partial i^2}\, \xi_0^2 + 2\, \frac{\partial^2 S}{\partial i\, \partial v}\, \xi_0 \xi_1 + \frac{\partial^2 S}{\partial v^2}\, \xi_1^2$$

evaluated at an equilibrium point is positive definite.

EXERCISE 2.1. Interpret the above results for $f(v)$ having the shape shown in Figs. 2.2a,b.

X.3. Sufficient Conditions for Stability in Nonautonomous Systems

In this section, we state some extensions of the results in Section 1 to nonautonomous systems

$$(3.1) \qquad\qquad \dot{x} = f(t, x)$$

where $f \colon [\tau, \infty) \times R^n \to R^n$, τ a constant, is smooth enough to ensure that solutions exists through every point (t_0, x_0) in $[\tau, \infty) \times R^n$, are unique and depend continuously upon the initial data. Let Ω be an open set in R^n containing zero. A function $V \colon [\tau, \infty) \times \Omega \to R$ is said to be *positive definite* if V is continuous, $V(t, 0) = 0$, and there is a positive definite function $W \colon \Omega \to R$ such that $V(t, x) \geq W(x)$ for all (t, x) in $[\tau, \infty) \times \Omega$. $V(t, x)$ is said to possess an infinitely small upper bound if there is a positive definite function $W(x)$ such that $V(t, x) \leq W(x)$, (t, x) in $[\tau, \infty) \times \Omega$. If $V \colon [\tau, \infty) \times \Omega \to R$ is continuous, we define the derivative $\dot{V}(t, \xi)$ of V along the solutions of (3.1) as

$$\dot{V}(t, \xi) = \overline{\lim_{h \to 0^+}} \frac{1}{h}\, [V(t+h, x(t+h, t, \xi)) - V(t, \xi)],$$

where $x(t, \sigma, \xi)$ is the solution of (3.1) passing through $(\sigma, \xi) \in [\tau, \infty) \times R^n$. If $V(t, \xi)$ has continuous partial derivatives with respect to t, ξ, then

$$\dot{V}(t, \xi) = \frac{\partial V(t, \xi)}{\partial t} + \frac{\partial V(t, \xi)}{\partial \xi} f(t, \xi).$$

THEOREM 3.1.   If $V \colon [\tau, \infty) \times \Omega \to R$ is positive definite and $\dot{V}(t, x) \leq 0$, then the solution $x = 0$ of (3.1) is stable. If, in addition, V has an infinitely small upper bound then the solution $x = 0$ is uniformly stable. If, furthermore, $-\dot{V}$ is positive definite, then the solution $x = 0$ of (3.1) is uniformly asymptotically stable.

PROOF.   Since $V(t, x)$ is positive definite, there exists a positive definite function $W(x)$ such that $V(t, x) \geq W(x)$ for all $(t, x) \in [\tau, \infty) \times \Omega$. Let $B(r)$

be the ball in R^n of radius r with center at the origin. There is an $r > 0$ such that $B(r) \subset \Omega$. For $0 < \varepsilon < r$, $0 < k = \min_{|x|=\varepsilon} W(x)$. For any $t_0 \in [\tau, \infty)$, let $0 < \delta < \varepsilon$ be a number such that $V(t_0, x) < k$ for $|x| \leq \delta$. Such a δ always exists since $V(t_0, 0) = 0$ and V is continuous. If x_0 is in $B(\delta)$, then the solution $x(t)$ of (3.1) with $x(t_0) = x_0$ is in $B(\varepsilon)$ for $t \geq t_0$ since $\dot{V}(t, x) \leq 0$ implies $V(t, x(t)) \leq V(t_0, x_0)$, $t \geq t_0$. This proves stability of $x = 0$.

If, in addition, V has an infinitely small upper bound, then there is a positive definite function $W_1(x)$ such that $V(t, x) \leq W_1(x)$ for all $(t, x) \in [\tau, \infty) \times \Omega$. With ε, k as above, choose $\delta > 0$ with $W_1(x) < k$ for $|x| \leq \delta$. Then, for any $t_0 \in [\tau, \infty)$, x_0 in $B(\delta)$, the solution $x(t)$ of (3.1), $x(t_0) = x_0$, is in $B(\varepsilon)$ for $t \geq t_0$ since $V(t, x(t)) \leq V(t_0, x_0) \leq W_1(x_0)$ for all t. This proves uniform stability of $x = 0$.

If, furthermore, $-\dot{V}(t, x)$ is positive definite, then there is a positive definite function $W_2(x)$ such that $-\dot{V}(t, x) \geq W_2(x)$ for all $(t, x) \in [\tau, \infty) \times \Omega$. The proof now proceeds in a manner similar to the proof of Theorem 1.1.

Let $R^+ = [0, \infty)$, $V(t, x)$: $R^+ \times R^n \to R$ be continuous, G be any set in R^n and $\bar{G}$ be the closure of G. We say V is a *Liapunov function of* (3.1) *on* G if

(i)   given x in $\bar{G}$ there is a neighborhood N of x such that $V(t, x)$ is bounded from below for all $t \geq 0$ and all x in $N \cap G$.

(ii) $\dot{V}(t, x) \leq -W(x) \leq 0$ for (t, x) in $R^+ \times G$ and W is continuous on $\bar{G}$.

If V is a Lyapunov function for (3.1) on G, we define

$$E = \{x \text{ in } \bar{G}: W(x) = 0\}.$$

THEOREM 3.2.   Let V be a Liapunov function for (3.1) on G and let $x(t)$ be a solution of (3.1) which is bounded and remains in G for $t \geq t_0 \geq 0$.

(a)   If for each p in $\bar{G}$, there is a neighborhood N of p such that $|f(t, x)|$ is bounded for all $t \geq 0$ and all x in $N \cap G$, then $x(t) \to E$ as $t \to \infty$.

(b)   If W has continuous first derivatives on $\bar{G}$ and $\dot{W} = (\partial W / \partial x) f(t, x)$ is bounded from above (or from below) along the solution $x(t)$, then $x(t) \to E$ as $t \to \infty$.

PROOF.   Let p be a finite positive limit point of $x(t)$ and $\{t_n\}$ a sequence of real numbers, $t_n \to \infty$ as $n \to \infty$ such that $x(t_n) \to p$ as $n \to \infty$. Conditions (i) and (ii) in the definition of a Liapunov function imply that $V(t_n, x(t_n))$ is nonincreasing and bounded below. Therefore, there is a constant c such that $V(t_n, x(t_n)) \to c$ as $n \to \infty$ and since $V(t, x(t))$ is nonincreasing, $V(t, x(t)) \to c$ as $t \to \infty$. Also, $V(t, x(t)) \leq V(t_0, x(t_0)) - \int_{t_0}^{t} W(x(s)) \, ds$ and hence

$$\int_{t_0}^{\infty} W(x(s)) \, ds < \infty.$$

Part (a). Assume p is not in E. Let $\delta > 0$ be such that $W(p) > 2\delta > 0$. There is an $\varepsilon > 0$ such that $W(x) > \delta$ for x in $S_{2\varepsilon}(p) = \{x \colon |x - p| < 2\varepsilon\}$. Also ε can be chosen so that $S_{2\varepsilon}(p) \subset N$, the neighborhood given in (a). If $x(t)$ remains in $S_{2\varepsilon}(p)$ for all $t \geq t_1 \geq t_0$, then $\int_{t_0}^{\infty} W(x(s))\, ds = +\infty$ which is a contradiction. Since p is in the limit set of $x(t)$, the only other possibility is that $x(t)$ leaves and returns to $S_{2\varepsilon}(p)$ an infinite number of times. Since $|f(t, x)|$ is bounded in $S_{2\varepsilon}(p)$, this implies each time that $x(t)$ returns to $S_{2\varepsilon}(p)$, it must remain in $S_{2\varepsilon}(p)$ at least a positive time τ. Again, this implies $\int_{t_0}^{\infty} W(x(s))\, ds = +\infty$ and a contradiction. Therefore $W(p) = 0$ and E contains all limit points.

Part (b). Since $\int_{t_0}^{\infty} W(x(s))\, ds < \infty$ and $\dot{W}(x(t))$ is bounded from above (or from below), it follows that $W(x(t)) \to 0$ as $t \to \infty$. Since W is continuous $W(p) = 0$ and this proves (b).

Example 3.1. Consider the equation

(3.2)
$$\dot{x} = y,$$
$$\dot{y} = -x - p(t)y,$$

where $p(t) \geq \delta > 0$. If $V(x, y) = (x^2 + y^2)/2$, then

$$\dot{V} = -p(t)y^2 \leq -\delta y^2,$$

and V is a Liapunov function on R^2 with $W(x, y) = -\delta y^2$. Also, $\dot{W} = -2\delta(xy + p(t)y^2) \leq -2\delta xy$. Every solution of (3.2) is clearly bounded and, therefore, condition (b) of Theorem 3.2 is satisfied. The set E is the x-axis and Theorem 3.2 implies that each solution $x(t)$, $y(t)$ of (3.2) is such that $y(t) \to 0$ as $t \to \infty$. On the other hand, if $p(t) = 2 + e^t$, then there is a solution of (3.2) given as $x(t) = 1 + e^{-t}$, $y(t) = -e^{-t}$. Since the equation is linear, every point on the x-axis is a limit point of some solution. This shows that the above result is the best possible without further restrictions on p.

Notice that the condition in (a) of Theorem 3.2 is not satisfied in Example 3.1 unless $p(t)$ is bounded.

A simple way to verify that a solution $x(t)$ of (3.1) remains in G for $t \geq t_0$ is given in the following lemma whose proof is left as an exercise.

LEMMA 3.1. Assume that $V(t, x)$ is a continuous scalar function on $R^+ \times R^n$ and there are continuous scalar functions $a(x)$, $b(x)$ on R^n such that $a(x) \leq V(t, x) \leq b(x)$ for all (t, x) in $R^+ \times R^n$. For any real number ρ, let $A_\rho = \{x \text{ in } R^n \colon a(x) < \rho\}$ and let G_0 be a component of A_ρ. If G is the component of the set $B_\rho = \{x \text{ in } R^n \colon b(x) < \rho\}$ contained in G_0 and $\dot{V} \leq 0$ on G_0, then any solution of (3.1) with $x(t_0)$ in G remains in G_0 for all $t \geq t_0$.

Notice that $a(x) \to \infty$ as $|x| \to \infty$ in Lemma 3.1 implies boundedness of the solutions of (3.1).

X.4. The Converse Theorems for Asymptotic Stability

In this section, we show that a type of converse of Theorem 3.1 for uniform asymptotic stability is true. If $V(t, x)$ is a continuous scalar function on $R^+ \times R^n$ and $x(t, t_0, x_0)$ is the solution of (3.1) with $x(t_0, t_0, x_0) = x_0$, we define the derivative $\dot{V}_{(3.1)}(t, \xi)$ along the solutions of (3.1) as in Section 3; namely,

$$\dot{V}_{(3.1)}(t, \xi) = \overline{\lim_{h \to 0^+}} \frac{1}{h} [V(t + h, x(t + h, t, \xi)) - V(t, \xi)] \ .$$

Our first converse theorem deals with the linear system since the essential ideas emerge without any technical difficulties.

THEOREM 4.1. If $A(t)$ is a continuous matrix on $[0, \infty)$ and the linear system

$$(4.1) \qquad\qquad \dot{x} = A(t)x$$

is uniformly asymptotically stable, then there are positive constants K, α and a continuous scalar function V on $R^+ \times R^n$ such that

$(4.2) \qquad$ (a) $\quad |x| \leqq V(t, x) \leqq K |x|,$

$\qquad\qquad$ (b) $\quad \dot{V}_{(4.1)}(t, x) \leqq -\alpha V(t, x),$

$\qquad\qquad$ (c) $\quad |V(t, x) - V(t, y)| \leqq K |x - y|,$

for all t in R^+, x, y in R^n.

PROOF. From Theorem III.2.1, there are positive constants K, α such that the solution $x(t, t_0, x_0)$ of (4.1) with $x(t_0, t_0, x_0) = x_0$ satisfies

$$(4.3) \qquad\qquad |x(t, t_0, x_0)| \leqq K e^{-\alpha(t-t_0)} |x_0|$$

for all $t \geqq t_0 \geqq 0$ and all x_0 in R^n. Define for $t \geqq 0$, x_0 in R^n

$$(4.4) \qquad\qquad V(t, x_0) = \sup_{\tau \geqq 0} |x(t + \tau, t, x_0)| e^{\alpha\tau}.$$

It is immediate from (4.3) that $V(t, x_0)$ is defined for all t, x_0 and satisfies (4.2a).

To verify (4.2c) observe that

$$|V(t, x_0) - V(t, y_0)| \leqq \sup_{\tau \geqq 0} |x(t + \tau, t, x_0) - x(t + \tau, t, y_0)| e^{\alpha\tau}$$

$$= \sup_{\tau \geqq 0} |x(t + \tau, t, x_0 - y_0)| e^{\alpha\tau}$$

$$= V(t, x_0 - y_0) \leqq K |x_0 - y_0|.$$

The proof of (4.2b) proceeds as follows:

$$\dot{V}_{(4.1)}(t, x_0) = \varlimsup_{h \to 0^+} \frac{1}{h} \left[\sup_{\tau \geq 0} |x(t + \tau + h, t + h, x_0)| e^{\alpha \tau} - \sup_{\tau \geq 0} |x(t + \tau, t, x_0)| e^{\alpha \tau} \right]$$

$$= \varlimsup_{h \to 0^+} \frac{1}{h} \left[\sup_{\tau \geq h} |x(t + \tau, t, x_0)| e^{\alpha(\tau - h)} - \sup_{\tau \geq 0} |x(t + \tau, t, x_0)| e^{\alpha \tau} \right]$$

$$\leq \varlimsup_{h \to 0^+} \frac{1}{h} \left[\sup_{\tau \geq 0} |x(t + \tau, t, x_0)| e^{\alpha(\tau - h)} - \sup_{\tau \geq 0} |x(t + \tau, t, x_0)| e^{\alpha \tau} \right]$$

$$\leq \varlimsup_{h \to 0^+} \left[\frac{1}{h} \sup_{\tau \geq 0} |x(t + \tau, t, x_0)| e^{\alpha \tau} (e^{-\alpha h} - 1) \right.$$

$$= -\alpha V(t, x_0).$$

It remains only to show that $V(t, x_0)$ is continuous. Notice that

$$|V(t + s, x_0 + y_0) - V(t, x_0)| \leq |V(t + s, x_0 + y_0) - V(t + s, x(t + s, t, x_0 + y_0))|$$
$$+ |V(t + s, x(t + s, t, x_0 + y_0)) - V(t, x_0 + y_0)|$$
$$+ |V(t, x_0 + y_0) - V(t, x_0)|.$$

The fact that the first and third terms can be made small if s, y are small follows from (4.2c). The continuity of the solution of (4.1) in t and an argument similar to that used in proving that $\dot{V}_{(4.1)}$ existed shows the second term can be made small if s is small. This proves right continuity. Left continuity is an exercise for the reader.

Theorem 4.1 is a converse theorem for exponential asymptotic stability. In fact, if there is a function V satisfying (4.2), and $x(t) = x(t, t_0, x_0)$ is a solution of (4.1), then

$$V(t, x(t)) \leq e^{-\alpha(t - t_0)} V(t_0, x_0) \leq K e^{-\alpha(t - t_0)} |x_0|.$$

The fact that $V(t, x) \geq |x|$ implies that $x(t)$ satisfies (4.3).

The basic elements of the proof of the above theorem are the estimate (4.3) and the lifting of the norm of x in the definition of V. If the lifting factor $e^{\alpha \tau}$ had not been applied one would have only obtained $\dot{V} \leq 0$.

Our next objective is to extend Theorem 4.1 to the nonlinear equation (3.1). We need

LEMMA 4.1.   Suppose $f(t, 0) = 0$. The solution $x = 0$ of (3.1) is uniformly stable if and only if there exists a function $\rho(r)$ with the following properties:

(a) $\rho(r)$ is defined, continuous and monotonically increasing in an interval $0 \leq r \leq r_1$,

(b) $\rho(0) = 0$,

(c)   For any x in R^n, $|x| < r_1$ and any $t_0 \geqq 0$, the solution $x(t, t_0, x_0)$, $x(t_0, t_0, x_0) = x_0$, of (3.1) satisfies

$$|x(t, t_0, x_0)| \leqq \rho(|x_0|) \qquad \text{for} \quad t \geqq t_0.$$

PROOF.   The sufficiency is obvious. To prove necessity, suppose $\varepsilon > 0$ is given and let $\hat{\delta}(\varepsilon)$ be the least upper bound of all numbers $\delta(\varepsilon)$ occurring in the definition of uniform stability. Then, for any x in R^n, $|x| \leqq \hat{\delta}(\varepsilon)$, and any $t_0 \geqq 0$, the solution $x(t, t_0, x_0)$ of (3.1) satisfies $|x(t, t_0, x_0)| \leqq \varepsilon$ for $t \geqq t_0$. For every $\delta_1 > \hat{\delta}$, there is an $\tilde{x}_0$ in R^n, $|\tilde{x}_0| < \delta_1$ such that $|x(t, t_0, \tilde{x}_0)|$ exceeds ε for some value of t. Consequently, $\hat{\delta}(\varepsilon)$ is nondecreasing, positive for $\varepsilon > 0$, and tends to zero as ε tends to zero. However, $\hat{\delta}(\varepsilon)$ may be discontinuous. Now choose $\hat{\delta}(\varepsilon)$ continuous, monotonically increasing and such that $\hat{\delta}(\varepsilon) < \hat{\delta}(\varepsilon)$ for $\varepsilon > 0$. Let ρ be the inverse function of b. For any x_0 in R^n, $|x_0| \leqq \hat{\delta}(\varepsilon)$, there exists an ε_1 such that $|x_0| = \hat{\delta}(\varepsilon_1)$ and, thus, $|x(t, t_0, x_0)| \leqq \varepsilon_1 = \rho(|x_0|)$. This proves the lemma.

LEMMA 4.2.   The solution $x = 0$ of (3.1) is uniformly asymptotically stable if and only if there exist functions $\rho(r)$, $\theta(r)$ with $\rho(r)$ satisfying the conditions of Lemma 4.1 and $\theta(r)$ defined, continuous and monotonically decreasing in $0 \leqq r < \infty$, $\theta(r) \to 0$ as $r \to \infty$ such that, for any x_0 in R^n, $|x_0| \leqq r_1$, and every $t_0 \geqq 0$, the solution $x(t, t_0, x_0)$ of (3.1) satisfies

$$|x(t, t_0, x_0)| \leqq \rho(|x_0|)\theta(t - t_0), \qquad t \geqq t_0.$$

The proof of this lemma is left as an exercise.

THEOREM 4.2.   If $f(t, 0) = 0$, $f(t, x)$ is locally lipschitzian in x uniformly with respect to t and the solution $x = 0$ of (3.1) is uniformly asymptotically stable, then there exist an $r_1 > 0$, $K = K(r_1) > 0$, a positive definite function $b(r)$, a positive function $c(r)$ on $0 \leqq r \leqq r_1$ and a scalar function $V(t, x)$ defined and continuous for $t \geqq 0$, x in R^n, $|x| \leqq r_1$, such that

(4.5)   (a) $|x| \leqq V(t, x) \leqq b(|x|)$,

             (b) $\dot{V}_{(3.1)}(t, x) \leqq -c(|x|)V(t, x) \leqq -|x| c(|x|)$,

             (c) $|V(t, x) - V(t, y)| \leqq K |x - y|$,

for all $t \geqq 0$, x, y in R^n, $|x| \leqq r_1$, $|y| \leqq r_1$.

PROOF.   From Lemma 4.2, there exist functions $\rho(r)$, $\theta(\tau)$, defined, continuous and positive on $0 \leqq r \leqq r_1$, $0 \leqq \tau < \infty$, such that $\rho(r)$ is strictly increasing in r, $\rho(0) = 0$, $\theta(\tau)$ is strictly decreasing in τ, $\theta(\tau) \to 0$ as $\tau \to \infty$ and for any x in R^n, $|x| \leqq r_1$, the solution $x(t, t_0, x_0)$ of (3.1) satisfies

$$|x(t, t_0, x_0)| \leqq \rho(|x_0|)\theta(t - t_0), \qquad t \geqq t_0 \geqq 0.$$

One can also assume that $\theta(\tau)$ has a continuous negative derivative, and $\theta(0) = 1$. From the properties of $\theta(\tau)$, there exists a function $\alpha(\tau)$ defined, continuously differentiable and positive on $\tau > 0$, $\alpha(0) = 0$, $\alpha(\tau)$ strictly increasing, $\alpha(\tau) \to \infty$ as $\tau \to \infty$ such that

$$\theta(\tau) = e^{-\alpha(\tau)}, \qquad \tau \geqq 0.$$

Suppose $q(\tau)$ is a bounded continuously differentiable function on $0 \leqq \tau < \infty$, such that $q(0) = 0$, $q(\tau) > 0$, $q'(\tau) < \alpha'(\tau)$ for $\tau > 0$, q' monotone decreasing, and define

$$V(t, x_0) = \sup_{\tau \geqq 0} |x(t + \tau, t, x_0)| e^{q(\tau)}.$$

The function $V(t, x_0)$ is defined for $t \geqq 0$, x_0 in R^n, $|x_0| \leqq r_1$ and satisfies

$$|x_0| \leqq V(t, x_0) \leqq \rho(|x_0|) \sup_{\tau \geqq 0} e^{-\alpha(\tau) + q(\tau)} = \rho(|x_0|) \overset{\text{def}}{=} b(|x_0|).$$

Furthermore, since there exists a continuous positive nondecreasing function $P(r)$, $0 < r \leqq r_1$ such that

$$\rho(|x|) e^{-\alpha(\tau) + q(\tau)} \leqq |x| \qquad \text{for} \quad \tau \geqq P(|x|),$$

it follows that

$$V(t, x_0) = \sup_{0 \leqq \tau \leqq P(|x_0|)} |x(t + \tau, t, x_0)| e^{q(\tau)}.$$

Consequently, for any $h > 0$, there is a τ_h^*, $0 \leqq \tau_h^* \leqq P(|x(t + h, t, x_0)|)$ such that τ_h^* is continuous in h and

$$V(t + h, x(t + h, t, x_0)) = |x(t + h + \tau_h^*, t, x_0)| e^{q(\tau_h^*)}.$$

If $h + \tau_h^* = \tau_h$, then

$$V(t + h, x(t + h, t, x_0)) = |x(t + \tau_h, t, x_0)| e^{q(\tau_h)} e^{-q(\tau_h)} e^{q(\tau_h - h)}$$
$$\leqq V(t, x_0) e^{-q(\tau_h)} e^{q(\tau_h - h)}.$$

Therefore,

$$\dot{V}_{(3.1)}(t, x_0) \leqq -V(t, x_0) \overline{\lim_{h \to 0^+}} \frac{1}{h} [e^{q(\tau_h)} - e^{q(\tau_h - h)}] e^{-q(\tau_h)}$$

$$= -V(t, x_0) q'(\tau_0) \leqq -V(t, x_0) q'(P(|x_0|))$$

$$\overset{\text{def}}{=} -c(|x_0|) V(t, x_0)$$

$$\leqq -|x_0| c(|x_0|),$$

which proves (4.5b). Since f in (3.1) is locally Lipschitzian in x uniformly in t, for any $r_0 > 0$, there is a constant $L = L(r_0)$ such that

$$|x(t, t_0, x_0) - x(t, t_0, y_0)| \leqq e^{L(t - t_0)} |x_0 - y_0|,$$

for all $t \geq t_0$ and x_0, y_0 for which $|x(t, t_0, x_0)| \leq r_0$, $|x(t, t_0, y_0)| \leq r_0$. Choose $r_0 = \rho(r_1)$. Since $P(r)$ is nondecreasing in r, it follows that

$$|V(t, x_0) - V(t, y_0)| \leq \sup_{0 \leq \tau \leq P(r)} |x(t + \tau, t, x_0) - x(t + \tau, t, y_0)| e^{q(\tau)}$$

$$\leq \left[\sup_{0 \leq \tau \leq P(r)} e^{L\tau} e^{q(\tau)} \right] |x_0 - y_0|$$

$$\stackrel{\text{def}}{=} K(r) |x_0 - y_0|,$$

for $|x_0|$, $|y_0| \leq r$. For $r = r_1$, $K = K(r_1)$, this proves V satisfies the inequalities of the theorem.

To prove $V(t, x_0)$ is continuous in t, x_0 we observe that

$$|V(t + h, x_0 + y_0) - V(t, x_0)|$$
$$\leq |V(t + h, x_0 + y_0) - V(t + h, x(t + h, t, x_0 + y_0))|$$
$$+ |V(t + h, x(t + h, t, x_0 + y_0)) - V(t, x_0 + y_0)|$$
$$+ |V(t, x_0 + y_0) - V(t, x_0)|.$$

The fact that the first and third terms can be made small if h, y_0 are small follows from the lipschitzian property of V and the continuity of the solution of (3.1). The second term can be seen to be small if h is small by an argument similar to that used in showing that $\dot{V}$ existed.

This shows right continuity. Left continuity is an exercise for the reader.

In Section 2, it was shown that a function V satisfying (4.5) implies that the solution $x = 0$ of (3.1) is uniformly asymptotically stable. Therefore, Theorem 4.2 is the converse theorem for uniform asymptotic stability.

X.5. Implications of Asymptotic Stability

In most of the applications that arise in the real world, the differential equations are not known exactly. Therefore, specific properties of the solutions of these equations which are physically interesting should be insensitive to perturbations in the vector field with, of course, the types of perturbations being dictated by the problem at hand. Therefore, one reasonable criterion on which to judge the usefulness of a definition of stability is to test the sensitiveness of this property to perturbations of the vector field. In this section, we use the converse theorem of Section 4 to discuss the sensitiveness of uniform asymptotic stability to arbitrary perturbations as long as they are "small." We need

LEMMA 5.1. Suppose f satisfies the conditions of Theorem 4.2, V is the function given in that theorem, $g(t, x)$ is any continuous function on $R^+ \times R^n$

into R^n and consider the equation

(5.1) $$\dot{x} = f(t, x) + g(t, x).$$

Then

(5.2) $$\dot{V}_{(5.1)}(t, x) \leqq -c(|x|) V(t, x) + K |g(t, x)|$$

for all $t \geqq 0$, $|x| \leqq r_1$.

PROOF. Let $x^*(t, t_0, x_0)$, $x^*(t_0, t_0, x_0) = x_0$, be a solution of (5.1) and $x(t, t_0, x_0)$, $x(t_0, t_0, x_0) = x_0$, a solution of (3.1). Using the definition of $\dot{V}_{(5.1)}$ and relations (4.5b), (4.5c), one obtains

$$\dot{V}_{(5.1)}(t, x_0) = \varlimsup_{h \to 0^+} \frac{1}{h} [V(t + h, x^*(t + h, t, x_0)) - V(t, x_0)]$$

$$= \varlimsup_{h \to 0^+} \frac{1}{h} [V(t + h, x(t + h, t, x_0)) - V(t, x_0)$$

$$+ V(t + h, x^*(t + h, t, x_0)) - V(t + h, x(t + h, t, x_0))]$$

$$\leqq \dot{V}_{(3.1)}(t, x_0) + K \varlimsup_{h \to 0^+} \frac{1}{h} |x^*(t + h, t, x_0) - x(t + h, t, x_0)|$$

$$\leqq \dot{V}_{(3.1)}(t, x_0) + K |g(t, x_0)|$$

$$\leqq -c(|x_0|) V(t, x_0) + K |g(t, x_0)|.$$

This proves the lemma.

Many results now follow directly from Lemma 5.1. In fact, if (3.1) is linear, then equation (5.1) is

(5.3) $$\dot{x} = A(t)x + g(t, x),$$

and Theorem 4.1 states that $c(|x|)$ in (4.5) can be chosen as $\alpha > 0$. Therefore, the inequality (5.2) becomes

(5.4) $$\dot{V}_{(5.3)}(t, x) \leqq -\alpha V(t, x) + K |g(t, x)|.$$

If $\omega(t, r)$ is a continuous function on $R^+ \times R^+$, nondecreasing in r, such that

(5.5) $$|g(t, x)| \leqq \omega(t, |x|),$$

then one sees that (4.2a) and (5.5) imply

(5.6) $$\dot{V}_{(5.3)} \leqq -\alpha V + K\omega(t, V).$$

Using our basic result on differential inequalities in Section I.6 and relation (4.2), one can state

THEOREM 5.1.   Suppose $\omega(t, u)$ in (5.5) is such that for any $t_0 \geqq 0$, $u_0 \geqq 0$, the equation

(5.7) $$\dot{u} = -\alpha u + K\omega(t, u)$$

has a solution passing through t_0, u_0 which is unique. If $u(t, t_0, u_0)$, $u(t_0, t_0, u_0) = u_0$, is a solution of (5.7) which exists for $t \geqq t_0$ and $u_0 \geqq K|x_0|$, x_0 in R^n, $|x_0| \leqq r_1$, then the solution $x(t, t_0, x_0)$, $x(t_0, t_0, x_0) = x_0$, of (5.3) satisfies

$$|x(t, t_0, x_0)| \leqq u(t, t_0, u_0), \qquad t \geqq t_0.$$

Another application of Lemma 5.1 is

THEOREM 5.2.   If f satisfies the conditions of Theorem 4.2 and the solution $x = 0$ of (3.1) is uniformly asymptotically stable, then there is an $r_1 > 0$ such that for any ε, $0 < \varepsilon < r_1$ and any r_2, $\varepsilon < r_2 \leqq r_1$, there is a $\delta > 0$ and a $T > 0$ such that for any $t_0 \geqq 0$ and any x_0 in R^n, $\varepsilon \leqq |x_0| \leqq r_2$, the solution $x(t, t_0, x_0)$, $x(t_0, t_0, x_0) = x_0$ of (5.1) satisfies $|x(t, t_0, x_0)| < \varepsilon$ for $t \geqq t_0 + T$ provided that $|g(t, x)| < \delta$ for all $t \geqq 0$, $|x| \leqq r_1$.

PROOF.   The hypotheses of the theorem imply there is a function $V(t, x)$ satisfying the conditions (4.5) of Theorem 4.2. Choose r_1 as in Theorem 4.2. Without loss of generality, we may assume the functions $b(s)$, $c(s)$ in (4.5) are nondecreasing. For any ε, r_2 as in the theorem, we know that $\varepsilon \leqq |x| \leqq r_2$ implies $c(\varepsilon) > 0$ and $\varepsilon < V(t, x) < b(r_2)$. Furthermore, Lemma 5.1 implies

$$\dot{V}_{(5.1)}(t, x) \leqq -c(\varepsilon)V(t, x) + K|g(t, x)|$$
$$\leqq -c(\varepsilon)\varepsilon + K|g(t, x)|$$

for all $t \geqq 0$, $\varepsilon \leqq |x| \leqq r_2$. Consequently, if $x = x(t)$ represents a solution of (5.1) with $x(t_0) = x_0$, $\varepsilon \leqq |x_0| \leqq r_2$ and $K\delta < \varepsilon c(\varepsilon)/2$ with $|g(t, x)| < \delta$ for all $t \geqq 0$, $|x| \leqq r_1$, then $\dot{V}_{(5.1)}(t, x(t)) \leqq -\varepsilon c(\varepsilon)/2$ as long as $\varepsilon \leqq |x(t)| \leqq r_2$. This clearly implies that $|x(t_0 + t, t_0, x_0)| < \varepsilon$ for $t \geqq T > 2b(r_2)/\varepsilon c(\varepsilon)$. This completes the proof of the theorem.

One could continue to deduce other more specific results but Theorems 5.1 and 5.2 should indicate the usefulness of the converse theorems of Liapunov.

X.6.   Wazewski's Principle

The purpose of this section is to give an introduction to a procedure known as Wazewski's principle for determining the asymptotic behavior of the solutions of differential equations. This principle is an important extension of the method

of Liapunov functions. To motivate the procedure and to bring out some of the
essential ideas, we begin with a very elementary example.

Let us consider the scalar equation

(6.1) $\dot{x} = x + f(t,x)$

where $f(t,x)$ is continuous, smooth enough to ensure the uniqueness of solutions and there is a continuous, nondecreasing function $K(\sigma), \sigma \geq 0$, constants
$k \in [0,1), \sigma_0 > 0$, such that

(6.2)
$$\sigma K(\sigma) \leq k\sigma^2 \qquad \text{for } \sigma \geq \sigma_0$$
$$|f(t,x)| \leq K(\sigma) \qquad \text{for } t \geq 0, |x| \leq \sigma.$$

The objective is to show that hypotheses (6.2) imply that Eq. (6.1) has a
solution which exists and is bounded for $t > 0$.

If $V(x) = x^2/2$, then the derivative $\dot{V}$ of V along the solutions of Eq. (6.1)
satisfy

(6.3) $\dot{V}(x) = x^2 + xf(t,x) \geq (1-k)|x|^2$

if $|x| \geq \sigma_0$. Consequently, on the set $|x| = \sigma_0$, the vector field behaves as
depicted in the accompanying figure.

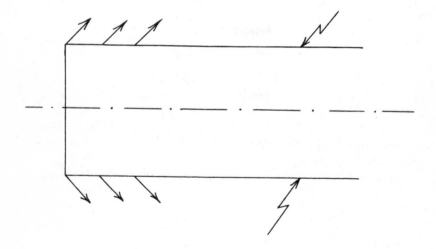

If $x(t,x_0)$ designates the solution of Eq. (6.1) satisfying $x(0,x_0) = x_0$, let

$\quad S_+^* = \{x_0 \in R : |x_0| \leq \sigma_0 \text{ and there is a } \tau \text{ with } x(\tau,x_0) = \sigma_0\}$

$\quad S_-^* = \{x_0 \in R : |x_0| \leq \sigma_0 \text{ and there is a } \tau \text{ with } x(\tau,x_0) = -\sigma_0\}$

$\quad S^* = S_+^* \cup S_-^*.$

The set S^* represents the initial values of all those solutions with initial value x_0 in $\{|x| \leq \sigma_0\}$ which leave the region $\{|x| \leq \sigma_0\}$ at some time, S_+^* the ones which leave at the top of the rectangle and S_-^* those that leave at the bottom of the rectangle.

Clearly $\sigma_0 \in S_+^*$, $-\sigma_0 \in S_-^*$. It is also intuitively clear that $S^* \neq [-\sigma_0, \sigma_0]$ by the way some paths want to go to the top and some to the bottom. Let us make this precise.

We prove first that $S_\pm^*$ are open in $[-\sigma_0, \sigma_0]$. If $x_0 \in S^*$, let τ be chosen so that $|x(\tau, x_0)| = \sigma_0$. Inequality (6.3) implies there is an $\epsilon > 0$ such that $|x(t, x_0)| > \sigma_0$ for $\tau < t \leq \tau + \epsilon$. Choose a neighborhood U of $x(\tau + \epsilon, x_0)$ such that $y \in U$ implies $|y| > \sigma_0$. By continuous dependence on initial data, there is a neighborhood V of x_0, $V \subseteq [-\sigma_0, \sigma_0]$ of x_0 such that, for every z in V, there is a $\bar\tau(z)$ such that the solution $x(t,z)$ satisfies $x(\bar\tau(z), z)$ in U. In particular, $|x(\bar\tau(z), z)| > \sigma_0$. Consequently, there is a $\tau(z)$ such that $|x(\tau(z), z)| = \sigma_0$. This proves $S_\pm^*$ are open in $[-\sigma_0, \sigma_0]$.

Obviously, $S_+^* \cap S_-^* = \phi$. Thus, $[-\sigma_0, \sigma_0] \backslash S^*$ is not empty. This is equivalent to saying that there is an $x_0 \in [-\sigma_0, \sigma_0]$ such that the solution $x(t, x_0)$ exists and satisfies $|x(t, x_0)| < \sigma_0$ for $t \geq 0$, as was to be shown.

Let us now generalize these ideas to n-dimensions. Suppose Ω is an open set in $R \times R^n$, $f : \Omega \to R^n$ is continuous and the solution $\phi(t, t_0, x_0), \phi(t_0, t_0, x_0) = x_0$, of the n-dimensional system

$$(6.4) \qquad\qquad \dot{x} = f(t, x)$$

depends continuously upon (t, t_0, x_0) in its domain of definition.

For notation, let P designate a representative point (t, x) of Ω, $(\alpha(P), \beta(P))$ designate the maximal interval of existence of the solution $\phi(t, P)$ through $P, \Phi(t, P) = (t, \phi(t, P))$ be the point on the trajectory through P at time t, and for any interval $I \subset (\alpha(P), \beta(P))$, let $\Phi(I, P) = \{\Phi(t, P) = (t, \phi(t, P)), t \in I\}$ be that part of the trajectory through P corresponding to t in I. Let $\omega \subset \Omega$ be a fixed open set, $\bar\omega$ be the closure of ω in Ω, and let $\partial\omega$ denote the boundary of ω in Ω.

Definition 6.1. A point $P_0 \in \partial\omega$ is a point of egress from ω with respect to Eq. (6.4) and the set Ω if there is a $\delta > 0$ such that $\Phi([t_0 - \delta, t_0), P_0) \subset \omega$. An egress point P_0 is a *point of strict egress* from ω if there is a $\delta > 0$ such that $\Phi((t_0, t_0 + \delta], P_0) \subset \Omega \backslash \bar\omega$. The set of points of egress is denoted by S and the points of strict egress by S^*.

Definition 6.2. If $A \subset B$ are any two sets of a topological space and $K : B \to A$ is continuous, $K(P) = P$ for all P in A, then K is said to be a *retraction* from B to A and A is a *retract* of B.

With these definitions, we are in a position to prove the following result known as the *principle of Wazewski*.

THEOREM 6.1. If $S = S^*$ and there is a set $Z \subset \omega \cup S$ such that $Z \cap S$ is a retract of S but not a retract of Z, then there exists at least one point P_0 in $Z \backslash S$ such that $\overline{\Phi([t_0, \beta(P_0)), P_0)} \subset \omega$.

PROOF. For any point P_0 in ω for which there is a $\tau \in [t_0, \beta(P_0)]$ such that $\Phi(\tau, P_0)$ is not in $\bar{\omega}$, there is a first time t_{P_0} for which $\Phi(t_{P_0}, P_0)$ is in S, $\Phi(t, P_0)$ is in ω for t in $[t_0, t_{P_0})$. The point $\Phi(t_{P_0}, P_0)$ is called the *consequent* of P_0 and denoted by $C(P_0)$. The set of points in ω for which a consequent exists is designated by G, the *left shadow of S*.

Suppose now $S = S^*$ and define the map $K : G \cup S \to S$, $K(P) = C(P)$ for $P \in G$, $K(P) = P$ for $P \in S$. We prove K is continuous. If $P \in \omega$ and $C(P) = (t_P, \phi(t_P, P))$, then $S = S^*$ implies there is a $\delta > 0$ such that $\Phi((t_P - \delta, t_P), P) \subset \omega$, $\Phi((t_P, t_P + \delta), P) \subset \Omega \backslash \bar{\omega}$. Since $\phi(s, P)$ is continuous in (s, P), for any $\epsilon > 0$, there is an $\eta > 0$ such that $|\phi(s, Q) - \phi(s, P)| < \epsilon$ for $s \in (t_P - \delta, t_P + \delta)$ if $|Q - P| < \eta$. This clearly implies that $C(Q) \to C(P)$ if $Q \to P$. If P is in $S = S^*$, then one repeats the same type of argument to obtain that K is continuous.

Since K is continuous, K is a retract of $G \cup S$ into S.

If the conclusion of the theorem is not true, then $Z \backslash S \subset G$, the left shadow of S. Thus, $Z \subset G \cup S$. Since $Z \cap S$ is a retract of S, there is a mapping $H : S \to Z \cap S$ such that $H(P) = P$ if P is in $Z \cap S$. The map $HK : G \cup S \to Z \cap S$ is continuous, $(HK)(P) = P$ if P is in $Z \cap S$. Thus $G \cup S$ is a retract of $Z \cap S$. Since $Z \subset G \cup S$, the map $HK : Z \to Z \cap S$ is a retraction of Z onto $Z \cap S$. This contradiction proves the theorem.

As an example, consider the second order system

$$(6.5) \qquad \begin{aligned} \dot{x} &= f(t, x, y) \\ \dot{y} &= g(t, x, y) \end{aligned}$$

for $(t, x, y) \in \Omega = \{t \geq 0, x, y \text{ in } R\}$. Let $\omega = \{(t, x, y) : t \geq 0, |x| < a, |y| < b\}$, where $a > 0, b > 0$ are fixed constants. Suppose

$$(6.6) \qquad \begin{aligned} xf(t, x, y) &> 0 \qquad \text{on } |x| = a, |y| < b, \\ yg(t, x, y) &< 0 \qquad \text{on } |x| < a, |y| = b. \end{aligned}$$

For the set ω, we have $S = S^* = \{(t, x, y) : t \geq 0, |x| = a, |y| < b\}$. Let $Z = \{(t, x, y) : t = 0, y = 0, |x| \leq a\}$. Then $Z \cap S$ is clearly a retract of S. However, $Z \cap S$ is not a retract of Z. Thus, Theorem 6.1 implies there is a solution of Eq. (6.5) which remains in ω for $t \geq 0$.

EXERCISE 6.1. Generalize the previous example by replacing the conditions (6.6) on f, g by the following:

$$\omega = \{(t, x, y) : t \geq 0, |x| < u(t), |y| < v(t), u, v \text{ continuous}$$
$$\text{together with their first derivatives } u', v'\}$$

$$xf(t,x,y) > u(t)u'(t) \qquad \text{on } |x| = u(t), |y| < v(t)$$

$$yg(t,x,y) < v(t)v'(t) \qquad \text{on } |x| < u(t), |y| = v(t).$$

Give some examples of u,v,f,g which satisfy these conditions and interpret your results in these special cases.

EXERCISE 6.2. Let $H_j : R \times R^n \to R$, $j = 1,2,\ldots,m$ be continuous together with fheir first derivatives and let $\dot{H}_j(t_0,x_0)$ be the derivative of $H_j(t,x(t,x(t,t_0,x_0)))$ at (t_0,x_0) along the solutions of the Eq. (6.4). Define

$$\omega = \{(t,x) \text{ in } R^{n+1} : H_j(t,x) < 0, \qquad\qquad j = 1,2,\ldots,m\}$$

$$\Gamma_k = \{(t,x) \text{ in } R^{n+1} : H_k(t,x) = 0, H_j(t,x) \leq 0, \qquad j = 1,2,\ldots,m\}.$$

The Γ_k are called the faces of ω. Suppose for each $k = 1,2,\ldots,m$, and each $(t,x) \in \Gamma_k$, either

$$\text{(a)} \quad \dot{H}_k(t,x) > 0$$

or

$$\text{(b)} \quad (t,x) \text{ is not a point of egress.}$$

Let $L_k = \{(t,x) \text{ in } \Gamma_k \text{ satisfying (a)}\}$, $M_k = \{(t,x) \text{ in } \Gamma_k \text{ satisfying (b)}\}$. Prove $S = S^* = \cup_{k=1}^m L_k \backslash \cup_{k=1}^m M_k$.

EXERCISE 6.3. One can use the results in Exercise 6.2 to prove results on the asymptotic equivalence of systems. Consider the two systems

(6.7) $$\dot{x}_j = f_j(t)x_j + g_j(t,x), \qquad j = 1,2,\ldots,n,$$

(6.8) $$\dot{y}_j = f_j(t)y_j, \qquad j = 1,2,\ldots,n.$$

Suppose $g = (g_1,\ldots,g_n)$, $f = (f_1,\ldots,f_n)$ are continuous for $t \geq T, (t,x) \in \Omega = \{(t,x) : T \leq t < \infty, x \text{ in } R^n\}$ and a unique solution of Eq. (6.7) exists through each point in Ω. If there exists a continuous function $F : [T,\infty) \to [0,\infty)$ and constant K such that

$$K \leq \int_t^v f_j(s)ds, \qquad v \geq t \geq T,$$

$$\int^\infty F(t)\exp\left(\int_T^t f_j(s)ds\right)dt < \infty, \qquad j = 1,2,\ldots,n,$$

$$|g(t,x)| \leq |x|F(t) \text{ on } \Omega$$

then, for every solution y of Eq. (6.8), there is a solution x of Eq. (6.7) such that

$$\lim_{t \to \infty} [x(t) - y(t)] = 0.$$

EXERCISE 6.4. Consider the second order equation

(6.9) $$\overset{..}{x} + f(t,x,\dot{x}) = 0$$

where $f(t,x,y)$ is continuous for $t \geq 0$, x, y in R^n and a uniqueness result holds for the initial value problem for Eq. (6.9),

$$x \cdot f(t,x,y) + y \cdot y > 0 \qquad \text{for all } t \geq 0, x \text{ in } R^n, y \neq 0 \text{ in } R^n.$$

Then there exists a family of solutions of Eq. (6.9) depending on at least n parameters such that, for any solution x in this family, there is a $t_0 > 0$ such that $x(t) \cdot x(t)$ is nonincreasing for $t \geq t_0$. Hint: Let $b > 0$, $\omega_b = \{(t,x,y) \text{ in } R^{2n+1} : \ell(t,x,y) < 0, m(t,x,y) < 0\}$ where $\ell(t,x,y) = x \cdot y - b$, $m(t,x,y) = -t$. Let Z_b be a line segment joining points $(t_0,\xi_1,\eta_1),(t_0,\xi_2,\eta_2)$ in distinct components of the set $\{(t,x,y) \text{ in } R^{2n+1} : x \cdot y = b, t > 0\}$ with $(t_0,0,0)$ not in Z_b. The principle of Wazewski yields a solution such that $x(t)\dot{x}(t) < b$ for $t > t_0$. To show $x(t)\dot{x}(t) \leq 0$, take a sequence $b_m \to 0$.

X.7. Remarks and Suggestions for Further Study

The main topic in the previous pages has been the application of the direct method of Liapunov to the determination of the stability of solutions of specific differential equations as well as the application of this method to the deduction of qualitative implications of the concept of stability. This chapter should serve only as an introduction and more extensive discussions may be found in Hahn [1], Krasovskii [1], LaSalle and Lefschetz [1], Halanay [1], Yoshizawa [2], Zubov [1]. The definition of Liapunov function used in Sections 1 and 3 may be found in LaSalle [2]. Theorem 1.1, Lemma 1.5 and Theorem 3.1 are due to Liapunov [1], Theorem 1.2 is essentially due to Cetaev [1], Exercise 1.5 to Hartman [2]. Section 2 is based on the paper of Moser [1]. The idea for the proofs of the converse theorems in Section 4 is due to Massera [1].

The basic idea of the direct method of Liapunov is applicable to systems described by functional differential equations as well as partial differential equations. For a discussion of functional differential equations as well as further references, see Krasovskii [1], Halanay [1], Hale [8], Cruz and Hale [1]. For partial differential equations, see Zubov [1], Infante and Slemrod [1].

The formulation of the topological principle in Section 6 was given by Wazewski [1] although the idea had been used on examples before (see Hartman [1] for more details). Exercise 6.2 is due to Onuchic [1], Exercise 6.3 to Onuchic [2]. Further extensions of the ideas of Wazewski and a more general theory of isolating blocks may be found in Conley [1].

Appendix. Almost Periodic Functions

In this appendix, we have assembled the information on almost periodic functions which is relevant for the discussion in the text. Only the most elementary results are proved. More details may be found in the following books.

[A.1] H. Bohr, Almost Periodic Functions, Chelsea, New York, 1947.
[A.2] A. S. Besicovitch, Almost Periodic Functions, Dover, New York, 1954.
[A.3] J. Favard, Fonctions presque-periodiques, Gauthier-Villars, Paris, 1933.
[A.4] A. M. Fink, Almost Periodic Differential Equations, Lecture Notes in Math., Vol. 377(1974), Springer-Verlag.
[A.5] B. M. Levitan, Almost Periodic Functions (Russian) Moscow, 1953.
[A.6] T. Yoshizawa, Stability Theory and the Existence of Periodic Solutions and Almost Periodic Solutions, Applied Mathematics Sciences, Vol. 14(1975), Springer-Verlag.

The books of Fink and Yoshizawa are very good presentations of the theory of almost periodic functions and differential equations. The presentation below follows these two books.

Notation. $\alpha, \beta, \ldots$ will denote sequences $\{\alpha_n\}$, $\{\beta_n\}$, ... in R^n. The notation $\beta \subset \alpha$ denotes β is a subsequence of α, $\alpha + \beta = \{\alpha_n + \beta_n\}$, $-\alpha = \{-\alpha_n\}$. We say α, β are *common subsequences* of α', β' if $\alpha_n = \alpha'_{n(k)}$, $\beta_n = \beta'_{n(k)}$ for some function $n(k)$, $k = 1, 2, \ldots$. If f, g are functions on R and α is a sequence in R, then $T_\alpha f = g$ means $\lim_{n \to \infty} f(t + \alpha_n)$ exists and is equal to $g(t)$ for all $t \in R$. The type of convergence (uniform, pointwise, etc.) will be specified when used. For example, we will write $T_\alpha f = g$ pointwise, $T_\alpha f = g$ uniformly, etc. All functions considered will be continuous and complex valued functions on R unless explicitly stated otherwise.

Definition 1. f is *almost periodic* if for every α' there is an $\alpha \subset \alpha'$ such that $T_\alpha f$ exists uniformly.

If $AP = \{f : f \text{ is almost periodic}\}$, $|f| = \sup_t |f(t)|$ if $f \in AP$, then AP with this norm is a normed linear space. It will be shown to be complete later.

LEMMA 1. Every periodic function is AP.

PROOF. If T is the period of f and α' is a sequence, then there exists an $\alpha \subset \alpha'$ such that $\alpha(\bmod T) = \{\alpha_n(\bmod T)\}$ converges to a point δ. Then $T_\alpha f = f(\cdot + \delta)$ uniformly.

LEMMA 2. Every almost periodic function is bounded.

PROOF. If not, there is a sequence $\alpha' \subset R$ such that $|f(\alpha'_n)| \to \infty$ as $n \to \infty$. But $f \in AP$ implies there is an $\alpha \subset \alpha'$ such that $T_\alpha f$ exists uniformly. In particular, $f(\alpha_n)$ converges which contradicts our supposition.

THEOREM 1. (Properties of almost periodic functions)

(1) AP is an algebra (closed under addition, product and scalar multiplication).

(2) If $f \in AP$, F is uniformly continuous on the range of F, then $F \circ f \in AP$.

(3) If $f \in AP$, $\inf_t |f(t)| > 0$, then $1/f \in AP$.

(4) If $f, g \in AP$, then $|f|, \bar{f}, \min(f,g), \max(f,g) \in AP$.

(5) AP is closed under uniform limits on R.

(6) $(AP, |\cdot|)$ is a Banach space.

(7) If $f \in AP$ and df/dt is uniformly continuous on R, then $df/dt \in AP$.

PROOF. We prove this theorem in detail since it is simple and illustrates the way subsequences are used in Definition 1.

(1) If $f, g \in AP$ and $\alpha'' \subset R$, there exists $\alpha' \subset \alpha''$ such that $T_{\alpha'} f$ exists uniformly, $T_{\alpha'}, af$ exists uniformly for any complex number a. Also there exists $\alpha \subset \alpha'$ such that $T_\alpha g$ exists uniformly. Thus, $T_\alpha(f + g)$ exists uniformly, $T_\alpha(fg) = (T_\alpha f)(T_\alpha g)$ exists uniformly.

(2) If $f \in AP$ and F is uniformly continuous on the range of f, then, for any $\alpha' \subset R$, there exists $\alpha \subset \alpha'$ such that $T_\alpha(F \circ f) = F \circ T_\alpha f$ uniformly.

(3) If $\inf_t |f(t)| > 0$, then $1/z$ is uniformly continuous on the range of f and $1/f \in AP$.

(4) The same argument proves $|f|, \bar{f} \in AP$ if $f \in AP$ by using the function $F(z) = |z|, z$, respectively. Since

$$\min(f,g) = \frac{1}{2}[(f+g) - |f-g|], \qquad \max(f,g) = \frac{1}{2}[(f+g) + |f-g|],$$

the remainder of (4) is proved by using (1).

(5) Suppose $\{f_n\} \subset AP$, f continuous, $|f_n - f| \to 0$ as $n \to \infty$. For any $\alpha' \subset R$, there exists $\alpha' \supset \beta' \supset \cdots \supset \beta^n \supset \cdots$ such that $T_{\beta^n} f_n$ exists uniformly, $n = 1, 2, \ldots$. Use the diagonalization process to obtain $\beta \subset \alpha'$ such that $T_\beta f_n$ exists uniformly for all $n = 1, 2, \ldots$.

For any $\epsilon > 0$, three is a $k_0(\epsilon)$ such that, for $n = 1, 2, \ldots, k \geq k_0(\epsilon), t \in R$

$$|f(t + \beta_n) - f_k(t + \beta_n)| < \epsilon/3.$$

For a fixed $k \geq k_0(\epsilon)$ choose $N(k,\epsilon)$ such that, for $n, m \geq N(k,\epsilon), t \in R$

$$|f_k(t + \beta_n) - f_k(t + \beta_m)| < \epsilon/3.$$

Then, for $n, m \geq N(k,\epsilon), t \in R$,

$$|f(t + \beta_n) - f(t + \beta_m)| \leq |f(t + \beta_n) - f_k(t + \beta_n)|$$
$$+ |f_k(t + \beta_n) - f_k(t + \beta_m)| + |f_k(t + \beta_m) - f(t + \beta)| < \epsilon.$$

Thus, $T_\beta f$ exists uniformly and $f \in AP$.

(6)   This is obvious from (1) and (5).

(7)   If $f_n(t) = n[f(t + 1/n) - f(t)]$, then the mean value theorem implies

$$f'(t) - f_n(t) = f'(t) - f'(t + \theta(t, n))$$

for some $t \leq \theta(t, n) \leq t + 1/n$. Since $f'(t)$ is uniformly continuous, we have $f_n(t) \to f'(t)$ uniformly on R. Since each $f_n \in AP$, we have from (5) that $f' \in AP$. This completes the proof of the theorem.

The space AP contains all periodic functions. Theorem 1 shows that it contains all functions which are sums of periodic functions and thus all trigonometric polynomials. It also contains the uniform limit of trigonometric polynomials. One can actually show (but the proof is not trivial) that AP consists precisely of functions which are uniform limits of trigonometric polynomials.

To obtain other characterizations of almost periodic functions, we need the following concept.

Definition 2.   The hull of $f, H(f)$, is defined by

$$H(f) = \{g : \text{there is an } \alpha \subset R \text{ such that } T_\alpha f = g \text{ uniformly}\}$$

If f is periodic, then $H(f)$ consists of all translates of f; that is,

$$H(f) = \{f_\tau : f_\tau(t) = f(t + \tau), t \in R, \tau \in R\}$$

and thus, $H(f)$ is a closed curve in AP if f is periodic. If $f \in AP$, then $H(f)$ may contain elements which are not translates of f. In fact, if $f(t) = \cos t + \cos\sqrt{2} t$, then $f \in AP$ since it is the sum of periodic functions, but is not periodic since it takes the value 2 only at $t = 0$. Also, $f(t) > -2$ for all t and there is an $\alpha' = \{\alpha'_n\} \subset R$ such that $f(\alpha'_n) \to -2$. If $\alpha \subset \alpha'$ and $T_\alpha f = g$ uniformly, then $g(0) = -2$ and g cannot be a translate of f.

THEOREM 2. $f \in AP$ if and only if $H(f)$ is compact in the topology of uniform convergence on R. Furthermore, if $f \in AP$ then $H(g) = H(f)$ for all $g \in H(f)$.

PROOF. If $H(f)$ is compact, then for any $\alpha' \subset R$, there is an $\alpha \subset \alpha'$ such that $T_\alpha f$ exists uniformly and, thus $f \in AP$.

Conversely, if $f \in AP$, $\{g_n\} \subset H(f)$, then there is an $\alpha' = \{\alpha_n'\}$ such that $|f(\cdot + \alpha_n') - g_n| < 1/n$ (we have used the diagonalization process here). Also, there is an $\alpha \subset \alpha'$ such that $T_\alpha f = g$ exists uniformly. If $\alpha = \{\alpha_n\} = \{\alpha_{k_n}'\}$, then $|f(\cdot + \alpha_n) - g_{k_n}| \to 0$ as $n \to \infty$. Thus,

$$|g - g_{k_n}| \leq |g - f(\cdot + \alpha_n)| + |f(\cdot + \alpha_n) - g_{k_n}| \to 0$$

as $n \to \infty$. Thus, $H(f)$ is compact.

To prove the last part of the theorem, if $g \in H(f)$ and $h \in H(g)$; that is, there is an $\alpha = \{\alpha_n\} \subset R$ such that $T_\alpha g = h$ uniformly, then there is a $\beta = \{\beta_n\}$ such that $||f(\cdot + \beta_n) - g(\cdot + \alpha_n)| < 1/n$. Thus, $T_\beta f = h$ and $h \in H(f)$; that is, $H(g) \subseteq H(f)$. Conversely, if $g \in H(f)$, $T_\alpha f = g$, then $T_{-\alpha} g = f$ by the change of variables $t + \alpha_n = s$. Thus, $f \in H(g)$. Since $H(g)$ is compact, this implies $H(f) \subseteq H(g)$. Finally, we have $H(f) = H(g)$ and the theorem is proved.

To obtain another characterization of an almost periodic function, we need the following definitions.

Definition 3. A subset $S \subset R$ is *relatively dense* if there is an $L > 0$ such that $[a, a + L] \cap S \neq \phi$ for any $a \in R$.

Definition 4. If f is any continuous complex valued function on R, the *ϵ-translation set* of f is defined as

$$T(f, \epsilon) = \{\tau : |f(t + \tau) - f(t)| < \epsilon \qquad \text{for } t \in R\}.$$

Definition 5. A continuous complex valued function f on R is *Bohr almost periodic* if for every $\epsilon > 0$, $T(f, \epsilon)$ is relatively dense.

Lemma 3. A Bohr almost periodic function is bounded and uniformly continuous.

PROOF. If $f(t)$ is almost periodic, then for any $\eta > 0$, there exist an l and a τ in every interval $[t - l, t]$ such that

$$|f(t) - f(t - \tau)| < \eta \qquad \text{for all } t \text{ in } (-\infty, \infty).$$

If we let $t' = t - \tau$ then $0 \leq t' \leq l$ and $|f(t) - f(t')| < \eta$. If $|f(t')| \leq B$ for $0 \leq t' \leq l$ then $|f(t)| \leq B + \eta$ for all t in $(-\infty, \infty)$. This proves boundedness.

Suppose η, l are as above. Since f is continuous, it is uniformly continuous on $[-l, l]$ and we can find a $\delta < l$ such that

$$|f(t_1') - f(t_2')| < \eta \qquad \text{if} \qquad |t_1' - t_2'| < \delta, t_1', t_2' \in [-l, l].$$

For any t_1, t_2 with $t_1 < t_2$, $|t_1 - t_2| < \delta$, consider the interval $[t_1, t_1 + l]$. Let τ be an almost period relative to η in this interval. Then $t_1' = t_1 - \tau$, $t_2' = t_2 - \tau$ are in $[-l, l]$ and $|f(t_1) - f(t_1')| < \eta$, $|f(t_2) - f(t_2')| < \eta$. The triangle inequality shows that $|f(t_1) - f(t_2)| < 3\eta$ and this proves uniform continuity.

THEOREM 3.   The following statements are equivalent:

  (i) $f \in AP$
  (ii) $H(f)$ is compact in the topology of uniform convergence on R.
  (iii) f is Bohr almost periodic.

PROOF.   We have already proved (i) $\Leftrightarrow$ (ii). We next prove (iii) $\Rightarrow$ (i). Suppose $\alpha' \subset R$. We first show that, for any $\epsilon > 0$, there is an $\alpha \subset \alpha'$ such that

$$|f(t + \alpha_n) - f(t + \alpha_m)| < \epsilon \qquad \text{for all } n, m$$

Let L be as in the definition of relatively dense for $T(f, \frac{\epsilon}{4})$, and choose δ from Lemma 1 so that $|f(t) - f(s)| < \epsilon$ if $|t - s| < 2\delta$. We may write the sequence $\alpha = \{\alpha_n\}$ as $\alpha_n = \beta_n + \gamma_n$, $\beta_n \in T(f, \frac{\epsilon}{4})$, $0 \leq \gamma_n \leq L$, $\gamma_n \to \gamma$ as $n \to \infty$, $\gamma - \delta < \gamma_n < \gamma + \delta$. Then

$$|f(t + \alpha_n) - f(t + \alpha_m)| \leq |f(t + \beta_n - \beta_m + \gamma_n - \gamma_m) - f(t - \beta_m + \gamma_n - \gamma_m|$$
$$+ |f(t - \beta_m + \gamma_n - \gamma_m) - f(t + \gamma_n - \gamma_m|$$
$$+ |f(t + \gamma_n - \gamma_m) - f(t)| < \epsilon \qquad \forall t \in R.$$

Now use the diagonalization process to get $\alpha'' \subset \alpha \subset \alpha'$ such that $T_{\alpha''}f$ exists uniformly.

We now prove (ii) $\Rightarrow$ (iii). $H(f)$ compact is equivalent to $H(f)$ totally bounded is equivalent to $\{f(t + \tau) : \tau \in R\}$ totally bounded is equivalent to, for any $\epsilon > 0$, there are $a_1, \ldots, a_n$ and a function $i(t)$ from R to $\{1, \ldots, n\}$ such that $|f(t + a_{i(\tau)}) - f(t + \tau)| < \epsilon$ for all t, τ. This is equivalent to

$$|f(t) - f(t + \tau - a_{i(\tau)})| < \epsilon \qquad \text{for all } t.$$

If $L = \max|a_i|$, then $\tau - L \leq \tau - a_{i(\tau)} \leq \tau + L$ and $\tau - a_{i(\tau)} \in T(f, \epsilon)$. Thus, $T(f, \epsilon)$ is relatively dense with constant $2L$.

This completes the proof of the theorem.

THEOREM 4. $f \in AP$ if and only if

  (iv) For any $\alpha', \beta' \subset R$, there are common subsequences $\alpha \subset \alpha'$, $\beta \subset \beta'$ such that

$$T_{\alpha+\beta}f = T_\alpha T_\beta f \qquad \text{pointwise.}$$

PROOF.   Suppose $f \in AP$, $\alpha', \beta' \subset R$ are given and choose $\beta'' \subset \beta'$ such that $T_{\beta''}f = g$ uniformly. Since $g \in AP$, for $\alpha'' \subset \alpha'$ common with β'', there is an $\alpha''' \subset \alpha''$ such that $T_{\alpha'''}g = h$ uniformly. Let $\beta''' \subset \beta''$ be common with α'''. Then we can find common subsequences, $\alpha \subset \alpha'''$, $\beta \subset \beta'''$ such that $T_{\alpha+\beta}f = k$

uniformly and also note $T_\beta f = g$, $T_\alpha g = h$ uniformly. If $\epsilon > 0$ is given, then, there is a n_0 such that, for all $n \geq n_0$ and all t,

$$|k(t) - f(t + \alpha_n + \beta_n)| < \epsilon, \; |g(t) - f(t + \beta_n)| < \epsilon, \; |h(t) - g(t + \alpha_n)| < \epsilon$$

Thus, for $n \geq n_0$,

$$|h(t) - k(t)| \leq |h(t) - g(t + \alpha_n)| + |g(t + \alpha_n) - f(t + \alpha_n + \beta_n)|$$
$$+ |f(t + \alpha_n + \beta_n) - k(t)| < 3\epsilon$$

Since ϵ is arbitrary, $h = k$; i.e. $T_{\alpha + \beta} f = T_\alpha T_\beta f$ uniformly and, in particular, pointwise.

To prove (iv) implies $f \in AP$, suppose $\gamma' \subset R$. Then there is a $\gamma \subset \gamma'$ such that $T_\gamma f$ exists pointwise. In fact, for $\alpha' = \{0\}$, $\beta' = \gamma'$, we know there are common subsequences $\alpha \subset \alpha'$, $\gamma \subset \gamma'$ such that $T_{\alpha + \gamma} f = T_\alpha T_\gamma f$ pointwise. Since $\alpha = \{0\}$, this proves the assertion. Suppose that $T_\gamma f$ does not exist uniformly. Then there exist $\alpha' \subset \beta$, $\beta' \subset \gamma$, $\tau' \subset R$, $\epsilon > 0$

(*) $$|f(\alpha'_n + \tau'_n) - f(\beta'_n + \tau'_n)| \geq \epsilon.$$

Apply condition (iv) to α', τ' to obtain common subsequences $\alpha'' \subset \alpha'$, $\tau'' \subset \tau'$ such that $T_{\tau'' + \alpha''} f = T_{\tau''} T_{\alpha''} f$ pointwise. Choose $\beta'' \subset \beta'$ common to α'', τ'' and apply (iv) to β'', τ'' to find $\beta \subset \beta''$, $\tau \subset \tau''$ such that $T_{\tau + \beta} f = T_\tau T_\beta f$ pointwise. Choose $\alpha + \alpha''$ common to β, τ. Then $T_{\tau + \alpha} f = T_\tau T_\alpha f$ pointwise from above. But $T_\alpha f = T_\gamma f = T_\beta f$ pointwise, since $\alpha \subset \gamma$, $\beta \subset \gamma$. Thus, $T_{\tau + \alpha} f = T_\tau T_\alpha f = T_\tau T_\beta f = T_{\tau + \beta} f$ pointwise. At $t = 0$, this contradicts (*) since $\tau + \beta \subset \tau' + \beta'$, $\tau + \alpha \subset \tau' + \alpha'$. Thus, the convergence is uniform.

This completes the proof of the theorem.

A remarkable consequence of Theorem 3 is that almost periodicity can be decided by discussing only pointwise convergence as in (iv). This is generally easier to check than uniform convergence on R.

From Proposition 1 and Theorem 1, the space AP contains all trigonometric polynomials and all uniform limits of trigonometric polynomials. The converse is also true and is stated without proof.

THEOREM 5. $f \in AP$ if and only if f can be uniformly approximated by trigonometric polynomials.

It is natural to investigate the problem of the existence of a theory of trigonometric series for almost periodic functions analogous to the theory of Fourier series for periodic functions. The first fundamental result is

THEOREM 6. If $f \in AP$, then the mean value

$$M[f] = \lim_{T \to \infty} \frac{1}{T} \int_t^{t+T} f(s) ds$$

exists, is independent of t and the convergence is uniform in t.

If $f \in AP$, then $f \exp(-i\lambda \cdot) \in AP$ for any real number λ. Define

$$a(\lambda, f) = M[fe^{-i\lambda \cdot}] = \lim_{T \to \infty} \frac{1}{T} \int_0^T f(t)e^{-i\lambda t}dt.$$

The set

$$\Lambda(f) = \{\lambda \in R : a(\lambda, f) \neq 0\}$$

is called the set of *Fourier exponents* of f, the numbers $a(\lambda, f)$, $\lambda \in \Lambda(f)$, the *Fourier coefficients* and the *Fourier series* of f is designated by

$$f(t) \sim \Sigma_{\lambda \in \Lambda(f)} a(\lambda, f) e^{i\lambda t}.$$

Then, one can prove

THEOREM 7. $\Lambda(f)$ is denumerable. If $f, g \in AP$, then $f = g$ if and only if $\Lambda(f) = \Lambda(g)$ and $a(\lambda, f) = a(\lambda, g)$ for $\lambda \in \Lambda(f)$. If $f \sim \Sigma_{\lambda \in \Lambda(f)} a(\lambda, f) \exp(i\lambda t)$, then $\Sigma_{\lambda \in \Lambda(f)} |a(\lambda, f)|^2 < \infty$. The usual operations on Fourier series are valid.

For any sequence $\{\lambda_n\} \subset R$, the *module* of $\{\lambda_n\}$ is the set consisting of all real numbers which are finite linear combinations of the $\{\lambda_n\}$ with integer coefficients. We say $\{\alpha_j\} \subset R$ is *linear independent over the rationals* if, for every $N \geq 1$, $\Sigma_{k=1}^N r_k \alpha_{j_k} = 0$ for each r_k rational implies $r_k = 0$, $k = 1, 2, \ldots, N$. The sequence $\{\alpha_j\}$ is said to be an *integral base* for $\{\lambda_n\}$ if $\{\alpha_j\}$ is linearly independent over the rationals and if each element of $\{\lambda_n\}$ is a finite linear combination of the $\{\alpha_j\}$ with integer coefficients. If $f \in AP$, the *module* $m[f]$ of f is the module of $\Lambda(f)$. A function $f \in AP$ is called *quasiperiodic* if there exist an integer $N \geq 1$, positive real numbers $T_1, \ldots, T_N$ and a function F on R^N such that $F(x_1, \ldots, x_N)$ is periodic in x_j of period T_j, $j = 1, 2, \ldots, N$ such that $f(t) = F(t, \ldots, t)$.

Example 1. If f is periodic of period $2\pi/\omega$, then $m[f] = \{n\omega, n = 0, \pm 1, \ldots\}$ and an integral base for $m[f]$ in $\{\omega\}$.

Example 2.   If $f(t) = \cos t + \cos 2t + \cos \sqrt{2}t$, then $\Lambda(f) = \{1, 2, \sqrt{2}\}$, $m[f] = \{n + m\sqrt{2}, n, m = 0, \pm 1, \ldots\}$ and an integral base for $m[f]$ is $\{1, \sqrt{2}\}$. This function f is quasiperiodic with

$$f(t) = F(t, t), F(x_1, x_2) = \cos x_1 + \cos 2x_1 + \cos \sqrt{2} x_2.$$

Example 3. If $\Sigma_{n=1}^\infty |a_n| < \infty$, $a_n \neq 0$ for all n, then

$$f(t) = \Sigma_{n=1}^\infty a_n e^{it/n}$$

is in AP, $\Lambda(f) = \{1/n, n = 1, 2, \ldots\}$. What is $m[f]$ and what is an integral base for $m[f]$?

The following result is basic for differential equations. The proof is complicated and contained in Fink, p. 61. In the statement of the theorem, "convergence in any sense" may be interpreted as "pointwise-convergence," "uniform convergence on compact sets," "uniform convergence" or "mean convergence"; that is, f_n converges to f in the mean if $M(|f_n - f|^2) \to 0$ as $n \to \infty$.

THEOREM 8. For $f, g \in AP$, the following statements are equivalent:

(i) $m[f] \supset m[g]$

(ii) for any $\epsilon > 0$, there is a $\delta > 0$ such that $T(f, \delta) \subset T(g, \epsilon)$

(iii) $T_\alpha f$ exists implies $T_\alpha g$ exists (any sense)

(iv) $T_\alpha f = f$ implies $T_\alpha g = g$ (any sense)

(v) $T_\alpha f = f$ implies there exists $\alpha' \subset \alpha$ such that $T_{\alpha'} g = g$ (any sense).

In the study of almost periodic differential equations, one encounters frequently the simple equation $\dot{y} = f(t)$ where $f \in AP, M[f] = 0$. For f periodic, this equation always has a periodic solution. However, when $f \in AP, M[f] = 0$, this may not be the case. In fact,

$$f(t) \sim \sum_{n=1}^{\infty} \frac{1}{n^2} e^{it/n^2}$$

is in $AP, M[f] = 0$. However, if $\int^t f$ is in AP, then

$$\int^t f \sim (\text{constant}) + \Sigma_{n=1}^{\infty}(-i)e^{it/n^2}.$$

However, this latter funtcion is not in AP since the sum of squares of the Fourier coefficients does not exist.

The following result is known about the integral of an almost periodic function.

THEOREM 9. If $f \in AP$, then $\int^t f$ is in AP if and only if it is bounded.

In the applications to differential equations, this proposition generally cannot be applied.

The next result due to Bogoliubov [1] is concerned with an AP approximation to the solution of the equation $\dot{y} - f(t) = 0$ when f is AP, has $M[f] = 0$ and does not necessarily have a bounded integral.

LEMMA 4. Suppose f is in AP, $M[f] = 0$. For every $\eta > 0$, there is a continuous scalar function $\zeta(\eta)$, $0 < \eta < \infty$, $\zeta(\eta) \to 0$ as $\eta \to 0$ such that the almost periodic function f_η defined by

(1)
$$f_\eta(t) = \int_{-\infty}^{t} e^{-\eta(t-s)} f(s) \, ds$$

satisfies

(2)   (a) $|f_\eta(t)| \leq \eta^{-1}\zeta(\eta)$,

             (b) $|df_\eta(t)/dt - f(t)| \leq \zeta(\eta)$,

for all t in R. Also, $m[f_\eta] \subset m[f]$.

PROOF.   Since $M[f] = 0$, there is a continuous nonincreasing scalar function $\varepsilon(T)$, $0 < T < \infty$, $\varepsilon(T) \to 0$ as $T \to \infty$ such that $|T^{-1} \int_t^{t+T} f(s)\, ds| \leq \varepsilon(T)$ for all t in R. Furthermore, for any $T > 0$,

$$f_\eta(t) = \int_0^\infty e^{-\eta s} f(t-s)\, ds$$

$$= \sum_{k=0}^\infty e^{-\eta k T} \int_{kT}^{(k+1)T} f(t-s) e^{-\eta(s-kT)}\, ds.$$

If B is a bound for $|f(t)|$ on $(-\infty, \infty)$, then the above relation yields

$$|f_\eta(t)| \leq \sum_{k=0}^\infty e^{-\eta k T} \left| \int_{kT}^{(k+1)T} f(t-s)\, ds \right|$$

$$+ B \sum_{k=0}^\infty e^{-\eta k T} \int_{kT}^{(k+1)T} (1 - e^{-\eta(s-kT)})\, ds$$

$$\leq \sum_{k=0}^\infty e^{-\eta k T} \varepsilon(T) T + B \int_0^T (1 - e^{-\eta s})\, ds \sum_{k=0}^\infty e^{-\eta k T}$$

$$\leq \varepsilon(T) T [1 - e^{-\eta T}]^{-1} + BT.$$

Since $\varepsilon(T) \to 0$ as $T \to \infty$ and is nonincreasing, there always exists a unique solution to the equation $1 - e^{-\eta T} = \varepsilon(T)$. Suppose T_η is chosen to be this solution. Since $\varepsilon(T) > 0$ for all T, it is clear that $T_\eta \to \infty$ as $\eta \to 0$ and this together with the fact that $\varepsilon(T) \to 0$ as $T \to \infty$ imply that $\eta T_\eta \to 0$ as $\eta \to 0$. The solution T_η is obviously continuous in η. If we let $\zeta(\eta) = (B+1)\eta T_\eta$, then relation (2a) is proved.

From (1), it follows that

(3) $$\frac{df_\eta(t)}{dt} - f(t) = -\eta f_\eta(t),$$

and therefore, (2b) follows from (2a).

It remains to show that $f_\eta \in AP$. Suppose $\alpha' \subset R$. Since $f \in AP$, there is a sequence $\alpha \subset \alpha'$ such that $T_\alpha f$ exists uniformly. Thus, for any $\epsilon > 0$, there is an $N(\epsilon)$ such that

$$|f_\eta(t + \alpha_n) - f_\eta(t + \alpha_m)| \leq \int_{-\infty}^0 e^{\eta s} |f(t + s + \alpha_n) - f(t + s + \alpha_m)|\, ds$$

$$\leq \epsilon \int_{-\infty}^0 e^{\eta s} = \epsilon/\eta,$$

for $n, m \geq N(\epsilon)$. This shows $f_\eta \in AP$. The fact that $m[f_\eta] \subset m[f]$ follows from Theorem 8. This proves the lemma.

We want to study almost periodic solutions of an almost periodic differential equation $\dot{x} = f(t,x)$. Thus, we need to know when $f(t,\phi(t))$ is almost periodic if $\phi \in AP$. In general, this is not true. One can see this from the example $f(t,x) = \sin xt$ by showing that $f(t,\sin t) = \sin(t \sin t)$ is not uniformly continuous. The difficulty arises from the fact that the ϵ-translation set for $\sin xt$ is not well behaved in x.

An appropriate generalization of the definition of an almost periodic function depending on parameters is the following one.

Definition 6. A continuous function $f:R \times D \to C^n$, where D is open in C^n, is said to be *almost periodic in t* uniformly for $x \in D$ if for any $\epsilon > 0$, the set $\mathcal{T}(f,\epsilon) = \underset{x \in K}{\cap} T(f(\cdot,x),\epsilon)$ is relatively dense in R for each compact set $K \subset D$. Equivalently, if for any $\epsilon > 0$, and any compact set $K \subset D$, there exists $L(\epsilon,K) > 0$ such that any interval of length $L(\epsilon,K)$ contains a τ with

$$|f(t + \tau, x) - f(t,x)| < \epsilon \qquad \text{for all } t \in R, x \in K.$$

One can then prove the following results (see Fink or Yoshizawa).

THEOREM 10. If $f(t,x)$ is AP in t uniformly for $x \in D$, then f is bounded and uniformly continuous on $R \times K$ for any compact set $K \subset D$.

THEOREM 11. If $f(t,x)$ is AP in t uniformly for $x \in D$, then, for any $\alpha' \subset R$, there is an $\alpha \subset \alpha'$ and a function $g(t,x)$ which is AP in t uniformly for $x \in D$ such that $f(t + \alpha_n, x) \to g(t,x)$ uniformly on $R \times K$ for any compact set $K \subset D$.

Conversely, if $f:R \times D \to C^n$ is continuous and for any $\alpha' \subset R$, there is an $\alpha \subset \alpha'$ such that, for any compact set $K \subset D$, $f(t + \alpha_n, x)$ converges uniformly on $R \times K$, then $f(t,x)$ is AP in t uniformly for $x \in D$.

In general, functions AP in t uniformly for $x \in D$ satisfy the same properties as before if convergence is always uniform on compact sets $K \subset D$.

THEOREM 12. If $f(t,x)$ is AP in t uniformly for $x \in D$, and $\xi \in AP$, $\xi(t) \in K$, a compact set of D, then $f(\cdot)) \in AP$.

In the applications to differential equations, it is desirable to have an improvement of Lemma 4 in this more general situation. More specifically, one wants an approximating solution of the equation

$$\frac{\partial y}{\partial t} - f(t, x) = 0,$$

which is very smooth in x. One can apply Lemma 4 and more or less classical smoothing operators to achieve this end result. If $M[f(\cdot, x)]$ denotes the mean value of $f(t, x)$ with respect to t, then we have

LEMMA 5.   Suppose $f(t,x)$ is AP in t uniformly with respect to x in a set D containing the closed ball $B_\sigma = \{x$ in $C^k : |x| \leq \sigma\}$. If $M[f(\cdot, x)] = 0$ for x in B_σ then for any $\sigma_1 < \sigma$, there are an $\eta_0 > 0$ and a function $w(t, x, \eta)$ defined and continuous for t in R, x in B_{σ_1}, $0 < \eta < \eta_0$, which is almost periodic in t uniformly with respect to x in B_{σ_1} and η in any compact set of $(0, \eta_0)$, $m[w(\cdot, x, \eta)] \subset m[f(\cdot, x)]$, $w(t, x, \eta)$ has derivatives of any desired order with respect to x and is such that, if

$$g(t, x, \eta) = \frac{\partial w(t, x, \eta)}{\partial t} - f(t, x),$$

then $g(t, x, \eta)$, $\eta w(t, x, \eta)$ and $\eta \partial w(t, x, \eta)/\partial x$ approach zero as $\eta \to 0$ uniformly with respect to t in R, x in B_{σ_1}. If, in addition, $f(t, x)$ has continuous first partial derivatives with respect to x, then $\partial g(t, x, \eta)/\partial x \to 0$ as $\eta \to 0$ uniformly with respect to t in R, x in B_{σ_1}.

PROOF.   From Lemma 4, it follows that there is a function $f_\eta(t, x)$ defined by (1) which is a.p. in t uniformly with respect to x in B_σ and η in any compact set such that

$$(4) \qquad\qquad |f_\eta(t, x)| \leq \eta^{-1}\zeta(\eta),$$

$$\frac{\partial f_\eta(t, x)}{\partial t} - f(t, x) = -\eta f_\eta(t, x),$$

where $\zeta(\eta) \to 0$ as $\eta \to 0$.

For a fixed $a > 0$ and some fixed integer $q \geq 1$, consider the function $\Delta_a(x)$ defined by

$$\Delta_a(x) = \begin{cases} d_a(1 - a^{-2}|x|^2)^{2q} & \text{for } |x| \leq a \\ 0 & \text{for } |x| > a \end{cases}$$

where the constant d_a is determined so that $\int_{B_\sigma} \Delta_a(x)\, dx = 1$. Define $w(t, x, \eta)$ by

$$(5) \qquad\qquad w(t, x, \eta) = \int_{B_\sigma} \Delta_a(x - y) f_\eta(t, y)\, dy.$$

It is easy to see that $w(t, x, \eta)$ is a.p. in t uniformly with respect to x in B_σ and η in any compact set since $f_\eta(t, x)$ has the same property.

The function $\Delta_a(x - y)$ possesses continuous partial derivatives up through order $2q - 1$ with respect to x which are bounded in norm by a function $G(a)/(\text{area of integration})$ where $G(a)$, $0 < a < \infty$, is continuous. The function $G(a)$ may approach ∞ as $a \to 0$. Therefore, from (4), the function $w(t, x, \eta)$ defined by (5) has partial derivatives with respect to x up through order $2q - 1$ which are bounded by $G(a)\zeta(\eta)\eta^{-1}$. Since q is an arbitrary integer, the number of derivatives with respect to x may be as large as desired. Choose $a = a_\eta$ as a function of η in such a way that $a_\eta \to 0$, $G(a_\eta)\zeta(\eta) \to 0$

as $\eta \to 0$. The conclusion of the lemma concerning $\eta w(t, x, \eta)$ is therefore valid since ηw, $\eta \partial w / \partial x$ are bounded by $G(a_\eta) \zeta(\eta)$.

For any $\sigma_1 < \sigma$, choose η_0 so small that $\sigma_1 + a_\eta < \sigma$ for $0 < \eta < \eta_0$. From the definition of $\Delta_a(x)$, it follows that $\int_{B_\sigma} \Delta_{a_\eta}(x - y)\, dy = 1$ for every x in B_{σ_1}, $0 < \eta < \eta_0$. Therefore, from relations (4) and (5), it follows that

$$(6) \qquad \bar{g}(t, x, \eta) = \int_{B_\sigma} \Delta_{a_\eta}(x - y)[f(t, y) - f(t, x)]\, dy,$$

where $\bar{g} = g + \eta w$. From the definition of $\Delta_a(x)$, it follows that

$$|g(t, x, \eta)| \leq \sup_{0 \leq |x-y| \leq a_\eta} |f(t, y) - f(t, x)|.$$

Since $a_\eta \to 0$ as $\eta \to 0$ and $f(t, x)$ is uniformly continuous for t in R, x in B_{σ_1}, there is a continuous function $\delta(\eta)$, $0 < \eta < \eta_0$, $\delta(\eta) \to 0$ as $\eta \to 0$ such that

$$\sup_{0 \leq |x-y| \leq a_\eta} |f(t, y) - f(t, x)| < \delta(\eta).$$

Consequently, $|\bar{g}(t, x, \eta)| < \delta(\eta)$ and $|g(t, x, \eta| < \delta(\eta) + G(a_\eta)\zeta(\eta)$ and g satisfies the properties stated in the first part of the lemma.

If, in addition, $f(t, x)$ has continuous first partial derivatives with respect to x, then an integration by parts in relation (6) yields

$$\frac{\partial \bar{g}(t, x, \eta)}{\partial x} = \int_{B_\sigma} \Delta_{a_\eta}(x - y) \left[\frac{\partial f(t, y)}{\partial x} - \frac{\partial f(t, x)}{\partial x} \right] dy.$$

The same type of argument as before shows that this expression is bounded by a function $\delta_1(\eta)$ which approaches zero as $\eta \to 0$. This completes the proof of the lemma.

If, in Lemma 5, the function $f(t, x)$ is periodic in some of the components of x, then the function $w(t, x, \eta)$ can also be chosen periodic in these components with the same period.

If g is an n-vector and ϕ is an r-vector, then $g(\phi)$ is said to be *multiply periodic* in ϕ with vector period $\omega = (\omega_1, \ldots, \omega_r)$, $\omega_j > 0$, $j = 1, 2, \ldots, r$, if the function g is periodic in the jth component of ϕ with period ω_j. If $g(\phi)$ is multiply periodic, then the function $g(\phi_1 + t, \ldots, \phi_r + t)$ is AP in t uniformly with respect to ϕ. It is therefore meaningful to define the mean value of this function with respect to t. This concept of mean value was first used in differential equations by S. Diliberto [1]. To simplify notations, let $\phi + t = (\phi_1 + t, \ldots, \phi_r + t)$ and designate the above mean value by

$$(7) \qquad M_\phi[g(\phi)] = \lim_{T \to \infty} \frac{1}{T} \int_0^T g(\phi + t)\, dt.$$

To understand this concept a little better, consider the case where ϕ is a two-vector, $\phi = (\phi_1, \phi_2)$ and

$$g(\phi) \sim \sum_{j,\,k} a_{jk}\, e^{i(k\mu\phi_1 + j\omega\phi_2)}.$$

In usual treatises on multiply periodic functions, the mean value of g is the constant term in the Fourier series of g; that is, a_{00}. According to definition (7),

$$M_\phi[g(\phi)] = \sum_{j,\,k:\,k\mu + j\omega = 0} a_{jk}\, e^{i(k\mu\phi_1 + j\omega\phi_2)}$$

and this could be a function of ϕ if the frequencies μ, ω are such that μ/ω is rational.

Lemma 5 has an appropriate generalization to the case of functions of the form $g(t, \phi, x)$ which are multiply periodic in the vector ϕ, AP in t, uniformly with respect to ϕ in C^r and x in B_σ and $M_{t,\phi}[g(t, \phi, x)] = 0$. Under these conditions, for any $\sigma_1 < \sigma$, there are an $\eta_0 > 0$ and a function $W(t, \phi, x, \eta)$, $0 < \eta < \eta_0$, multiply periodic in ϕ, AP in t such that the function

$$G(t, \phi, x, \eta) = \frac{\partial W}{\partial t} + \sum_{j=0}^{r} \frac{\partial W}{\partial \phi_j} - g(t, \phi, x)$$

as well as ηW, $\eta \partial W/\partial\phi$, $\eta \partial W/\partial x \to 0$ as $\eta \to 0$ uniformly with respect to t in R, ϕ in C^r, x in B_{σ_1}. If $g(t, \phi, x)$ has continuous first partials with respect to x, then the partial derivatives of $G(t, \phi, x, \eta)$ with respect to ϕ, x also approach zero as $\eta \to 0$ uniformly with respect to t, ϕ, x. The proof of this fact is very similar to the proof of Lemma 5 and may be found in Hale [3].

REFERENCES

André, J., and P. Seibert [1]. On after-endpoint motions of general discontinuous control systems and their stability properties. *Proc. 1st Intern. IFAC Congr.* Moscow 1960, **II**, 919–922.

Andronov, A. A., S. E. Khaikin, and A. A. Witt [1]. *Theory of Oscillators*, Pergamon Press, New York, 1966.

Andronov, A. A. Leontovich, E. A., Gordon, I. I. and A. G. Maier [1], *Theory of Bifurcations of Dynamic Systems on a Plane*, John Wiley and Sons, 1973.

Antosiewicz, H. [1]. Boundary value problems for nonlinear ordinary differential equations. *Pac. J. Math.* **17** (1966), 191–197. [2]. Un analogue du principe du pointe fixe de Banach. *Ann. Mat. Pura. Appl.* (4) **74** (1966), 61–64. [3]. On an integral inequality. To appear.

Arnold, V. I. [1]. Small denominators. I. Mapping the circle onto itself. *Izv. Akak. Nauk SSSR Ser. Mat.* **25** (1961), 21–86. [2]. Small denominators and problems of stability of motion in classical and celestial mechanics. *Uspehi Mat. Nauk* **18** (1963), no. 6(114), 91–192.

Arscott, F. M. [1]. *Periodic Differential Equations*. Pergamon Press, New York, 1964.

Auslander, J. and W. H. Gottschalk [1]. *Topological Dynamics*. W. A. Benjamin, New York, 1968.

Bailey, H. R. and L. Cesari [1]. Boundedness of solutions of linear differential equations with periodic coefficients. *Arch. Rat. Mech. Ana.* 1 (1958), 246–271.

Bancroft, S., J. K. Hale and D. Sweet, [1]. Alternative problems for nonlinear functional equations. *J. Differential Equations* 4 (1968), 40–56.

Bartle, R. G. [1]. Singular points of functional equations. *Trans. Am. Math. Soc.* 75 (1953), 366–384.

Bass, R. W. [1]. Equivalent linearization, nonlinear circuit analysis and the stabilization and optimization of control systems. *Proc. Symp. Nonlinear Circuit Anal.* 6 (1956). [2]. Mathematical legitimacy of equivalent linearization by describing functions. *IFAC Congress*, 1959.

Bellman, R. [1]. *Stability Theory of Differential Equations*, McGraw-Hill, New York, 1953. [2]. *Introduction to Matrix Analysis*. McGraw-Hill, New York, 1960.

Bendixson, I. [1]. Sur les courbes définies par des équations differéntiélles. *Acta. Math.* 24 (1901), 1–88.

Birkhoff, G. D. [1]. Dynamical Systems. *Am. Math. Soc. Colloq. Publ.* New York, 1927.

Bogoliubov, N. N. [1]. On some statistical methods in mathematical physics. *Akad. Nauk. Ukr. R.S.R.* 1945.

Bogoliubov, N. N. and Y. A. Mitropolskii [1]. *Asymptotic Methods in the Theory of Nonlinear Oscillations*. Gordon and Breach, New York, 1961. [2]. The method of integral manifolds in nonlinear mechanics. *Contributions to Differential Equations* 2 (1963), 123–196.

Bogoliubov, N. N., Jr., and B. I. Sadovnikov [1]. On periodic solutions of differential equations of nth order with a small parameter. *Symp. Nonlinear Vibrations*, Kiev, USSR, Sept., 1961.

Borg, G. [1]. On a Liapunov criteria of stability. *Am. J. Math.* 71 (1949), 67–70.

Borges, C. [1]. Periodic solutions of nonlinear differential equations: existence and error bounds. Ph.D. dissertation. University of Michigan. 1963.

Cartwright M., [1]. Forced oscillations in nonlinear systems. Contributions to the Theory of Nonlinear Oscillations, 1 (1950), 149–241, *Annals Math. Studies*, No. 20, Princeton.

Cesari, L. [1]. *Asymptotic Behavior and Stability Problems*. Springer, 1959; Second edition, Academic Press, New York, 1963. [2]. Proprietà asintotiche delle equazioni differenziali lineari ordinaire. *Rend. Sem. Mat. Roma* 3 (1939), 171–193. [3]. Sulla stabilitá delle soluzioni dei sistemi di equazioni differenziali lineari a coefficienti periodici. *Atti Accad. Mem. Classe Fis. Mat. e Nat.*, (60) 11 (1940), 633–692. [4]. Existence theorems for periodic solutions of nonlinear differential systems. *Boletin Soc. Mat. Mexicana* 1960, 24–41. [5]. Functional analysis and periodic solutions of nonlinear differential equations. *Contr. Diff. Equations* 1 (1963), 149–167. [6]. Functional analysis and Galerkin's method. *Michigan Math. J.*, 11 (1964), 385–414. [7]. Existence in the large of periodic solutions of hyperbolic partial differential equations. *Arch. Rat. Mech. Anal.* 20, (1965), 170–190. [8]. A nonlinear problem in potential theory. *Michigan Math. J.*, 1969.

Cesari, L. and J. K. Hale [1]. A new sufficient condition for periodic solutions of weakly differential systems. *Proc. Am. Math. Soc.* 8 (1957), 757–764.

Cetaev, N. G. [1]. Un théorème sur l'instabilité. *Dokl. Akad. Nauk SSSR* 2 (1934), 529–534.

Chafee, N. [1]. The bifurcation of one or more closed orbits from an equilibrium point of an autonomous differential system. *J. Differential Equations*, 4 (1968), 661–679.

Chen, K. T. [1]. Equivalence and decomposition of vector fields about an elementary critical point. *Amer. J. Math.* 85 (1963), 693–722.

Coddington, E. A. and N. Levinson [1]. *Theory of Ordinary Differential Equations.* McGraw-Hill, New York, 1955.

Conley, C. [1], Isolated invariant sets and the Morse index. CBMS number 38, Am. Math. Soc., Providence, R. I.

Coppel, W. A. [1] *Stability and Asymptotic Behavior of Differential Equations. Heath Mathematical Monographs,* 1965.

Coppel, W. A. and A. Howe [1]. On the stability of linear canonical systems with periodic coefficients. *J. Austral. Math. Soc.* 5 (1965), 169–195. [2]. Corrigendum: On the stability of linear canonical systems with periodic coefficients. *J. Austral. Math. Soc.* 6 (1966), 256.

Coppel, W. A. and K. J. Palmer [1], Averaging and integral manifolds. *Bull. Australian Math. Soc.* 2(1970), 197–222.

Cronin, J. [1]. Branch points of solutions of equations in a Banach space. *Trans. Am. Math. Soc.* 69 (1950), 208–231.

Cruz, M. A. and J. K. Hale [1]. Stability of functional differential equations of neutral type. *J. Differential Equations.* 7(1970), 334–355.

Denjoy, A. [1]. Sur les courbes définies par les équations différentielles a la surface du tore. *J. de Math.* 11 (1932), 333–375.

Diliberto, S. P. [1]. Perturbation theorems for periodic surfaces. I, II. *Rend. Circ. Mat. Palermo* 9 (2) 1960, 265–299; 10 (2)(1961), 111. [2]. On stability of linear mechanical systems. *Proc. Internat. Sympos. Nonlinear Vibrations. Izdat. Akad. Nauk Ukrain. SSR,* Kiev. 1, 1963, pp. 189–203. [3]. New results on periodic surfaces and the averaging principle. *U.S.—Japanese Seminar on Differential and Functional Equations,* pp. 49–87. W. A. Benjamin, 1967.

Fillipov, A. F. [1]. Differential equations with discontinuous right hand side. *Mat. Sbornik (N.S.)* 51 (1960), 99–128.

Flugge-Lotz, I. [1]. Discontinuous automatic control. Princeton, 1953.

Friedrichs, K. [1]. *Special Topics in Analysis.* Lecture Notes, New York University, 1953–1954. [1]. *Advanced Ordinary Differential Equations.* Lecture Notes, New York University, 1956.

Gambill, R. A. and J. K. Hale [1]. Subharmonic and ultraharmonic solutions of weakly nonlinear systems. *J. Rat. Mech. Ana.* 5 (1956), 353–398.

Golomb, M. [1]. Expansion and boundedness theorems for solutions of linear differential equations with periodic or almost periodic coefficients. *Arch. Rat. Mech. Ana.* 2 (1958), 284–308.

Gottschalk, W. H. and G. A. Hedlund [1]. *Topological dynamics. Amer. Math. Soc. Colloq. Publ.* 36 (1955).

Graves, L. M. [1]. Remarks on singular points of functional equations. *Trans. Am. Math. Soc.* 79 (1955), 150–157.

Hahn, W. [1]. *Theory and Application of Lyapunov's Direct Method.* Prentice-Hall, New York, 1963. [2]. *Stability of Motion,* Springer-Verlag, New York, 1967.

Halanay, A. [1]. *Differential Equations—Stability, Oscillation, Time Lags.* Academic Press, New York, 1966.

Hale, J. K. [1]. Periodic solutions of nonlinear systems of differential equations. *Riv. Mat. Univ. Parma* 5 (1954), 281–311. [2], Sufficient conditions for the existence of periodic solutions of first and second order differential equations. *J. Math. Mech.* 7 (1958), 163–172. [3]. Integral manifolds of perturbed differential systems. *Ann. Math.* 73 (1961), 496–531. [4]. On differential equations containing a small parameter. *Contr. Diff. Equations.* 1 (1962) [5]. Asymptotic behavior of the solutions of differential difference equations. *Proc. Intern. Symp. Nonlinear Oscillations, Kiev.* Sept. 1961, II (1963), 409–426. [6]. On the behavior of solutions of linear differential equations near resonance points. *Contr. Theory Nonlinear Oscillations* **5** (1960). [7]. *Oscillations in Nonlinear Systems.* McGraw-Hill, New York, 1963. [8]. Sufficient conditions for stability and instability of autonomous functional differential equations. *J. Diff. Equations* **1** (1965), 452–482. [9]. Periodic solutions of a class of hyperbolic equations containing a small parameter. *Arch. Rat. Mech. Anal.* **23** (1967), 380–398. [10], *Theory of Functional Differential Equations.* Appl. Math. Sciences, Vol. 3, Springer-Verlag, 1977. [11], Bifurcation near families of solutions. Proc. Int. Conf. Diff. Eqns., Uppsala (1977).

Hale, J. K. and A. Stokes [1]. Behavior of solutions near integral manifolds. *Arch. Rat. Mech. Ana.* **6** (1960), 133–170.

Hale, J. K. and H. M. Rodriques [1], Bifurcation in the Duffing equation with independent parameters, I. Proc. Royal Soc. Edinburgh, 78A(1977), 56–65; [2] II. Ibid 79A(1977), 317–322.

Hale, J. K. and P. Z. Táboas [1], Interaction of damping and forcing in a second order equation. Nonlinear Analysis, 2(1978), 77–84; [2] Bifurcation near degenerate families, Applicable Analysis. To appear.

Hall, W. S. [1]. Periodic solutions of a class of weakly nonlinear evolution equations. Ph.D. Thesis. Brown University, June, 1968.

Harris, W. A., Jr., Y. Sibuya, and L. Weinberg [1]. Holomorphic solutions of linear differential systems at singular points. To appear.

Hartman, P. [1]. *Ordinary Differential Equations,* Wiley, New York, 1964. [2]. On stability in the large for systems of ordinary differential equations. *Canad. J. Math.* **13** (1961), 480–492.

Henry, D. [1], Geometric Theory of Semilinear Parabolic Equations, Springer-Verlag, 1980.

Hopf, E. [1], Abzweigenemer periodischen Lösung von einer stationaren Lösung eines Differentialsystems. Berich. Math. Phys. Sachlischen Akad. Wursenchaften Leipzig 94(1942), 3–22.

Howe, A. [1]. Linear canonical systems with periodic coefficients. Ph.D. Dissertation, University of Canberra, Australia, 1967.

Infante, E. F. and M. Slemrod [1]. An invariance principle for dynamical systems on Banach spaces, *Proc. IUTAM Symp. on Stability* Springer, 1970.

Kelley, A. [1]. The stable, center-stable, center, center-unstable, unstable manifolds. *J. Differential Equations,* **3** (1967), 546–570.

Knobloch, H. W. [1]. Remarks on a paper of Cesari on functional analysis and nonlinear differential equations. *Michigan Math. J.* **10** (1963), 417–430. [2]. Comparison theorems for nonlinear second order differential equations. *J. Diff. Equations* **1** (1965), 1–25.

Kolmogorov, A. N. [1]. Théorie générale des systèmes dynamiques et mécanique classique. *Proc. Int. Cong. Math.* 1954, **1**, 315–333. Noordhoff, 1957.

Krasovskii, N. N. [1]. *Stability of Motion.* Stanford University Press, 1963.

Krein, M. G. [1]. The basic propositions in the theory of λ-zones of stability of a canonical system of linear differential equations with periodic coefficients (Russian), Pamyati A. A. Andronova, *Izdat. Akad. Nauk. SSSR,* Moscow, 1955, pp. 413–498.

Krylov, N. and N. N. Bogoliubov [1]. The application of methods of nonlinear mechanics to the theory of stationary oscillations. *Publication 8 of the Ukrainian Academy of Science,* Kiev, 1934. [2]. Introduction to nonlinear mechanics, *Annals Math. Studies, No. 11,* Princeton University Press, Princeton, N.J., 1947.

Kruzweil, J. [1]. Invariant manifolds for flows, *Proc. Sym. Diff. Equations and Dynamical Systems.* Academic Press, New York, 1967. [2]. The averaging principle in certain special cases of boundary problems for partial differential equations. *Casopis Pěst.*

Mat. **88** (1963), 444–456. [3]. van der Pol perturbation of the equation for a vibrating string. *Czechoslovak Math. J.* **17** (92) (1967), 558–608.

Kyner, W. T. [1]. Invariant manifolds, *Rend. Circ. Mat. Palermo*, **10** (2) (1961), 98–110.

Laksmikantham, V. and S. Leela [1]. *Differential Inequalities.* Academic Press, New York, 1969.

Laloy, M. [1], On the inversion of Lagrange-Dirichlet theorem in the case of an analytic potential. Rpt. 107, December 1977, Louvain-la-Neuve.

LaSalle, J. P. [1], Relaxation oscillations. *Quart. Appl. Math.* **7** (1949), 1–19. [2]. An invariance principle in the theory of stability. *Int. Symp. on Diff. Eqs. and Dyn. Sys.*, p. 277, Academic Press, New York, 1967.

LaSalle, J. P. and S. Lefschetz [1]. *Stability by Liapunov's Direct Method.* Academic Press, New York, 1961.

Lee, B. and L. Markus [1]. *Optimal Control Theory.* Wiley, New York, 1967.

Lefschetz, S. [1]. *Differential Equations; Geometric Theory.* Second Edition. Interscience, New York, 1959. [2]. *Stability of Nonlinear Control Systems.* Academic Press, New York, 1965.

Levinson, N. [1]. Transformation theory of nonlinear differential equations of the second order. *Ann. Math.* **45** (1944), 723–737. [2]. Small periodic perturbations of an autonomous system with a stable orbit. *Ann. Math.* **52** (1950), 727–738.

Lewis, D. C. [1]. On the role of first integrals in the perturbation of periodic solutions. *Ann. Math.* **63** (1956), 535–548. [2]. Autosynartetic solutions of differential equations. *Am. J. Math.* **83** (1961), 1–32.

Liapunoff, A. [1]. *Probléme Géneral de la Stabilité du Mouvement. Annals Math. Studies*, No. 17, Princeton University Press. 1949.

Lillo, J. C. [1]. A note on the continuity of characteristic exponents. *Proc. Nat. Acad. Sci. U.S.A.* **46** (1960), 247–250.

Locker, J. S. [1]. An existence analysis for nonlinear equations in a Hilbert space. *Trans. Am. Math. Soc.* **128** (1967), 403–413.

Lykova, O. B. [1]. Investigation of the solution of a system of differential equations with a small parameter on a two-dimension local integral manifold in the resonance case. *Ukrain. Mat. Z.* **10** (1958), 365–374.

Loud, W. [1]. Periodic solutions of $x'' + cx' + g(x) = \varepsilon f(t)$. *Mem. Am. Math. Soc..* No. 31, 1959, 58 pp. [2], Branching of solutions of two-parameter boundary value problems for second order differential equations. Ingenieur-Archiv 45(1976), 347–359.

Magnus, W. and S. Winkler [1]. *Hill's Equation.* Interscience, New York, 1966.

Malkin, J. G. [1]. Theory of Stability of Motion. AEC-tr-3352. [2]. *Some Problems in the Theory of Nonlinear Oscillations.* Moscow, 1956.

Markus, L. [1]. Asymptotically autonomous differential systems. *Contr. Theory Nonlinear Oscillations*, **3** (1956), 17–29.

Markus, L. and H. Yamabe [1]. Global stability criteria for differential systems. *Osaka Math. J.* **12** (1960), 305–317.

Marsden, J. E. and M. McCracken [1], The Hopf Bifurcation and its Applications. Springer-Verlag, 1976.

Massera, J. L. [1]. Contributions to stability theory. *Ann. Math.* **64** (1956), 182–206.

Massera, J. L. and J. J. Schäffer [1]. *Linear Differential Equations and Function Spaces.* Academic Press, New York, 1966.

McCarthy, J. [1]. The stability of invariant manifolds, I, Stanford University, Appl. Math. Stat. Lab., *ONR Tech. Rep.* **36**, 1956.

McGarvey, D. [1]. Linear differential systems with periodic coefficients involving a large parameter. *J. Differential Equations.* 2 (1966), 115–142.

McLachlan, N. W. [1]. *Theory and Application of Mathieu Functions.* The Clarendon Press, Oxford, 1947.

Minorsky, N. [1]. *Nonlinear Oscillations.* van Nostrand, Princeton, N.J., 1962.

Minorsky, N. [2], *Introduction to Nonlinear Mechanics,* J. W. Edwards, Publisher, Ann Arbor, 1947.

Mishchenko, Z. and L. S. Pontrjagin [1]. Differential equations with a small parameter attached to the higher derivatives and some problems in the theory of oscillation. *IRE Trans.* CT-7 (1960), 527–535.

Mitropolski, Y. A. [1]. *Problèmes de la théorie asymptotique des oscillations non stationaires.* Gauthier-Villars, Paris, 1966.

Montandon, B. [1], Almost periodic solutions and integral manifolds for weakly nonlinear nonconservative systems. J. Differential Eqns. 12(1972). 417–425.

Morrison, J. A. [1]. An averaging scheme for some nonlinear resonance problems. *SIAM J. Appl. Math.* 16 (1968), 1024–1047.

Moser, J. [1]. On invariant curves of area-preserving mappings of an annulus. *Nachr. Akad. Wiss. Göttingen Math.-Phys. Kl. II* 1962, 1–20. [2]. Bistable systems of differential equations with applications to tunnel diode circuits. *IBM J. Res. Develop.* 5 (1961), 226–240. [3], Stable and Random Motions in Dynamical Systems. Princeton University Press. 1973.

Nemitskii, V. V. and V. V. Stepanov [1]. *Qualitative Theory of Differential Equations.* Princeton University Press, 1960.

Newhouse, S. [1], Lectures on Dynamic Systems. CIME, Summer Session on Dynamical Systems, June, 1978.

Nirenberg, L. [1]. *Functional Analysis.* Lecture Notes, New York University, 1960–1961.

Nitecki, Z. [1], Differentiable Dynamics. M.I.T. Press, Cambridge, Mass., 1971.

Olech, C. [1]. On global stability of an autonomous system on the plane. *Contributions to Differential Equations* 1 (1963), 389–400.

Onuchic, N. [1], Applications of the topological method of Ważewski to certain problems of asymptotic behavior in ordinary differential equations. Pacific J. Math. 11(1961), 1511–1527; [2] On the asymptotic behavior of the solutions of functional differential equations, pp. 223–233 in Differential Equations and Dynamical Systems. Ed. J. K. Hale and J. P. LaSalle, Academic Press. 1967.

Opial, Z. [1]. Sur la dépendance des solutions d'un système d'équations différentielles de leurs seconds membres. Applications aux systèmes presques autonomes. *Ann. Polon. Math.* 8 (1960), 75–89.

Palamadov, V. P. [1], Stability of equilibrium in a potential field. Functional Analysis and its Applications (1978), 277–289.

Palmore, J. [1], Instability of equilibria, Preprint.

Peixoto, M. [1]. Structural stability on two-dimensional manifolds. *Topology* 1 (1962), 101–120.

Perelló, C. [1]. Periodic solutions of differential equations with time lag containing a small parameter. *J. Diff. Equations.* 4(1968), 160–175 [2]. A note on periodic solutions of nonlinear differential equations with time lags, *Differential Equations and Dynamical Systems.* Academic Press, 1967, 185–188.

Perron, O. [1]. Die Stabilitätsfrage bei Differentialgleichungssysteme. *Math. Zeit.* 32 (1930), 703–728.

Pliss, V. A. [1]. On the theory of invariant surfaces. *Differentialniye Uravnaniya* 2 (1966), 1139–1150. [2]. *Nonlocal Problems of the Theory of Oscillations.* Academic Press, New York, 1966.

Poincaré, H. [1]. Sur les courbes définies par les équations différentielles. *J. de Math.*, (3) **7** (1881), 375–422; (3) **8** (1882), 251–296; (4) **1** (1885), 167–244; (4) **2** (1886), 151–217. [2]. *Les méthodes nouvelles de la mécanique celeste.* Paris, 3 vols., 1892, 1893, 1899.

Rabinowitz, P. H. [1]. Periodic solutions of a nonlinear wave equation. *Comm. Pure Appl. Math.* **20** (1967), 145–205.

Reuter, G. E. H. [1]. Subharmonics in a nonlinear system with unsymmetric restoring forces. *Quart. J. Mech. Appl. Math.* **2** (1949), 198–207.

Rouche, N. Habets, P. and M. LaLoy [1], *Stability Theory by Liapunov's Direct Method.* Appl. Math. Sci. Vol. 22, 1977, Springer-Verlag.

Sacker, R. J. [1]. A new approach to the perturbation theory of invariant surfaces. *Comm. Pure Math.* **18** (1965), 717–732.

Sansone, G. and R. Conti [1]. *Nonlinear Differential Equations*, Pergamon Press, New York, 1964.

Schwartz, A. J. [1]. A generalization of a Poincaré-Bendixson theorem to closed two-dimensional manifolds. *Am. J. Math.* **85** (1963), 453–458; errata, ibid, **85** (1963), 753.

Sethna, P. R. [1]. An extension of the method of averaging. *Quart. Appl. Math.* **25** (1967), 205–211.

Siegel, C. L. and J. K. Moser [1], *Lectures on Celestial Mechanics*, Grundlehren, Vol. 187, Springer, 1971.

Smale, S. [1]. Differentiable dynamical systems. *Bull. Amer. Math. Soc.* **73** (1967), 747–817.

Sotomayor, J. [1], Generic one parameter families of vector fields, Publ. Math. IHES, 43(1973), 5–46.

Sternberg, S. [1]. Local contractions and a theorem of Poincaré. *Am. J. Math.* **79** (1957), 809–824. [2]. On the structure of local homeomorphisms of euclidean n-space, II. *Am. J. Math.* **80** (1958), 623–631. [3]. III, Ibid, **81** (1959), 578–604.

Szarski, J. [1]. *Differential Inequalities.* Warszawa, 1965.

Urabe, M. [1]. *Nonlinear Autonomous Oscillations.* Academic Press, New York, 1967. [2]. Galerkin's procedure for nonlinear periodic systems and its extension to multipoint boundary value problems for general nonlinear systems. *Numerical Solutions of Nonlinear Differential Equations* (*Proc. Adv. Sympos.* Madison, Wis., 1966), pp. 297–327. Wiley, New York, 1966.

Vainberg, M. M. and V. A. Tregonin [1]. The methods of Lyapunov and Schmidt in the theory of nonlinear differential equations and their further development. *Mathematical Surveys*, **17** (1962), 1–60.

van der Pol, B. [1]. Forced oscillations in a circuit with nonlinear resistance. *Phil. Mag.* **3** (1927), 65–80; *Proc. Inst. Radio Eng.* **22** (1934), 1051–1086.

Wasow, W. [1]. *Asymptotic Expansions of Ordinary Differential Equations.* Interscience, New York, 1965.

Ważewski, T. [1], Sur un principe topologique de l'examen de l'allure asymptotique des integrales des équations différentielles ordinaires. Ann. Soc. Polon. Mat. 29(1947), 279–313.

Williams, S. [1]. A connection between the Cesari and Leray-Schauder methods. *Michigan Math. J.*, 1969.

Yakubovich, V. A. [1]. Structure of the group of symplectic matrices and the unstable canonical systems with periodic coefficients. *Mat. Sbornik N.S.* **44** (86) (1958), 313–352. [2]. Critical frequencies of quasi-canonical systems. *Vestnik Leningrad Univ.* **13** (1958), 35–63.

Yoshizawa, T. [1]. Asymptotic behavior of solutions of a system of differential equations. *Contributions to Differential Equations 1* (1963), 371–387. [2]. *Stability Theory of Liapunov's Second Method.* Mathematical Society of Japan, 1966.

Yoshizawa, T. and J. Kato [1]. Asymptotic Behavior of Solutions near Integral Manifolds. *Differential Equations and Dynamical Systems*, J. K. Hale and J. P. LaSalle,
Eds., Academic Press, New York, 1967, pp. 267–275.

Zubov, V. I. [1]. *Methods of A. M. Lyapunov and their Applications*. Noordhoff. 1964.

Index